George A. Bubenik   Anthony B. Bubenik
Editors

Horns, Pronghorns, and Antlers

Evolution, Morphology, Physiology, and Social Significance

With 204 Illustrations

Springer-Verlag
New York Berlin Heidelberg
London Paris Tokyo Hong Kong

George A. Bubenik
Associate Professor
Department of Zoology
University of Guelph
Guelph, Ontario N1G 2W1
Canada

Anthony B. Bubenik
Consultant in Big Game Research
& Management
Thornhill, Ontario L3T 3X7
Canada

Cover illustrations: Descriptions can be found in the legends to the following figures in the text—Chapter 1: Figure 1 (page 6); Figure 4 (page 11); Figure 26 (page 57). The illustration of the deer with antlers does not appear in this book. It was prepared by and is copyrighted by Anthony B. Bubenik.

Library of Congress Cataloging-in-Publication Data
Horns, pronghorns, and antlers: evolution, morphology, physiology, and social significance/George A. Bubenik, Anthony B. Bubenik, editors.
p. cm.
Includes bibliographical references.
ISBN-13:978-1-4613-8968-2
1. Horns. 2. Antlers. I. Bubenik, George A. II. Bubenik, Anton B.
QL942.H77 1990
599.73'504185–dc20 89-29975

Softcover reprint of the hardcover 1st edition 1990

Typeset by Publishers Service of Montana, Bozeman, Montana.

9 8 7 6 5 4 3 2 1

ISBN-13:978-1-4613-8968-2 e-ISBN-13:978-1-4613-8966-8
DOI: 10.1007/978-1-4613-8966-8

Horns, Pronghorns, and Antlers

Preface

Since the first drawings left on walls of ancient caves, human beings have been fascinated with that unique phenomenon of the animal kingdom, the presence of horns and antlers.

From the mythical "unicorn" exercising the power over life and death to the perceived aphrodisiacal and other medical properties of rhinoceros horns and growing antlers, these conspicuous protuberances have had a significant place in the history of mankind.

Part of that ancient interest in antlers and horns was due to their value as symbols of masculinity; this interest persists today in trophy hunting, an honorable tradition carried on for centuries in many countries of the world.

This book, which deals with evolution, morphology, physiology, and behavior, has not been devised as a comprehensive review of the subject of horns, pronghorns, and antlers; rather, it is a series of chapters stimulating thoughts, discussions, and initiation of new studies.

As editors, we did not interfere with the content of articles nor with the opinions and interpretations of our contributors, and we left them to decide whether to accept the suggestions of our reviewers. Despite the fact that various aspects of cranial appendages have been studied since the end of the eighteenth century, many controversial views still exist, as witnessed in various chapters of this book. We do not consider the presence of contradictory views as a shortcoming but rather as an advantage demonstrating where our knowledge needs improvement, where conformity exists, or where future studies should be aimed.

Finally, both of us would like to thank all contributors to this volume for their collaboration on this rather unique project, which brings together endocrinologists, taxonomists, paleontologists, experimental morphologists, neuroendocrinologists, behaviorists, and physiologists. It was a great challenge, and we hope that we met it "head on."

GEORGE A. BUBENIK
ANTHONY B. BUBENIK

Acknowledgments

The scientific accuracy as well as the organization and flow of individual chapters were greatly enhanced by the generous help of our panel of reviewers. Some of them were acknowledged by individual contributors, but others were not. Therefore, we as editors would like to express our gratitude to our following colleagues for their generous effort in reviewing various chapters of this book: E.D. Balon, C. Barrette, R.C. Bigalke, J.P. Bogart, D.E. Brown, N. Chapman, T.H. Clutton-Brock, R.J. Goss, P. Heizmann, H.A. Jacobson, D.W. Kitchen, R.L. Marchinton, C.D. McInnes, S.R. Scadding, D. Schams, J.M. Suttie, E. Thenius, D.E. Ullrey, A.C.V. Van Bemmel, and S.D. Webb.

In addition, we would like to thank Mary Bubenik, Carolyn Pollard, Yoko Imai, and Jane Taylor for their excellent technical assistance.

George A. Bubenik
Anthony B. Bubenik

Contents

Preface v
Acknowledgments vii
Contributors xi

I. Evolution and Morphology

1. Epigenetical, Morphological, Physiological, and Behavioral Aspects of Evolution of Horns, Pronghorns, and Antlers 3
ANTHONY B. BUBENIK

2. Correlation of Reproductive and Digestive Strategies in the Evolution of Cranial Appendages 114
CHRISTINE M. JANIS

3. *Muntiacidae* 134
COLIN P. GROVES AND PETER GRUBB

4. *Cervidae* of Southeast Asia 169
PETER GRUBB

5. Cranial Appendages of Giraffoidea 180
CHARLES S. CHURCHER

6. Evolution and Dispersal of African Bovidae 195
ALAN W. GENTRY

II. Physiology, Genetics, and Behavior

7. The Pronghorn (*Antilocapra americana*) 231
BART W. O'GARA

8. Neuroendocrine Regulation of the Antler Cycle 265
George A. Bubenik

9. Of Antlers and Embryos 298
Richard J. Goss

10. Antler Regeneration: Studies with Antler Removal, Axial Tomography, and Angiography 313
James M. Suttie and Peter F. Fennessy

11. The Role of the Nervous System in the Growth of Antlers 339
George A. Bubenik

12. Experimental Manipulation of the Neural Control of Antler Growth 359
James M. Suttie

13. Experimental Induction of Antler Growth 371
Zbigniew Jaczewski

14. The Annual Antler Cycle of the European Roe Deer (*Capreolus capreolus*) in Relation to the Reproductive Cycle 396
Antoine J. Sempéré

15. Endocrine Mechanisms and Antler Cycles in Rusa Deer, *Cervis (Rusa) timorensis* 416
Simone van Mourik and Teodor Stelmasiak

16. Nutrition and Antler Development 426
Robert D. Brown

17. Social Status and Antler Development in Red Deer 442
Luděk Bartoš

18. Genetic Variability and Antler Development 460
Kim T. Scribner and Michael H. Smith

19. The Antler as a Model in Biomedical Research 474
George A. Bubenik

References 488
Index of Extant and Extinct Species of Pecora 545
Index of General Terms 551

Contributors

LUDĚK BARTOŠ, Group of Ethology, Institute of Animal Production, CS-104 00, Praha 10-Uhrineves, Czechoslovakia.

ROBERT D. BROWN, Department of Wildlife and Fisheries, Mississippi State University, Mississippi State, Mississippi 39762-5917 U.S.A.

ANTHONY B. BUBENIK, Consultant in Big Game Research & Management, 10 Stornoway Crescent, Thornhill, Ontario L3T 3X7 Canada.

GEORGE A. BUBENIK, Department of Zoology, University of Guelph, Guelph, Ontario N1G 2W1 Canada.

CHARLES S. CHURCHER, Department of Zoology, University of Toronto, Toronto, Ontario M5S 1A1; and Department of Vertebrate Palaeontology, Royal Ontario Museum, Toronto, Ontario M5S 2C6 Canada.

PETER F. FENNESSY, MAF Technology, South Invermay Agricultural Centre, Private Bag, Mosgiel, New Zealand.

ALAN W. GENTRY, Department of Palaeontology, The Natural History Museum, Cromwell Road, London SW7 5BD England.

RICHARD J. GOSS, Division of Biological and Medical Sciences, Brown University, Providence, Rhode Island 02912 U.S.A.

COLIN P. GROVES, Department of Prehistory and Anthropology, The Australian National University, GPO Box 4, Canberra ACT 2601 Australia.

PETER GRUBB, 35 Downhills Park Road, London N17 5PE England.

ZBIGNIEW JACZEWSKI, Institute of Genetics and Animal Breeding, Polish Academy of Sciences, Popielno, 12-222 Wejsuny Poland.

CHRISTINE M. JANIS, Division of Biology and Medicine, Brown University, Providence, Rhode Island 02912 U.S.A.

BART W. O'GARA, U.S. Fish and Wildlife Service, Montana Cooperative Wildlife Research Unit, University of Montana, Missoula, Montana 59812 U.S.A.

KIM T. SCRIBNER, The University of Georgia, Savannah River Ecology Laboratory, Drawer E, Aiken, South Carolina 29802; and Department of Zoology, The University of Georgia, Athens, Georgia 30602 U.S.A.

ANTOINE J. SEMPÉRÉ, Centre d'Etudes Biologiques des Animaux Sauvages (CNRS), Villiers-en-Bois, F-79360 Beauvoir-sur-Niort, France.

MICHAEL H. SMITH, The University of Georgia, Savannah River Ecology Laboratory, Drawer E, Aiken, South Carolina 29802; and Department of Zoology, The University of Georgia, Athens, Georgia 30602 U.S.A.

JAMES M. SUTTIE, MAF Technology, South Invermay Agricultural Centre, Private Bag, Mosgiel, New Zealand.

TEODOR STELMASIAK, Biocene International (Vic) Pty Ltd, Veterinary Precinct, VRI Building, Parkville 3052 Australia.

SIMONE VAN MOURIK, University of Melbourne, School of Agriculture and Forestry, Parkville 3052 Australia.

I
Evolution and Morphology

1
Epigenetical, Morphological, Physiological, and Behavioral Aspects of Evolution of Horns, Pronghorns, and Antlers

ANTHONY B. BUBENIK

"Prediction is the very Hallmark of Science."

Eldredge, 1985

Part A. Terms—Hypothesis—Controlling Mechanisms—Function

Development of Terms for Cranial Appendages

The fact that the "true" horns are a product of the skin, and "'horns' of deer" develop as bone protuberances, and thus both are not homologous by nature was already known to Aristotle (Peck 1965). Despite this, both antlers and horns are frequently considered homologous terms (Frick 1937; Geist 1966a, 1974a; Janis 1982; Kiltie 1985). Moreover, the common view that antlers and horns have been developed by natural selection primarily as weapons—a view we find in Democritos (Mulachius 1843) and supported by E. Darwin's (1794) notion (adopted by C. Darwin 1859, 1871)—persists in the present zoological systematic and behavioral interpretations (e.g., Geist 1966a, 1974a; Clutton-Brock 1982; Janis 1982; Kiltie 1985).

Perhaps it was this ancient notion and the lack of a proper term for antlers in Latin (both antlers and horns are *cornus*) that led Linnaeus (1758) to put the horned and antlered ruminants into one suborder *Pecoran* of the *Artiodactyla*; this false meaning and lack of behavioral knowledge might also be behind Darwin's (1871) statement that "when males are provided with weapons which, in the females, are absent there can hardly be any doubt that these serve for fighting with other males, that they were acquired through sexual selection, and were transmitted to the male sex alone."

In spite of the incorrect conclusion of the Darwins, we have to stress that the broad context of the term *Pecora* has its meaning because of our long-standing and intuitive conceptualism (Carr 1972). *Pecora* is derived from the Latin *pecus*, i.e., horned livestock. Thus, should "pecora" be used correctly, i.e., in the sense of Linnaeus (1758), then the suborder would encompass only the true-horned ruminants. Horns, of course, are not the only character of bovids or Cavicornia, because, e.g., pronghorns of *Antilocarpra americana* also are cavicorn, but morphologically not analogous to bovid's horns.

Thus, the suborder Pecora should not encompass 1) cervoids whose skulls were or are equipped with bony protuberances (e.g., Frick 1937; Simpson 1945), and 2) the "sinecorn" propagules, like the extinct blastomerycides, and early paleomerycids, moschids, and hydropotins as well. The morphophysiological adaptations necessary for development of cranial appendages are so profound that we agree with Webb & Taylor (1980) [contrary to Groves & Grubb (1987); Scott & Janis (1987)] that it is necessary to separate the *Pecora* into two distantly related lineages:

1. The one that includes all sinecorn ruminants and their extant relatives; these we call *Sinecornua*.
2. The rest with any kind of cranial appendages; these should be called *Eupecora* (Webb & Taylor 1980).

Therefore, it remains a question why these basic differences are overlooked also in the most recent papers (e.g., Clutton-Brock 1982; Geist 1968, 1978a; McFarland et al. 1985; Scott & Janis 1987).

In our view, one of the reasons why the classification of pecoran in general and that of cervoids in particular is still in stage of development, is the disregard of processes which control and form the shape of the cranial appendages and their histology, as indicator of developmental stage. This, inevitably must lead, on the one hand, to overlooking of convergences and morphological flexibilities and, on the other, to erroneous conclusions about mimicked relationships or diminutive tendencies as adaptive processes (Gray 1821; Brooke 1878; Frick 1937; Hershkowitz 1969; Sondaar 1977; Leinders 1979, 1984; Vrba 1980; Groves & Grubb 1987; Janis & Lister 1985; Eisenberg 1987).

An important step forward is the consensus that cranial appendages with anlagen in dermis are not homologous to the cranial apophyses, which should have terminated the long-lasting dispute (e.g., Cunningham 1900; Rörig 1900; Brandt 1901; Fambach 1909; Atzkern 1923; Duerst 1926; George 1956; Bubenik, A. 1982a; Goss 1983).

Due to this consensus, Pecora should not be called "horned ruminants," and students working in this field should know that either the term *antlers* or *horns* is delineative enough to avoid confusion, because not every apophysis is an antler, and not every horn-sheath indicates bovid taxon. Thus, new terms must be defined to let the reader know what the author means.

Proper Terms for Cranial Appendages

Names which are used in this paper were developed with the aim of having symbolic and specific words for appendages for both bovoids and cervoids as well.

Appendages of Epiphyseal Nature

The Ossicones. Historically (Atzkern 1923), all bony cores of pecoran which are of dermal origin are called ossicones. In that case it is too broad a term because it does not allow one to distinguish between both the loose and ankylosed ossicones. Kolda (1951) calls the latter *os cornu*. Then for the former, which apparently represents a more primitive stage, the term proto-ossicones seems suitable (Fig. 1, v). In all known bovids the os cornu can be protected either by skin or horn or cavicorn horn. Hence, we need to speak about the former as *velericorn proto-ossicone*, e.g., in giraffe genus *Giraffa* and okapi genus *Okapia*. The latter is a *horned ossicone* of bovids.

The base of the os cornu is frequently covered by skin and may be wider than the part covered by horn. This is called a boss or stalk (see Gentry, Chapter 6).

Spur. In many extinct eupecoran and in extant giraffe, a protrusion of occipital bone will be termed here the spur. The morphophysiology of their development, microstructure, vascularization, and innervation need to be investigated, because we do not know whether all spurs were built as a compact bone (osteon) or whether they also have a spongy core (tuber).

Horn Sheath. Keratinized epidermis covering the os cornu is called horn. It consists of many periodically developed horn cones clinging tightly together. The cavicorn horn differs from compact scurs or skin horns, which can develop at any site of the body and are not homologous to the pronghorn of antilocaprins.

Appendages of Apophyseal Nature—Bony Protuberances

Cranial Protuberances. They were developed on different bones as their protrusion or apophyses. Their compact cortex has a spongious core which is connected with that of the corresponding cranial bone.

(a) *Peduncles* or *processus conuiformis*: These are perennial, conically shaped, and eventually may become bifurcated. In some extinct lineages they might be velericorn or protected by a deciduous horn sheath, as in the pronghorn *Antilocapra americana*.

(b) *Pedicles* or *processus cornu cervi*: They are the proximal parts of cranial apophyses whose distal, mostly cervicorn branched apices were facultatively perennial, and later became deciduous (Kolda 1951). It would seem reasonable that the pedicles of North American merycodonts should be termed pseudopedicles and that of the Eurasiatic protocervids protopedicles. Depending on the developmental stage, the discarded part of the pedicle may belong to one of three different types (Figs. 2, 3):

Pseudoantlers were the first known cervicorn apophyses of facultatively perennial or aperiodically deciduous character used by the North American merycodontins.

Protoantlers were also facultatively perennial distal parts of protopedicles of protocervids of Eurasia which appeared a few MYA (million years ago) later.

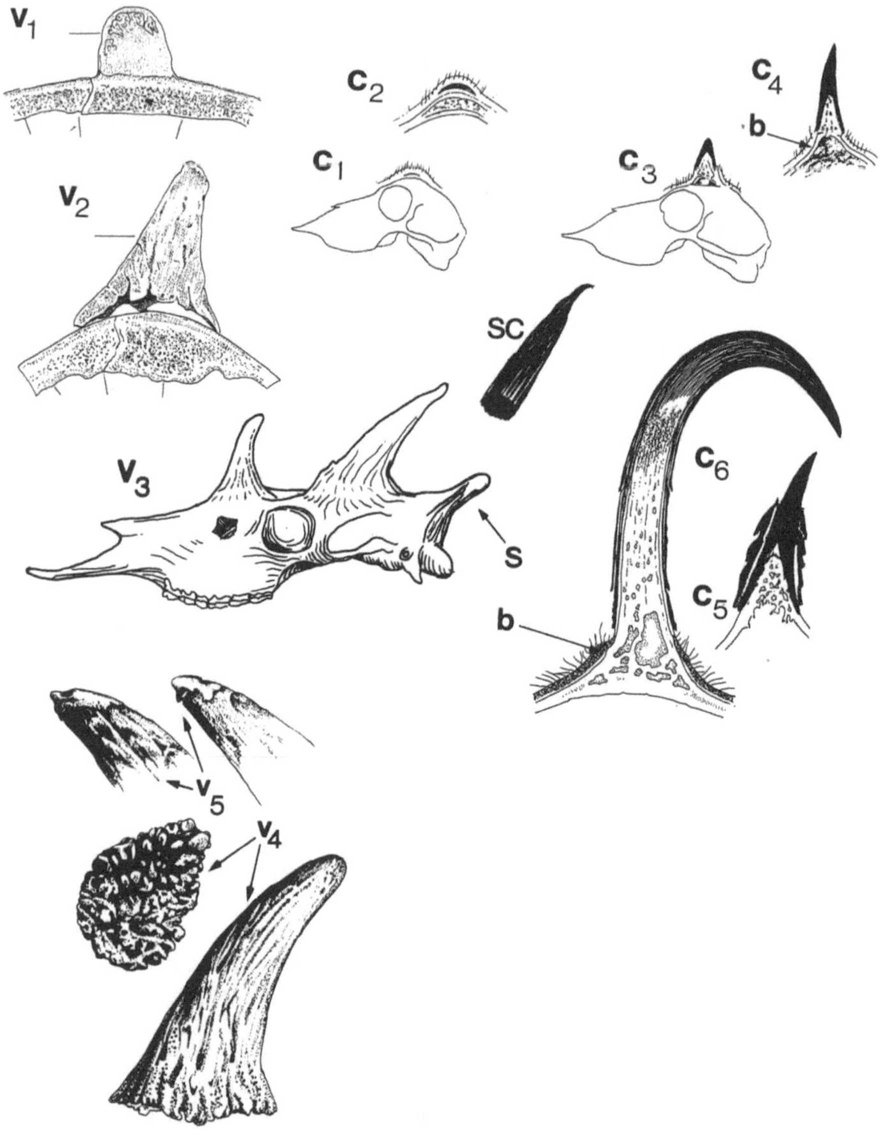

FIGURE 1. Terminology of cranial appendage of epiphyseal nature, or dermal origin: *v* velericorn, not fused proto-ossicone of a giraffe and okapi, in different stages of development: v_1 after birth, v_2 juvenile, v_3 mature, v_3 possibly also in the tylopod *Paratoceras wardi*, of which cranial appendages are yet of unknown origin, *s* occipital spur; v_4 not fused proto-ossicone of okapi: lateral view and base, v_5 the deciduous caps. *c* Cavicorn ossicone: development and fusion with the cranial boss in chamois: *1* dermal anlage; *2* enlarged from *1*; *3, 4* the first horn sheath in developing stage; *5* the first horn cap pierced by the permanent horn. *6* chamois horn in sagittal section; *b* the boss or stalk as cranial protrusion, covered by skin; *sc* scur or skin horn of a chamois. (Figs. $v_{1.2.4.5}$ after Lankester 1907, 1910; Fig. v_3 after Patton & Taylor 1973; Courtesy, American Museum of Natural History.)

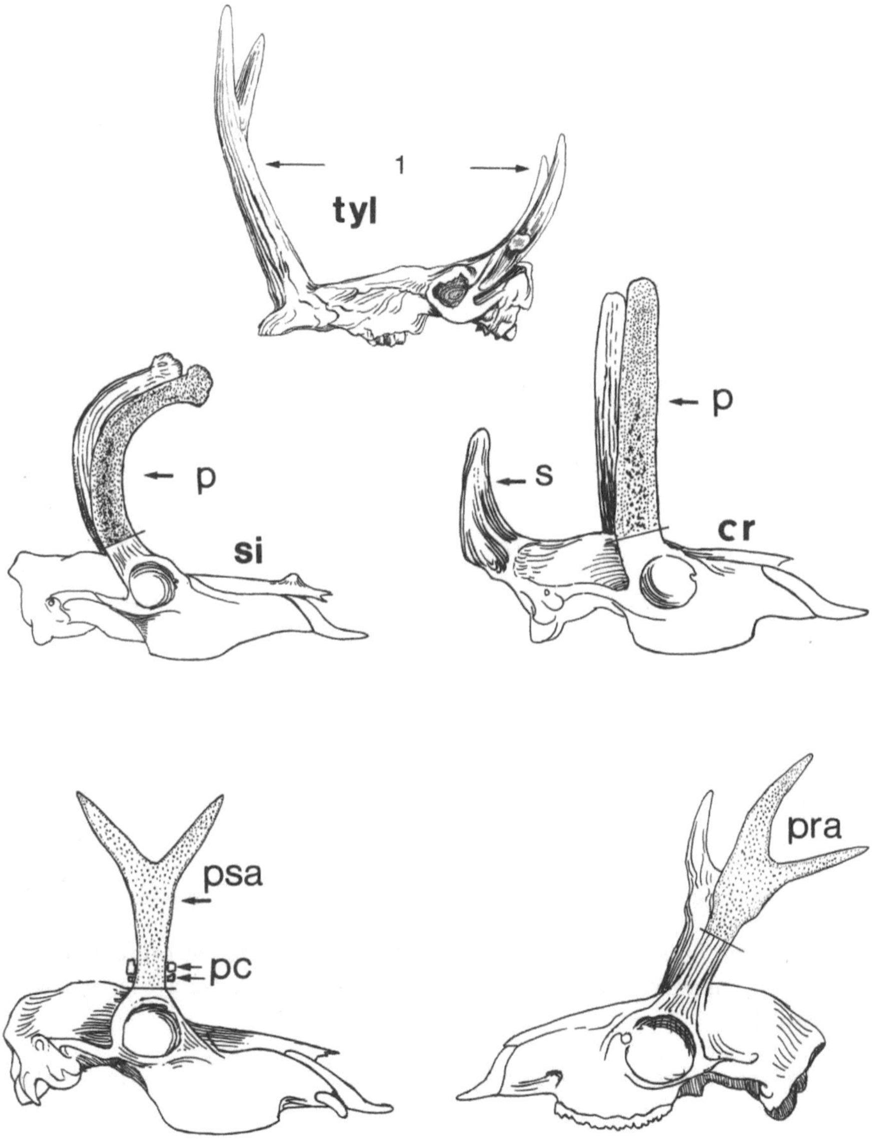

FIGURE 2. Peduncles and pedicles in sagittal section. Compare the peduncles *tyl* of a noneupecoran tylopod *Synthetoceras*, with those of two cervoids: *si* in *Sinclairomeryx* and *cr Cranioceras*. *s* is a spur as protrusion of the occiput. *p* shows the pedicle-like apophyses in longitudinal section. *psa* is a pseudoantler of the merycodont *Meryceros* with pseudocoronets *pc*. *pra* is a protoantler of *Dicroceros elegans*, which did not develop a coronet. If the distal part died, it was sequestered below the swollen base.

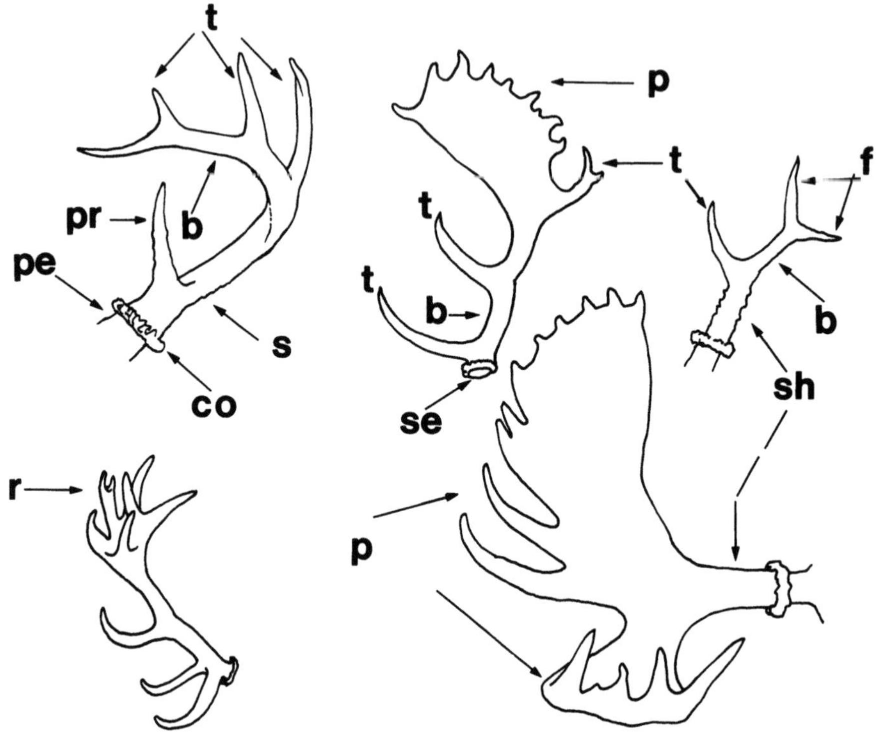

FIGURE 3. Description of antlers: *pe* pedicle, *co* coronet or burr, *sh* shaft, *b* beam, *t* tine, *pr* sprout, called prong, *f* terminal fork, *r* multipointed royal, *p* palm, resp. a double palm.

Antlers are the circannually cast and regenerated apices of pedicles.

Each of the three types of appendages have quite different origins, history, morphophysiological control of growth, and regeneration processes, as will be discussed later.

Pseudocoronets were developed by pseudocervids, originally as dermal bones.

Coronet or *burr* is the bony rim of the antlers base which seals the pedicle skin and appears with the first subsequent antler.

Shaft is used for the not yet ramified part of the cervicorn apices of any kind of pedicles.

Beam is that part of cervicorn apophysis which has the potential to develop two types of branches: sprouts (pseudotines) or tines.

Sprouts or *pseudotines*: Sprouting seems to be the pristine type of branching, used also in cervicorn apophyses of climacocerids consider either as giraffoids (MacInnes 1936), or protocervids (Bubenik, A. 1982c). Sprouts develop as exostoses of the apophyses' cortex. Sprouts can mimic tines in length, location, and shape, and may be the origin for future genotypic tines.

Tines develop by splitting the beam's apex into unequally long forks. The shorter end is termed tine, the longer is the beam. Any tine can bifurcate, but its growth is limited. Tines may originate by genocopy from sprouts.

Seal is the base of a cast antler, which plugs the dead antler from the core of the living pedicle.

Theories of Development of Cranial Appendages

C. Darwin's concept (1859), taken from E. Darwin (1794)—that antlers and horns have developed as weaponry by sexual selection—is a hypothesis that persists in recent papers—(e.g., Geist 1978a; Clutton-Brock 1982). However, there are also other views which are worth mentioning:

Duerst (1926) considered injury of exposed skin or bone loci as the primary vector of the development of cranial appendages.

Janis (1982) describes the development of cranial appendages as the best tool for the defense of new resources connected to a more fibrous diet, which is close to Cope's (1887) idea that inheritance of new characters should be ascribed to changes of nutritional conditions.

Augusta (Vodička & Augusta 1942), Kurtén (1968), and other scholars considered the cranial appendages to be species-specific, sex-dimorphic cues which have changed size in concert with changes of habitat and body size.

Krieg (1936a) and Beninde (1937) regarded antlers as luxurious head adornments.

Fisher (1958) considered cranial appendages as a "handicap," selected to attract female interest but shortening life expectancy.

Geist (1978a, 1983, 1986a) ascribes the trend toward variability in size and shape of the body and cranial appendages to the pulsation between the efficiency- or dispersal-phenotype. This, of course, is only an analogy to the theory of *ecomorphs* (Aleyev & Bubrak 1984; White & Keller 1984). Aleyev (1986) defines the ecomorph as "an integral system of the inter-conditioned ecological-morphological adaptations determining a general structure of the organism body in accordance with a concrete trend of the species evolution under all living systems of the organismic rank." White & Keller (1984) note that "Some morphological characters imply a common ancestry whereas others correlate with habits and are the result of convergence. A distinction needs to be made between a taxon and an ecomorph."

Both dispersal phenotypes and ecomorphs have similar aspects, but the ecomorph concept differs from that of dispersal in being general and having some elements of epigenetic restructuring. The importance of migration into vacant niches is not stressed, and pedo- and hypermorphism are not considered as the main vectors for further phyletic development.

Geist's theory (1986a) has too many weak points to be considered, in his own words, as "the best" explanation of eupecoran evolution. First, there is enough evidence that new lineages of eupecoran have developed, not on the periphery, but in the centrum of the ancestor's origin (Matthew 1908; Stehlin 1937; Colbert

1940; Azzaroli 1948, 1979; Kahlke 1951; Thenius 1980; Groves & Grubb, Chapter 3). There also was an attempt to prove invariable tendency for diminution, e.g., in *Capreolus* by Lehmann (1960), which was easy to deny (Bubenik, A. 1984).

Second, the fluctuations between hypermorphic and pedomorphic pulses in relation to food abundance or social disorder are common events in ungulates in all climatic zones (Kingdon 1979; Bubenik, A. 1984; Vrba 1985; Stringham & Bubenik 1974; Heimer 1987). These pulses are well documented in fossils (Azzaroli 1948, 1979; Sher 1986), historical records (Erbach-Erbach 1986), and nutritional studies (Krieg 1937; Vogt 1947; Bubenik, A. 1959a; Bayern, A.V. & J.V. Bayern 1975).

Third, the insular dwarfism (Thaler 1973) of deer cannot be ascribed to pedomorphic changes, as documented by Sondaar (1977). The extinct dwarfs of Cretan deer of the genus *Candiacervus* sp., referred to by Geist (1986a), and particularly the smallest of them, *Candiacervus repalophorus*, retained antlers longer than their much larger relatives (De Vos & Dermitzakis 1986).

Therefore, there is a good reason to question whether, in Geist's hypothesis, the terms hypermorphic, pedomorphic, and neoteny (Gould 1977) are used in a proper context, an objection also expressed for human neoteny by Ewer (1960). Thus, the maintenance phenotype should be better termed retarded or suppressed. Hypermorphic cranial appendages can appear and disappear in any population without leading to new restructured lineages.

Finally, there is a legitimate question as to why the switches between the maintenance and dispersal phenotypes cannot be applied, for example, to African bovids, whose pulses are well documented (Kingdon 1979; Sinclair 1979; Vrba 1984, 1985). Similar lack of change can be found in the protocervids of the Eurasiatic Miocene such as dicrocerids, heteroproxins, stephanocematids, triceromerycins, or the Pliocene rusin deer, which spread across the whole of Eurasia from east to west and vice versa, being exposed to enormous climatic and environmental changes (Colbert 1936; Bohlin 1935, 1953; Thenius 1980; Janis & Lister 1985).

Cranial Appendages as Epigenetic Induced and Co-Opted Exaptations

The terms exaptations and adaptations have their proponents and opponents (Krimbas 1984; Wallace 1984). However, it was difficult to find any better English word, which in its symbolism could replace exaptation and adaptation. Therefore, both exaptation and adaptation will be used in this paper.

The great variety of shapes of cranial appendages, and their appearance in unrelated taxa in quite distinct epochs (Frick 1937; Patton & Taylor 1971, 1973; Webb & Taylor 1980; Janis 1982) (Fig. 4), are intriguing and difficult to explain by random mutations in view of the extremely low probability of success (Schmidt 1985).

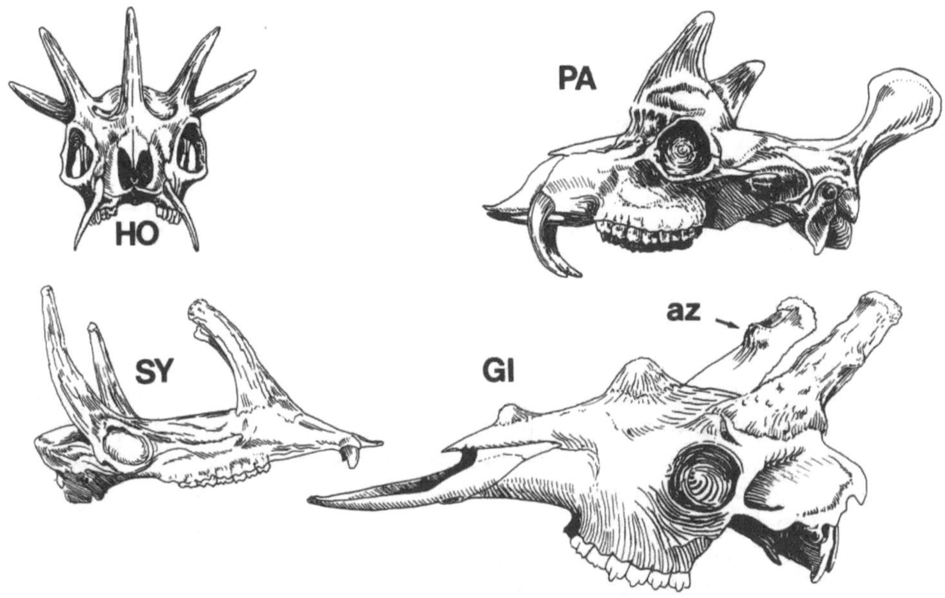

FIGURE 4. Cranial appendages of different origin in unrelated taxa: *HO Hoplitomeryx* (Leinders 1984), *PA Paleomeryx tricornis* (Zhanxiang et al. 1985), *SY* protoceratid *Synthetoceras* (Webb 1981), *GI* giraffe *Giraffa*; *az* azygous exostosis.

The hypothesis of Gould & Vrba (1982) that new organs may develop from random exaptations can be used as an analogy for the evolution of cranial appendages. We can assume that cranial exaptations as tissue defense against permanent injury led to a socially important structure that was co-opted and further developed by either activating dormant DNA sequences or sexual selection.

According to Gould & Vrba (1982), exaptations are morphological changes arising randomly from the "enormous pool of non-aptations, the wellspring of most evolutionary events." To induce their activation, we have to presume that specific sites on heads of eupecoran propagules become programmed for their manifestation by some corresponding neuronal messengers (Davies et al. 1987; Lockerbie 1987; Pollard 1987; Thaller & Eichele 1987), and their appearance unfolded other pliomorphic and physiological changes. To develop a new messenger, it is assumed that the population–and, concretely, the tissues concerned–must be exposed to stressors to which the previous generations were never exposed (Waddington 1961; Bateson 1963; Belyaev 1979; Wallace 1982).

The most logical way the messenger could reach the genome seems by epigenetic mechanisms, i.e., the new messenger can penetrate the placental barrier and activate the "dormant" or "unemployed" plasmatic or nonplasmatic DNA sequences and stimulate the corresponding anlagen of the exposed cranial loci in either the dermis or the cranial bones. If the new sequence or allele is linked with the female chromosome, the development of new exaptation will spread over the daughters' populations faster than a random mutation of individuals and their selection by contest.

Offensive Defense in Response to Inexperienced Stress as Vector for Exaptations

The hypothesis that inexperienced stressors may be vectors for new messengers and exaptations is supported by the fact that virtually only socially stressed deer populations (those in disorganized status) (Bubenik, A. 1986a) exhibit high frequencies of genetic atavisms such as upper canines, doubled or missed molars, and/or nonspecific antler anomalies (e.g., Brandt 1901; Rhumbler 1911, 1932; Bubenik, A. 1966; Robinette & Jones 1959; Reyel 1963; Thing et al. 1986).

In our concept, a population exposed to an inexperienced stress has to respond by "offensive defense" as the ultimate way to survive. It is open to question whether the offensive defense leads to a "bottleneck" crisis with only few survivals, which microevolutive changes allow further promising evolution. From an epigenetic view it is important whether most of the surviving individuals develop new DNA messengers (Adams, D.B. 1979) which may activate some dormant or silent alleles on the one hand, as for example, for the cranial exaptations and induction of neurons for the stimulation of their development and growth control on the other. Thus, a population under inexperienced stress should have a better chance to develop a quite different lineage in a saltatory way (Hardy 1965) than would a population experiencing well-being in vacant niches with only a few new mutants.

The Offensive-Defensive Mechanisms of Cranial Loci Exposed to a Permanent Injury

Physiologically, offensive defense is a normal mechanism to fight diseases or to cope with long-lasting trauma. The response of permanent stress on bone results in the development of exostosis. This is a general phenomenon in humans as well as other mammals. Injury of the forehead bone, leading to development of pedicles and velvet antlers, was first noticed by Blasius (1903). The defense of injured forehead skin in bovids, by thickening and keratinizing certain cranial angles, was presumed by Cope (1887) and proved by the studies of Duerst (1926).

Duerst (1926) not only speculated that these anlagen must be developed long before (i.e., were dormant, author's remark), but he also speculated (p. 5) that there might be a psychological background for the epidermal protection which may precede the development of horned ossicone: "It seems that the psychological tendency of the epidermal protection is running ahead of that of the bony one" (free translation by the author). A similar view about behavioral stimulation was expressed, in a more modern interpretation, by Waddington (1957), Bateson (1979), and Ho & Saunders (1980); Ewer (1958) termed it the "habit a jump ahead of the structure."

The opposition toward the validity of behaviorally "enforced" macrochanges (Maynard-Smith et al. 1985) may rely on the developmental duality of the "Januslike" mechanism (Bateson 1979; Balon 1988).

In the Neo-Darwinian concept, they are the randomly developed mutations, whose origin is questioned (e.g., Charlesworth et al. 1988) and which are selected if useful. However, new behavioral innovations may activate, as mentioned earlier, via new messengers the silent or silenced alleles which later are assimilated as genocopy (Bateson 1979) into the genome. To illustrate this point, we have not only the classical case of the African suids (Ewer 1958), or the spotted hyena *Crocuta crocuta* (Racey & Skinner 1979), but also on the one hand, the evolution of horns and antlers, and, on the other hand, the co-optation of exapted voices into verbal language (Sebeok 1985).

Presently, it is difficult to say how many sequences of homeostatic thresholds (Balon 1985a,b) were necessary for either a genomic transcription or expression of other yet dormant DNA sequences to develop and fix genetically new forms of horns or antlers. The response of roe deer antlers architecturally to habitat in general (Bubenik, A. & König 1985) and the speed of architectural remodeling to other texture of the understory in particular (Markowski 1987), or the size of os cornu as response to thermal pressure (see p. 32), may serve as evidence for the impact of extrinsic stressors and epigenetic pathways for achievement of corresponding restructuring of the organ in question. They also support, as the most plausible explanation, the notion of a vault of stored programs and their activation by umwelt stimuli (Von Uexküll 1982).

More interesting, however, are the rare incidents of atypic antler constructions, which may mimic either a more primitive or a more advanced ontogenic stage or even develop yet unknown antler forms (Bubenik, A. 1966; Bubenik, G. et al. 1982). Noteworthy is the fact that appendages with construction that match the evolutionary, predictable design develop only on intact cranial loci (e.g., pedicles or frontals) where a direct connection to the corresponding neuron via its axon could be presumed (e.g., Landois 1904; Jaczewski 1967; Nellis 1965; Bubenik, A. 1966; Hartwig & Schrudde 1974; Vandal et al. 1986).

Thus, it seems plausible to hypothesize that the atypical shape of ectopic antlers may result either from lack of a neuronal connection (see pp. 17–18), or from a nonspecific messenger which may activate other DNA sequence (Fig. 5).

Support for this hypothesis may be seen in the space- and time-independent appearance of cranial appendages of the same nature (epiphyseal or apophyseal) and almost identical construction on the same cranial loci (Figs. 1,2) in very distant and often obviously genetically divergent lineages—e.g., in tylopods, cervoids, giraffoids, and bovids (Pilgrim 1941a; Bohlin 1953; Bubenik, A. 1962, 1966; Webb 1973, 1981; Patton & Taylor 1971; Webb & Taylor 1980; Janis 1982; Scott & Janis 1987). This enables us to hypothesize that these lineages might have a common but very distant ancestor with, in their effect, the same or very similar DNA sequences which are dormant (or cannot develop) as long as they are not activated by corresponding messenger. This is an idea close to that expressed long before modern genetics by Tandler & Grosz (1913): "The sexual characters did not develop by sexual selection as 'Nova,' but were acquired by ways and means within the range of phyletic processes" (free translation by author).

By and large, the dormant allelles could activate the development of the dermal or cranial exaptations. After they were co-opted as adaptations, their form may

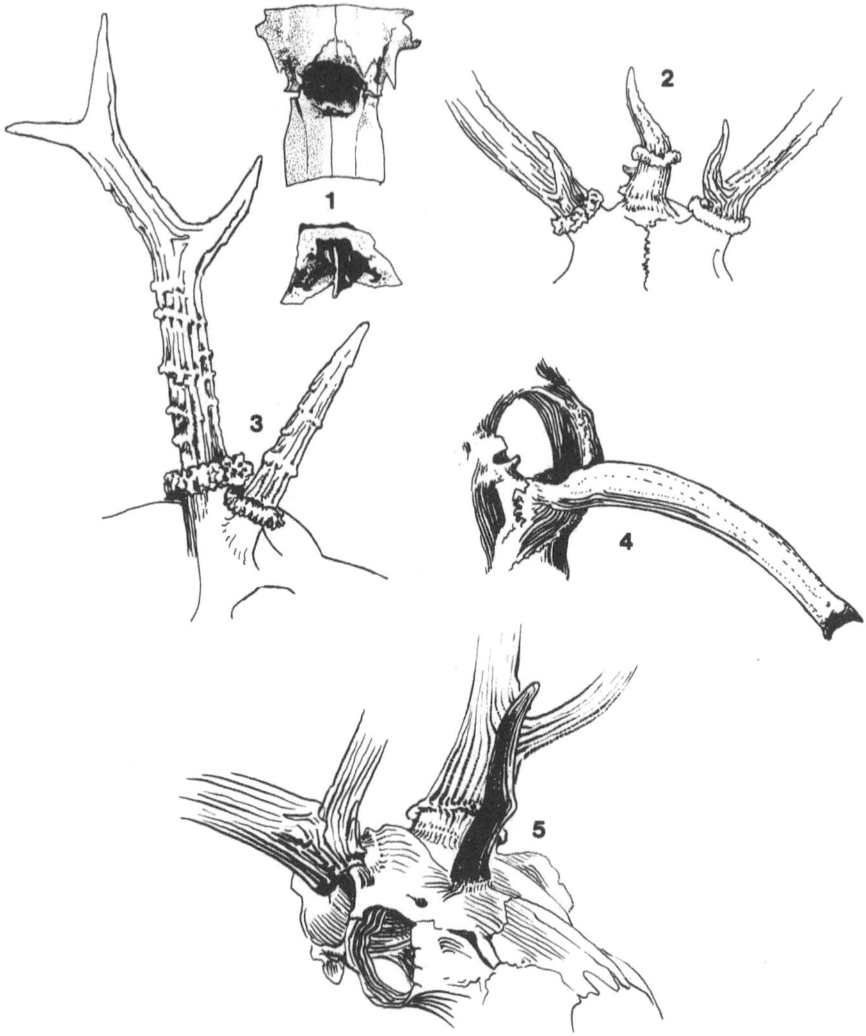

FIGURE 5. Examples of ectopic antlers developed on improper loci: *1* antler rudiment over the nasals in white-tailed deer *Od. virginianus* (Wislocki 1952); *2* antler on the parietal bone of a red deer stag *C. elaphus hippelaphus* (Landois 1904). *3* ectopic antler on the pedicle of a roe deer buck *Capreolus capreolus* (collection A. Bubenik); *4* antler on the zygomatic arch of a white-tailed deer (Nellis 1965); *5* ectopic antler over the junction of frontal bones of a caribou male *R. tarandus caribou* (redrawn from Vandal et al. 1986).

be selected from the available programs, or according to new transcriptions due to the inevitable pliomorphic changes. Using this hypothesis, eupecoran lineages with three different programs were possible:

1. Lineages carrying both velericorn or cavicorn ossicones and apophyses, located between rostrum and occipital bone. It must be confirmed whether

such complicated developments were possible, as might be the case with *Paleomeryx tricornis* (Zhanxiang et al. 1985) or some extinct giraffoids (Churcher, Chapter 5). In view of present knowledge such dual mechanism seems improbable.

2. Lineages possessing only proto-ossicones which may be the propagules of giraffes and okapi, and may lead to bovids. If the cranial appendages of *Paleomeryx kaupi*, shown by Ginsburg & Heintz (1966) were loose proto-ossicones, then some paleomerycins and giraffins may be very close relatives.
3. Lineages with only cranial apophyses which are represented by the different families of cervoids.

Origin of Sexual Dimorphism of Cranial Appendages and Tusks Regression

Because species-specific behavior, physiology, and form are interdependent, and their remnants can be found in neonates or as recapitulation in ontogeny, they should be some of the best cues for tracing their evolutionary pathway back to the extinct propagules.

The first firm support that horns are activated exaptations co-opted as useful organs was proposed by Duerst (1926). His remark, that it would be a *vicious circle* to assume that horns appeared prior to the drive to use them, hits the nail on the head.

The responsiveness of the cranial appendages to extrinsic and intrinsic "instructions," and their subsequent saltatory ontogeny, points to recorded "information" of past environments, behavioral traits, structural improvement, and their assimilation by the genome (Waddington 1961; Bubenik, A. 1982a). They even seem to be true phyletic recapitulations (Gould 1977; Alberch et al. 1979). Ontogenetic changes of antlers and horns (Rörig 1900; Rhumbler 1911; Bubenik, A. 1966; Vrba 1984; Jehenne 1986) show how the development might progress and theoretically, how extrinsic information could lead, via new messengers, to morphophysiological changes with hereditary effect.

Sex Monomorphic Nature of Embryonal Anlagen of Cranial Appendages

The sex dimorphism of cranial appendages of most cervid and many bovid lineages attested to the notion that in monomorphic lineages the antlers and horns of females are latterly acquired organs (Prevost 1869; Rörig 1899; Colbert 1958; Clutton-Brock 1982), or evolutive reversal or "quasi-independent" selection (Lewontin 1978; Kiltie 1985). However, there is no proof for such a hypothesis. On the contrary, there is enough evidence that the anlagen for cranial appendages were from the beginning monomorphic and their sex dimorphism is acquired.

The evidence that the anlagen for cranial apophyses are present, but dormant, or silenced in female of sexually dimorphic genera is the fact that they appear occasionally in females, or it is not difficult to reactivate them (Prevost 1869; Alston 1879; Landauer 1925; Dixon 1927; Bubenik, A. 1966; Jaczewski 1976; Bubenik, G. 1982; Kierdorf 1985). For example, the pristine pseudo-ossicones of giraffe (Lankester 1902; Spinage 1968b) are monomorphic as are ossis cornu of early caprins (Gentry 1978) and also of some early bovids. In deer the problem of sexual monomorphism of pedicle anlagen was first discussed by Berthold (1831), then by Cameron (1892) and Lydekker (1898). However, it seems that it was Rörig (1900, 1907) who first pointed to the fact that in the more primitive telemetacarpal deer the intact, but "antlered" females appear more frequently than in the more advanced plesiometacarpal taxa.

Frankenberger (1951) was the first to discover that early embryos of cervid species have monomorphic anlagen of pedicles. However, their postpartum development needs stimulation by testicular testosterone in all genera except rangiferins. Only the rangiferins can use other androgenic sources (e.g., from adrenals), which explains why both sexes can carry antlers. Without an adequate androgen level, both pedicles and antlers may accidentally disappear, not only in males but also in both sexes of rangiferins, e.g., on Slate Islands, Ontario (Bubenik, A., own records).

In some extinct antilocaprins, the peduncles were sex monomorphic (Webb 1973), and many females of extant pronghorn "antelope" are carrying velericorn or cavicorn peduncles (O'Gara 1969a).

Therefore, the horn dimorphism must be considered as a secondary adjustment—e.g., when only the much larger males needed better thermoregulation, or where the females also must defend calving territories and/or food resources. In cervoids, the silencing of alleles for antlerogenesis in females must have other reasons, because velvet-antlers do not have thermoregulatory effect and sexual antagonism (except during breeding period) excludes the need for resources defense.

If we accept as legitimate that cranial exaptations have developed by activation of dormant defense anlagen of tissues exposed to injury, then it is rightful to hypothesize why the traguloids or at least the closest sinecorn lineages, such as moschids and hydropotids, remained equipped with tusks (Aitchison 1946; Dubost 1971, 1975; Kingdon 1979) but did not develop cranial appendages to protect the head against possible injuries.

Theoretically, the answer is simple. They did not find it necessary to protect head angles, and/or for some reason they belong to lineages which never developed any dormant alleles which might be activated if a tissue defense became necessary. If behavior can be made responsible for the development of alleles for protective cranial exaptations, then a possible explanation of why cranial exaptations were not a primary concern of their evolution is the fact that both traguloids and sinecorn pecoran endanger mostly the neck, torso or pelvic area by slashes with tusks or kicking with hind legs.

In context with this notion it is worthwhile to mention that among the tylopods, a distant relative of sinecorn pecoran, only the Protoceratidae have developed cranial appendages. They lost the upper incisors, the upper canines were absent or reduced (Fig. 4, s), and molars were hypsodont (Patton & Taylor 1973). By contrast, the camelids have never developed cranial appendages. They use their sharp, pointed teeth, i.e., third incisors, first premolars, and canines in both upper and lower jaw, (Scott 1937), for biting, or eventually kick each other (Grzimek 1968). Head-butting is not in their fighting repertoire.

The Relationship Between Both Tusks and Cranial Appendages as Weaponry

The coincidence that in some ancestral lineages of eupecoran the tusks begun to regress with development of cranial appendages led to the notion that the tusks regression must be ascribed to a "compensatory reduction in favor of antlers" as better weaponry (Beninde 1937). This myth has been continually repeated, despite it was refuted by Obergfell (1957).

According to her, the upper canines began to develop into long protruding tusks in the Lower Oligocene, e.g., in the sinecorn *Eumeryx*. At that time the tusks were as large in males as in females. The plausible reason for their reduction should be the preference for more fibrous food, which may support Janis's (1982) theory that development of cranial appendages correlates with preference of that kind of forage. According to Obergfell (1957) that food needed further development of molarization and more intense chewing. This caused a shift in the chewing closer to the esophagus, because a higher pressure of the jaws is a logical prerequisite for digestion of woody food. Because intense chewing necessitates great sideward movement of the lower jaw, the long tusks have to be shortened to stumps. At this stage they became unemployed organs and the alleles for their development were silenced—up to oblivion in some lineages.

Other evidence toward the compensatory hypothesis is the fact that it cannot be implied for bovids and muntjacs. The latter are still equipped with tusks and use them as offensive weapons in intraspecific contests. The divergent pedicles and antlers serve as a shield that protects the face and the neck (Barrette 1977). However, in interspecific defense (e.g., toward dogs), the antlers are used as offensive weaponry (Bubenik, A., unpublished observation).

Finally, the frequency and length of rudimentary tusks presented in many cervid lineages also does not support the notion of compensatory regression. For example, the roe deer lost the upper canine despite very short and primitive antlers. In contrast, the red deer or reindeer with elaborated antlers have upper canines present—in the former as dull stumps, in the latter as small but sharp teeth. Thus, the tusks' regression in cervoids cannot be considered as the result of compensatory pressure toward better weaponry for a more important reason: the developmental control of pedicles and antlers is a complicated neurophysiological mechanism which has nothing in common with dentition.

Neuronal Control of Organ Development

The evolutionary pathways by which exaptations were developed and co-opted as periodically growing adaptations presuppose that these organs grow under neuronal control. A short review of mechanisms that control ontogeny of organs will show how far they coincide with those controlling ontogeny of cranial appendages.

First we have to emphasize that the neuronal and biochemical pathways through which the growth of organs is initiated and controlled in vertebrates require that the organ-anlage becomes activated "just on time," in the so-called "sensitive period" to produce an attractant for the axon of the neuron "in charge." To achieve this aim, specific messengers are used.

Neuronal Messengers

Both defense and offense (Lorenz 1964) are monitored primarily by the gray midbrain sets of neurons (Adams, D.B. 1979) and mediated by neurotransmitters (e.g., catecholamines). These and other substances can then guide the growth cones of axons to the local exaptation-anlagen and sensitize the new target tissue to produce nerve growth factors. If the new site is under the control of an allele, then the target tissue must produce neuronal attractants (Davies et al. 1987; Lockerbie 1987), and receptors for corresponding enzymes, hormones, and retinoic acid (Thaller & Eichele 1987; Umesono et al. 1988) to secure the development and growth of an adaptation as a "brand new" organ on the particular locus.

Thus, new messengers of extrinsic or intrinsic nature may be the primary vectors of reactivation of dormant alleles or development of new DNA sequences and organs. In that case the epigenetic approach (Waddington 1957; Ho & Saunders 1980; Lockerbie 1987; Pollard 1987; Løvtrup 1988) to evolution in general and that of eupecoran appendages in particular, seems more logical than any form of dispersal hypothesis, or the random mutations in the Darwinian or Neo-Darwinian concept of natural selection.

Assuming that specific dormant alleles need specific messengers, we can hypothesize that the same messengers can develop under the same or similar stressors, which up to now were not experienced. Because these crucial stressors depend on local situations, we may have the explanation for the space- and time-independent expression of same or almost similar cranial appendages. The accidental appearance of cranial appendages, in a form which is used by another genus, is additional support for the notion that the genome must have programs for more than one construction. This hypothesis is close to that of Roth (1984), who presumes that common developmental pathways are governed by the same batteries of genes, which makes these structures homologous. This concept finds also support in Todd's (1975) karyotype analyses of Artiodactyla and criticism of impact of mutation in evolution (Charlesworth et al. 1988; Ho 1988) or genetic imprinting (Weiss 1989).

A certain analogy about the link between chemistry of the messenger and dormant allele offers the adaptive mimicry of the caterpillar *Nemoria arizonia* discovered by E. Greene (Wicke 1989). The caterpillar changes its external coloration according to the chemistry of the food eaten.

Neuronal Connection

The connection between the target tissue and the corresponding neuron can be established only after the particular neuron can be established only after the particular neuron has migrated into its final position (Heumann 1987; Lockerbie 1987). Generally, this may occur in a relatively short period, depending when synaptogenesis occurs (Davies et al. 1987); in respect to appendages this can happen during fetal development, as in the giraffe, or postpartum, as in other eupecoran.

This principle is matched by the fact that generally the development of both anlagen, (i.e., pedicles and ossicones) needs more time than the fetal period. In deer, the velvet, as the most densely innervated tissue in mammals (Vacek 1955), may be the sensory receptor and transmitter of extrinsic tactile information via the parasympathetic nerves to the hypothetical brain centers for antler growth (see p. 78) (Bubenik, A. & Pavlansky 1965; Bubenik, G. et al. 1982b; G. Bubenik, Chapter 8).

Role of Chemotaxis in Neuronal Connections

Pathfinding of the target tissue by the corresponding axon is eased because the target tissue produces a diffusing attractant for the growth cone of the axon of a particular neuron (Davies et al. 1987; Lockerbie 1987). For pedicles, this chemoattractant is presumably the nerve growth factor (NGF) (Suttie & Fennessy 1987). Evidence that transection of the sciatic nerve raises the NGF level (Heumann 1987) may point to the reason why after a deep cut into the cranial locus of the pedicle, its growth can be initiated without hormonal booster (Jaczewski 1982; Bubenik, G. et al. 1982b; Lincoln 1984).

This means that as long as a certain anlage is incapable of producing a chemoattractant for the growth cone of the axon, it remains dormant. In deer these chemical activators are androgens, specifically testosterone. It is unknown whether in giraffoids and bovids it is also the sex or other hormone, or some presently unknown factors.

Neurotransmitters

Activation and arrest of growth cone advancement is promoted by neurotransmitters. One of them is serotonin, the effect of which is reversible (Lockerbie 1987).

Alpha-adrenergetic receptors for adrenaline or noradrenaline have been found on blood vessels of growing antlers (Wika 1978). The negative impact on the species-specific antler form by interruption of neuronal connection to the target field has been corroborated by transplantation of the pedicle anlage far away from the corresponding axon (Jaczewski 1967; Hartwig 1968; Hartwig & Schrudde 1974; Goss 1987). Such transplants are incapable of developing a species-specific form of the appendage and respond only to the seasonal levels of testosterone and their impact on antler cycle (Jaczewski, Chapter 13; Goss, Chapter 9, respectively). In contrast to this, transplanted anlagen of ossicones retain the species-specific form, and there are no records of eventual impact of neuronal control.

Zone of Polarizing Activity (ZPA)

Assuming that the growth pattern and differentiation of the cranial appendages are monitored by similar biochemical mechanisms as in other organs, the appendage-anlage should also have ZPA, where the *trans*-retinoic acid (RA) may operate as the local permissive agent (Thaller & Eichele 1987). Its concentration in these cranial loci may be monitored neuronally (G. Bubenik, Chapter 8).

Epidermal Growth Factor (EGF)

EGF stimulates the fast tissue growth. It has been found in relatively high concentrations in antler velvet (Ko et al. 1986). The action of hormones on the pre- or postsynaptic membrane, which alters the permeability to neurotransmitters or their precursors, and/or the function of a neutrotransmitter's receptors, is considered nongenomic (McEwen et al. 1979). Because the timing of the antler growth is also neurohormonally controlled, we hypothesize that the beginning of the mechanism which induces pedicle and antler development was also originally nongenomic. Thus the androgens stop and initiate antler growth by altering the permeability of the pre- or postsynaptic membranes of the corresponding axons and dendrites. This dual mechanism of induction and discontinuing of antler growth proceeds over complicated feedback mechanisms acting on the receptors in the brain and/or the target tissue (see G. Bubenik, Chapter 8; Suttie, Chapter 12).

Electrical Fields

Another prerequisite for the guidance of the growth of the neuron's cone is the extracellular electrical field (Lockerbie 1987), which can attract the cone to, but also deflect it from, the original locus. In growing antler the most negative potential was recorded from the apex. Electronegativity increased with the rate of antler growth. Long-term application of direct current to the supraorbital branch of the trigeminal nerve accompanying the superficial arteries of the growing antler produced uni- or bilateral acceleration of antler growth, or foreign and even up to now unknown branching pattern (Bubenik, G. et al. 1982; Bubenik, G. & Bubenik, A. 1978; Lake et al. 1982; G. Bubenik, Chapter 8). Again, we do not have records about the impact of electrical fields on cornuogenesis. Summing up, the physiological processes controlling antlerogenesis and cornuogenesis seem to go along with all the known pathways by which development of other organs is controlled. However, the independence in neuronal control of both pedicles and behavior of transplanted anlagen in cervids, in comparison to bovids, constitutes the line of evidence for the different origins and growth control of both antlers and horns. The species-specific shape of cranial appendages, its ontogeny and correlations with body size, other visual releasers, habitat texture, and architectural ritualization point strongly to epigenetic influence.

Behavioral Impact on Development of Cranial Appendages

In the epigenetic concept, the extrinsic or intrinsic information leading to activation or restructuring of new expression of genome is introduced during fetalization or neogeny (Bolk 1926; Waddington 1957; Balon 1986). The observed stepwise expressions during ontogeny may be recapitulations of evolutionary important changes (Alberch et al. 1979; Stanley 1979; Umesono 1988).

The fact that in extant eupecoran the embryonic anlagen of cranial appendages are monomorphic supports the assumption that the injuries of the corresponding loci which elicited the defensive exaptations were of an "every-day nature." The most probable injuries may originate from sparring strategies; these can be either head-toward-head clashes ritualized from greeting rituals on the one hand, or fighting with forelegs (flailing), or kicking with hind legs toward the head.

Sex Monomorphic Differences in Head Display of Neonates

Support for sex monomorphic nature of cranial appendages in eupecoran deliver the greeting or sparring pattern of neonates. They are monomorphic regardless of the presence or absence of cranial appendages, and different from family to family, with specialized tuning in the genera, species, and even subspecies.

Newborn giraffes use the lateral blow with their small "hornets" (Innis 1958), or offer them for licking (Bubenik, A., unpublished observation). Newborn bovids, which in general are "followers" (Walther 1984) and within hours socialize with peers, have the forehead exposed to trauma just after birth as they run behind their mothers through the dense undergrowth or spar with forehead or upward or lateral head blows toward the torso (Schaffer & Reed 1972; Bubenik, A. 1982b; Kingdon 1979; Walther 1984; Byers 1987). In contrast, the newborn deer, which by and large are "hiders" (Lent 1974), socialize weeks after birth and do not perform frontal butting, flailing, and kicking prior to pedicle growth (Barrette 1977; Bubenik, A. 1982b; Byers 1987; MacNamara & Eldridge 1987).

However, there are some interesting "coincidences": the only follower among deer, the antler-monomorphic reindeer, is also the only deer species in which pedicles and antlers begin to grow just after birth and their velvet is shed about 90 days afterward (Bubenik, A. 1966). Among bovids the interesting exception are the paired anlagen of the anterior ossicones of the four-horned antelope *Tetracerus quadricornis*. The posterior pair of horns begin to grow soon after the birth (Duerst 1926). The anterior pair begin to develop after sparring with posterior horns has become a daily ritual (Bubenik, A., unpublished observations). Unfortunately, there are no records showing whether the asynchronous ontogeny of horns of the four-horned antelope is related to differential timing of ossicone development or other unknown factors. The lack of knowledge of greeting or sparring patterns in neonates of sinecoran does not allow verification of the above hypothesis of the possible importance of behavioral patterns in eupecoran neonates for tracing their evolutionary history.

The co-optation of cranial appendages as more or less conspicuous organs and sparring tools cannot be ascribed only to their defensive attributes. The earliest stages of the pre- and postnatal development of cranial appendages corroborate that their development was accompanied by other pliomorphic changes.

Primary and Secondary Function of Cranial Exaptations

Prior to the local keratinization of the skin or protrusion of pedicles, we can notice a restructuring of the skin covering these locations. The integument becomes not only thicker, but also equipped with numerous sebaceous and/or apocrine glands (Schaffer 1940).

The sebum helps to grease the surface of the hair and, later, of the exaptations, making them more slippery and more resistant to abrasion or injury. Residues of this primordial skin restructuring and function are still preserved in eupecoran: both sexes of newborns have on the forehead two rut wrinkles [Brunstfalten (Schaffer 1940)] (Lankester & Ridewood 1910; Spinage 1968b). With development of cranial appendages, the scent-producing skin areas develop either glandular field around the basis of ossicones or expand with the skin of the horn stalks or into the velvet (Schuhmacher 1939; Schaffer 1940; Vacek 1956).

Thus, these areas are also important scent glands which dissipate and leave individually tuned odors in the optimum horizon of olfactory reception (Bubenik, A. 1962, 1971, 1982b); they ease not only the following of conspecifics but also the pleiotropic discrimination of males' preferences by the female.

Therefore, regardless of breeding strategy (Kirkpatrick 1987), the olfactory function of cranial appendages cannot be overlooked when searching for models of sexual selection by female's choice in Eupecora (Bubenik, A. 1985). Thus, the chemoreceptive importance of the cranial appendages may stimulate their growth to magnify the surface and odor production (Bubenik, A. 1971; Walther 1984). As soon as the appendages grew over the upright-held ears they became more and more important visual releasers, which accentuated the head movement as they became longer and more branched. Hence, both natural and sexual selection may contribute to their further development.

Role of Cranial Appendages in Visual Perception of Eupecoran

Due to the increase of cranial appendages, the front-pole—which, contrary to Portmann's notion (Portmann 1970), is not just the head-pole but involves also the neck—became a severe competitor to the optical accents of the caudal part of the body (i.e., the rump patch and the tail). In deer, the cryptic coloration of the torso faded out, and the rump patch and the tail became low-key releasers; in giraffoids the overall cryptic coloration pattern remained; in many bovids the cryptic coloration of the torso, on the one hand, and a long tail on the other, were retained despite well-developed horns. In other bovids only the torso became semantic by conspicuously contrasting colors; horns almost disappeared in the head profile, and tails were shortened to a visually unimportant cue (Fig. 6). By and large the polarization of bodily accents with emphasis on the front-pole was achieved only in cervids (Fig. 7).

►

FIGURE 6. Polarization of optical accents from torso in bovids and cervids: *1* Gemsbok *Gazella oryx*, *2* white-tailed gnu or black wildebeest *Connochaetes gnu*, *3* roan antelope *Hippotragus equinus*, *4* water buck *Kobus defassa ellipsiprymnus*, *5* mouflon *Ovis musimon*, *6* Thompson's gazelle, *Gazella thompsoni*, *7* sika deer *Cervus nippon*, *8* barren ground caribou *Rangifer t. caribou*.

1
2
3
4
5
6
7
8

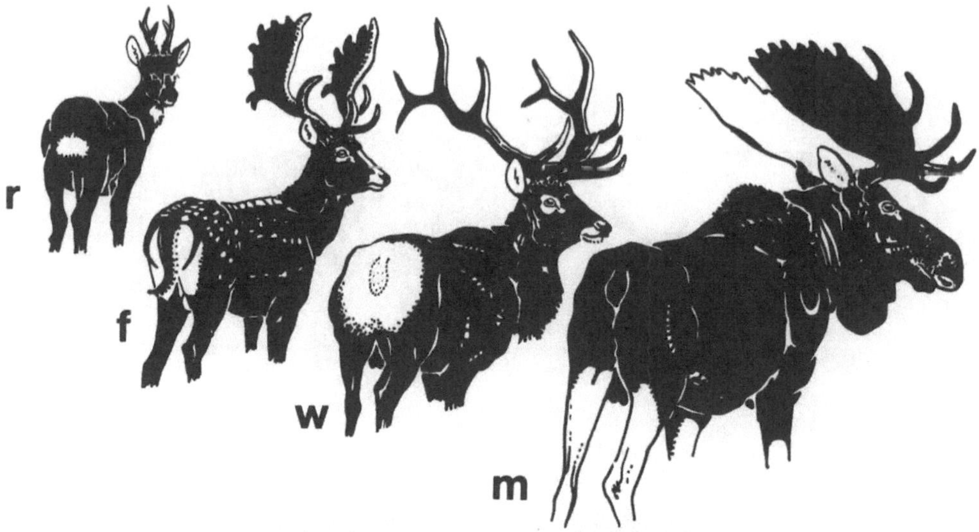

FIGURE 7. Polarization of optical accents in cervids. The front-pole is more and more accentuated at the expense of the caudal-pole in: *r* roe deer *Capreolus capreolus*, *f* fallow deer *Dama dama*, *w* wapiti or elk *Cervus e. canadensis*, *m* Alaskan moose *Alces a. gigas*.

The rank-order in which these optical accents operate or to which the conspecifics respond, seems to depend on the evolutionary stage of the species, role of horns in female's choice, visibility in the original habitat, and texture of the original understory which must be penetrated by flight (Bubenik, A. 1982b; Walther 1984). For us as observers, the use of cranial appendages in visual communication bears one tempting error, ignored in all papers dealing with this topic (e.g., Walther 1966; Geist 1978b, 1987a): the eye-angles of eupecoran are much more narrow than those of humans, and their optics differ from those of primates.

Visual Perception in Eupecoran

Basically, the eye of ungulates has an oblique pupil and retina (Johnson 1901), which remain permanently horizontal (Schneider 1930) regardless of the position of the head; the lens cannot accommodate, the retina is ramp and not spherical (Prince et al. 1960); when the eupecoran look straight forward, most of their eye-angles lie between 32° and 55°; occasionally they can reach 85°. Therefore, when the ungulate does not converge the eyes in order to improve the stereoscopy, the vision is merely monoscopic (Jackson 1977; Bubenik, A. 1987).

Thus, from a distance the animals see the conspecifics in one plain. In other words, from behind the rump patch is projected as part of the head and does not operate as an independent stimulus. Equally, the tridimensional cranial appendages are projected in one plain regardless of their angle toward the observer. The apparent monoscopy and difficulty in assessment of distances appears as a handicap for the human observer. For the eupecora it is an advantage: it allows a control of a horizon of more than 300° (Bubenik, A. 1987a). For location of suspicious or even dangerous objects, the stereophonic and olfactory location is much better than the visual.

When it is necessary to discriminate spatial configuration and distance of close objects (e.g., the horns or antlers of a combatant, or a creeping predator), the stereoscopy can be temporarily improved by turning the eyes forward, backward, or downward (Bubenik, A. 1987a). Due to the specialized optics, the front-pole and caudal-pole, the coloration and posture of the torso, and the other visual stimuli represent very pleiotropic, but species- or even subspecies-specific codes for which both the transmitter and receiver are programmed (Bubenik, A. 1982b, 1987).

The disregarding of the specificity of quite different optics of eupecoran on the one hand, and the evolutionary importance of the body poles and torso coloration on the other, led to erroneous conclusions since the classical studies in bovids of Walther (1966, 1984), for whom torso posture was an important communicative cue. This, a priori, should not mean that for cervids, just because they are ungulates and ruminants as well, the torso must have the same importance as it has for bovids. This basic difference could be confirmed by the impact of artificial front-pole carried on the chest of a man, whose bipedal torso does not play any role, as will be discussed later.

It is interesting that the male cervids and bovids do not discriminate between species-specific and species-atypic antlers and horn shapes. However, the male cervids know each other's antler shape and rank in the sense of "proper names" (Hediger 1976), apparently by frequent sparring. It is unknown if such recognition of partners exist in bovids.

Male and female bovids discriminate between different horn size as a rank symbol. However, male bovids with horns of different rank like to spar in a playful manner whenever they are in the mood to do so. Horns are a good fighting tool, and male bovids are ever ready to use their horns toward antlered deer in defending food resources, etc. In sheep the female discriminates and recognizes species-specific head and horn size (Collias 1956; Kendrick & Baldwin 1987), and it prefers mates with the highest-ranking horns (Geist 1971). In contrast, the female deer seems to have innate discrimination of the species and even subspecies shape of antlers (Bubenik, A. 1982a) and their scent (Bubenik, A. 1987a).

Species-Specific Design of Cranial Appendages

Which design of cranial appendage is selected among the stored, dormant, or temporarily silenced construction programs seems to depend on the optimum utility. Growth and shaping may proceed to better accentuate the front pole, as an olfactory and optical cue and/or as a better permanent or temporary sparring tool. Also, the texture of the habitat which must be penetrated seems to exert pressure for selection of the angle of the appendages in relation to the profile of the head and the body axis. In that case, again, the habit may jump ahead of the new structure. Being pushers, twisters, or head-clashers, "slippers" (slipping through the understory, the Schlüpfers in German), duikers, gallopers, or runners (Bubenik 1982b), necessitated not only appropriate design of the appendages (Bubenik, A. 1982b; Walther 1984; Kitchener 1988), but also anatomical and morphological restructuring of the skull (Schaffer & Reed 1972), the occipital area (Kobrin-Morizi 1984), the postcranial skeleton (Scott 1987), and eventually a temporary higher density of fibers of the neck muscles (Leinders 1984; Schnare & Fischer 1987) to withstand the pressure exerted on the appendages during sparring or contest.

Cervicorn appendages, as more olfactory and intimidating organs, were aimed more towards surface expansion than toward length increase (Chard 1958; Bubenik, A. 1982b); horns, by contrast, were directed primarily towards shapes suitable as a battering tool (Walther 1966) and less as visual releaser. Constructive or architectural reshaping to higher stages with eventual reduction or even disappearance of no longer important tines can be followed in evolution of many cervid lineages, as will be discussed later.

Polymorphism of Cranial Appendages

Astonishing similarities exist in the construction and architecture of cranial appendages of extant and extinct North American and Eurasiatic eupecoran. This points to some genetical links; however, their proper connection is difficult to establish due to the fact that the giraffoids and paleomerycids as ancestors either were polymorphic in regard to their cranial appendages or represent groups of different lineages with analogous characters. For example, the North American cervoid dromomerycids are considered an offshoot of paleomerycids (Webb 1977). This might be possible if we can link them to the Eurasiatic cervoid *Paleomeryx simplicicornis* (Schlosser 1924) with simple spike-like apophyses. All other paleomerycids, e.g., the Eurasiatic *P. kaupi* or Chinese *P. tricornis* were equipped with ossicones, a specific bovoid feature. Similarly, the earliest Eurasiatic cervoids seem to be offshoots of both giraffoids and paleomerycids (Webb 1977; Churcher 1978; Webb & Taylor 1980). But it is difficult to imagine that true giraffoids could have a lineage with cervicorn apophyses like climacocerids, or that the *Procervulus aurelianensis*, with points developed by sprouting like the dromomerycid *Triceromeryx*, is the ancestor of *Procervulus dichotomus*, a typical protocervid with forked beam; questionable is whether cervoids like hoplitomerycids (Leinders 1984) could produce ossicones [Bubenik, A., unpublished study], or whether the cervoid *P. simplicicornis* (Schlosser 1924) could be a relative of the other paleomerycids with ossicones (Fig. 8).

A plausible explanation of these morphological discrepancies within one lineage is difficult with the present knowledge of mechanisms controlling both cornuogenesis and antlerogenesis. The similarities in skeleton and dentition (Teilhard de Chardin & Trassaert 1937; Sewertsow 1951; Thenius & Hofer 1960; Hamilton 1978a,b; Janis & Lister 1985) among the pristine giraffoids and paleomerycids and the diversity of their cranial appendages allow us to question

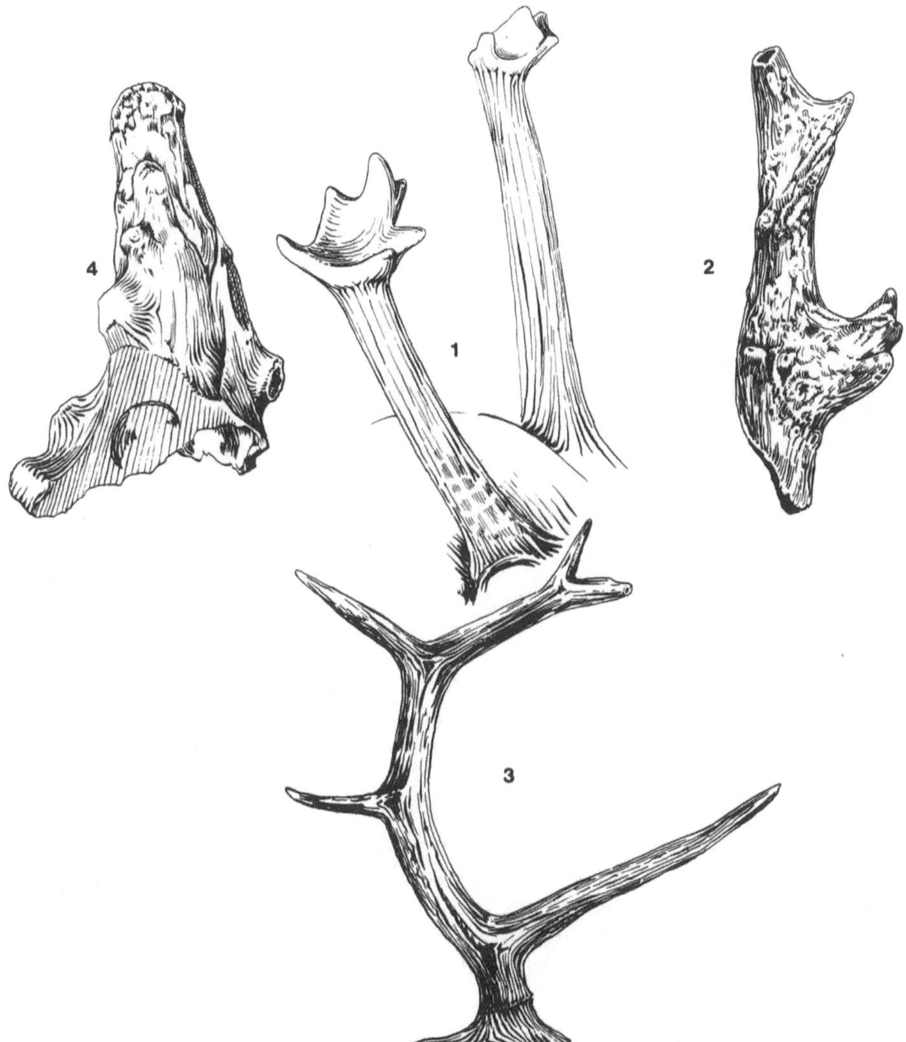

FIGURE 8. The pristine taxa among giraffoids-paleomerycids: *1 Lagomeryx praestens* with supraorbital apophyses, *2 Procervulus aurelianensis*, (both Museum of Natural History, Basel), *3 Climacoceras* (Hamilton 1978a), and *4 Triceromeryx*, (Bohlin 1953). All are ramified by sprouting.

whether paleomerycids were closer to cervoids, as has been proposed by Hamilton (1978a,b) and Groves & Grubb (1987) or to giraffoids (Bubenik, A. 1982c), or whether they were an independent clade, as proposed by Janis & Lister (1985) and Scott & Janis (1987).

Besides this, some of the velericorn appendages seem to be lifelong, and others facultatively perennial, i.e., aperiodically deciduous. When the distal part of the latter was rejected, the base lacked a compact bridge and coronet and had a concave profile. The sequestration often occurred asynchronously, biased to the axis and sometimes far below the "swollen" shaft (Bubenik, A. 1982c).

In extinct paleomerycids, giraffoids, and bovoids, the locations and numbers of cranial loci varied from two to three and even to four or five. Examples of this are the extant giraffe (Spinage 1968b), the extinct *Giraffokeryx* (Colbert 1933; Churcher, Chapter 5), the *Dromomerycinae* and *Cranioceratidae* (Webb 1983a), the four-horned antelope *Tetraceros* (Grzimek 1968), and the *Hoplitomeryx matthei* with five appendages (Fig. 4).

In modern eupecoran, the cranial appendages are either epiphyseal or apophyseal, but not of both natures. The exception is the giraffe, in which azygous exostoses mimic sprouts of climacocerids (MacInnes 1936; Urbain et al. 1944).

Part B. Horned Eupecoran

The "Horned" Eupecora: Giraffoids, Protobovids, Eubovids

The anlagen of proto-ossicones (in the author's term) (Lankester 1907a,b) and ossicones as well (Dove 1935; George 1956) are in species-specific dermal loci of covering the frontal and parietal bones. In contrast, ectopic-horned – "exaptation" or, precisely, the scurs (Fig. 9) can develop at any skin location of the body, but scurs close to specific loci have the shape most similar to the specific horn (Schönberg 1928; Meile & Bubenik, A. 1979).

The dermal embryonal anlagen of ossicones unite the giraffoids with the bovids. The main superficial difference is the velericorn nature of the former and the cavicorn of the latter. Morphologically, of course, the cavicorns are more advanced taxa. Unfortunately, we do not have reliable information about the neurophysiological growth control of both.

Ossicones of Giraffids and Paleotragins

The embryonal anlagen of the proto-ossicones of the giraffe *Giraffa* sp. develop into small, not ankylosed velericorn appendages during the fetal life of both sexes (Lankester 1907a). They begin to ankylose during adulthood (Spinage 1968b). Surprisingly, the proto-ossicones appear only in males in the more primitive okapi *Okapia*, subfamily Paleotraginae, superfamily Giraffoidea (Geraads 1986). It would be worthwhile to know whether these anlagen are also sex dimorphic in embryos and newborns. The residues of rut wrinkles of calves of both sexes (Lankester 1907a), and the reported but questionable ossicups of an okapi female (Lankester 1902) might be examples of the presence of "dormant" anlagen in females.

The integument of the apices of a giraffe's proto-ossicones is subjected to only slight keratinization. This produces a horny disk, which is discarded probably due to permanent regeneration of the underlying skin (Lankester 1907b;

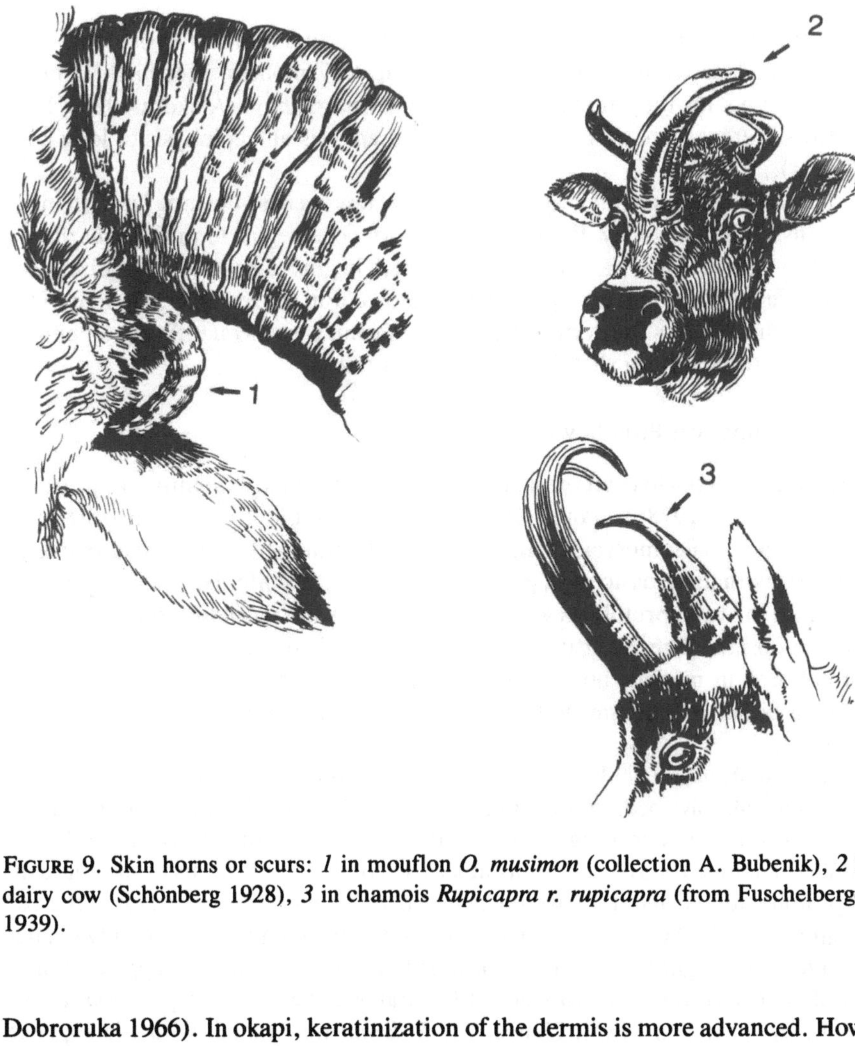

FIGURE 9. Skin horns or scurs: *1* in mouflon *O. musimon* (collection A. Bubenik), *2* in dairy cow (Schönberg 1928), *3* in chamois *Rupicapra r. rupicapra* (from Fuschelberger 1939).

Dobroruka 1966). In okapi, keratinization of the dermis is more advanced. However, the small horned caps can be rejected temporarily and rebuilt subsequently, as confirmed by Grzimek (1968).

In maturing and mature giraffes, the top of the skull and the forehead bones are exposed to powerful blows to which they respond by permanent reinforcement. According to Spinage (1968b), the superficial bone growth is so dynamic that during aging the exposed facial veins are finally completely embedded into new bone layers. Perhaps, the long and strong hair around the thick, very poorly keratinized "horned" apices and the cartilaginous "cushions" between the proto-ossicones and the skull serve as elastic buffer which may reduced the effect of horn blows on skull and brain, and prevent severe injuries in juveniles (Innis 1958). The additional reinforcement of the skull by supraorbital and azygous exostosis (Spinage 1968b) may also prevent subperiosteal hemorrhages and produce better physical conditions.

Bovid Horns

The embryonal ossicone anlagen develop postpartum or a few days before birth. They ankylose completely, very soon or up to 7 months after birth (Kingdon 1979). In bovids they ankylose with the protruding *boss* (Atzkern 1923), but in a way different from that known between diaphysis and epiphysis of long bones (Fig. 1).

The first horn cup may appear within a few weeks, but not later than 6–7 months. Simultaneously, the base of the horn core begins to proliferate through the periosteum above the cranial locus, which begins to protrude, building a boss which relatively soon ankyloses with the os cornu (Dove 1935; George 1956). During this process the dermis begins to keratinize, subsequently developing the typically formed horn sheaths at specific intervals.

The Hypothetical Protobovids

Assuming that bovid cavicornia have developed not independently—as presumed by Hamilton (1978b); Janis (1987) and others—but from some offshoot of giraffoids or paleomerycids, then they should emerge as a cavicorn lineage or lineages with loose ossicones, poorly ankylosed late in life. For such a hypothetical stage the term protohorn and for their bearers the term protobovids seems to be appropriate. The incentive for this hypothesis is the accidental occurrence of protohorns in modern bovids, which appear as scurs (Dove 1935) or may be related either to atavism or to fetal malnutrition (Sandiford 1829; Rhumbler 1932).

At present, we do not have evidence of any extinct or extant protobovid (albeit not all bovids have been studied in this regard). The fact that the "most primitive" bovid—*Eotragus* (Ginsburg & Heintz 1966), from the Mid-Helvetian (19 to 17 MYA) (Thenius 1983)—has firmly ankylosed ossicones, leads us to speculate that if protobovids existed, they would have evolved much earlier, perhaps at the very beginning of the Miocene or towards the end of the Oligocene. The only connection to them might be the aforementioned *Paleomeryx kaupi*. It appeared in the Burdigaliean (lower Mid-Miocene of Central and Western Europe, about 23.8 to 17 MYA) (Thenius 1983). Hence, the *P. kaupi* may match with the supposed evolutionary line of protobovids derived from some paleomerycids.

The Eubovids

If the protobovids did indeed exist, the modern bovids with horned ossicones fused with the skull after birth (Atzkern 1923; Dove 1935; George 1956; Kiltie 1984) deserve the term eubovids. The primitive caprids and antelopes, including gazelles and ovibovins (Thenius 1980), such as *Pseudotragus* and *Oioceros*, which appeared at the end of the Tertiary are considered as the next evolutionary steps, above that of *Eotragus* have had still short and slightly curved ossicones (Gentry, Chapter 6). However, the explosive diversification of horn spirals erupted during the Pliocene.

Activation and Control of Os Cornu Growth

We do not know whether what messenger is necessary to activate both the dermal anlage of os cornu, the cranial boss and the keratinization of the epidermis. Whether or not male or female sex hormones are involved in activation of anlagen, as is the case in deer pedicles, is still unsolved. Landauer (1925) considers sex hormones as facultatively necessary. Duerst (1926) and Goss (1969c) are convinced that horn growth is independent on pubertal testosterone, because horns of juvenile castrates of boreal bovids grow faster than those of intact males and females. This, of course, does not prove that the cranial loci over which species-specific horns or scurs grow (Long & Gregory 1978) do not need a booster of fetal sex hormones.

The noninvolvement of sex hormones in horn sheath growth is evident in tropical bovids, which are of Miocene or Pliocene origin. The keratinization of the epidermis of the ossicone does not stagnate seasonally or periodically. This may be due to the fact that the sex hormone secretion, at least of males, only undulates seasonally, and females may experience many estrous cycles. Therefore, it would be worthwhile to study the physiology of cornuogenesis in sex dimorphic genera with extremely delayed os cornu and horn development, such as *Tragelaphus*, *Redunca*, and *Kobus* (Kiltie 1985).

However, in boreal bovids of Pleistocene origin, which have recurrent puberty and therefore a great annual differential between nadir and zenith of sex hormone levels, the development of horn cones is stopped seasonally during the period of high sex hormone levels. Hence, with the achievement of full maturity the annual gain of horn sheaths is only a fraction of that of juveniles (e.g., in sheep, goat, or cattle) and just smaller (>2 mm) in the rupicaprins.

The mentioned differential in involvement of sex hormones in horn growth is corroborated by the records of Khan et al. (1936), Dobroruka (1960), or Mohr (1965). According to these records, the regeneration of horn sheaths is possible in many bovids of Miocene-Pliocene origin but is rare in Pleistocene bovids, which shows that the sex hormone involvement might correlate with different evolutionary age of the genera.

Genetics of Cavicorn Ossicones

The interesting studies with transplanted os cornu cast more light on the inheritance of horns. Duerst (1902) emphasized that artificially separated os cornu cannot develop horn sheath and vice versa, but Dove (1935) proved that skin-horn or scur may have a loose ossicone. In the relatively primitive goats (Gentry 1978), a horn core transplanted above the frontomedial suture produces a horn sheath and ankyloses with the parietal bone. In goats, surgically polled individuals regenerate the ossicone (Dove 1935). In some modern bovids (e.g., cattle), the transplanted os cornu anlage does not ankylose and the os cornu does not regenerate after polling. Beside this, regeneration of an os cornu in boreal bovids is possible only before the first puberty (Mohr 1965).

It is important to note that a horn core transplanted on the occipital bone fuses completely with the cranium and is rebuilt as a solid bone (Dove 1935), analogous to a compact spur. Thus, I conclude that the occipital protuberance might not be morphophysiologically homologous to other cranial protuberances, i.e., pedicles. In that case, indeed, the development of occipital spurs in the pristine cupcoran (for example, in some paleomerycids, giraffoids, cervoids, and tylopods) may also be ascribed to a common allele, which expresses itself under the same stimulus.

Therefore, Dove (1935) speculated that there is only a single genetic factor controlling the character of both os cornu and horn sheath; he considered the keratinized sheath as a product of interacting tissues. Asdell (1944) pointed out that intersexes are found only among polled goats. Hutt (1964) had shown that homozygotes in goats are polled in both sexes, but intersexes only in females. Thus, genetically polled goats are the product of selection of a dangerous mutation. Long & Gregory (1978) established four independent loci for inheritance of horns and scurs in races of cattle and found all cattle homozygous for the gene for horns, which is epistatic to the gene for scurs.

By and large, the horn symmetry is almost perfect, and both horn cores and horn sheath develop synchronously despite the fact that the left horn seems to be used more frequently in sparring (Ludwig 1970). Evident horn asymmetry in bovids is extremely rare. There is also no record that horn symmetry is affected by skeletal injuries or any kind of stress. Anomalies of horn spiral, such as reverse-turned horns and horns growing into cheeks or neck, are also symmetrical.

The almost perfect horn symmetry may point to one controlling mechanism for both horns; however, horns of hybrids in cattle, goat, and sheep which we encountered in India are asymmetrical. Horn asymmetry is known in the continental mouflons *Ovis musimon*, which were crossed with domestic sheep 100 years ago. In contrast, the horn asymmetry in the oryx, *Oryx gazella*, and kudu, *Tragelaphus strepticeros*, seen in Namibia must be ascribed to injuries caused by slipping through the numerous fences of farmlands.

Shaping of Horns and Thermoregulatory Function of Os Cornu

Principally, the horn grows as a spiral (Cook 1914), with almost infinite radius as in oryx, alternating radii as in chamois (Bubenik, A. et al. 1977), or very small and compressed as in goats and sheep. The fact that the os cornu of boreal bovids is substantially shorter than the horn cavity, and that the scurs have shapes of specific horns, led Duerst (1926) to the assumption that the species-specific horn shape is monitored by the species-specific property of the stratum corneum and the torsion effect of the hair medulla, and not by the horn core alone.

However, the morphological properties of hair seem not to be the single determinant of horn shape, as corroborated by the dramatic ontogenetical changes in horn spiral in gnu *Connochaetes gnou* or muskox *Ovibos moschatus*, and the sex dimorphic horns in ibex *Capra ibex* or bighorn, *Ovis canadensis*. According to R. Köenig (personal communication, 1986), the horn spiral together with the sulci and ridges are the result of the unequal thickness and growing speed of the keratinizing epidermis around the horn core (Fig. 10). However, why the spirals are so manifold is difficult to explain. Sometimes it seems that the primary aim was optical accentuation of the head-pole; other times the horns are shaped to a smooth penetration of the understory or reinforcement and reshaping of the horn surface to a better buffering effect in violent contests (Walther 1966; Vrba 1984).

Last but not least, we must take into consideration the ritualization of horn shape, aimed toward the optimum expression of the status with a minimum offensiveness in the form (Walther 1966). Synchronously with spiral growth and torsion also axes of the os cornu lamellae must be permanently rearranged in order to adjust to the torsion effect of the horn spiral (Atzkern 1923), a histologically important feature.

Thermoregulatory Function of Os Cornu

The fact that variation in horn spiral was more diversified and the os cornu is much longer in the tropics than in cold latitudes may be a result of their thermoregulatory function (Taylor 1966). This, of course, does not mean that horns have developed only as thermoregulatory organs.

The extremely long horn cores which reach the bottom of the horn apex [for example, in oryx *Gazella oryx*, sable antelope *Hippotragus nigger*, and the extinct *Pelorovis* and *Homoioceros* (Thenius 1980)] probably could never develop under the severe climatic conditions of the Pleistocene tundra and forest-tundra, even though the horns, (for example, of *Ovis dalli* or *Ovis ammon polii*) are respectably long. Thus, horn length and ossicone length seem to be two independent variables, adapting opportunistically to two different conditions: on the one hand, for visual assessment of conspecificity (Kendrick et al. 1987), maturation class, or behavioral reasons (Bubenik, A. 1984), and on the other hand, for ambient temperature of the habitat. The former determines the horn core as thermally sensitive bone (Table 1).

Unfortunately, it was difficult to obtain enough specimens with uncut ossicones to verify the hypothesis statistically. However, the very small differences in measurements of the specimens we used may be representative for a rough comparison of the mentioned responsiveness of os cornu to ambient temperature.

All measurements of horns and ossis cornu of the subarctic sheep have energetically the best parameters, i.e., they protect from great losses through os cornu. In contrast, the sheep and ibex of more moderate climate have relative larger surface of os cornu and narrow relationship between horn and os cornu length. Of course, we have to admit that the calculation of surfaces are simplified, reducing the core to a straight cone and neglecting the unknown thickness of horn sheath and sulci, which may change the energetic losses in one or other direction.

The Nature of Sexual Dimorphism and Social Significance of Horns

The evolutionary trend toward small horns or total disappearance of horns in females may be monitored by their social role and energetics. The weight of

A
B
C
1
2
3

TABLE 1. Average horn sheath and hore core measurements of mountain sheep and ibex.

Name	HL mm	BCH mm	OL mm	OL:HL %.	S cm^2	Age year	N
Capra ibex ibex Alpine ibex	930.0	250.0	570.0	61.3	723	11–14	5
Capra ibex sibirica Mongolian ibex	1200.0	210.0	373.0	31.5	429	?	4
O. nivicola Kamchatka sheep	740.0	335.0	228.0	30.8	455	?	2
Ovis dalli White sheep	1110.0	367.0	275.0	24.8	399	9–11	2
Ovis ammon polii Pamir argali	1078.0	275.0	404.0	25.4	622	?	2
Ovis canadensis nelsoni Desert sheep	1109.0	394.0	422.0	38.0	740	13–16	5
Ovis c. canadensis Bighorn sheep (Colorado)	1230.0	395.0	387.5	32.6	890	7–10	8
Ovis catclawensis[a]		363.0	335.0		713	?	?

Measurements of intact horn cores of mature individuals, which horn length did not differ more than ±2.5 cm from the presented mean. The significance of these values must be questioned due to the low number of specimens, and some other parameters. HL = horn length, BHC = basal horn circumference, OL = ossicone length, S = surface of the ossicone (calculated as a straight cone with circular basis).
Measurements are from collections of A.B. Bubenik, J.P. Boone, Denver, Dr. Tikhonov, Zool. Instit. USSR Academy of Science Leningrad, and data from Schaffer & Reed (1972).
[a]Corner 1977.

horns, possibly the high metabolic claims on horn growth, and the entropy may be energetically either expensive or advantageous, depending on the ambient temperature and abundance of food resources. Generally, the long-horned females may be socially disadvantageous during the rut. Females in the just attractive, but not proceptive or receptive, phase of estrus (Beach 1975), do not tolerate the penetration of their "intimate zone" by the male. Well-horned females may respond to a male's harassment by offensive use of horns which would bring havoc in mating herds and eventually handicap the male's sexual performance.

◄

FIGURE 10. The form of the horn spiral: *A* as a visual cue in *1* great kudu *Tragelaphus strepticeros*; *2* Dall sheep *Ovis dalli*; *3* sable antelope *Hippotragus nigger*; *B* as a "streamlined" organ for easy penetration of understory in *1* giant duiker *Cephalophus sylvicultor*, *2* Nile-Lechwe *Onototragus megaceros*, *3* sitatunga *Tragelaphus spekii*; *C* as a battering tool in frontal contest, in *1* muskox *Ovibos moschatus*, *2* kaama *Alcelaphus buselaphus caama*, *3* bison *Bison b. bison*.

In contrast, the well-horned males and small-horned, or even hornless females ease sex discrimination and male-mate selection. Well-horned females may be necessary only in habitats where food scarcity affords general competition for and defense of resources.

Socially, the perennial character and relative insensitivity of horns make them into a year-round sparring tool (Walther 1984; Byers 1987). The routine use of horns may also be the reason that only a large size difference is respected as a rank indicator. The detailed observations of Walther (1966, 1984) corroborate that horns are used not as deadly weapons but primarily as a tool in playful combat or very sophisticated sparring rituals of males and intimidating organs in females. Packer's (1983) view that in African antelopes only the male horns have been developed as ritualized duel-weaponry, in contrast to the stabbing (offensive) character of female horns, does not find the support of other students (e.g., Leuthold 1977).

In boreal bovids the situation is rather complicated. Despite the fact that female horns may look more offensive (e.g., in Alpine chamois *Rupicapra rupicapra*), there is no evidence of their offensive use (Meile & A. Bubenik 1979). In contrast, the females of the Abruzzo chamois, *Rupicapra ornata*, fight with their horns as vigorously as the males do (Lovari 1984/85). Thus female horns, which are not used as weaponry may only be designed to mimic males, but in neither case they can represent a "pseudoneotenic" stage.

Virtual Horn Length and Ritualized Shaping of Horn Spiral

In bovids the relationship between body size or body weight and horn size is not well expressed. The absolute length of the horn spiral and the visual one, i.e., the length of the chord between the horn apex and its base, seem to be two intradependent variables (Duerst 1926). On the one hand, the more compressed the spiral and the smaller its radius, the shorter is the horn chord and the more effectively is defused the offensive feature of horns as injury inflicting weaponry.

On the other hand, the spiral enables all kinds of architectural ritualization and adaptation to the habitat and both visual and physical impact are of social importance (Kitchener 1988). The tendency to ritualize the shape during phylogeny and ontogeny to less offensive and more defensive spiral is obvious.

In accordance with the importance of the torso and the diversity of visual accentuation of the frontal- and caudal-pole of the body, there is not a general tendency to correlate body size or shoulder height with horn length. The different slopes of regression lines between horn-length against shoulder height in African antelopes according to their breeding strategy (Popp 1985) may be rather a coincidence (Weiner 1985) despite the statistical significance (Fig. 11). Otherwise there should be one regression line for all bovids, as it is in cervids (Bubenik, A. 1985).

Olfactory Function of Horns

Although there is no doubt about the importance of pheromones in bovid communication and sexual stimulation (Johnston 1983), there seem not to be reliable

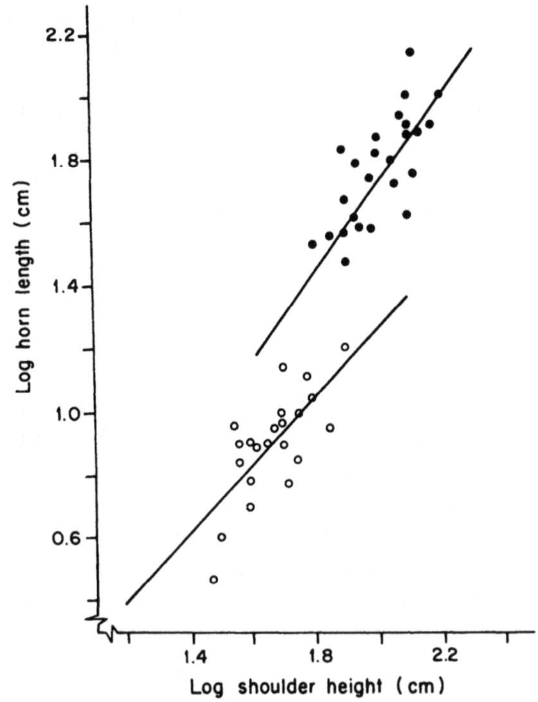

FIGURE 11. Double logarithmic representation of the relationship between horn length and shoulder height for 47 antelope species. Closed circles represent polygynous species, open circles monogamous species (Popp 1985).

records that the females discriminate between males according to the odor of their horns as if might be the case in deer (Bubenik, A. 1982b, 1989).

Part C. Cranial Apophyses of Cervoids

Cranial Appendages of Cervoids as Important Taxonomical Character

The metatarsal gully and the special configuration of lacrimal orifices are considered as the most reliable taxonomical characters of cervoids (Leinders 1979); Leinders & Heintz 1980; Scott & Janis 1987). This can be questioned, at least as it concerns the *Hoplitomeryx matheii* (Leinders 1984; Fig. 4) whose cranial appendages have the microstructure of os cornu (Bubenik, A., unpublished).

Webb & Taylor (1980) and A. Bubenik (1982c) emphasized the fact that cranial appendages in cervoids are the primary characters for cervoid origin or alliance. Similar to ossicones, the programmed defense toward injuries of periosteum and underlying bones of the most exposed skull angles – i.e., from nasals over orbital rims and zygomatic arch, to forehead and parietal bone – seem to be the plausible explanation for development of cranial exaptations, which were co-opted as peduncles or cervicorn pedicles.

However, their species-specific shaped pedicles can be developed only on few cranial loci which are close enough to the supraorbital branches of the axons of the trigeminal nerve. The anlagen for development of the species-specific protuberances from their embryonal bases has the inductive periosteum (Hartwig 1968a; Goss 1987; Goss, Chapter 9). In pristine cervoids and probably in the Late Oligocene tylopods (Patton & Taylor 1971; Webb 1981) as well, this inductive periosteum was over the supraorbital rims and nasal bones.

As a few accidental examples corroborate, the periosteum which surrounds the inductive periosteum must also have a defensive potential for development of ectopic protuberances. However, normally the corresponding alleles seem to be silenced, unless an accidental or artificial stressor, generally a properly timed injury, will reactivate them. All ectopic protuberances are as more atypical as further they are from the species-specific pedicle bases (Filhol 1890; Nitsche 1898; Landois 1904; Frick 1937; Wislocki 1952; Jaczewski 1967; Hartwig 1968a; Bubenik, A. 1975; Goss 1984; Vandal et al. 1986).

Because the cranial appendages of cervoids have appeared independently of space and time over a period between 35-5 MYA, we have to accept the notion that they remained dormant for millions of years.

In the lower Mid-Miocene, i.e., at the very beginning of cervoid history, the small supraorbital protuberances were velericorn in both North America and Eurasia. Since the Mid-Miocene they began to increase in length and points and have developed into facultatively deciduous pseudoantlers in North America, and protoantlers in Eurasia. The former, called merycodonts were equipped with pseudocoronets, the latter termed dicrocerids, lacked any bony protruding rim between the perennial and facultatively deciduous part. Both merycodonts and the early dicrocerids were facultatively perennial. Their distal parts mineralized centrifugally, allowing apposition of new subperiostal osteon layers on their surfaces.

The dicrocerids achieved their maximum development in the Mid-Miocene with *D. elegans* who was of the size of fallow deer. In a few specimens from Sansan, France, the protoantlers were three-pointed, with a basal and a straight forward oriented tine (Stehlin 1939).

Another new lineage which developed at the Mid-Miocene of North America were the antilocaprins, with supraorbital peduncles protected by regularly deciduous pronghorns (Frick 1937; Pilgrim 1941b; Webb 1973), with which the evolutionary history of original North American cervoids ended. Meanwhile, towards the end of the Miocene the first presently known antlered deer with pedicles on frontal cristae emerged (Thenius 1948; Bubenik, A. 1962, 1982c). The oldest known is the small spiker, *Euprox minimus* (Thenius 1950).

The relocation of the pedicle bases from orbits to the frontals happened only in Eurasia, where a dynamic diversification of deer is apparent since the Upper Miocene and the early Pliocene. However, an antler gigantism erupted after the Mid-Pliocene and culminated in the Mid Pleistocene. The North American cervicorn cervoids died out during the early Pliocene except the pronghorned "antelope," *Antilocapra americana*. It took almost 2 MY (million years) until at the end of the Pliocene the Eurasiatic deer invaded that continent.

Activation of Pedicle Anlagen and Their Cervicorn Development

The postnatal ontogeny of the originally monomorphic embryonal pedicle anlagen in red deer correlates (Lincoln 1973) with the rise of fetal testosterone. Then the bases almost disappear, remaining dormant. The actual development begins only postpartum, during a genus-specific period. Frankenberger (1951) reported the monomorphic pedicle anlagen in red deer and fallow deer, and showed them to A. Bubenik (1958) also in early embryos of reindeer from Sweden. However Wika (1980, 1982) did not find any embryonal anlagen, but reported about the epidermal infolding, over future pedicle bases.

In contrast to ossicones which cannot regenerate after removal, the pedicle regrowth and regeneration of its apex cannot be stopped as soon as it was developed. Then annually a large scale of different hormones and enzymes is involved in their growth and cycle (Suttie & Fennessy 1987; G. Bubenik, Chapter 8).

However, the recent hypothesis of van Jaarsveld & Skinner (1987) that the androgenization of female genitalia in spotted hyena, *Crocuta crocuta* (Racey & Skinner 1979), may happen due to translocation of the segment of the Y chromosome responsible for this role, or its regulator, to an autosome or X-chromosome, might also be applicable to female *Rangifer.* Similar situation may exist in the pronghorned females of the pronghorned antelope. Otherwise, any antlerless cervid female can develop pedicles if she is supplied with testosterone during the sensitive phase. Thus hermaphrodites (Wislocki 1954), or females with adrenal tumors or with ovarian cysts, both of which produce large amounts of androgens (Cowan 1946; Donaldson & Doutt 1965; Crispens & Doutt 1970) can develop pedicles and velvet antlers. Even an accidental or artificial injury of the pedicle locus during the sensitive phase can induce pedicle growth, possibly by mobilizing adrenal androgens due to the neuronal signals transferred from the velvet (Bubenik, A. et al. 1976; Morris & Bubenik, G. 1982; Kierdorf 1985).

As mentioned above antlers, as a dead distal part of pedicles, appeared as a relative late developmental stage or a new program parallel to the miocene protoantlers. A good model for tracing their development are the velericorn antlers of hypogonadal deer (Thomas et al. 1970; Bubenik, A. & Weber-Schilling 1987) and castrated males of higher developed deer (Bubenik, A. & Pavlansky 1965), and hermaphrodites (Wislocki 1954).

Cervicorn Development of Pedicle's Apex—Tines as Genocopies of Sprouts

Evolutionary the branching of cranial apophyses by sprouting seems to be the pristine way for increasing their surface. This mechanism is still preserved in antlers of rangiferins (Bubenik, A. 1956, 1959c), in the first antlers of elaphins, genus *C. elaphus*, a basal long sprout and a short one mimic the brow-tine and bez-tine, in second antler of North American moose and as a prong on the shaft of odocoileins of the genus *O. virginianus* and *O. hemionus* (Fig. 12).

The fact that sprouts appear on the location of tines and prongs developed later in evolution of odocoileins supports the hypothesis that tines developed from

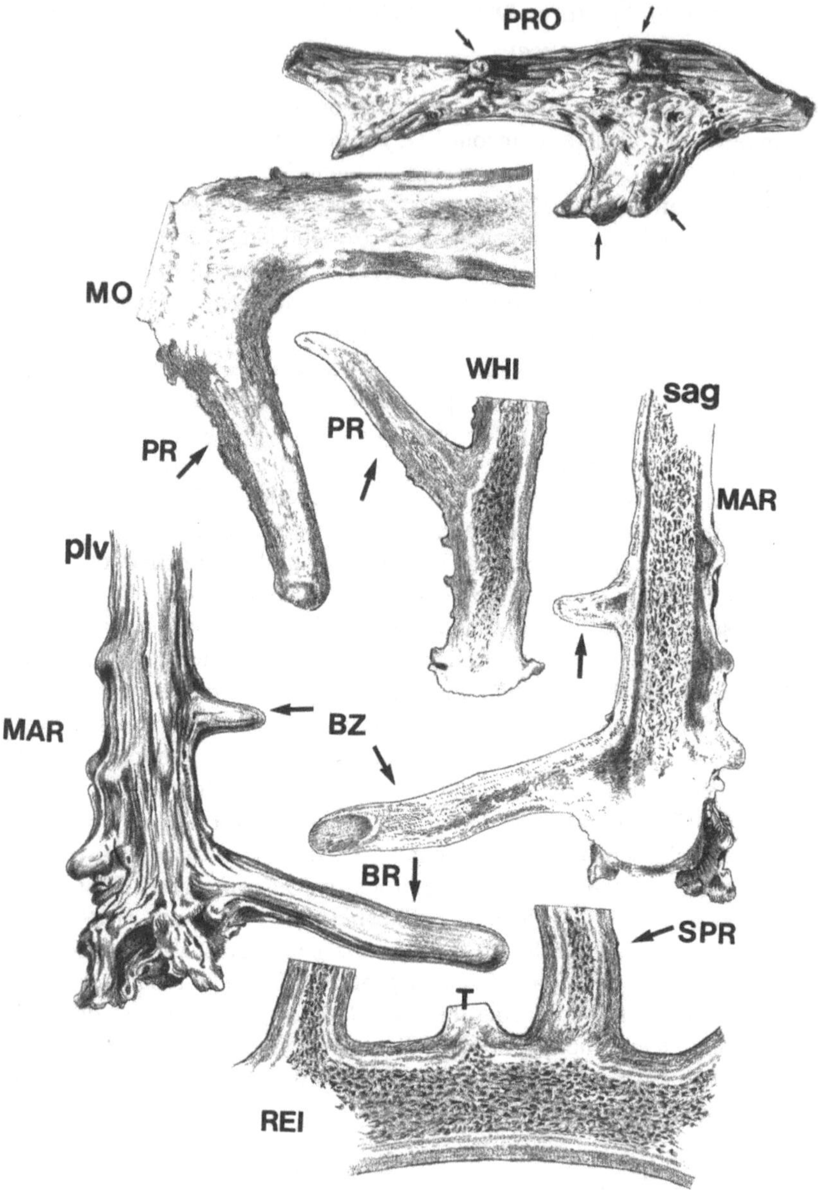

FIGURE 12. Sprouts marked with arrows: in the first antlers of maral *MAR Cervus elaphus maral* which mimic the *BZ* (bez-tine) and *BR* (brow-tine), *plv* in plain view, *sag* in sagittal section. Sagittal section of reindeer beam *REI* which amputated tine *T* was regenerated as a sprout *SPR* The prong *PR* as sprout in white-tailed deer *WHI*, *O. virginianus*; *PR* in second antler of the moose *MO Alces* sp. The core of any sprout is separated from that of the beam by the compact cortex. Sprouting knobs of the *Procervulus aurelianensis PRO* considered as probable giraffoid.

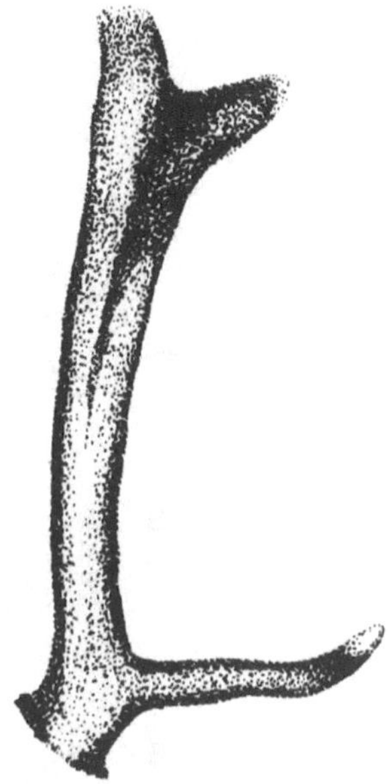

FIGURE 13. Autoradiograph of antler of a red deer with incorporated radiophosphorous (P^{32}) shows upward proceeding intensified development of bone matrix up into the new tine. The denser the dots, the higher the radioactivity of the tissue.

sprouts as epigenetical genocopy in order to improve the defensive architecture when the antler length needed a better defensive shield of the eye-parietal area. Two facts seem to corroborate the "young age" of both bez-tines and prongs: (1) They are not fixed in the genome as tines which developed originally by beam's splitting. They may or may not appear regularly, or they are either developed on one beam only, or alternate from one to the other side in subsequent antlers. (2) None of the bez-tine and prongs match the "bauplan" (Gould & Levontin 1979) of the "one plain" hypothesis of the antler construction developed by Thompson D'Arcy (1940).

The successive splitting of the pedicle's apex into tines is a process that begins to develop far below the actual point of ramification. An autoradiograph of an antler of a red deer stag with incorporated radioactive phosphorous P^{32} cut off just at the stage of splitting (Fig. 13) (Bubenik, A. et al. 1956) shows increasing production of bone matrix in the beam's core up to a stage that leads to splitting of the beam. Thus the ramification of vessels into the tine and beam, as studied first by Rhumbler (1911), seems to be a morphological consequence and not the primary impetus for beam splitting as argued by Suttie et al. (1985b). Hence, sprouts are not homologous, but only analogous to tines.

The epigenetical copying of tines in genome can have also a reverse trend. Tines that become "unemployed" due to better defensive properties of new tines in the more distal part of the beam may gradually disappear or be architecturally reshaped. This trend is well documented in the phylogeny of antler construction of many genera (Kahlke 1951; Bubenik, A. 1966) (Fig. 14).

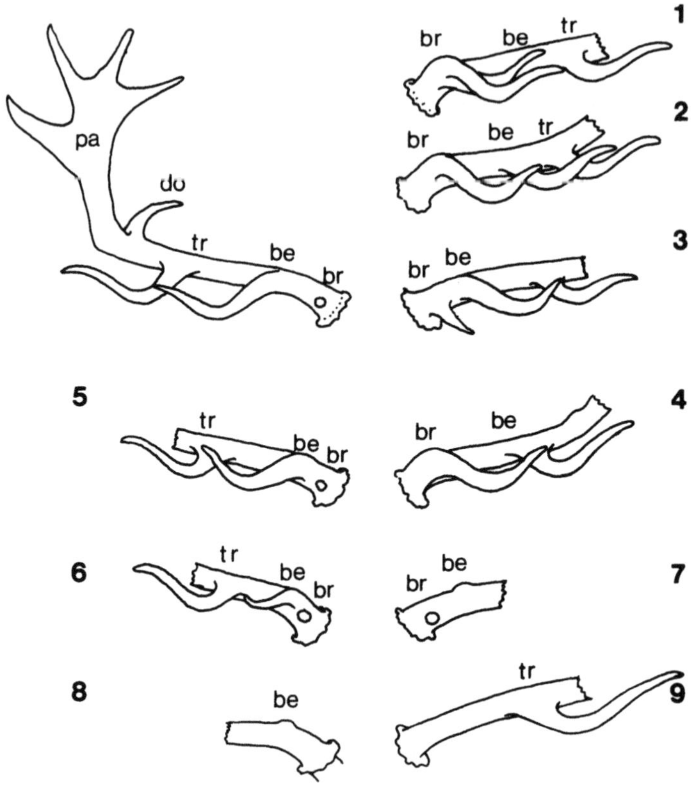

FIGURE 14. The beginning of tine's regression of "unemployed" tines is indicated by reduction of their size until to the point when their new size and shape, or their disappearance is fixed in the genome. Good example of the regression of unemployed tines was shown by Kahlke (1956) in the giant fallow deer of the *Orthogonoceras* alliance. The sequence of the antlers was rearranged by A. Bubenik, according to the presumed subsequent process of regression of the basal tines: *br* brow-tine, *be* bez-tine, *tr* trez-tine, *do* dorsal tine, *pa* terminal palm. The numbers in parenthesis correspond to those of Kahlke (1956). *Orthogonoceras verticornis* Dawk. *1* (24), *2* (23), *4* (22), *7* (3), *8* (2). *C. belgrandi* Harmer *9* (1), *6* (13). *C. mosbachensis* Soerg. *5* (7), *3* (10).

Genetical explanation of such a process is not easy unless we accept epigenetical effects by gradual silencing of corresponding DNA sequences by extrinsic messengers or their epigenetical effect on the trophic memory for antlers construction (see p. 78). In this regard the cybernetical theory of evolution (Schmidt 1985) gives many interesting ideas.

Facultatively Perennial Nature of Protoantlers and Pseudoantlers

The protoantlers of eurasiatic protocervids (Matthew 1904; Dehm 1944; McFarland 1985) or preantlers (Leinders 1984), and the pseudoantlers of merycodonts, also termed 'nondeciduous antlers' by Voorhies (1969) and Goss (1983) are considered either as lifelong or facultatively perennial appendages (Lartet 1839; Filhol 1890; Hilzheimer 1922; Frick 1937; Bubenik, A. 1962), or annually deciduous antlers (Lartet 1851; Kiernik 1913/14; Stehlin 1928, 1937; Dehm 1944; Ginsburg 1968; Kretzoi 1974; Ginsburg & Crouzel 1976).

Branching by sprouting presumes a cortex which remains highly active, long before the antler construction is finished and a centripetal mineralization sets in (Bubenik, A. 1966). This is also the case in the rare "velveted-antlers" (see p. 46) of extant deer (Bubenik, A. 1962, 1982c), and can be seen well on the osteon of the extinct proto- and pseudocervids. The peripheral layers of the tissue of their appendices remain active because mineralization progresses centrifugally: either from the center of the core as in dicrocerids and heteroproxins (Fig. 15) or just from the inner wall of the cortex, as was the case in climacocerids, lagomerycids, merycodontins and possibly dromomerycins, and is still preserved in deer pedicles.

The centrifugal mineralization and development of new layers of osteon is well documented histologically in the pedicle of extant deer and fossil pseudoantlers (Fig. 16) and protoantlers as well; macroscopically, we can distinguish the zones of additive regrowth on both proto- and pseudoantlers (Fig. 17).

Symptomatical for the facultatively perennial nature of proto- and pseudoantlers was the sequestration of the distal parts, e.g., of the protoantlered 'giraffoid' *Lagomeryx prestans*, the protocervid *Dicroceros elegans*, in contrast to the antler's casting, shown in the Miocene cervulids *Euprox furcatus*, which occurs below the well developed coronet (Fig. 18). We use the term sequestration for the rejection of the dead, or dying, distal part of the pedicle without the prior development of a protective bony bridge, which separates the living pedicle. There is no evidence of Looser's rebuilding zone as speculated for "frostbite" antlers sequestered in velvet above the coronet. Casting below the coronet is a process typical for antlers whose base is plugged with compact bone, in contrast to both or sealed from the living pedicle. The shedding means the peeling of dying velvet, or the rejection of the death pronghorn sheath.

Hypothesis on Gradual and Polymorphic Development of Skin Covering Cranial Appendages of Cervoids

In living deer the skin covering the pedicle is the same as that of the surrounding forehead area. In contrast, the skin covering the growing antler has different structure and properties. We call it velvet (Vacek 1955; Goss 1983, 1987). The other important character of velvet, with possibly epigenetic impact on antlers shape and construction is the density of endings of sensitive nerves which is the highest known among mammalian skin (Vacek 1955). Velvet also has much more sebaceous glands than pedicle skin; its short hair lacking *arectores pilli* stay perpendicular to the surface of the skin (Goss 1983). From an evolutionary view it seems important that well organized velvet develops only after the first casting of antlers. The seemingly not well transformed velvet of first antlers looks similar to the skin covering the perennial velericorn antlers (see below).

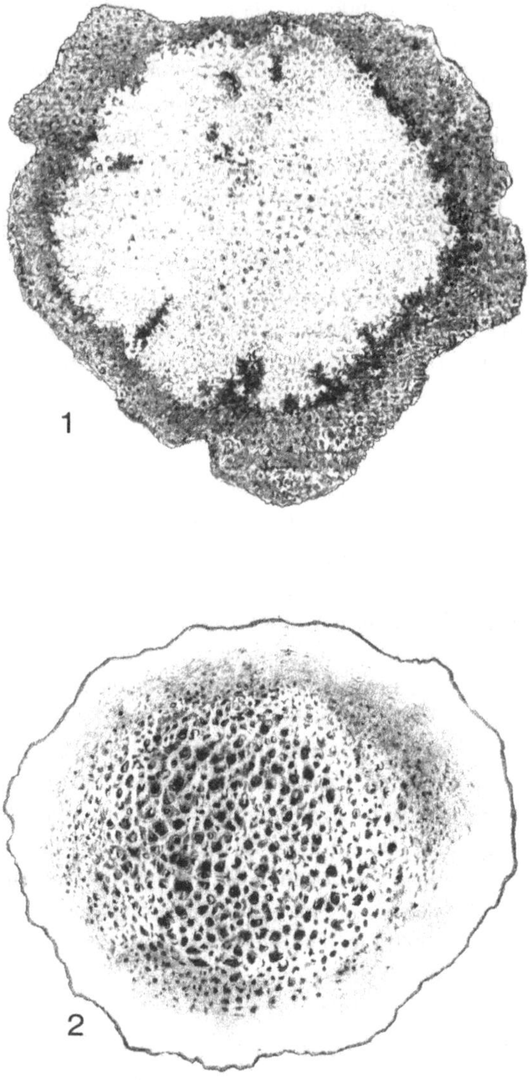

FIGURE 15. Cross-section of a protoantler of *Dicroceros elegans* with centrifugal mineralization *1* in comparison with the centripetal mineralization of the cortex of red deer antler *C. elaphus* sp. *2*.

▶

FIGURE 16. Microphotograph of the pseudoantler of the merycodont *Cosoryx* in cross-section: *a* outer layer of the still parallel oriented primary osteons. *b* inner part of the cortex, with mineralized osteon; *c* center of the core with wide channels and signs of osteon rebuilding.

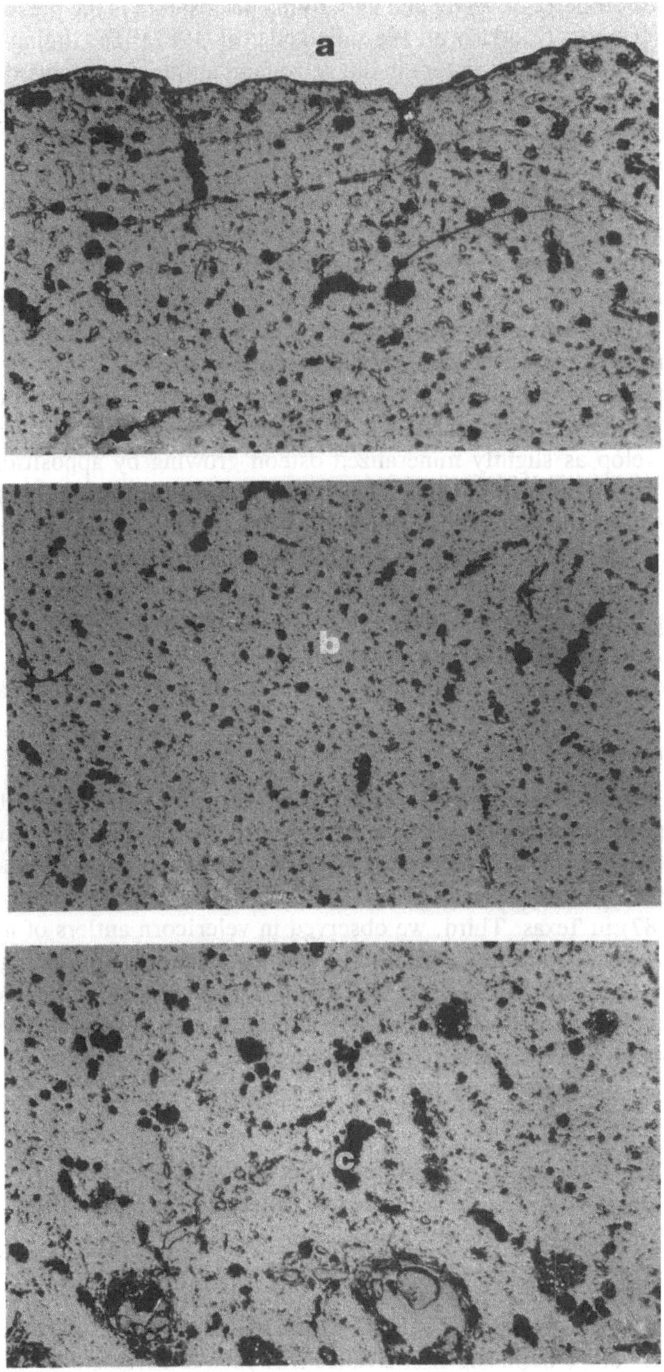
a
b
c

The other property of velvet and underlying periosteum is the presence of sex-hormone receptors (Plotka et al. 1983, Barrell et al. 1987). The timing of the activation of the androgen receptors in the velvet and antler tissue toward the end of antlers growth when testosterone begins to rise seem to be responsible for the short period of antler development and the centripetal mineralization (G. Bubenik, Chapter 8).

Because the pristine facultatively perennial appendages of the Miocene cervoids mineralized centrifugally (like the pedicle) we hypothesize that their skin was less susceptible to androgens than velvet, and androgen level was much lower as in deer. Another kind of morphological transformation may experience the skin of the pseudocervids, which underwent periodical keratinization, characterizing the antilocaprids, as will be discussed later.

In order to understand better the hypothesis of stepwise transformation of forehead skin into velvet we must begin with the present stage of antlerogenesis. The antlers develop as slightly mineralized osteon growing by apposition of bone matrix on the apex of the pedicle. Thus, they cannot gain subsequently on circumference and the diameter of the shaft is that of the pedicle.

Normally, the velvet dies due to progressive mineralization of the cortex. The openings of the Haversian system become so narrow that the velvet's life-supporting blood exchange between the core and external veins is cut off. Thus Wika & Krog (1980) called it a disposable vascular bed. In that case it cannot regenerate, or it cannot cover a perennial cervicorn proto- or pseudoantler.

The Regenerating Velvet

The first time we observed velvet regeneration was when we temporarily blocked the testosterone receptors in growing antlers with cyproterone acetate (Bubenik, G. & A. Bubenik 1978). Next, were observations about velvet regeneration in sexually intact juvenile chital *Axis axis*, made by G. Bubenik (personal communication, 1987) in Texas. Third, we observed in velericorn antlers of moose bull *Alces alces* castrate a mummification of velvet's epidermis and its regeneration from the hypodermis (Bubenik, A. & Larsen, D., unpublished). Besides this we know that velvet can also survive on hard antlers as long as their surface remains porous enough and the antlers' core remains alive.

Velveted-Antlers

Hard antlers with partly mummified, and partly alive velvet are an unique phenomenon, for which a term velveted-antlers seems well suited. Velveted-

---

►

FIGURE 17. Proto- and pseudoantlers in the stage of additive regrowth: *A* the Miocene *Heteroprox larteti* at the beginning of bifurcation (two periods of regrowth) and bifurcated (three regrowing cycles); *B* the pseudo-antlers of the merycodont *Ramoceros (1)*, apparently in a stage of temporary quiescence, and *2* a pseudoantler of *Paracosoryx* in a stage of subsequent (additive) regrowth; *3* a close-up of the tip in stage of reactivated growth.

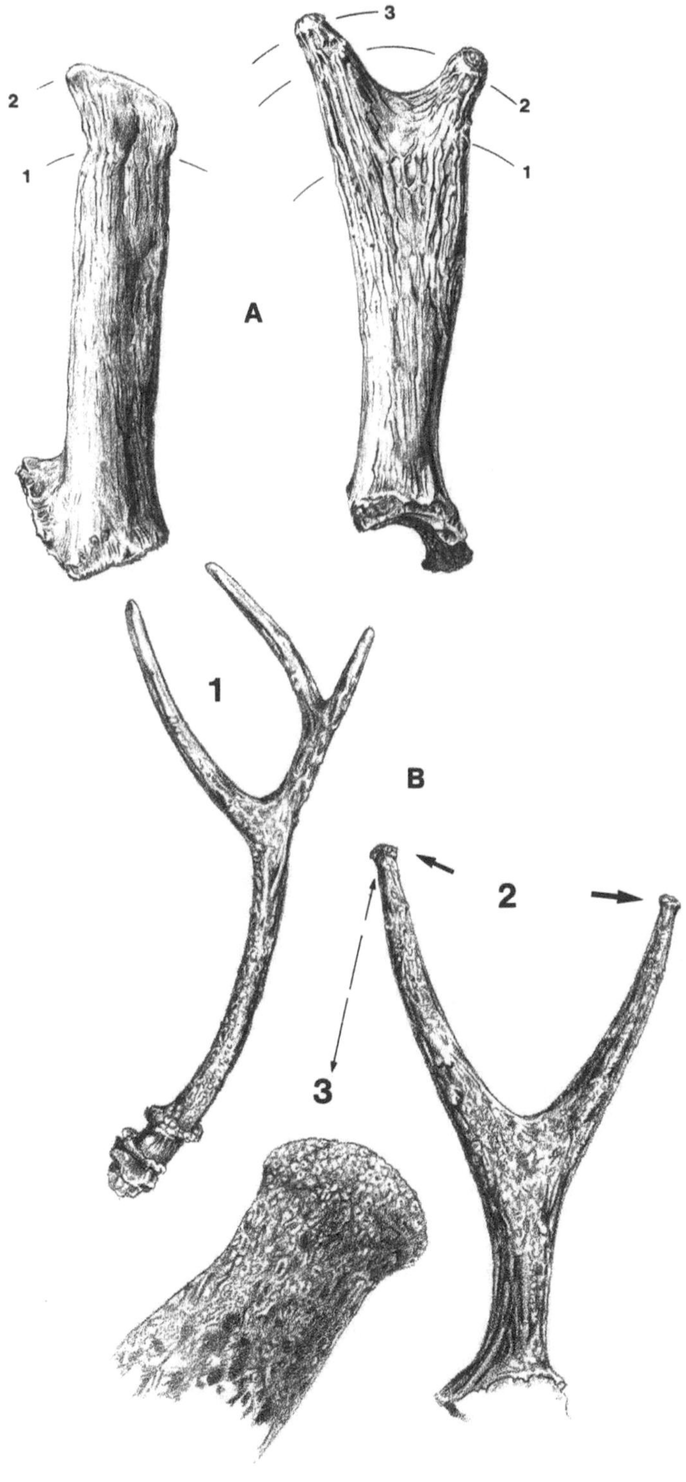

3
2
2
1
1
A
1
B
2
3

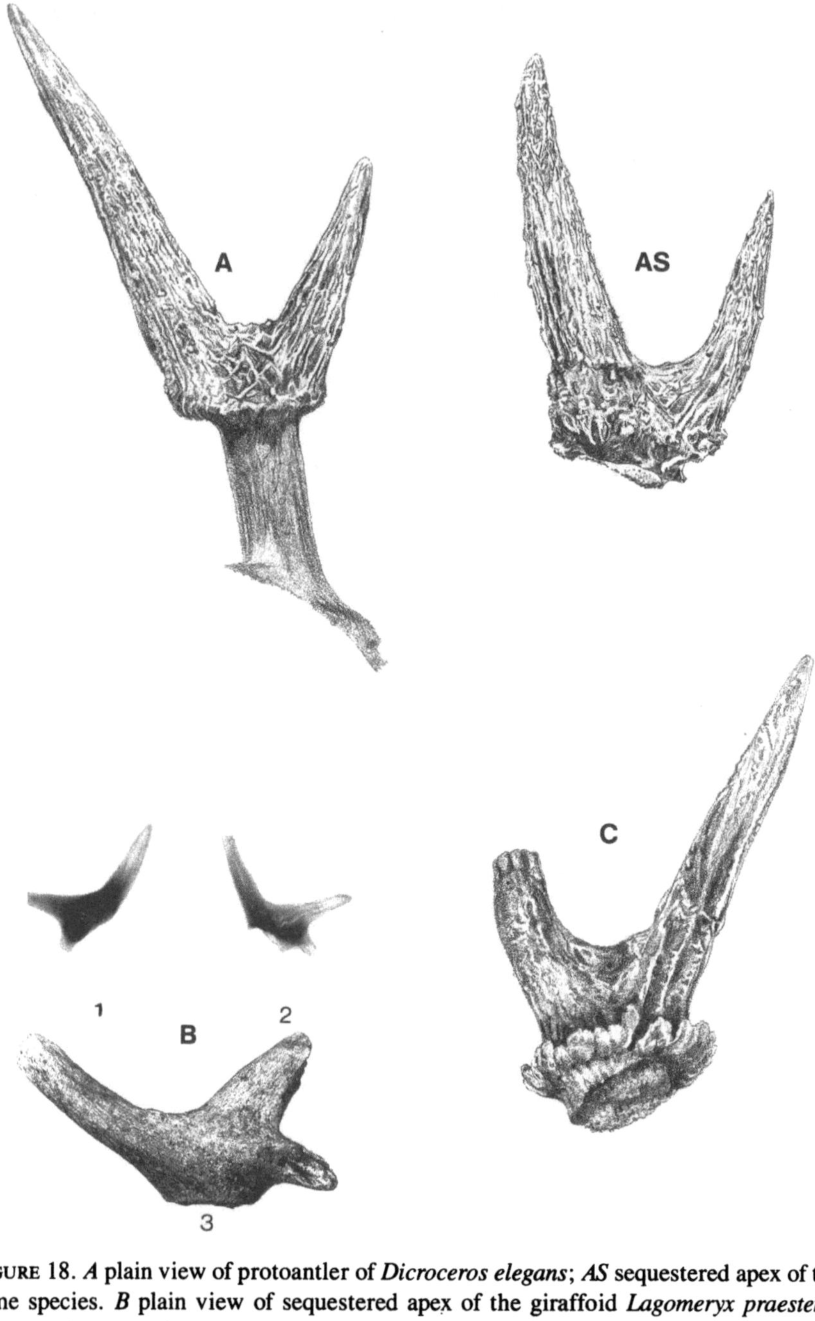

FIGURE 18. *A* plain view of protoantler of *Dicroceros elegans*; *AS* sequestered apex of the same species. *B* plain view of sequestered apex of the giraffoid *Lagomeryx praestens*, blown-up in comparison to *1, 2* which are positives of two radiographic projections of the same sequester. The blood vessels (white lines) and the poor mineralized points visible in *2* show that blood flow was still possible through the base up to the end of the apex. (*A, AS* and *B* from collection Museum of Natural History, Basel.) *C* The casted antler of the cervid-like *Dicroceros furcatus*; drawing made after a specimen in Museum d'Histoire Naturelle, Paris, courtesy Dr. Ginsburg.

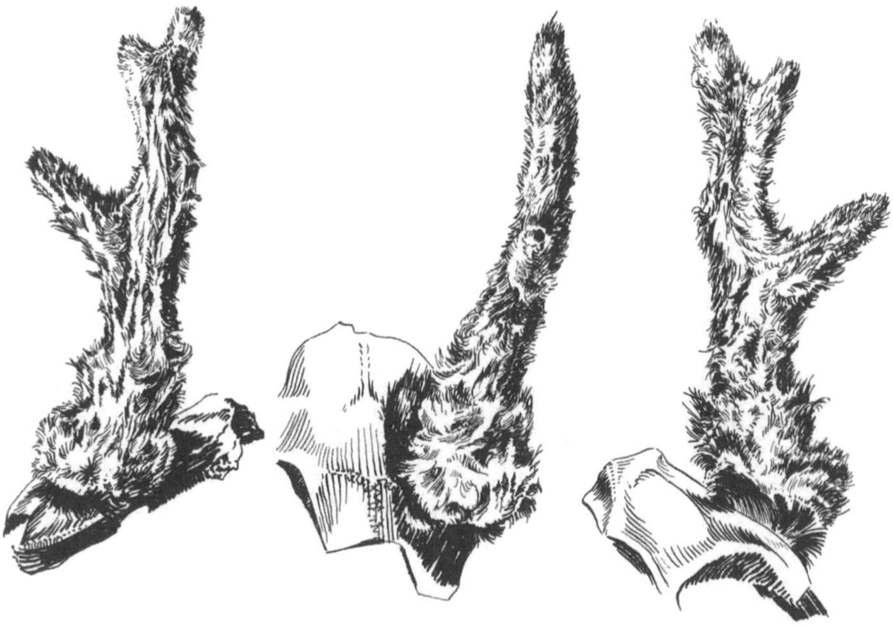

FIGURE 19. Velveted-antlers of hypogonadal white-tailed buck from Texas (after a specimen submitted for investigation by Dr. Thomas).

antlers are common in the hypogonadal white-tailed bucks from Texas (Fig. 19) (A. Bubenik, unpublished; see also Thomas et al. 1970), from female caribou *Rangifer r. caribou* (Thing et al. 1986; C. Barette, personal communication, 1987) and alive antler's core found Bouchud (1966). We suggest that the long lasting cleaning of antlers from velvet in muntjacs is due to velvet survival on pointed antlers, which we documented radiographically on the primitive muntjac *M. atherodes* (A. Bubenik, unpublished observation, 1985).

Paravelvet, Pseudovelvet, and Protovelvet

Superficially the velveted-antlers are not covered by well differentiated velvet. This is either due to the long survival of the velvet or it is a "primitive" stage of velvet, i.e., an intermediate product, equipped partly with typical long body-colored hair of the pedicle, growing among the short, not erectable, and distinctly colored velvet hair. Unfortunately there is no record that this kind of not-well-differentiated velvet was studied. Thus, as long as we do not know its morphology, we will call it paravelvet. Paravelvet is typical for hypogonadal or castrated males or "antlered" females of sex dimorphic species (Bubenik, A. & Weber-Schilling 1986).

Under the cover of paravelvet the speed of mineralization of the peripheral lamellae seems to be slowed down. Similar deceleration of the osteon's mineralization has to be presumed in the pseudo- and protoantlers, but due to quite

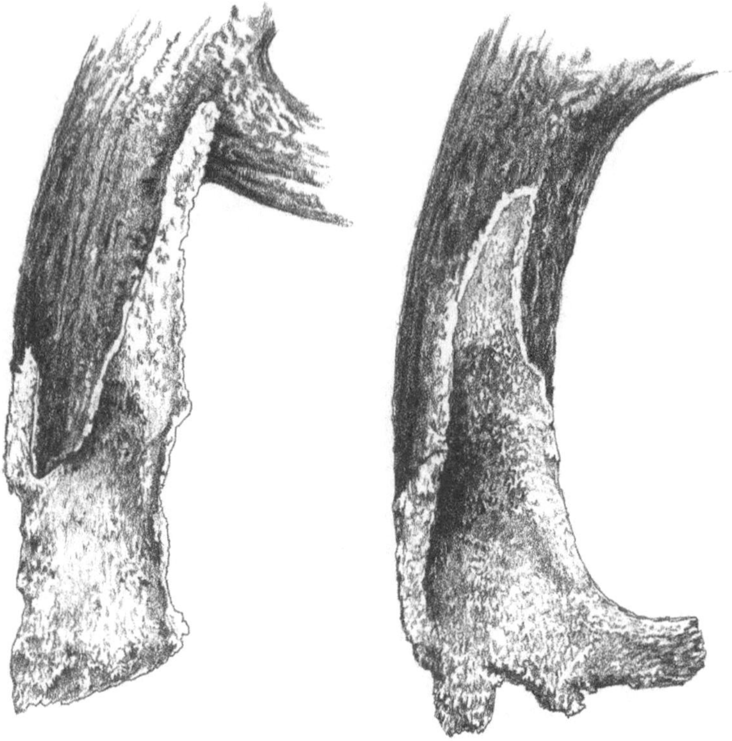

FIGURE 20. The uneven surface of sequestered antler of a cryptorchid red deer (Bubenik, A. & Weber-Schilling 1986).

different development of both pseudo- and protoantlers we have to assume that each was covered by special differentiated skin. Thus, we call the former pseudovelvet and the latter protovelvet.

Under the pseudovelvet the apposition of circumferal lamellae was slow, the lamellae were thin and the aperiodically sequestered distal parts were not delimited either from inside or outside. However, underneath the zone of sequestration dermal rings were developed which later fused with the pseudopedicle.

Underneath the protovelvet the periodically developed circumferential lamellae were coarser in the basal part, where later sequestration may occur. Under the protovelvet no coronet was developed.

The common character of both pseudo- and paravelvet and the paravelvet of velericorn antlers, is their regenerative potential. Their epidermis can die but the hypodermis remains alive, developing new velvet and thickening the surface by both apposition of new circumferential lamellae and osteochondritic ossification (Bubenik, A. & Larsen, D., unpublished).

The velericorn antlers of castrated or cryptorchid (hypogonadal) males seem to be always sequestered in a more or less diagonal direction to pedicle axis (Fig. 20). In contrast, the sequestration of protoantlers occurred through the proto-

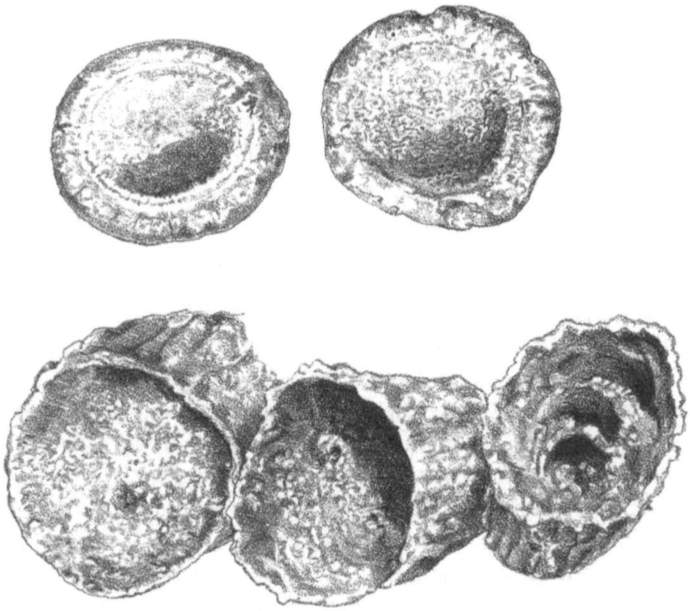

FIGURE 21. Seals of different degree of concavity in chital *Axis axis*. (Courtesy Dr. Acharjyo.)

pedicle in more or less one direction; only the multipointed apices of lagomerycids were sequestered perpendicularly to the long pedicles. Due to very few specimens of pseudopedicles and lack of sequestered pseudoantlers we do not know whether they were cast or sequestered.

The seal of the sequestered part of protoantlers has a concave, rarely a flat profile (Stehlin 1937, 1939; Ginsburg 1963, 1968), and the surface of the perennial part which has a porous surface remains convex (Ginsburg & Crouzel 1976).

A concave seal is typical for all deer of the Miocene and the Pliocene era, i.e., from tropical zones (Fig. 21). The concavity is as larger as younger is the male (Acharjyo & A. Bubenik 1982). The relationship with age correlates with the acme of plasma testosterone level which is highest in prime age. However, in deer of pleistocene origin the acme is high and nadir below the level necessary for testicular activity. In tropical deer (e.g., in chital), the plasma testosterone never drops so low that spermatogenesis is discontinued (Loudon & Curlewis 1988).

The opponents of the hypothesis of facultatively perennial nature of proto- and pseudoantlers argue that the well-pointed apices often have deep scratches that can originate only by fraying on hard objects or digging into soil. However, the sharp ridges below the apices look to be not worn at all in comparison to deer; in old protoantlers the surface of the base has Y-formed ridges which do not correspond to the normal pattern of subperiosteal blood vessels (Fig. 18A). In our view they were developed by subperiosteal osteochondritis. These differences in wear of apices and base show that apparently only the latter were temporarily without dermal cover or permanent skin as assumed by Teilhard de Chardin (1939).

This points to the probability that the paravelvet and also the pseudovelvet either survived intact below the bare apices, or experienced a partial desiccation with subsequent recovery as mentioned above for the rare cases of extant deer.

The Cervoids of North America: Cranioceratidae and Antilocapridae

The first cervoids of North America were the Cranioceratidae which appeared with the first chronofauna (the Early Miocene). They were equipped with two perennial, conically shaped, and probably velericorn peduncles located supraorbitally, and one occipital spur of variable length in both subfamilies cranioceratinae and dromomerycinae (Bohlin 1953; Webb 1977). Thenius (1980) considers the dromomerycins as an offshoot of giraffoids, but presently they are classified as palaeomerycids (Janis, Chapter 2). As we pointed out earlier such a relationship seems dubious unless there were two clades with paleomerycid characters but different cranial appendages, i.e., ossicones or peduncles.

Regardless as to who was the ancestor of cranioceratids, its pristine lineage entered Asia probably with the Early Miocene; the early Mid-Miocene *Triceromeryx* is considered as their offshoot (Bohlin 1953; Churcher 1970). The presence of *Triceromeryx pachecoi* (Crusafont Pairó 1952) in the Lower Miocene of Spain and Central Asia (Bohlin 1953) points to a very fast spread and possible further diversification. Its cranial apophyses covered by knob-sized sprouts, appear very similar to those of the *Procervulus aurelianensis*, whose distal parts of cranial apophyses might be facultatively perennial (Fig. 8).

Therefore, the relationship of triceromerycins with the dromomerycins (Webb 1983b) is not improbable. Part of the evidence is the morphology of their P_4, which according to Sewertsow (1951) and Janis & Lister (1985), is not specific only for giraffes. In that case of course, the connection to the Giraffoidea with ossicones (Webb 1977; Thenius 1979; Leinders 1984) is questionable. In North America the dromomerycins were soon accompanied by cervicorn merycodontins (Matthew 1904; Frick 1937; Webb 1973), with paired, supraorbital located pseudopedicles; their distal parts were facultatively perennial (Bubenik, A. 1982c).

Both velericorn peduncles and cervicorn pedicles are morphophysiologically not very distant, thus it is not impossible to assume as common ancestor paleomerycids or giraffoids, of course, with apophyseal appendages.

The Pseudocervidae–The Merycodontinae

Although we don't have evidence that the merycodontins ever entered Asia, their propagules, the paleomerycids and giraffoids, were established in West Europe since the Early Miocene (Thenius 1980). Therefore, it is not surprising that in Eurasia it is sometimes difficult to separate all the pristine lineages into well-defined taxa.

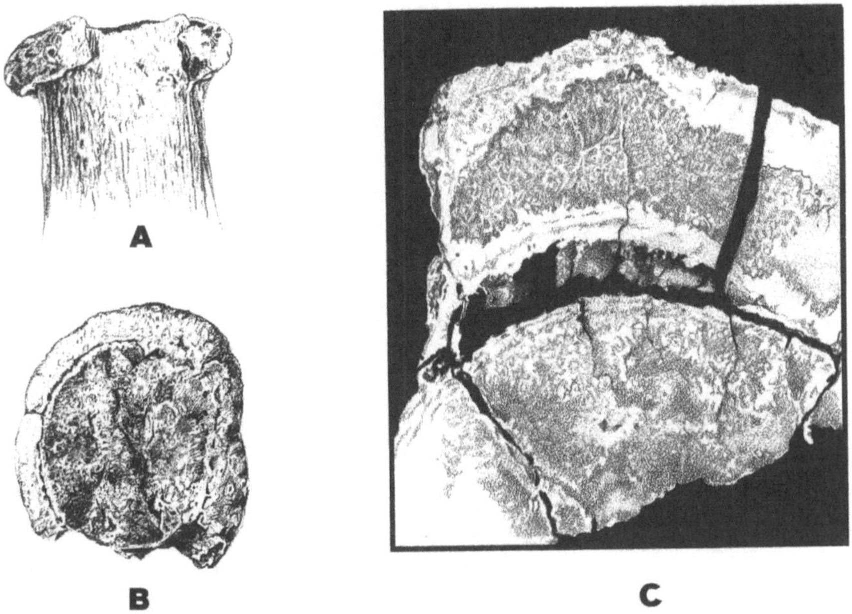

FIGURE 22. Merycodonts' pseudopedicle and pseudocoronet: *A* bare surface of a pseudopedicle of *Cosoryx*. *B* The same specimen from lateral view with the pseudocoronet partially lost due to incomplete fusion. *C* Cross-section of the subperiostally developed pseudocoronet, not well fused with the body of the pseudopedicle.

The pseudocervids or merycodontins have had only few common characters with the protocervids, i.e., early dicrocerids. Despite that they appeared almost 5 MYA before protocervids, their cervicorn appendages were more advanced, and the pedicles were adorned by pseudocoronets. Thus, we must conclude that they represent different, and on protocervids entirely independent clade. The immediate sinecorn ancestors of merycodontins are unknown. Taxonomically, the merycodonts are classified as a subfamily Merycodontinae of the family Antilocapridae, despite the fact that Antilocapridae appeared as late as with the 3rd chronofauna, i.e., almost 5 MY years later (Webb 1977).

Pseudocoronets

The basic feature which separates merycodontins from their latter contemporaries the protocervoids in Eurasia (Cope 1874 in Frick 1937, p. 282) are the pseudocoronets, which are not homologous to coronets of antlers because they did not develop by local "mushrooming" of the pedicle cortex as presumed by Voorhies (1969). The pseudocoronets seem to originate as subperiosteal osteochondritis. The cross-section of a pseudopedicle with the pseudocoronet shows that they are dermal bony rings, which ankylosed, point by point, with the pedicle's cortex (Fig. 22).

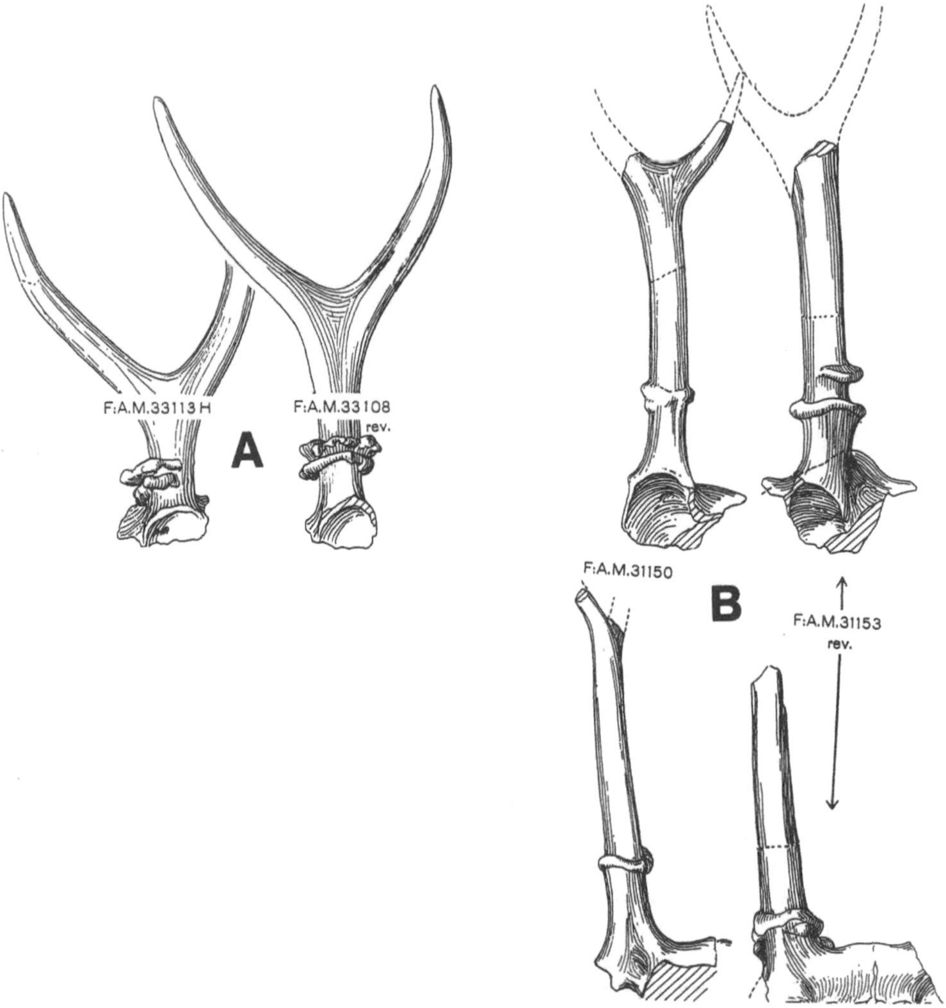

FIGURE 23. Partially lost pseudocoronets in: *A Cosoryx (Subcosoryx) cerronensis* (F:A.M. 33113H, 33108); and *B C. (Paracosoryx) alticornis* (F:A.M. 31150, 31153). (From Frick 1937.)

Also in contrast to Voorhies (1969), the youngest pseudocoronet cannot be the lowest, but must be the uppermost one (Bubenik, A. 1962, 1982c). The fact that the proximal coronets are sometimes only partly fused and/or partly lost points to a sequestration after the protecting skin was lacerated by rubbing and periosteum infected (Fig. 23).

It is interesting that pseudocoronets did not develop before the end of the second year of life, and no more than four pseudocoronets are found on one pseudopedicle. We also do not know whether the pseudocoronets were developed annually. Unfortunately we do not know about any sequestered pseudoantlers. In

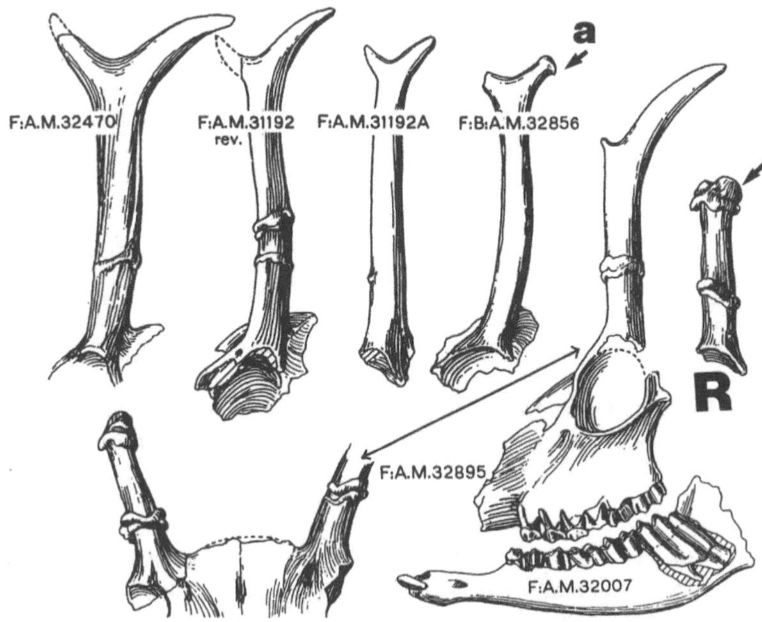

FIGURE 24. *Cosoryx (Paracosoryx) wilsoni* with asymmetrical branching, and asynchronous bilateral or unilateral development of pseudocoronets and asynchronous development of pseudocoronets (F:A.M. 32895); *Cosoryx (P.) dawesensis* (F:B:A.M. 32856) with dorsal point *a* and in *R* a pseudopedicle in regrowth of sequestered or lost apex. (From Frick 1937.)

contrast, few pseudopedicles with a bare surface above the uppermost pseudocoronet exist (Fig. 23). Assuming that the pseudocoronets were developed annually after the second or third year of life, the life expectancy of merycodont males was less than 7 years (Voorhies 1969). However, the irregular distances between the pseudocoronets may indicate an irregular timing of their development (Fig. 24).

Voorhies (1969) thought that the pseudocoronets had a protective function during rutting combats. Assuming that the core of the pseudoantlers remained alive as long as the distal part was not sequestered, it is questionable whether the sensitive pseudoantlers were a suitable weaponry. To substantiate this we have to emphasize the lack of a reliable number of broken pseudoantlers which would corroborate vigorous fighting. The few known cases of broken and subsequently healed tines (Voorhies 1969) show that the fracture happened when the tissue was alive and that the injury was by accident rather than by vigorous fighting. We assume that each of the new pseudocoronets might serve as a solid shield of the coronary ring of recurrent blood vessels. They might evolve when the distal part of the appendage was sequestered, or in a stage to be lost; this, of course would be a situation similar to that we know in deer prior to the velvet shedding (Frankenberger 1954b). This seems to be corroborated, e.g., by those specimens of *Meryceros (Submeryceros)* shown in Frick (1937) (Fig. 25). Their pseudocoro-

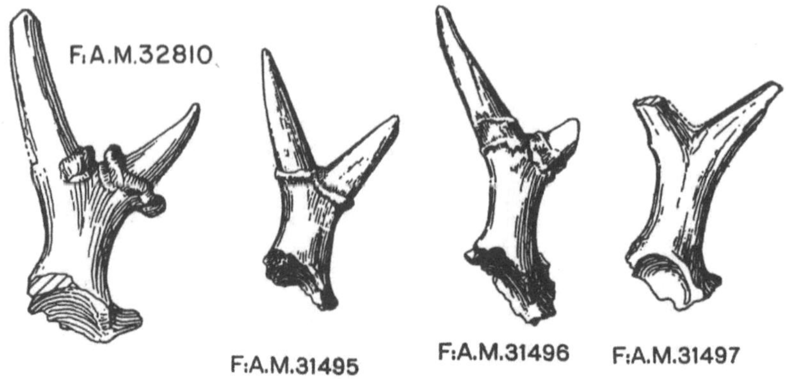

FIGURE 25. Pseudoantlers of *Meryceros (Submeryceros)* with coronets above the point of forking and asymmetric points. The coronets may develop after breakage. (From Frick 1937.)

nets are not on the shaft but on the forks. Thus we hypothesize that they have developed after accidental breakage and during healing of the wound.

The hypothesis that the youngest, i.e.. the lowest pseudocoronet marked the upper limit of the normal skin ("and flesh," Voorhies 1969) would presuppose that the pseudovelvet was present only during the period of additive horn growth (Voorhies 1969). In that case the pseudovelvet must regenerate over the whole and bare pseudoantler, which is difficult to substantiate.

Because merycodontins inhabited subtropical areas, we have to presume that they were in a permanently reproductive stage like the tropical deer (Brokx 1972; Loudon & Curlewis 1988). Under such condition only the apices might become bare. The lower part remained protected by pseudovelvet, which may temporarily die by infarction, or be desiccated or mummified. However, the hypodermis with periosteum seems to remain alive and regenerate as soon as an appropriate stimulus appeared. This would be a process we discovered in the velericorn antlers of castrated moose (Bubenik, A. & Larsen, D., unpublished).

Development of Horn-Protected Peduncles of Antilocaprins and Their Behavioral Significance

The deciduous horn-sheath of antilocaprins is a phenomenon among cervoids. Generally, it is believed that all extinct antilocaprins behaved like *Antilocapra americana*. The cervoid alliance of antilocaprins is shown by the cranial and metapodial characters (Leinders & Heintz 1980), and more importantly by the fact that the development of the cavicorn sheath and it regeneration is dependent on periodical sex-hormone secretion as in any cervid (Pocock 1905; Lyon 1908; Bubenik A. 1982c) (Fig. 26).

The development of antilocaprins from merycodontins seem to be gradual and must happen between the second and third chronofauna when some of their offshoots began to split the cervicorn construction into two peduncles as found,

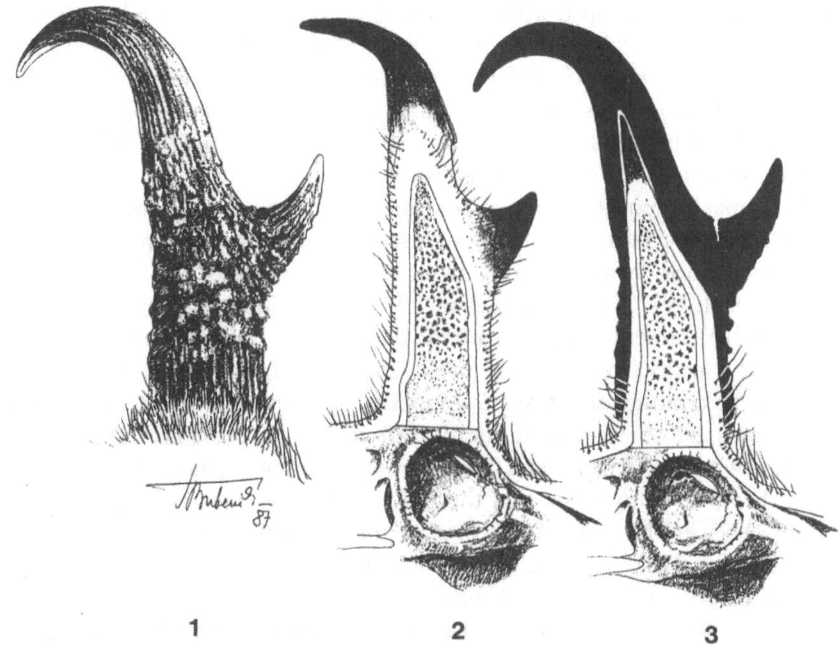

FIGURE 26. Schematic section through three phases of a pronghorn cycle: *1* mature pronghorn in plain view; *2* sagittal section of a pronghorn sheath in development and process of keratinization which begins at two points: on the apex and on the tip of the prong; *3* sequestration of dead pronghorn by keratinized apex of new pronghorn sheath.

e.g., in *Plioceros*, or reduce the number of points to a small fork, as in *Sphenophalos*. The regression progressed even further by developing a tall-based, laterally compressed horn core ending with a weak anterior and a strong posterior point in *Antilocapra* (*Subantilocapra*) *garciae* (Webb 1973). In Webb's view (personal communication, 1988), which we share against our previous view (Bubenik, A. 1982c), these extinct antilocaprins represent most probably the stages of transformation of pseudoantlers into conically shaped peduncles with a slight proximal boss covered by horned sheath.

The Keratovelvet

Thus there should be a distinct, close, or remote possibility that the facultatively perennial pseudovelvet of merycodontins could become keratinized, a notion expressed by Pilgrim (1941b). The deciduous character of the pronghorn sheath may be due to the perennial nature of the original pseudovelvet and its programming for periodical susceptibility for sex hormones as inductor of keratinization.

In that case the term keratovelvet seems to be appropriate for this type of dermis. The pronghorn sheath of *Antilocapra* is an unevenly growing and keratinized integument (Boas 1917; Duerst 1926; O'Gara & Matson 1975), whose growth cannot be stopped without testicular testosterone, which is a typical cervoid and not bovid mechanism. According to Boas (1917) and Duerst (1926), the behavior of keratovelvet is different from that of the keratinizing epidermis of the bovids' horn core. First, the distal, and then the more proximal layers of the horned sheath keratinize centripetally and so slowly and slightly, that the follicles of the hair can easily grow through the pronghorn.

The annually repeated cycle of keratinization of the dermis ends with atrophy of the uppermost portion of the hypodermis due to the loss of connection with the dead pronghorn sheath. Judging from the description of the pronghorn cycle (O'Gara et al. 1971; O'Gara & Matson 1975), it is evident that regeneration of the germinative epithelium underneath the old pronghorn sheath correlates with the drop in the testosterone after the peak of the rut. The regeneration process, and subsequent keratinization takes 3–4 months. The new rise of testosterone arrests the process and completes the keratinization. The old pronghorn sheath is lifted by the newly growing apex and frontal prong. The development of the new pronghorn sheath is so dynamic that the old sheath may crack between the tip and prong. O'Gara & Matson (1975) compare the loss of the pronghorn to the exfoliation of horns. However, in close-up view, the loss of the pronghorn looks quite different from that of a horn sheath.

The fact that prime bucks shed their pronghorns earlier than the bachelors (Kitchen 1974) goes along with the cycle of testes activation (O'Gara et al. 1971). It is an absolute analogy with prime and submature or "psychologically castrated" deer. The mature males cast the antlers earlier in accord with the differential in timing of testosterone secretion among premature and prime males (Bubenik, G. & A. Bubenik 1986).

Despite the parallelism between merycodonts and antilocaprins on the one hand and the protocervids on the other, the North American cervoids were apparently not programmed for a shift of the peduncle bases higher on the external cristae of frontal bones, as in true cervids. Both peduncles and pseudopedicles remained on the upper orbits akin to lagomerycids or protocervids. The only reinforcement occurred posteriorly by developing a crest toward the brain case. The clinging of the very exposed but fragile orbital rim seems to be the reason why the pseudoantlers and peduncles could never rise to the size of antlers.

Behavioral Significance of the Pronghorn Sheath

A mature pronghorn has an inward hooked tip, and a forward protruding prong (see O'Gara, Chapter 7). Thus the pronghorns are optically and functionally a rank indicator, and also a weaponry as any other cervicorn-like appendage. Even they operate as a scent dissipating organ, perfumed by the own sex-pheromones taken from vegetation on which the scent of the subauricular glands was smeared, and/or on which the buck had urinated (Bubenik, A., unpublished).

However, the possibility that the torsion of *Antilocapra*'s hair, or the hair's medulla contributes to the final pronghorn twist seems remote. Nobody has investigated the character of pronghorn hair-medullae and scurs' are questionable. The only published skin horn (Anonymous 1983) was too small to develop a mini-pronghorn and according to O'Gara (Chapter 7) longer scurs were apparently not thoroughly investigated. The relatively elastic tips of the pronghorn may function partly as a buffer when pronghorns are clinched and twisted during sparring. Thus pronghorns are a sophisticated armoire, combining the advantages of both horns and cervicorn antlers.

Of course, this speculation is based on the behavior of the apophyses of extant antilocapra. Whether the peduncles of the extinct illinogocerotins and stockocerotins (Frick 1937) were also protected by deciduous pronghorns or periodically atrophying pseudovelvet remains to be elucidated. The osteon of merycodontins and antilocaprins differs slightly as is shown in Fig. 27.

The Protocervids of Eurasia

The first protocervids of the family Dicroceridae appeared in the lower Mid-Miocene (the Burdigalian) as small ruminants with tusks, long supraorbital monomorphic pedicles and facultatively perennial protoantlers. According to Ginsburg (personal communication, 1986) the protoantlers of females were smaller.

Evidence of facultatively perennial nature gives their microstructure (Fig. 15/1). It shows subperiosteal apposition of lamellae and centrifugal mineralization of the cortex. Further indicators are the radiographs (Fig. 28) which show in well pointed protoantlers that open vessels penetrate from the protopedicles into the core of the protoantlers. This is another evidence for the irregularly deciduous character of proto- and pseudoantlers.

The dicrocerins developed mostly a fork, sometimes with (probably) accessoric sprout(s) between or around the fork. The protoantlers of the more or less affiliated genus *Heteroprox* ramified subsequently in many points laying almost parallel with the skull axis, as in dicrocerins. Their history covers 7–9 MA from the Helvetian into the Tortonian period.

All lineages of protocervids, in contrast to the North American pseudocervids, were also equipped with large tusks in males, smaller ones in females. The lack of broken protoantlers is more conspicuous than the scarcity of sequestered protoantlers. This may be strong evidence that true protoantlers were relatively sensitive organs inept for combat (Hilzheimer 1922; Bubenik, A. 1982b) and their sequestration may occur in more than annual intervals.

Muntiacidae, Fam. Nov. Bubenik 1982

In the Upper Miocene appeared a quite new lineage of cervoid characterized by short antlers with real coronets and long, streamlined pedicles, based still supraorbitally. The antlers' cortex was mineralized centripetally. This led to the speculation that they grew under velvet, have a short growing period, and were probably cast circannually. Because two genera of this large miocene family—the very small *Elaphodus* and different large muntjacs *Muntiacidae*—survived to our time, we can partly reconstruct their history.

All Muntiacidae were equipped with tusks in both sexes and used them instead of antlers in intraspecific contests. The timing of the antler cycle was individual

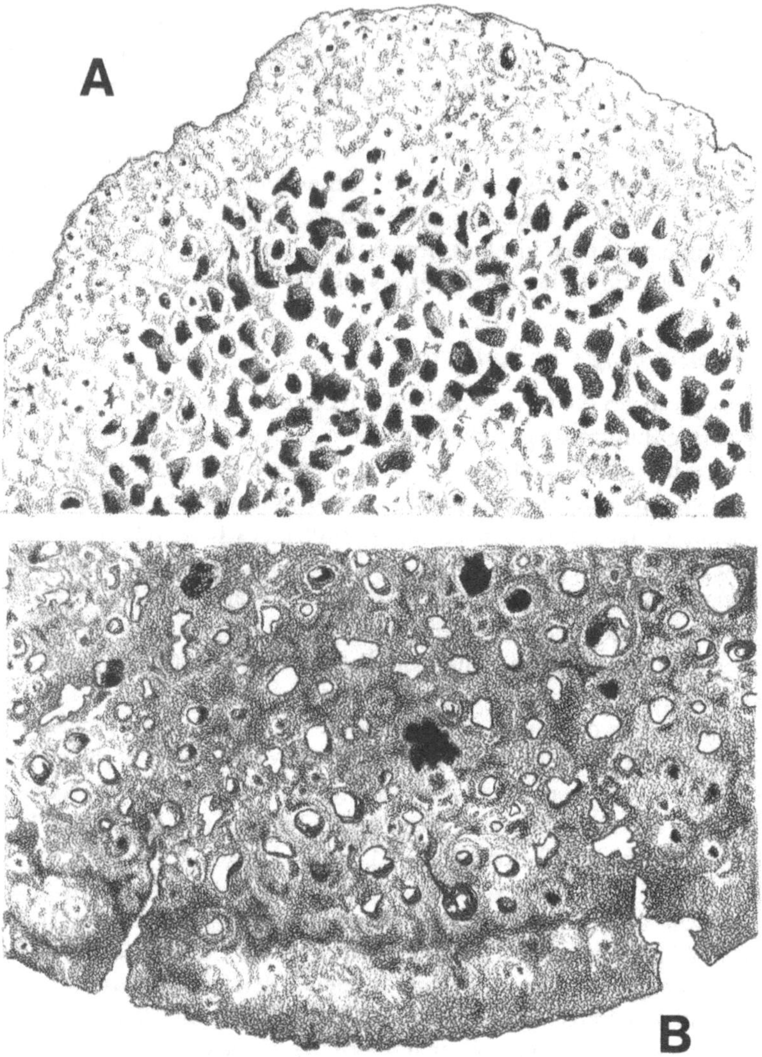

FIGURE 27. *A* A cross-section of the peduncle of a *Stokoceros* F:M.A. 51345, and *B* of *Merycodus leidy*. (Courtesy of Frick's laboratory 1965.)

(often asynchronous on each pedicle), the velvet cleaning lasts long, and the core of the first antlers, at least, may survive for months with bare surface (Fig. 29). The antler cycle was absolutely dependent on testicular testosterone. The diameter of the antler base and coronet could be much larger than that of the pedicle, which points to subperiosteal ossification of the surface. This notion is supported by the fact that the short points that occasionally develop on the base seem to be sprouts (Fig. 30). Casting occurred frequently, far below the coronet and in diagonal level.

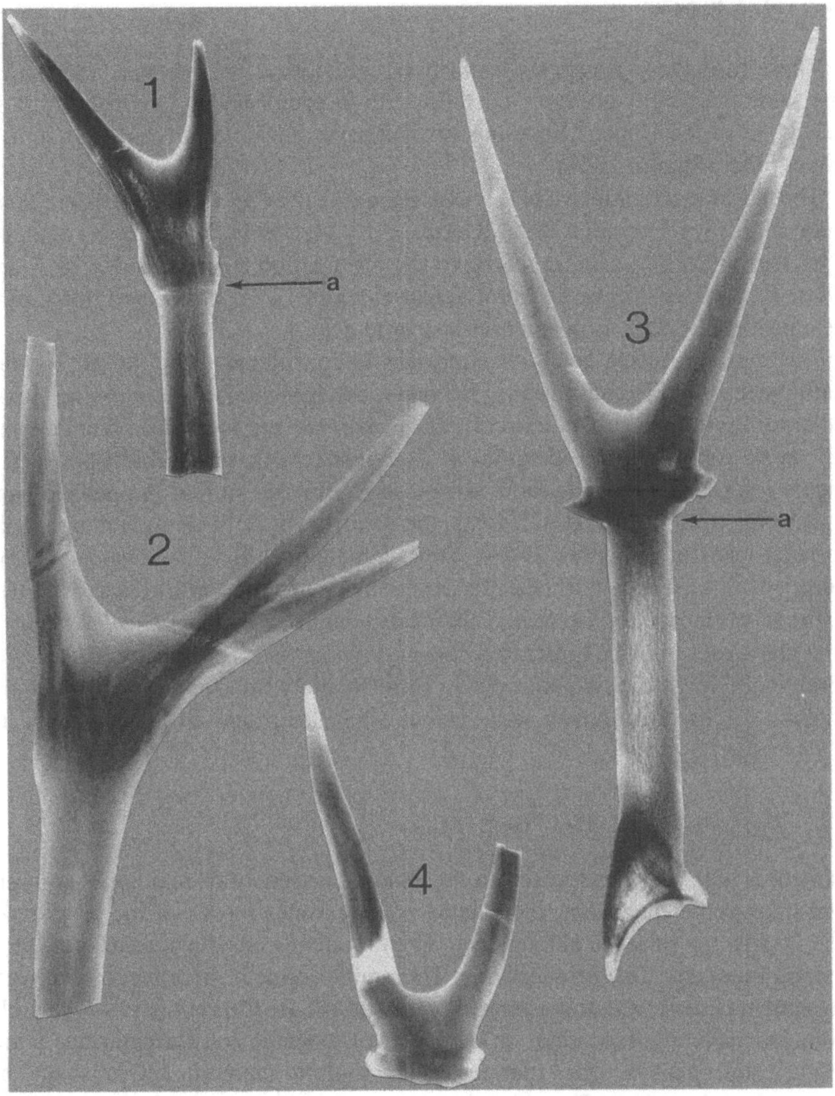

FIGURE 28. *1* Positive print of radiograph of *Dicroceros elegans*, and *2 Heteroprox larteti*: both do not show compact bridge between pedicle and the swollen base; thus the blood flow between pedicle and apex was still possible. The points in *1* are well petrified in contrast to the porous and worn tips in *2*. The protoantler of *1* is in stage of sequestration which proceeds centripetally (arrow *a*). Both *1* and *2* have no coronet. The dark area of the swollen basis only mimics a heavier mineralization due to the thickness of the bone. *3,4* are positive prints of radiographs of the more advanced, muntjac-like *Dicroceros furcatus*. *3* has well developed coronet, but not petrified apices. A faint, centripetally progressing crack around the pedicle below the coronet (arrow *a*), may indicate that rejection of the apex was in progress. *4* is positive print of radiograph of cast protoantler of another *D. furcatus*. The apices are petrified up to the seal. (Radiographs thanks to courtesy of Museum of Natural History, and Dr. Hügin, Basel, Museum of Natural History, Toulouse, and Museum of Natural History, Paris.)

These character—except the centripetal mineralization, coronet, and shaft of the same diameter—point to a relationship to protocervids, possibly to dicrocerids, in which Upper Miocene representatives began to develop first signs of the coronet (Stehlin 1939).

The oldest representative of this new lineage is the *Euprox furcatus* from Steinheim (Fig. 18c) (Hofmann 1893; Kiernik 1913/14; Stehlin 1939). Being equipped with true antlers, it cannot belong to the protoantlered dicrocerids like *D. elegans*. *E. furcatus* is close to Muntiacidae (Stehlin 1937, 1939), and thus cannot be considered either as protocervids or true deer.

The most primitive genus of euproxins is considered to be the *E. minimus* (Schlosser 1928; Thenius 1950). However, the characters of *E. minimus* justify quite different lineage. Pedicles of *Euprox minimus* are located on frontal bones cristae in contrast to the pedicles of *E. furcatus* with supraorbital bases. It is highly improbable that the more advanced *E. furcatus* shifted the pedicle bases back to the orbital rims. Besides this, the dentition of *E. furcatus* is more similar to heteroproxins (Thenius 1948). Thus, although *E. furcatus*, was a contemporary of *E. minimus*, both species cannot belong to the same genus, or at least to the same lineage. *E. furcatus* is obviously a muntiacid, while *E. minimus* is the first known cervid. The bifurcated *Euprox furcatus* was classified by Hensel 1859 (Stehlin 1928) as a new genus *Prox*. Unfortunately this term was preoccupied (Stehlin 1928), but Hensel was correct when he proposed a new genus for *E. furcatus*.

The Eucervidae—The True Deer

Only deer with pedicles located on the external cristae of frontal bones and relative short pedicles, absent, or reduced upper canines represent the progressive and flourishing lineages among deer. Therefore, we consider it as advantageous to term them true deer or eucervids. Their antler cycle is circannual merely and dependent of testicular androgens. The exceptions are Capreolidae, which antler cycle is absolutely dependent on testicular androgens, and Rangiferinae which can use androgens of other than testicular origin to maintain antlers shape, but cannot shed the velvet. The oldest known member of eucervids is the mentioned *Euprox minimus*, a small spiker described first by Toula 1884 (Thenius 1948) and possible propagule of the pliocervins of Khomenko 1913 (Thenius 1950). The common character of Muntiacidae and Eucervidae may be the qualitatively similar responsiveness of the velvet and antler tissue to sex hormones as activators of pedicle development and final mineralization.

The other (and flexible) character used in classification of cervoids is the anatomy of the metapodia, resp. the persistence or reduction of the metacarpi. The plesiometacarpal Muntiacidae and telemetacarpal Capreoloidae are older than the holometacarpal Pliocervinae. The Odocoileidae which appeared after the pliocervini are telemetacarpal, while the most advanced Cervidae are plesiometacarpal.

The anlagen cannot be activated on time without the input of the testosterone as a booster during the sensitive phase for pedicle development, for whatever rea-

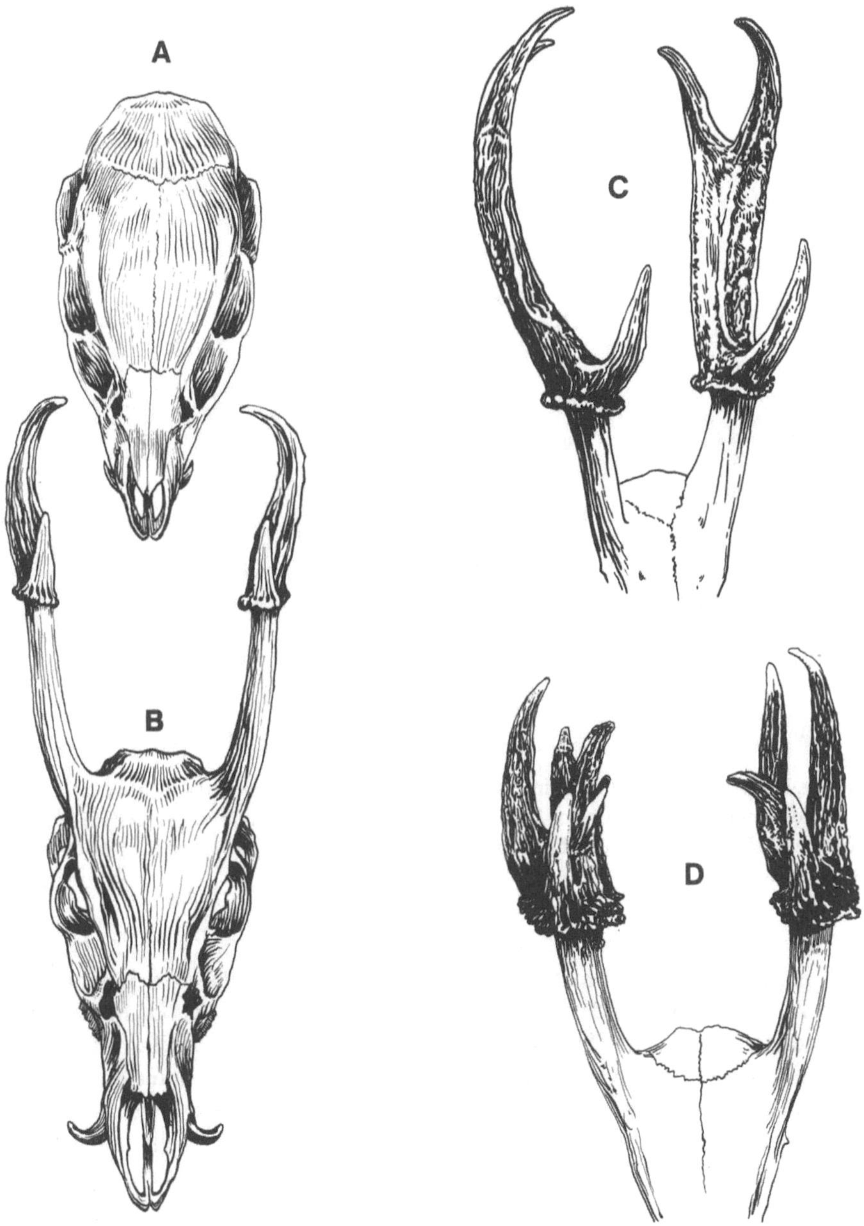

FIGURE 29. Skulls of the living fossil, *Muntiacus muntiacus* Zimmermann, all from Java (courtesy of Dubois Collection, Museum of Natural History, Leiden). *A* Infantile (few-weeks-old) skull before development of pedicles. Note that the nasosupraorbital bridges are just well developed. *B* A normal antler form. *C* Antlers with forked apex, almost identical with similar form in mazama deer *Mazama* sp. Fig. 46. *D* Multipointed antlers. The basal points seem to be developed as sprouts, which may be an atavistic process, known from dicrocerids.

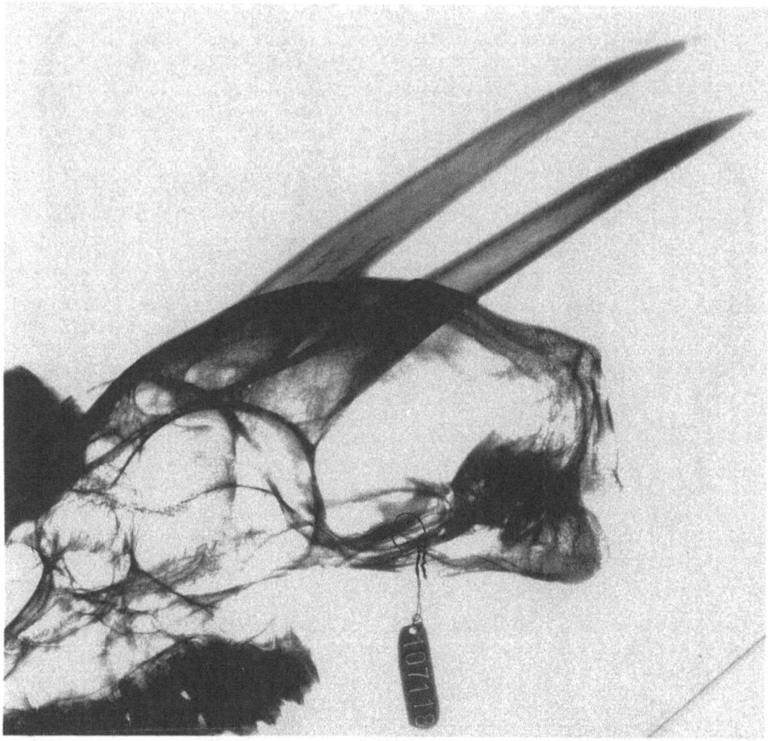

FIGURE 30. Radiograph of the brain case of *Muntiacus atherodes* with first antlers of a specimen older than 2 years according to the full developed dentition (courtesy British Museum of Natural History, London). The whole bony core from the diploe up to the tips of antlers is still alive.

son—e.g., due to a retarded development of the CNS-gonadal axis (Levine 1985; Moberg 1985) or a missed connection between the axon of corresponding neurit to the pedicle. Then, neither pedicles nor antlers are developed, until a hormonal or neuronal booster reactivates their growth (see G. Bubenik, Chapter 8).

For these or other not yet well-known reasons, only pedicle stumps may appear, or the skull remains bald. Such males are called retarded yearlings (knoblers) in roe deer (Bejšovec 1955), or hummels in intact males of mule and red deer (Robinette & Gashwiler 1955; Lincoln 1984; Lincoln & Fletcher 1984). Hummels also can appear in heavily stressed rangiferins, e.g., in the population on Slate Islands in Lake Superior (A. Bubenik, unpublished). The only safe way of permanent polling is the removal of the inductive periosteum prior to the pedicle development.

In all living deer except *Rangifer*, the activation of the inductive periosteum happens in the first 3-4 months of life. However, in odocoileins and *Capreolus*, the pedicle growth and mineralization of their distal part needs only 2-3 months. The first antlers are generally bare knobs, or at most small prickets or forklettes without coronets. In contrast, the pedicle anlagen in rangiferins are activated just after birth and the first, sometimes multipointed antlers are finished within 90 days. They are cast during the winter, and a second set of antlers is developed within the first year of life (Bubenik, A. 1966).

In Cervidae and Muntiacidae, the development of pedicles after priming of anlagen in autumn is slower. Their apices grow into the first antlers at the age of about 12 months, and the velvet period lasts 3-5 months.

Phases and Periods of Antler Cycle

During the velvet period the antler tissue experiences deep transformation from cartilaginous matrix over primary and tertiary bone up to petrification (Muir et al. 1987; G. Bubenik, Chapter 8). The transformation is gradual, but by and large we can differentiate one from the other as follows: The first phase lasts for about two thirds of the velvet period, during which the apposition proceeds very quickly (up to more than 20 mm/day), the osteon is fragile and poorly mineralized, almost without cortex. For this phase we do not have an appropriate term in English. In Chinese it is called 'pênts'ao' which means unossified antlers (Kong & But 1985). In Russian and other Slavic languages an abbreviated term panty is used. The second phase lasts about one third of the velvet stage during which the growth slows down until it stops, due to dynamic mineralization of the osteon which develops a compact cortex, petrified apices and tend to seal the base in order to plug it sooner or later from the core of the living pedicle. The velvet is shed with petrification of the antler points, but not always of the antler base; this is the third phase of antler cycle.

Velvet shedding is the last phase of the growth period. Generally it indicates the approaching or just occurred death of the antler core. However, the velvet also can die before the petrification of points is accomplished. This led G. Bubenik (Bubenik, G. & A. Bubenik 1986) to the hypothesis that velvet death is a separate process, whose direct cause may be infarction of its vessels or blockade of the panty's core sinuses by lipids, as observed by A. Bubenik (1959c, 1966). In rare cases, as mentioned earlier, the velvet can survive infarction and can regenerate by creeping regrowth, as can any skin.

Next is the period of hard, bare, and dead antlers attached to the pedicle. During this fourth phase the testosterone level declines to critical level below which the osteoclastic activity of pedicle's core is activated and leads to antler casting (G. Bubenik, Chapter 8). Then follows the fifth period of bare pedicles soon overgrown by skin which lasts up to new development of panty.

Antler Cast and Seal Profile

In boreal deer the survival of the antler's core after velvet shedding is short (from a few hours to few days at most) (Bubenik, A. 1966). Their mineralization and petrification are fast, due to the relatively sudden and steep rise of plasma testosterone (see G. Bubenik, Chapter 8).

In equatorial deer, the testosterone level does not drop so low in order to arrest the spermatogenesis (Loudon & Curlewis 1988). The antler core can survive the velvet death for at least several weeks, perhaps for months, being nourished through a central, wide blood vessel reaching into the pedicle (Acharjyo & A. Bubenik 1982; Bubenik, A., unpublished observation of antlerogenesis in muntjac). Similar protracted plugging of the antler base happens in the females of rangiferins (Bouchud 1966).

The period of hard and to the pedicles attached antlers correlates positively with the level of plasma testosterone at the beginning of the rut, and the speed of its decline. When the drop reaches a certain critical level, the petrified plug of the pedicle begins to demineralize (Frankenberger 1961). The process heads towards the edge of the pedicle, just beneath the coronet; however, in the axial direction it can be either retarded or accelerated (Fig. 31). In the former case the seal of the cast antler becomes convex, in the latter flat or concave (Bubenik, G. & Schams 1989). In hypogonadal deer, in deer with undulating testosterone secretion, as in tropical deer, and in proto- and pseudocervids, the seal is never convex (see also G. Bubenik, Chapter 8).

An interesting attribute of the antler cycle is the asynchronous timing of any phase and also, as a special event, in different height above the pedicles.

Basically, the antler cycle is a natural result of the dual interference of different levels of androgens: first the testosterone booster activates growth of the panty; then it drops to allow the panty to grow. When it starts to rise again, it operates via mineralization as a growth retardant. Finally, its drop causes the cast. Therefore, the attempt to explain the cast as an antipredator strategy to imitate antlerless females during the critical winter period (Geist & Bromley 1978) is an unrealistic hypothesis. In tropical deer with an aseasonal antler cycle, such mimicry would be meaningless. In seasonally breeding boreal deer, the timing of the critical phases for pedicle growth and antler cast is age and rank dependent (Bartoš & Hyánek 1982a; Bartoš, Chapter 17).

In boreal deer the timing of sex hormone secretion and cessation is so well tuned that it is only logical that prime males with the earliest acme of plasma testosterone level are the first in optimum reproductive performance.

Another erroneous theory about development of antler cast relies on the false notion that "apical growth [of velvet antlers] seems to be the only way to produce a branching structure" which must be replaced by a new, more complicated one (Kitchener 1986, p. 405). As has been emphasized here, the ramifying by sprouting seems to be pristine. Therefore, the annual antler cycle is a process which has developed as a physiological consequence.

Antler Size, Ramification Pattern, and Architectural Ritualization

The importance of the largest possible antler surface as a dissipator of sex pheromone and head accentuation on the one hand, and as a status symbol and sparring tool on the other, may be the two main reasons for favoring the ramification concept and positive correlation between antler mass (represented optically by the total antler length, Bubenik, A. 1982d) and body weight or shoulder height

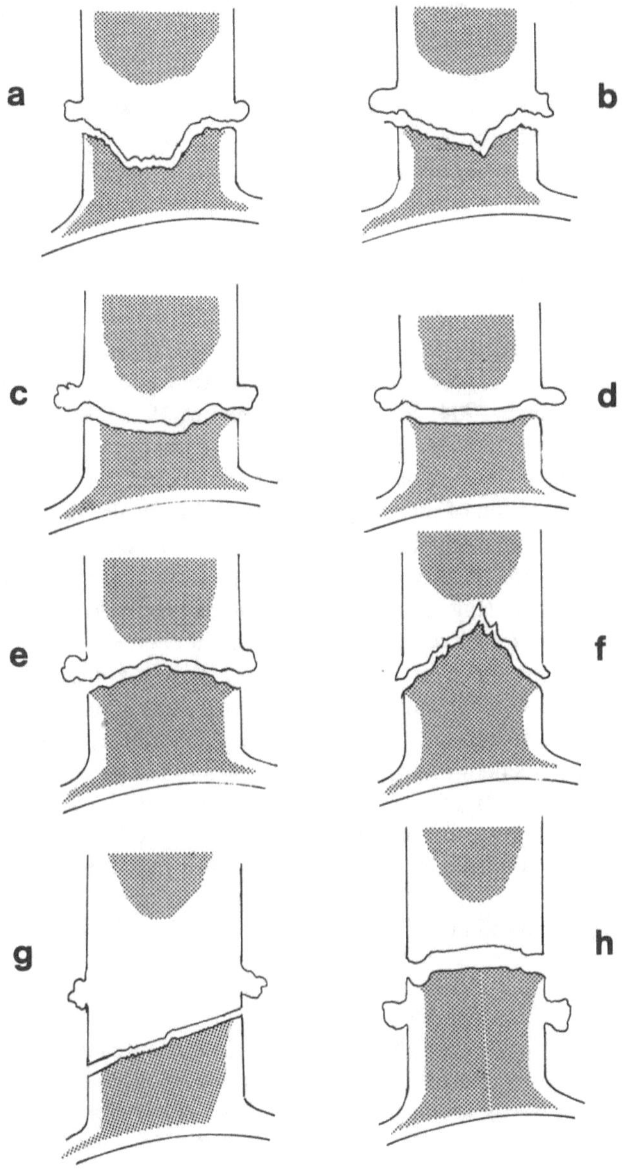

FIGURE 31. Schematic process of loosening of the compact connection between the dead antler and living pedicle depending on testosterone level at the rut and the speed of its decline, which leads to more or less convex, then flat and finally concave seal: *a* in high prime age, *b* juvenile or toward the end of adulthood, *c* to the end of prime age, *d* when senescence is approaching, *e*,*f* testosterone acme was too low, and resorbing process too long, especially in *f*; *g* cast below the coronet, typical in protocervids and in muntjacs. *h* cast above the coronet common in hypogonadal or castrated males.

(Huxley 1931; Clutton-Brock et al. 1980; Bubenik, A. 1985). The high significance of these correlations is shown in Fig. 32.

As mentioned earlier (p. 22, Fig. 7) the trend in permanent overaccentuation of the front-pole went obviously at the expense of the anal-pole. In the open-country species, the tail was shortened, and the contrast coloration of the rump patch lowered. This trend culminated apparently in alcins. In moose *Alces* spp. the rump patch is absent in males and almost invisible in females.

Hypothesis of Principles of Antler Ramification

The numerous attempts to design the principles of the development of antler construction were based partly on ontogenetical changes of antlers, partly on comparison of construction which appeared in different epochs of deer evolution (Hofmann 1901; Weber 1904; Kapherr 1924; Pocock 1933; Beninde 1937; Hoffmann 1956). None of these scholars tried to investigate the branching by experimental studies. The only known trial was done by A. Bubenik & Pavlansky (1965). They split the top of just-developing panty surgically in two ways: transversally (i.e., perpendicularly to) and sagittally (i.e., parallel with the body axis), which rendered quite different results in plesiometacarpal red deer and telemetacarpal roe deer (Fig. 33). Later A. Bubenik proceeded with unpublished study in fallow deer, *Dama dama*, and white-tailed deer, *O. virginianus*.

Sagittal splitting produced in any genera identical twin-antlers (Fig. 33). However, the transversal split showed basic difference of ramification potential of the anterior and posterior half of the beam in both telemetacarpal and plesiometacarpal deer. In the red deer and fallow deer, it was the anterior half of the beam which showed the potential to produce tines; in contrast, in roe deer and white-tailed deer, it was the posterior half of the beam which produced tines. Both other halves have produced only a long beam, which apex may bifurcate.

This may be considered as the evidence that the ramification principle of antlers is based on the following (Fig. 34): dichotomous split of the shaft into two

---

►

FIGURE 32. *A* Double logarithmic representation of the relationships between the $\log_e$ of an average antler length (a total of both beams and tines in cm, in case of moose the antlers surface in cm^2), and average shoulder height cm^{-1}. *B* Relationship between average antler length in cm^{-1} and average metabolic body weight $wkg^{.75}$ of 33 living species and subspecies of deer revealed almost absolute correlation regardless breeding strategy. Only three subspecies scattered so much that they were not included in calculation: No. *1* Tufted deer *Elaphodus cephalophus*, the most primitive muntjacids we know, No. *21 Mazama americana* a very primitive odocoilein about whose antlers length the records vary too much, and the cervicorn, primitive moose No. *31*, *Alces a. cameloides* with small cervicorn antlers. Note the fact, that using total antler length of representative, but not record specimens, one regression line was obtained in contrast to Clutton-Brock et al. (1980) who used beam length of record antlers and got different slopes of regression for serially monogamous, and polygynous deer breeding in small and large groups, or promiscuously.

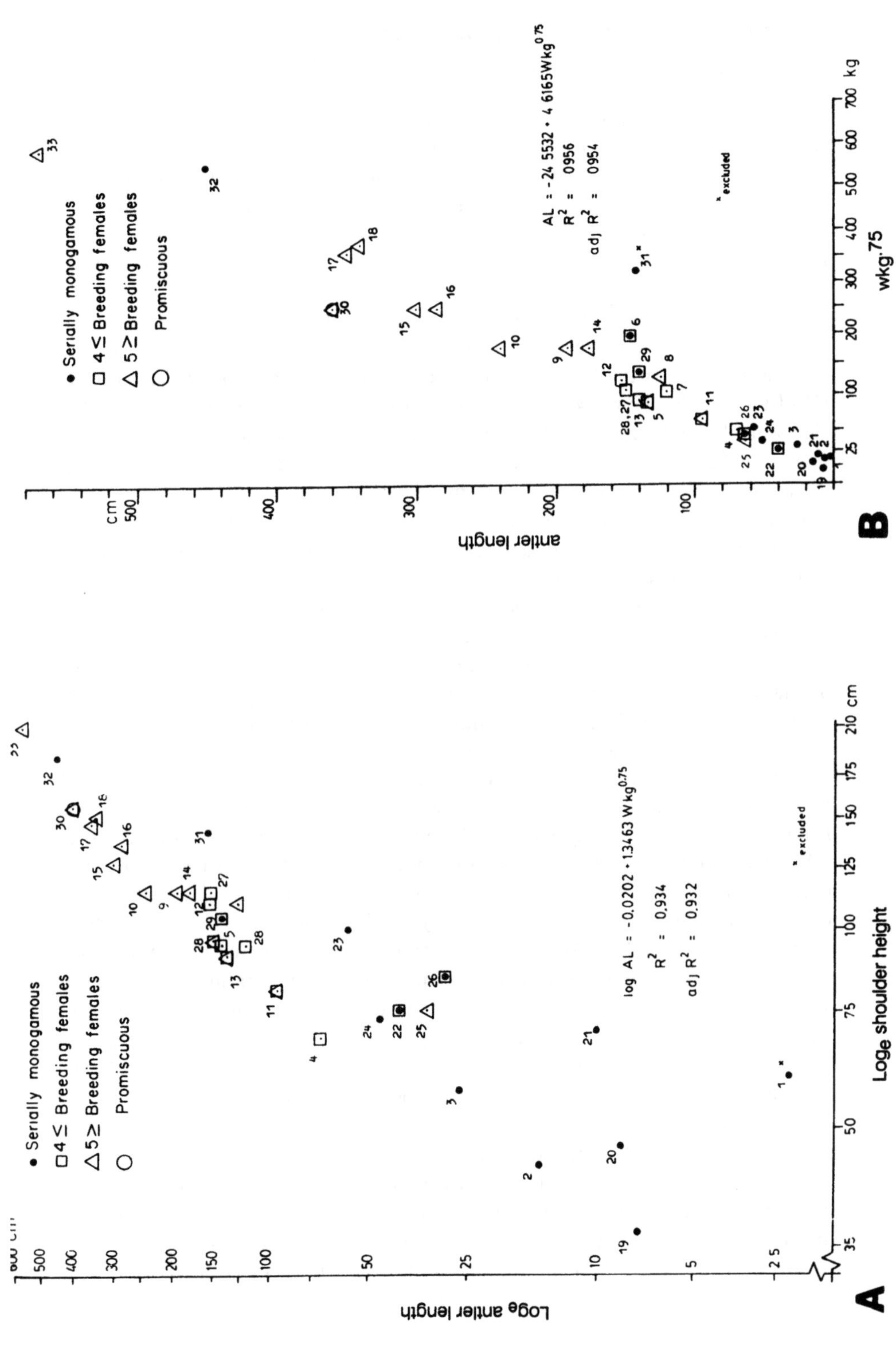
A
B
Serially monogamous
4 ≤ Breeding females
5 ≥ Breeding females
Promiscuous
Loge antler length
Loge shoulder height
log AL = -0.0202 + 1.3463 Wkg0.75
R2 = 0.934
adj R2 = 0.932
excluded
antler length
wkg.75
AL = -24.5532 + 4.6165Wkg0.75
R2 = 0.956
adj R2 = 0.954

equal parts or developing an inadaptive or adaptive fork, as was presumed for the 3-pointed construction of the primitive deer of the rusa-rucervus-axis group by Teilhard de Chardin & Trassaert (1937; Grubb, Chapter 4). In the former the terminal fork cannot produce tines but many points by further dichotomous splitting. In the latter case, the same half of the beam always retains the potential for tining, or the potential will switch to the other half, changing the tines' direction. Ultimately, the switching for tining can change after each ramification, producing tines on both sides of the beam, as in megacerins. Besides this, the beam can rotate and mimic the branching potential of each half, as suggested by Rhumbler (1913) for fallow deer antlers (Fig. 34).

The use of these principles lets us presuppose that also construction should be used where only the posterior half of the beam wil tine just above the coronet. This design was virtually used by the extinct *Algamaceros* (*Blastoceros*) *blicki* (Hoffstetter 1952) (Fig. 34/2), a very primitive deer from the last interglacial of Equador. The unbranched, dorsally bent beams with dorsally protruding tine may be used as good evidence for the notion that antler construction was not aimed toward development of an offensive weapon.

Basically, as long as the grooves on the beam's surface run parallel, the beam is growing in length. Convergent grooves point to terminating development, finished by a single point. Divergent grooves indicate splitting into a tine and a beam, or eventually into two beams. If the production of antler bone tissue becomes hypertelic, the nourishing vessels begin to cross each other erratically and in a different manner on both sides of the beam with resulting webbing or palmation (Fig. 35). Astonishingly enough, after this disarrayed period, a control over the vessels' course is regained and the vessels begin to converge into points protruding from the edge of the palm.

The anlage for primary dichotomous branching is hidden in the shaft (e.g., of mule deer, *Od. hemionus*, or marsh deer, *Dorcelaphus dichotomus*, or Père David's deer, the Mi-lu *Elaphurus davidianus*, where it is so much ritualized that it has been not recognized (Fig. 34). In general, the beam's apex carries also the potency for a secondary dichotomy, which is used, e.g., in the barasingha, *Rucervus duvauceli*. A tertiary dichotomy produces palms, as described above (Fig. 32). Under malnutrition or stress the tertiary dichotomy cannot develop and is replaced by basic ramification. The cervicorn antlers in the moose are good examples of palm regression. However, it is a false opinion that the basic number of tines in palmated antlers, (e.g., in moose) (Geist 1987b) can be estimated from even or odd number of the palm's points. The number varies not only between both palms but within any moose population of the world.

In the primitive antlers of the Miocene-Pliocene epoch, the spongy core was not always homogenous. Depending on construction type, the core also may have two or one large vessels, depending on whether the shaft had split in a dichotomous way or remained a monopodium with tines (Fig. 36). According to these different core patterns, also antlers fragment can be assessed to the corresponding deer lineage; e.g., in Trinil deposits of Java we can easy distinguish between

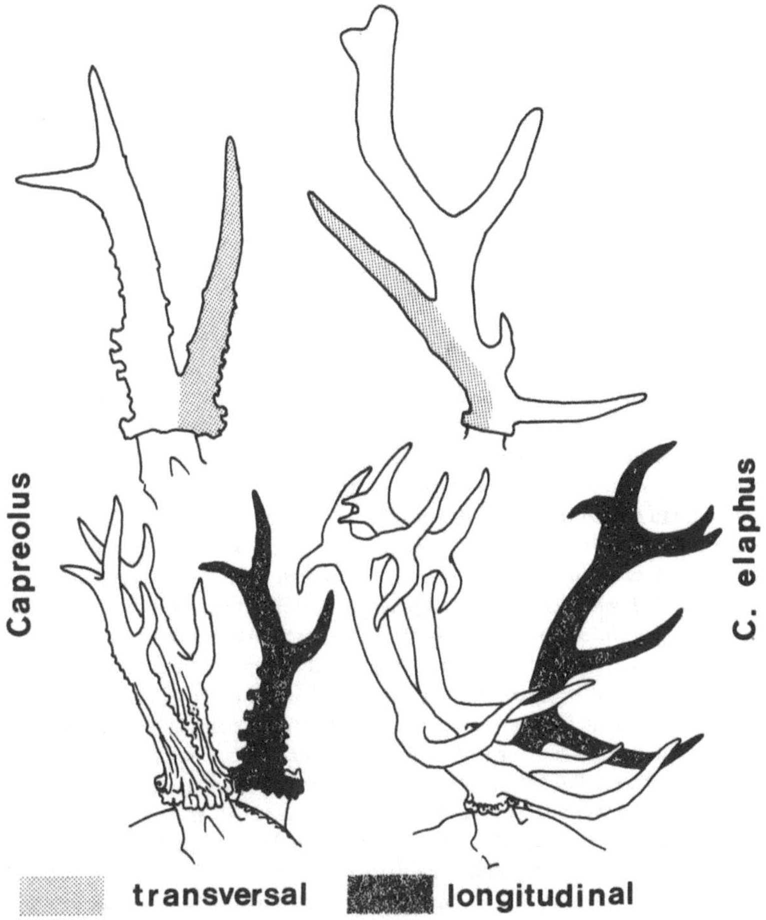

FIGURE 33. Experimental splitting of the developing panty in both telemetacarpal roe deer *Capreolus* sp. and plesiometacarpal red deer *C. elaphus* sp.: a split parallel to the body axis results into twin-identical antlers. However, a transverse split shows that in roe deer the ramification potential has only the posterior half of the beam, whereas in the red deer this potential seems to be inherent to the anterior half only.

antlers belonging to the sambar-ring *R. unicolor* with one huge central vessel and *R. timorensis* with a homogenous core as in red deer (Fig. 37). This method helped to estimate beyond a doubt that the highly ritualized antler construction of Père David's deer *Elaphurus*, is still dichotomous, as in its only forked pliocene ancestors (Otsuka & Hasegawa 1976).

In context with this, we consider it legitimate to question how fast new construction patterns have developed. Lister (1987) offers two possibilities: All

FIGURE 34. A scheme of antler ramification in telemetacarpal (*t*), and plesiometacarpal (*p*) deer. *Telemetacarpi*: first antlers are generally knobs. Next step goes over a spiker like pudu *Pudu* pudu or *Mazama 1*, or can produce different types of ramified constructions: *2* in extinct *Algamaceros (Blastoceros) blicki* a backward beam has dorsal tine, *3* primary dichotomy with two equal beams is known in marsh deer *Dorcelaphus dichotomus*, or mule deer *Od. h. hemionus*; the medial prong in mule deer seems to be developed later in order to close the wide gap between both beams, similarily to the white-tailed deer *Od. virginianus* (*4*) with tines on the dorsal edge of the beam; *5* roe deer *Capreolus* sp., *6* pampas deer *Ozotoceras bezoarticus* use the same basic model; *7* in huemul, *Hippocamelus bisulcus* the construction can vary from a fork up to a 4 pointed beam. Anterior (probably from sprouts) and posterior tines are in reindeer and caribou (*8*). The extinct broad fronted moose *Alces (Libralces) latifrons*, (*9*) shows symmetrical palms, developed probably from dichotomous branching. In extant moose *Alces alces* sp. (*10*), a palm is generally developed by secondary branching between the second tine and the beam's tip. However, also the first tine can bifurcate or palmate. In the extinct moose *Cervalces scotti*, the palmation produced three palms (*11*). *Plesiometacarpi*: *1* first antlers of all species are mostly spikers. In the second set the construction splits: the beam over the brow-tine can develop the inadaptive fork, like in sambar, *Rusa unicolor* (*2*), which can secondarily dichotomize like in barasinga deer *Rucervus duvaucelli* (*3*); or the terminal fork is adaptive. In that case the anterior fork takes over the role of beam with posterior tines (or ?) sprouts like in thamin *Rucervus eldi* (*4*). *5* in the axis deer the second tine is medially oriented; the adaptive fork can retain the posterior tip as beam; then all further tines are developed from the anterior edge as in *6* Java-rusa (*R. timorensis*). If the beam's end shall anew develop terminal fork, it can be again inadaptive or adaptive. This led to following constructions: the frontal fork becomes a beam (*7*) like in sika deer *Cervus nippon*, and *9* fallow deer *Dama dama*; in case of inadaptive fork a multipointed royal or crown can develop, like in red deer *Cervus elaphus hippelaphus* (*8*); however, an adaptive fork can proceed with forward branching and develop construction known in wapitis *C. elaphus canadensis* (*10*). An alternating adaptive fork was developed in the extinct giant megaceroids, like in the simple ramificated *11 Megaloceros (Dolichodoryceros) süssenbornensis*, and its palmated descendant, the Giant Irish elk *Megaloceros giganteus* (*12*). Some plesiometacarpi also have used dichotomous branching, typical e.g., for the pleistocene eucladocerids and elaphuroids, from which only the Mi-lu, or Père Davids' deer *Elaphurus davidianus* survived (*13*).

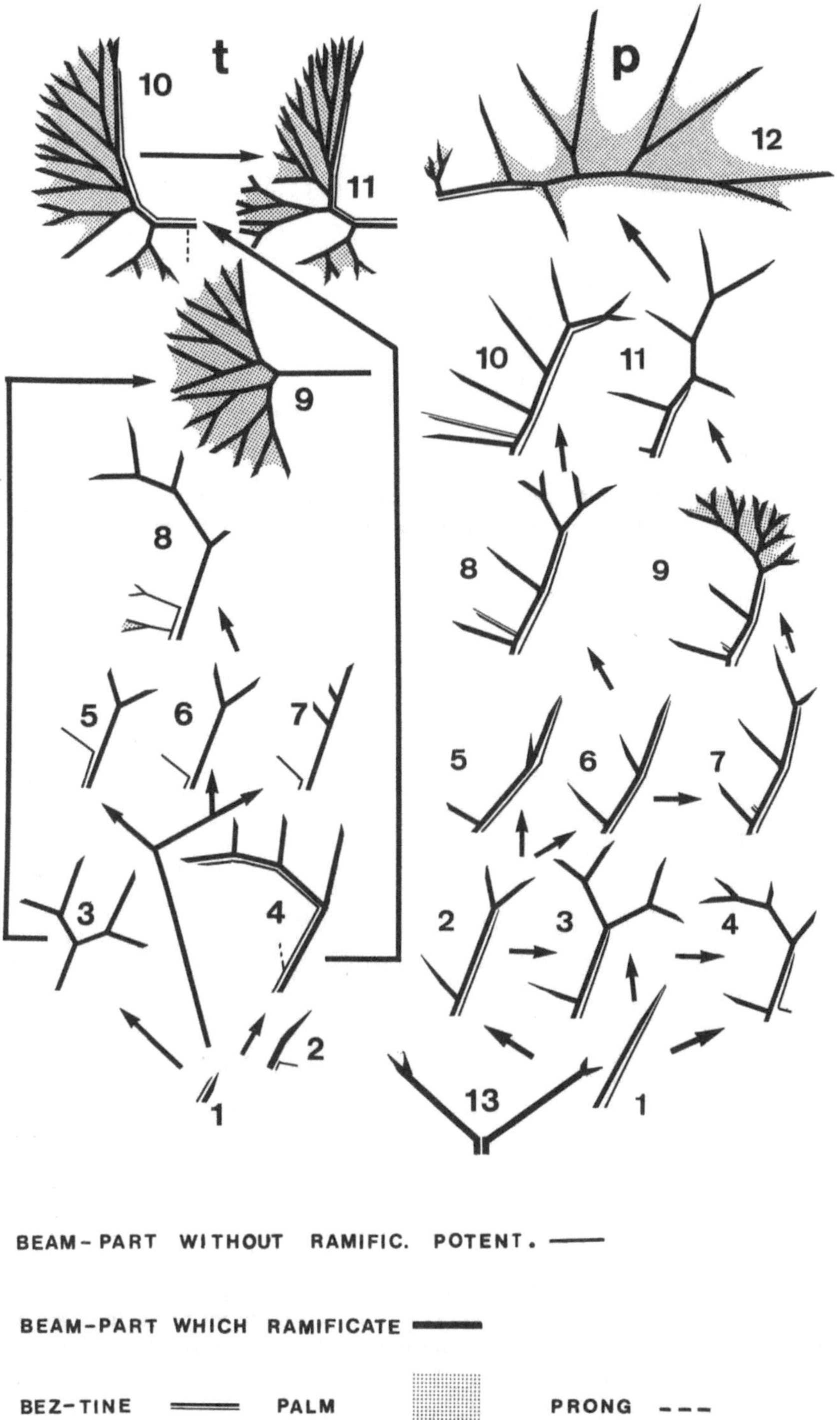
t
10
11
9
8
5
6
7
3
4
2
1
p
12
10
11
8
9
5
6
7
2
3
4
13
1
BEAM-PART WITHOUT RAMIFIC. POTENT.
BEAM-PART WHICH RAMIFICATE
BEZ-TINE
PALM
PRONG ---

FIGURE 35. Dorsal and ventral part of a brow-palm in moose. Note the erratic course of the blood vessels bed and their not identical part of both sides of the palm.

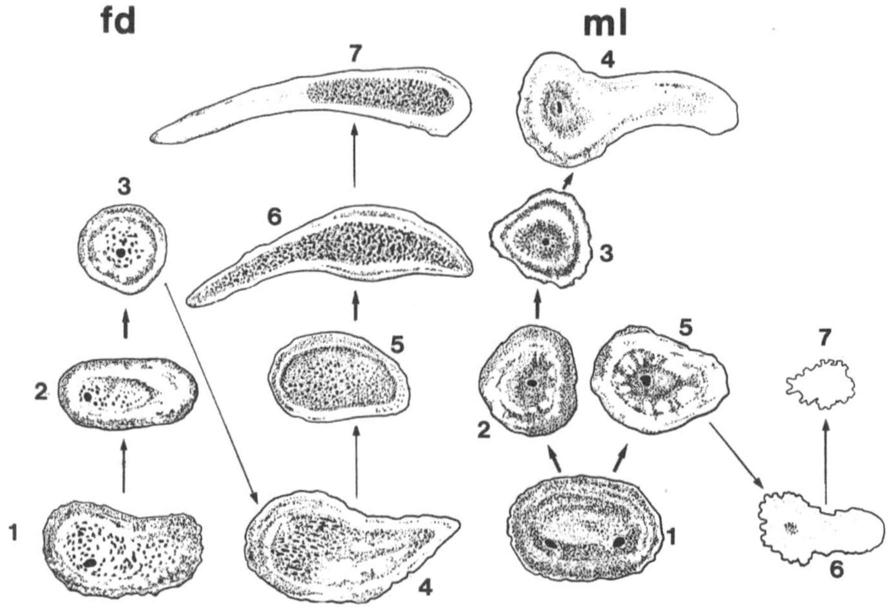

FIGURE 36. Macrostructure of the antler core in two different construction designs from the base to the top. *fd* fallow deer *Dama dama*: *1* shaft, *2* at the ramification into brow-tine, *3* before developing the palm, *4,5,6,7* sections up to the top of the palm. *ml* Mi-lu *Elaphurus davidianus*: *1* shaft, *2,3,4* anterior beam, *5,6,7* posterior beam. In *ml* are two eccentric channels, one for each beam whose architecture is highly ritualized, hiding the original dichotomous design.

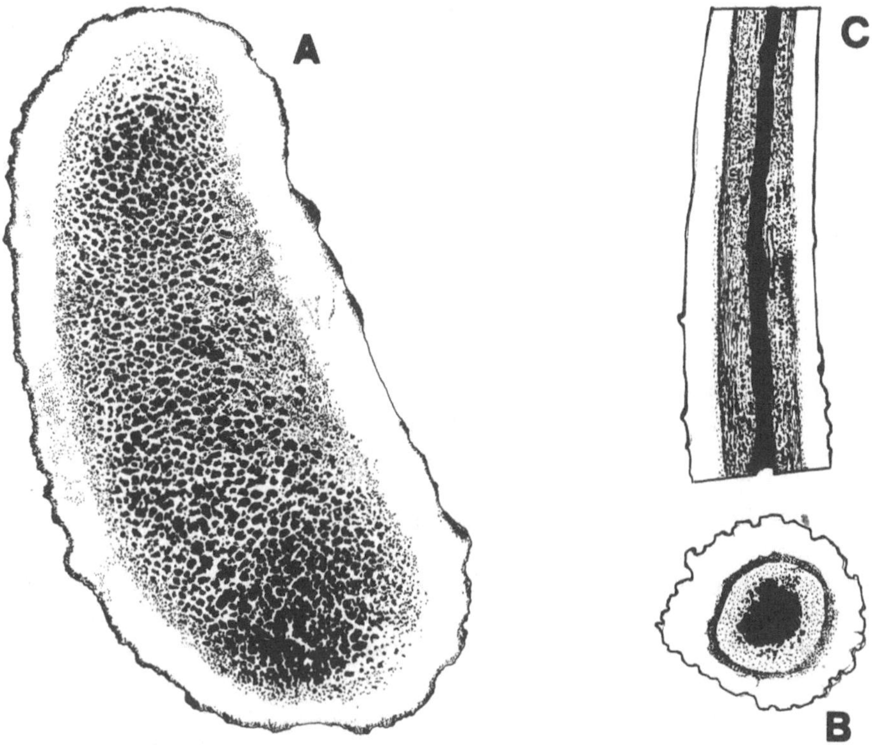

FIGURE 37. *A* transverse section through the antler of a red deer *C. elaphus*, *B* cross-section and *C* sagittal section through the antler of sambar *Rusa unicolor*.

levels of diversity are explicable by accumulated microevolution; or the basic structural types arise in larger steps whose direction contains a stochastic component, and these are subsequently refined by microevolution.

The epigenetic approach leads to another conclusion: the constructive changes went along with the body growth and changing behavior or habitat, or even with all of these variables simultaneously (Goss 1980; Bubenik, A. 1982b). Assuming that the genome stores alleles for many construction designs—of which only one is species-specific as long as the biochemical messenger is the same one—we have to conclude that the same antler pattern can survive in the same shape for millions of years, and even become dormant for different long periods until they are activated; pliomorphic changes which may accompany such a switch can of course induce transcription and enrich the number of construction alleles producing fast, yet unknown designs (Heintz 1970), or convergent constructions.

Antler architecture can change even faster than construction design, perhaps in only a few generations (Markowski 1987), or even in subsequent antlers (Bubenik, A. 1966; Bayern, A. v. & J. v. Bayern 1975). How fast such changes can occur was demonstrated by G. Bubenik et al. (1982) with electrical stimulation of axons innervating the pedicle (see G. Bubenik, Chapter 9).

Architectural Ritualization of Antler Shape

The diversity of antler shape relies not only on the design of ramification but also on architectural ritualization of the basic construction (Bubenik, A. 1982b, 1988). Geist (1974a) rejects the theory of ritualization of antler construction as obsolete myth, and Meunier (1981) dismisses any adaptive changes at all, comparing antlers with the random variability of mandibles of the stag-beetles (*Lucanidae*). Geist's negative attitude to antler ritualization is a mannerism. Meunier's attitude is out of line: comparing evolution of organs in such widely different phyla is too dangerous.

Antlers were, and probably still are, capable of being adapted architecturally to the visibility and density of the understory (Fig. 38). A good example is the roe deer. Woodland populations preferring dense forest edges have relatively short but robust antlers, stream-lined in the head profile, which do not hamper flight with the head kept low. Roe deer antler architecture can remain stable during prime age, but can also change from one subsequent antler to the other. The reason for this is unknown. However, it might be possible that when the buck changes his territory, then a quite different structure and texture of the understory may be the cause of neuronally induced architectural remodeling.

By and large, antlers can operate as short-distance or long-distance signals; simultaneously they must not impede flight through the cover or by high wind velocity at the level above the head (Fig. 38). Where the understory is dense, the

---

►

FIGURE 38. Architectural adjustment of antlers shape to the habitat conditions and signal function: *a* The streamlined antlers of the roe deer, running with head down. Visual function is minimal. *b* A plowshare shape of white-tailed deer antlers, whose beams are bent inward. We do not know about their importance as optical cue. *c* A bulldozing shape of red deer antlers, whose more or less heart-shaped beams with the royals go through any thicket with raised head. The antlers' length and the multipointed royals are intimidating signal by head nodding. *d* A twin-bladed antlers of a fallow deer palms, which do not hamper flight and indicate the position of the buck when his head is still hidden in the bush. By turning the head they reflect the light like mirror. *e* A weather-vane-shaped antler of barren ground caribou. Sideward pressure of a strong wind is so powerful that caribou prefer to go either with or toward the wind. In visual communication the palmated first tine is displayed by tilting the head. *f* A glider-wing architecture, such as of the tundra moose *Alces a. gigas*, seems to reduce the antlers' weight, if the wind is strong enough. For visual perception of antlers rank the tundra moose bends the head down, the woodland moose as e.g., *Alces a. americana* present the palms laterally. (Modified from W. Trense 1989.)

a b c d e f

antlers remain streamlined, regardless of the number of points or body size—for example in muntjac, mazama, roe deer, chital, and sambar—or they take the shape of plowshares. Due to their size, they are more or less only short-distance signals. In less dense habitats, antlers also become a visual long-distance signal. Depending on the habitat, they can work as bulldozers, or go easily through the thicket, with palms as two almost parallel blades. Long, harp-shaped antlers with many points on the top may operate (after experiences with dummies) like a weather-vane, which force the animals to go with the wind when its velocity is too high. The parallel palms are excellent long-distance signals when the head is turned. Antlers also may function like the wings of a glider, which reduces the weight of antlers, lifting the head (Bubenik, A. 1982b, 1989).

Hypothesis of Antler Growth Centers (AGC)

G. Bubenik (Chapter 8) refers in detail to the theory about AGC developed by A. Bubenik & Pavlansky (1965). However, for taxonomic reasons, the chapter on antler development cannot disregard some basic facts: Asymmetry of pseudoantlers, protoantlers, antlers, and pronghorns is almost a rule and symmetry an exception; also the asynchronous antler cycle is nothing exceptional, as is the frequency of many anomalies, such as casting and regeneration of antlers above or below the coronet on one or both pedicles.

However, an important phenomenon is the trophic memory for injuries, hormone boosters, and optimum food composition. The memory for injury that occurred just at the beginning of velvet-antler growth can produce malformed antlers for many subsequent cycles. The memory for hormones, when applied in appropriate periods, produced responses in the subsequent antlers (see G. Bubenik, Chapter 9). Ingesting optimum composed food over many years produced, for at least three subsequent hunger periods, voluminous antlers on which the tines and beams collapsed due to inadequate consumption of calcium and phosphorous, the builders of the bones' hydroxyapatite (Vogt 1947; Bubenik, A. 1966).

We also have evidence that the hypothetical AGC independently control each pedicle, as far as it concerns responsiveness to hormone receptors alongside the beam and its timing, as well as antler shape and weight.

Some support for the hypothesis of AGC is given by the elegant studies of S.W. Botjer on zebra finches (as referred to by Weisburd 1987). Botjer found that during the course of song learning the caudal nucleus of the ventral hyperstriatum increases 50% during the 70 days of the sensitive phase. At this time the magnocellular nucleus of the anterior neostriatum loses half of its cells. Morphologically this means that when one half of the brain is growing, the other is diminishing. Thus, the development of nuclei controlling the pedicles may be analogous to the development of nuclei for song learning, or to the importance of NGF during sensory interaction between the whisker pad in the mouse before and after the establishment of neuronal connection to the tactile whiskers field (Davis et al. 1987; Heumann 1987).

Perukes and Velericorn Antlers

The development and control of species-specific antler construction needs testosterone in order to arrest surface proliferation by centripetal mineralization of the velvet-antler tissue. Thus, castrated, cryptorchid, or hypogonadic male deer carry a variety of perennial appendages (Brandt 1901; Bubenik, A. 1963, 1966; Thomas et al. 1970; Bubenik, A. & Weber-Schilling 1986; Baber 1987). Their type correlates with the phyletic age of the species, genus, or family, and also with individual levels of plasma androgen of other than testicular origin, and eventually sensitivity of androgen receptors in the velvet and bony tissue. By and large, there are many types of proliferating appendages, some of unknown nature.

1. The permanently proliferating perukes (Fig. 39), of which osteon have a sarcoma-like character (Olt 1927a). The peruke tissue is benign and does not metastasize. But it has malignant impact on the calcium/phosphorus metabolism, forcing the body to utilize these elements from the skeleton. The perukes grow so fast that within 2 years their nodules grow over the whole head and together with the decalcification of the parietal bone contribute to the death of the castrate (Bubenik, A. 1963; Bubenik, A. & Larsen, D., unpublished).

2. Phenomenal variation is a peruke of intact male (collected by A. Bubenik, February 1967, in Bilje-Yugoslavia), with beginning spermatogenesis (Fig. 40). It allows us to speculate that the velvet did not have testosterone receptors or that they were silenced.

3. The cactus antlers (Baber 1987) (Fig. 41) are common in castrates of more highly evolved genera, but occasionally can appear in intact males with the apparently protracted phase of dynamic growth of velvet-antlers. They are characterized by numerous exostoses called pearls, which may cover the whole surface. However, they do not proliferate with a rate we know from peruke, and the velericorn antler remains in a more or less specific shape. The accidentally lost or naturally sequestered parts are regenerated (Fig. 42) (Bubenik, A. & Pavlansky 1965). Velericorn antlers are common in castrates of more highly developed lineages, such as odocoileins or cervins, in which the adrenals seem to produce enough corticoids to keep the shape under control (see G. Bubenik, Chapter 9).

4. Velvet-warts (Fig. 43) appears as large spherical appendices on the panty of intact reindeer males. They are shed without leaving any trace on the antler surface (Bubenik, A. 1966, 1975). Their origin is unknown.

5. Antler-osteoma (Fig. 44) seems to be caused by severe, but finally encapsuled and mineralized, hemorrhages of panty. It appears in intact males. A. Bubenik (1966) recorded osteomas in reindeer and red deer (Bubenik, A. 1966, 1975).

Although the disposition to develop perukes or velericorn antlers seems to be lineage-specific, some individuals can produce the opposite perennial features (i.e., a peruke instead of velericorn antlers and vice versa) (Murie 1928; Bubenik, A. 1963, 1966). This may depend on the individual potential for secretion of adrenal androgens.

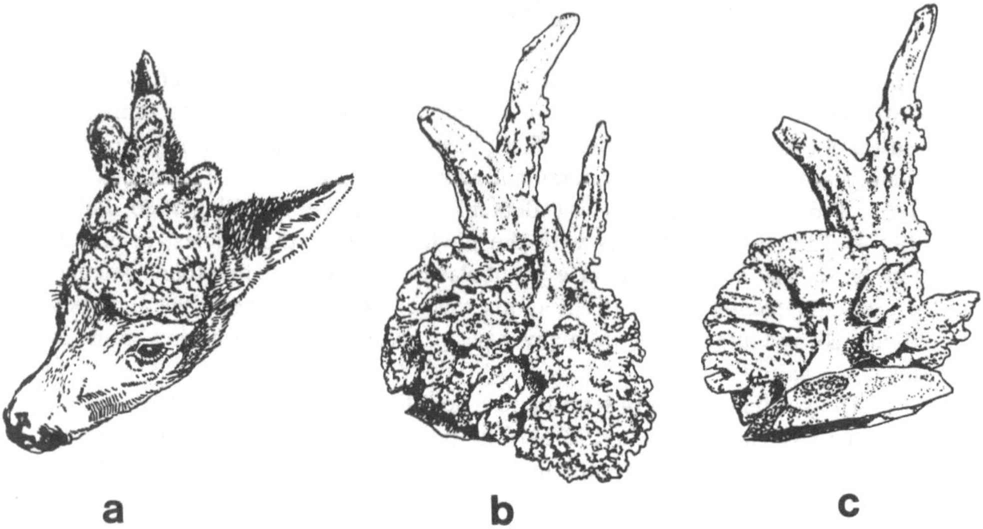

▲
FIGURE 39. Roe deer peruke: *a* plain view, *b* skinned, *c* the left half.

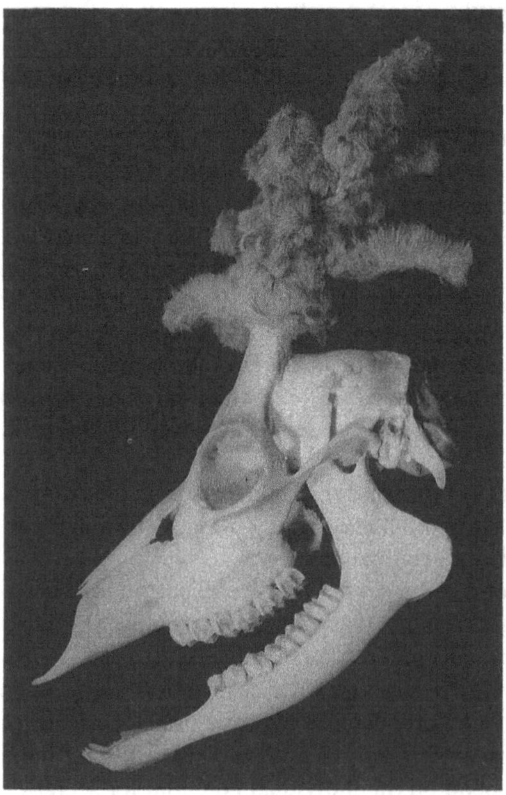

◄
FIGURE 40. Peruke of a sexually intact roe buck, shot in February.

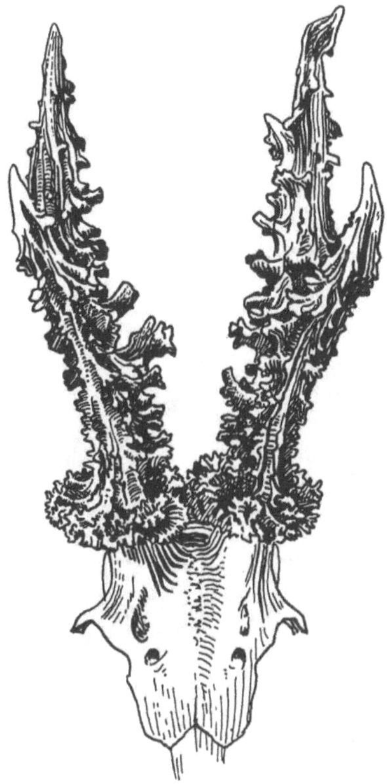

FIGURE 41. A typical cactus antlers of a roe deer. They are rare in roe deer castrates, but frequent in some roe deer populations, probably due to delayed mineralization of panty.

Cryptorchids and sometimes castrates may exhibit unilateral delay in casting above the coronet, which points to different zonal sensitivity of testosterone receptors.

Social Significance of Cervoid Cranial Appendages

As mentioned earlier, the development of cranial exaptations and their cooptations into appendages probably can be ascribed to permanent injuries of exposed skull angles. The social significance of all these appendages as scent dissipator and indicator of maturation status and individual rank may contribute with differential impact to the restructuring into larger, and in many aspects more effective, organs. Organs with such a polyvalent importance must be a target for both natural and sexual selection (Fisher 1958).

The perennial or facultatively perennial nature of proto- and pseudoantlers entitles us to the notion that they were organs almost as fragile and sensitive as panty, or velericorn antlers of living deer. The proto- and pseudoantlers were

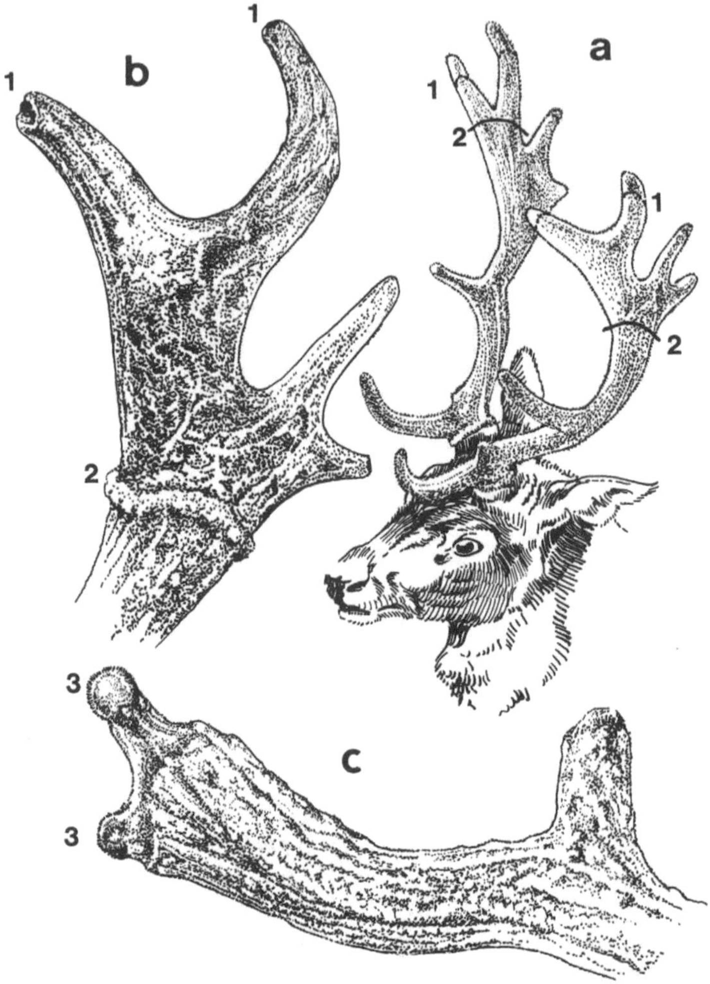

FIGURE 42. Velericorn antlers of a fallow deer castrate. *a,b* Prior sequestration of the distal parts. *1* The first lost tips, *2* the line along which half of the palm was sequestered. *b* shows in detail the left velericorn palm with developing coronet (*2*), above which the sequestration occurred. *c* Sequestered part in the stage of regrowth of new point (*3*).

handicapped as effective weapons by their exposed supraorbital location, possible fragility of the bone, and probable relatively low level of plasma testosterone, as is the case in tropical deer such as muntjac (N. Chapman, personal communication, 1987) and chital (Loudon & Curlewis 1988).

The low testosterone values may lower the aggressive behavior to such an extent that muntjacs prefer to fight with tusks (Barrette 1977) and sambar and chital by flailing despite the fact that their antlers are hard (F. Kurt and G. Bubenik, personal communication, 1969 and 1987, respectively). Flailing is an exclusively feminine fighting technique in deer. In boreal males it is used only during

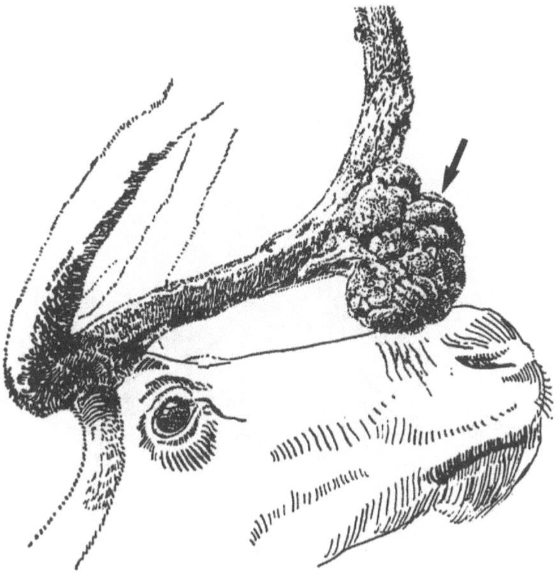

FIGURE 43. Velvet-wart on panty (arrow) of an intact reindeer bull looks like a huge skin wart. It can weigh over 1 kg and is shed without a trace on antler surface.

the period of testicular quiescence, i.e., from a few days prior to antler cast up to the beginning of dynamic mineralization (see Bartoš, Chapter 17). Therefore, the hypogonadal white-tailed bucks from Texas, equipped with velveted antlers, also use flailing in sparring and contests (Thomas et al. 1970).

Olfactory Function of Panty and Hard Antlers

Support for the notion that the proto- and pseudoantlers served primarily as scent dissipators is rendered by the fact that in all living deer, from the pudu to the moose, the panty and antlers are used for scent marking and pheromonal stimulation. During the panty phase, the sebum of the velvet is smeared daily over the whole body and vegetation, leaving a very broad scent track, most intense in the zone of antlers horizon (i.e., in the height of olfactory perception of moving animals) (Chard 1958; Bubenik, A. 1966).

It is obvious that the olfactory function of velvet is lost with its shedding. The fact that apparently all deer have developed sophisticated techniques to impregnate their hard antlers with their own sex pheromones from different glands and urine is good evidence for the importance of olfactory communication with antlers (Bubenik, A. 1982b, 1988). The scent glands in question are rubbed on vegetation, from which they are transferred to the antlers or to the pronghorns, mostly after careful olfactory checking. In deer species in which the penis is suited for jettisoning urine (Haigh 1982) towards the sideward tilted antlers, this strategy is preferentially used (Bubenik, A. 1982b).

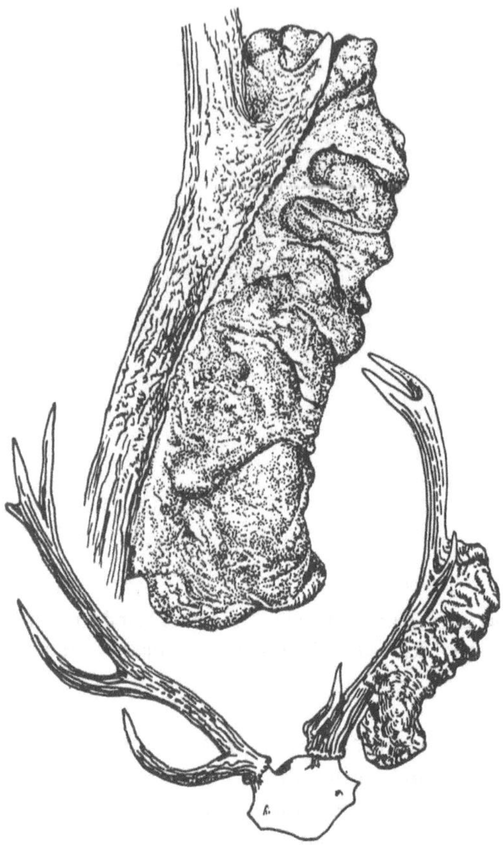

FIGURE 44. Antler osteoma on a red deer antler looks like a proliferated antler bone tissue.

As mentioned earlier, the importance of the scent of growing or dead cranial appendages may have led to the trend to enlarge their surface by branching, by palmation of the beam, or by paired, twisted, or horned-prong-equipped peduncles, as it happened in antilocaprins such as *Osbornoceros*, *Illingoceros*, or *Antilocapra* (Frick 1937). These observations led to speculation that the epigenetic vector of this development was primarily the olfactory function. The role of a visual status symbol may be of secondary importance. Thus velvet-antlers, hard-antlers, and pronghorns seem to play a role of sociophysiological chemocommunicators, monitoring the timing of puberty of subadult classes of males as hypothesized for red deer (Bubenik, A. 1986b), and as a primer of estrus, e.g., in moose (Bubenik, A. 1987a).

The importance of sex pheromones dissipated from the antlers is evident during the rut. Generally, the breeding male positions himself so that the wind can carry the scent of the antlers toward the female(s) and eventually to other males as well. Thus, the antler pheromones serve as informer of the status of the dissipator, and possibly enable the discrimination of his genomic quality. Their odor

FIGURE 45. Naso-nasal contact is an important ritual, as soon as both individuals displayed friendly moods. This picture, redrawn from Super 8 film, shows that the bipedal male is not distinguished from a quadrupedal peer.

helps to synchronize the females' estrus and intensify the rutting milieu. Thus, the intense pheromonal atmosphere possibly stimulates the gonadal activity of both sexes as well. Due to the fact that antlers or pronghorns are replaced annually and that the pheromones seem to change according to the fitness and rank of the male (Claus, Bubenik & Hamr, unpublished), the antlers and pronghorns appear to be more sophisticated organs than horns. In many cervids, the inferior male's naso-nasal checking (Fig. 45) or olfactory checking of all antler points of the superior male secures a safe presence in the closest proximity of the superior peer (Bubenik, A. 1987a).

Antlers as Status Symbol and Visual Supercue

Being of a deciduous nature and producing a distinct number of ontogenetic steps, antlers indicate annually the developmental status of the deer. Because they also serve as a sparring tool and organs of intimidation, their offensive or defensive shape should go along with the social status. This can be confirmed by development of the construction and architecture in any deer antlers of juvenile deer which are designed offensively, and defensively in prime individuals. However, during the adulthood the architecture can change back and forth, i.e., toward more or less advanced and intimidating stages, depending on the social status in the previous year (Bubenik, A. 1966). Approaching the postprime age, antler construction begins to regress, losing tines from the top down. Thus, the antlers regain again a more and more offensive and simple shape. With senescence, a collapse of antlerogenesis is inevitable. If the male survives, the last antlers are generally large, fist-sized stumps (Bubenik, A. 1982b). Therefore, the flexibility in ritualization of antler construction is not only an obsolete myth, but a remarkable phenomenon, whose logical explanation points again to epigenetic connection.

The studies with artificial front-poles of deer species whose lateral antler length is more than twice as long as the head (Fig. 46), have convincingly shown that antlers can raise the front-pole of the male deer to an optical supercue (Tinbergen 1951) and in caribou or moose to an absolute releaser which overwhelms any other cue, including human scent (Bubenik, A. 1982b, 1987a,b). The fact that a bipedal peer is accepted by large cervids as any quadrupedal conspecific corroborates that in large-antlered deer species the torso and its direction toward conspecifics does not play any important role. Thus, the role of the antlered front-pole is an important character that distinguishes deer from bovids and puts the function of large antlers far ahead of large horns.

It is interesting, and of evolutionary importance, that only the cervid females seem to have the inherited anlage for assessment of the species-specific range of antler shape and its rank. However, both sexes seem to have probably inherited if not acquired discrimination of the quality of the scent that the antlers dissipate (Bubenik, A. 1988). Therefore, in Kirkpatrick's term (1987), antlers and maybe also horns are a pleiotropic indicator of a male's preference. Because we cannot qualify the olfactory pattern with our nose, the use of antlers as only visual cues in simulated games or other theories is pointless. In this case, only a carefully monitored bioessay can be useful (Bubenik, A. 1982a).

Antlers as Weaponry, "Handicap" Organ for Female's Choice, and Thermoregulatory Agent

There is no doubt that, as soon as protoantlers have been replaced by hard, insensitive, and sharp-pointed antlers, they became theoretically extremely dangerous weapons. To an ethologist who "lives with deer," antlers appear as sophisticated sparring tools by means of which rank order can be assessed and overflow intrinsic tension can be calmed down by tiresome contest. Antlers also can be an excellent shield of face and body. Their use as a defensive shield is more important than the aim to inflict incapacitating injury, or even death of the contestant, as argued by Geist (e.g., 1974a). Routine merciless use of antlers is a behavioral anomaly, which is common in socially disorganized populations (e.g., on the Isle of Rhum) (Clutton-Brock et al. 1982), or by politically motivated mismanagement (Bubenik, A. 1984, 1986a, 1989).

Finally, there is the indisputable fact that during evolution and ontogeny, every antler construction experienced architectural ritualization. Its obvious aim is the

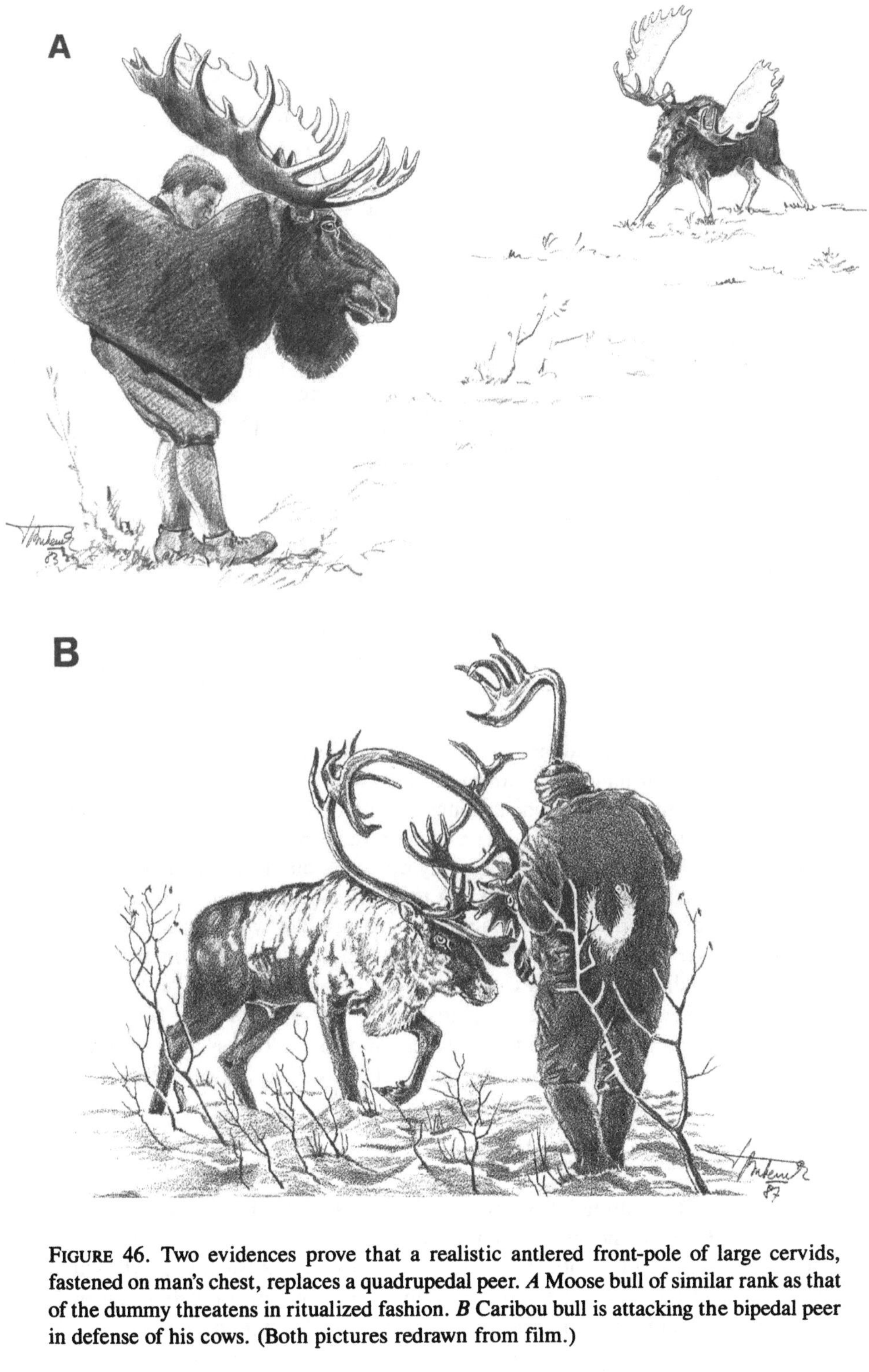

FIGURE 46. Two evidences prove that a realistic antlered front-pole of large cervids, fastened on man's chest, replaces a quadrupedal peer. *A* Moose bull of similar rank as that of the dummy threatens in ritualized fashion. *B* Caribou bull is attacking the bipedal peer in defense of his cows. (Both pictures redrawn from film.)

reduction of offensive design on the one hand, and improvement of defensiveness and visual impact on the other (Geist 1978a,b, 1982). However, the numerous skin injuries found, e.g., on hides of mule deer (Geist 1986b), do not imply the evidence that they were caused by antlers alone. Scholars familiar with year-round behavior of any deer know well that most of these skin scratches originate from hooves, inflicted either by flailing during antlerless periods, or on females' shoulders by mounting males. Most other scratches originate from collisions with hard objects in panic flight, and only a few from antlers blows.

Antlers and the Handicap Principle

Another aspect of the role of antlers opened the theory of selection for a handicap, i.e., females select males with the largest nuptial organs or ornaments like antlers, horns, or peacock feathers etc. (Fisher 1958).

The idea that antlers are a handicap must be rejected as a notion lacking a realistic background. The most objective paper of Kirkpatrick (1987) dealing with sexual selection by female choice in polygynous species including elaphins points to the complexity of this problem. It does not matter if it is considered as illustrative of the theory of "good genes" or of the "inadaptive school."

The notion that antlers or horns are a handicap that lowers the life expectancy (Geist 1966b) does not reflect conditions in well-organized populations. Sheep with optimum-sized horns and deer with optimum-sized antlers have low life expectancy only in disorganized populations. In contrast, under optimal social welfare, males with the largest horns or antlers experience the longest life (Bubenik, A. 1986a).

Of course, there is no doubt that antlers that exceed in size the sociobiological range are a virtual handicap. However, deer with antler size over the socially "acceptable" limit, which can be produced easily in deer farms, are not preferred as mates by females with free choice and are rare in free-living deer. The fact that female's choice qualifies antlers not only optically but also olfactorily does not allow their use in modeling, where just the optical impact can be used.

For a female in estrus, the males are chosen visually by antler size. However, to select a mate from two with the same antler rank, the female will check olfactorily the "quality" of the mate by the scent of his antlers. In such cases the dummy will fail, as well as a strange male who defeated her mate (Bubenik, A. 1985). This example shows that the concept of physical fitness in general and in antlers in particular is based on anthropomorphistic imagination and is far from the concept of actual selection rules.

Panty as Thermoregulatory Organs

In contrast to the perennial ossicones and horns, only the heavy vascularized panty could be considered as a thermoregulatory agent (Stonehouse 1968; Ohtaishi & Too 1974). With the death of the antler bone, such property will be lost. The fact that the antler cycle developed in a warm, probably seasonally humid and dry climate of the Mid-Miocene opposes the hypothesis that the antler cycle evolved as an adaptation to cold periods to avoid the possible freezing of the growing panty (Barrette 1977). An example are the antlers of roe deer, one of the oldest species whose ancestors, the procapreolids emerged at the end of the Miocene. Roe deer antlers grow from the December solstice to the vernal equinox, i.e., during the coldest months of the year. They may suffer frostbite only if the temperature drops below −30°C (Bubenik, A. 1984). The panty of reindeer does not show any thermoregulatory responses to ambient temperatures (Wika et al. 1975; Hove & Steen 1978), contrary to the hypothesis that velvet-antlers thermoregulate body temperature. The so-called frostbitten antlers (i.e., which sequestered the distal, velveted part) are known only in roe deer. J. Portmann (1970) ascribed this phenomenon not to the frost but to the development of the Looserian rebuilding zone known from animals. However, their prerequisite is not only a hunger osteopathy but also physical stress which is never exerted on growing antlers.

Synchronization of Velvet Shedding with the Rut

Chapman (1981) expressed the notion that the obvious timing of the period of hard antlers with the reproductive period may be more coincidental than selected. This might be true for the cervids of warm latitudes which experience just one, lifelong puberty. However, for the pleistocene cervids with annually recurrent puberty, the velvet shedding in the middle of the rut would be biologically pointless in general, and for prime males in particular. Despite this, Geist (1982, 1985) considers as evidence of Tule elk's (*C. e. canadensis nelsoni*) Siberian or Beringian ancestry the alleged (*sensu* McCullough 1971) velvet shedding after the beginning of the rut. Even though the subspecies status of the Tule elk was questioned (Hutton 1972; Schonewald-Cox et al. 1985), the alleged information cannot be found in the McCullough paper. The fact that males with first antlers or by and large most of the submature males may shed the velvet at the beginning of the rut or do not have petrified antler points may be considered a two-fold advantage: having well-pointed antlers too late means being inferior as mates and having less offensive impact in fights with peers.

Genetics of Antlers

A certain insight in the genetics of inheritance of antler form and size gives the intersubspecific of interspecific hybrids. In contrast to bovids, hybridization between deer does not produce either unilateral preference for the antler construction of parent (e.g., red deer x maral, or red deer x wapiti), as has been suggested by Sallač (1912), or intersex as in polled goats (Asdell 1944). Interspecific hybrids of deer quite frequently develop antlers of the more primitive parent (Gray 1954) or have their coloration pattern. Intergeneric hybrids are infertile. Thus the hypothesis that some new deer species may be hybrids between primitive and advanced lineages (Geist 1974b) seems questionable. Hybridization with a larger subspecies is successful only if the females have no other choice, i.e., a mate of the same subspecies is not available. Otherwise, hybridization resulting in larger antlers is possible (Winans 1913; Vogt 1947).

It would be very difficult to explain by random mutation the phenomenal diversity in the construction, and architecture of antlers in general and their reappearance in unrelated lineages independently of space and time in particular, as suggested by Lister (1987). In contrast, the very dense innervation of the velvet and the neuronal connection to the hypothetical centers for antler growth make epigenetic modeling and copying into genome and activation of silent or silenced DNA sequences more plausible.

The fast spread in time and space from ecotype to phenotype and the transcription of the genocopy into the genome's program point to fixation of these alleles on females chromosome. This notion is corroborated by Vogt's (1947) crossing experiments.

Presently we do not have any proof whether an individually programmed allele or gene for antler design and size exists, and/or what kind of quality range it may have. The theoretical paper of Sohr (1975) and the records concerning white-tailed deer bucks (Templeton et al. 1982) are not persuasive. The higher responsiveness to gonadotropin-releasing hormones in white-tailed deer with optimum antlers quality (Bubenik, G. et al. 1987) can be as inherited as socially acquired. Thus, when field data are compared, androgen levels in relation to age and rank are an important and not easily obtained factor. Rörig (1901) reports how the antler quality of roe buck, isolated from does for years, improved after the animal could mate. Similar was the experience with white-tailed deer bucks and castrates (G. Bubenik, Chapter 8).

Marchinton et al. (1987) questions the view that a high degree of isoenzyme heterozygosity is a prerequisite for overall excellence of antler quality (Smith et al. 1987; Scribner, Chapter 19). This criticism goes along with the experiences in the project "Achental," where red deer and roe deer deer populations are artificially held in social well-being and overall best fitness (Bubenik, A. 1984, 1986a). The isoenzyme heterozygosity of the resident Achental-population dropped to 75%, in contrast to more than 90% in 11 other red deer populations of Central Europe, living either in disorganized or socially organized conditions (Herzog 1986).

Part D. Timing of Growth in Relation to Zeitgebers

Timing of Reproductive Periods in Relation to Photoperiodicity and Climate

All known facts about horns, pronghorns, and antlers, and their pristine stages as well, leave no doubt that they are involved in the breeding strategy, mainly as informer, primer, and organs of female's choice. We know also that the cervids can be divided in three groups: photoperiodically independent, photoperiodically dependent, and photoperiodically flexible.

The Photoperiodically Independent Eupecora

The fact that eupecoran have evolved in a warm climate lets us presuppose that theoretically the timing of their most important life periods, such as breeding, calving, and maturity of cranial appendages, should have little or no dependence on photoperiodicity, or be independent of the length of gestation. The crucial timer could be only the food abundance, because offspring born during dry periods would hardly survive. This prediction goes along with experiences from the wilderness, where breeding periods are tuned to stay in accord with zeitgebers such as food abundance after calving, but are random or seasonally independent in zoos or in zones with year-round food abundance. This can also be confirmed by the uninterrupted puberty of many African antelopes (Bigalke 1963; Skinner 1971; Lincoln 1985; Skinner & van Jaarsveld 1987) and the timing of new estrus shortly after parturition. Hence eupecoran, with pregnancy around 5 months, can have two offspring annually (Asdell 1964). Climatic conditions which favored permanent reproduction persisted in most regions until the Late Miocene when episodes of aridity or cold seasons began to appear (Webb 1984). The savanna and steppe opened the formerly dense habitat, and large Eupecora of the runner-type were developed. Their greater body size demanded gestational periods around 7–8 months (or longer, as in the giraffe) and more intense lactation which often prolonged the anestrus period into weeks or months. Therefore, only one parturition period annually was possible.

The lifelong (even if undulating in intensity) puberty was also found in tropical deer of the genera *Muntjacus*, and *Axis*, and is suggested in *Rusa unicolor*, *Rucervus*, *Mazama*, *Ozotoceras*, or *Dorcelaphus* (Acharjyo 1982; Brokx 1972; Jackson 1985; Lincoln 1985). However, due to the individually tuned antler cycle of these genera, the female's choice is narrowed to the more competitive males with hard antlers. Based on this experience, we have to presume that the predecessors of deer, the protocervids and pseudocervids, and antilocaprins as well, were also aseasonal breeders, capable of tuning the breeding cycles.

The Photoperiodically Dependent Eupecora

During the second half of the Villafranchian, when the summer-winter seasonality north of the Tropic of Cancer began to be more severe, and especially when the glacial Pleistocene approached (Frenzel 1967; Buchardt 1978) a seasonally independent breeding cycle could not be maintained. Thus, the eupecoran which evolved during the Upper Villafranchian and the adjacent glacial Pleistocene (Frenzel 1967; Kowalski 1971) have to develop harmonious intradependence of breeding and calving seasons and gestational length to secure optimum survival of the offspring. The best solution to this problem was photoperiodically fixed timing of all these periods on the one hand, and recurrent puberty with recurrent quiescence of gonad on the other. Due to the food abundance during spring and summer, and 6-8 months' pregnancy, all eupecoran which evolved in the Pleistocene are invariable "short-day" breeders, theoretically between the autumnal equinox and winter solstice. Practically, due to long survival of sperm in the epididymis (Jaczewski & Jasiorowski 1974), and recurrent estrus sometimes up to the vernal equinox, a long parturition season as a "last reserve" is possible. However, the late-born offspring can only survive in temperate areas – seldom in boreal and never in subarctic climate.

The photoperiodical adjustments and their fixation in genome are easier to explain as a consequence of epigenesis than as a result of random mutation. The short period of spermatogenesis of short-day breeders between the end of summer and the winter solstice is also in accord with winter severity and the fitness of males, who with almost zero testosterone level become altruistic and are deprived of the weight of antlers during a long winter.

Thus, the timing of the horn growth, pronghorn cycle, and antler cycle in those eupecoran who evolved from the Miocene to Pleistocene cannot go unnoticed from a taxonomic view. Therefore, and according to the physiological mechanisms which correlate with the impact of the climate, the bovoids and cervoids can be categorized according to the epochs in which they evolved and which give evidence of their phyletic age (Bubenik 1986a). The fact that some of these epochs are still present in distinct geographical zones facilitates some insight into physiological processes which monitored the survival of extinct taxa.

The Photoperiodically Flexible Eupecora

With the cooling of Central and Southern European climates in the Pleistocene, the invariably aseasonal breeders had to emigrate into more southern warm zones, or perish. Only a few eupecoran that evolved in warm climates were genetically so flexible that they could adjust the estrus to photoperiodical control without giving up the more rigid program of breeding during long days, i.e., close to the equinocturnal light length of the tropics. It was this flexibility that enabled them to occupy enormously wide climatic zones and be successful aseasonal breeders in the tropical zone and short-day breeders in the boreal zone as well. Using the prefix mio-, plio-, or plei-, from the Miocene, Pliocene, and Pleistocene we can categorize the giraffoids, bovids and cervoids as follows.

Evolutionary Categories of Eupecoran

Mio-Giraffoids

Mio-giraffoids are characterized by straight velericorn proto-ossicones. The only survivor, the okapi, has estrus right after parturition, but due to a pregnancy period of 14–15 months the breeding season is not annual. The proto-ossicones never ankylose with the cranium.

Mio-Bovids

Using as a model the extant survivors (Asdell 1964; Anderson 1979), which are only in the tropical zone, we have to assume that the extinct mio-bovids would have bred any time, many twice a year, but that they also might "tune" the breeding period to optimal calving conditions (see the review by Spinage 1973). According to Vrba (1975), the extant are typical ecological generalists. The cornuogenesis is continuous; ossicones fuse with the cranium soon after birth.

Plio-Bovids

The extant plio-bovids still live in zones with tropical or subtropical climate. They are generalists, with permanent horn growth and ability to breed any time of the year; however, due to longer gestation, they have only one parturition annually (Kingdon 1979). Cornuogenesis is continuous; ossicone fusion may last up to 7 months. In the Late Villafranchian most of the Plio-bovids in Southern Europe perished. Whether the extinction can only be ascribed to the invariability of aseasonal breeding is questionable. Many of the obligate long-day breeders migrated from a cooling climate southward of Pamir and Tibet into Pakistan, India, and Indonesia (Heintz & Brunet 1982). Only a few chose the northern route and reached China, from where they migrated south when the climate began to cool down. Those with gestational periods of about 9 months could remain long-day breeders also in the boreal climate by tuning the breeding period so that it matched the photoperiodically optimum time for parturition.

Some surviving plio-bovids of Africa seem also to be energetically flexible (Skinner & van Jaarsveld 1987) in order to survive in subtropics as well. These species may create confusion for the behavioral ecologists and physiologists, because their physiological parameters depend on the latitude at which the population lives. One of these flexible species may be the South African blessbok, *Damaliscus albifrons*, and bontebok, *Damaliscus pygargus*, or in the northern hemisphere the serow, *Capricornis sumatranensis*.

Given the fact that the almost pliocene climate persists between the tropics of Cancer and Capricorn, it is natural that the species in concern proceed in their pliocene evolutionary trend. Thus we can follow for example the processes of remodeling of os cornu architecture in the early *Parmularius* lineage, and in the sister-groups, Alcelaphini and Aepycerotini. Of course, the evolution of other cooling organs (e.g., the ears) may also proceed, but is undetectable in fossils.

Pleisto-Bovids

Pleisto-bovids are invariably short-day breeders with periodic horn growth. Most of them seem to breed invariably with photoperiodic control. All are specialists (Vrba 1975) with very fast diversification and adjustability to almost any kind of habitat, and are a good example of flexible ecomorphs (Aleyev 1986). Ossicones ankylose within days or weeks after birth.

Pleisto-Bovids exhibit a great variation in maturation rates, depending on infrastructural conditions. Thus one population can appear typically pedomorphic or neotenic, while another, even a sympatric one, shows all the signs of a typical hypermorph (Bubenik, A. 1984).

Mio-Cervids

They are aseasonal breeders, with aseasonal and often asynchronous antler cycle. The long pedicles are located either supraorbitally or low on the frontal bone cristae; the antlers have one or at most three points, and a core which survives long after velvet shedding. With time it is solidly petrified along the whole length. The gestational period is about 5 months long, and overt estrus appears immediately or within 2 weeks after parturition.

Typical representatives are the plesiometacarpal *Muntiacidae*. However, there is a distinct probability that among the South American telemetacarpal deer, both the *Pudu* and the *Mazama* also may be mio-cervids (Garrod 1877; Spillmann 1931). For this conclusion we give evidence: the primitive skeleton of a slipper (after the German Schlüpfer; Krieg 1936b) or duiker, and the FMM breeding strategy [Female Monopolizes (one) Male] (Bubenik, A. 1985), the 5–6-months gestational period, and aseasonal breeding. The antlerogenesis is primitive, with often asynchronous cycle, antler asymmetry, and branching similar to that of dicrocerins or euproxins. Some mio-cervids are flexible enough to tune their breeding period according to climate. In temperate niches of Southern Chile, the pudu exhibits a seasonal breeding cycle (Vanoli 1967; MacNamara & Eldridge 1987), but in captivity it breeds year-round (Hershkowitz, 1969). Aseasonal breeders with aseasonal antler cycles are also the mazama deer, which in original habitat fawn in any month of the year (Krieg 1948; MacNamara & Eldridge 1987).

The progress in metapodial characters of the pudu deer (Hershkowitz 1982) may be a convergence. The idea of Eisenberg (1987) and Groves & Grubb (1987)—that pudu and mazama are probably secondarily small deer, derived from large ancestral forms with more complex antler—is questionable (Krieg 1948) in view of the necessity of reversal of all morphophysiological features. Such complex reversal cannot be documented for any taxa. The use of caryotype of pudu looks close to mule deer and white-tailed deer (Spotorno & Fernandez-Donoso 1975) may show the unreliability of karyotypes in comparing taxa very distant in time and lineage. Both Spillmann (1931) and Kraglievich (1932) consider pudu and mazama as the most primitive deer, close to procervulids, or according to Matthew (1908), as possible descendants close to the sinecorn Blastomerycids. However, the multipointed antlers of *Mazama simplicicornis* (Fig. 47), or (as a rare specimen) the multipointed antlers of pudu deer have all the characteristics of protocervids or procervulids, e.g., of the extinct *Cervulus cf. sinensis* (Teilhard de Chardin & Piveteau 1930).

Plio-Cervids

This term is not synonymous with the *Pliocervinae* Khomenko (1913). The plio-cervids, like the plio-bovids, do not represent a special taxon like pliocervins. The plio-cervids include not only cervids of the Pliocene, but also those of the adjacent and climatically similar Early Villafranchian. As inhabitants of mostly open habitat, the plio-cervids were large ungulates which experienced more rapid speciation (Teilhard de Chardin & Trassaert 1937; Stanley 1979) than Mio-cervids. Their larger bodies required a 7–9-month-long period of gestation, with

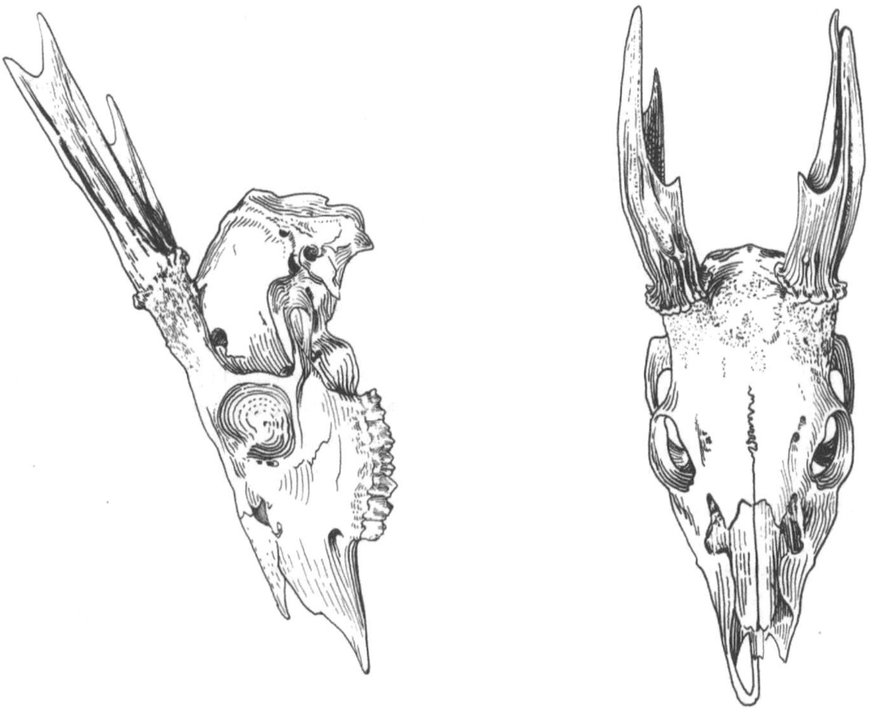

FIGURE 47. Multipointed antlers of *Mazama americana* (Collection H. Krieg). Very primitive ramification similar to multipointed muntjacs.

estrus appearing a couple of weeks or months after parturition. As there is a positive correlation between body and antler size, they must also have produced larger antlers. From the numerous taxa, those which survived include only members of the most primitive genera, such as the rusin-, rucervin-, axis-, and elaphuroid-lineages among the plesiometacarpi and the pampas deer (*Ozotoceras bezoarticus*) and marsh deer (*Dorcelaphus dichotomus*) among the telemetacarpi. It is questionable whether the two hippocamelids, the huemul and taruca *H. bisulcus* and *H. antisiensis* are plio-cervids.

The plio-cervids are irreversible aseasonal breeders, with individually tuned antler cycles and breeding periods adapted to the date of birth and to the more pronounced pluvial and dry seasons on the one hand, and continuous or long-lasting estrus periods on the other. The open mosaic-habitats favored a gregarious way of life and the MMF-s breeding strategy (Bubenik, A. 1985; one Male tries to Monopolize as many Females as he can herd and defend).

As soon as the European climate began to cool (Frenzel 1967), mainly due to the uplift of the rising Tibetan plateau (Kuhle 1986, personal communication, 1987), seasonable breeding became essential for survival. Pliocervids have had two choices: to escape from or to adapt to zones where breeding periods needed photoperiodic control. The former was certainly easier than the latter—i.e., to readjust the aseasonal cycles to seasonal breeding for which they might be not programmed (Kiltie 1984). Thus emigration was essential for lineages with invariable independence of photoperiodicity. Surprisingly, in contrast to bovoids, all moved eastward north of the Pamirs and Tibetan plateau into the still temperate or warm zones of China. When the climate there became extremely seasonal, just before the Mid-Pleistocene, the survivors migrated south and west to India and Malaysia, where they presently live (Thenius 1980).

In South America the telemetacarpal plio-cervids were established since the Mid-Pliocene (Hershkowitz 1982). There is no doubt that the pampas deer and the marsh deer (Mohr 1932, 1962; Jackson 1985; Lincoln 1985) are invariable aseasonal breeders. Their phyletic relationships are not clear. The pampas deer has all transferrin carried by the white-tailed deer, but also has eight other transferrin phenotypes (Quinteros et al. 1971). As mentioned above, the evolutionary age of both hypocamelids huemul and taruca is uncertain. Both are seasonal, short-day breeders (Povilitis 1983; Merkt 1987), a character typical for pleisto-cervids. However, according to their karyotype (Spotorno et al. 1988) they are close to mazama, which points to a pliocene origin. Unfortunately, we do not know how both will behave at low altitudes with temperate climate.

Their extinct relatives, the antiferins (genus *Antifer*), of temperate pampa east of the Andes lived together with the above-mentioned aseasonal odocoileins (Spillmann 1931; Kraglievich 1932). Therefore, both huemul and taruca may belong to plio-cervids whose genetical flexibility allows them to be short-day breeders in a strong seasonal environment.

Among the Old World plesiometacarpi, the most primitive of all were the pliocene rusin deer with adaptive three-pointed antler construction (Teilhard de Chardin & Trassaert 1937) whose many lineages stretched from Mediterranean Europe up to China (Heintz 1970; Groves & Grubb 1987; Grubb, Chapter 4). They also disappeared from European theater with the approach of the glacial Pleistocene and survived in China until the Mid-Pleistocene before the winter temperature reached the freezing point (Kuhle, personal communication 1987). They began to migrate southward. However, the genetic flexibility or adaptability to the pleistocene seasonal severity was also developed in some plesiometacarpal plio-cervids.

Plio-Cervids That Mimic Pleistocene Origin

Evidence of such flexibility exists. Each of these tele- and plesiometacarpal plio-cervids used a quite different and sophisticated strategy to thrive in the seasonally harsh climate.

The white-tailed deer above the Tropic of Cancer mimics with seasonal breeding and antler cycle a true pleisto-cervid (Bubenik, G. 1987); in the tropical zone it is a typical aseasonal plio-cervid (Brokx 1972; Branan & Marchinton 1981; Stüve 1985).

The roe deer is another flexible long-day breeder of miocene or pliocene ancestry. As an invariable long-day breeder, it has the rut in summer, but has only 5 months of actual pregnancy due to the delayed ovum implantation (the diapause), which can last up to the winter solstice, i.e., 4.5 months (Stieve 1950a,b). The bucks have the antler growth period between the winter solstice and the vernal equinox. The spermatogenesis lasts from March/April to the beginning of September. Antlers are cast between September and January. Estrus can repeat from July to January (Short & Mann 1966a; Schams et al. 1980). The winter antlerogenesis and the diapause may both be remnants of pristine living conditions in strongly pluvial and dry seasons. Delayed egg implantation might be the only gateway mechanism for unfavorable feeding conditions – as is known, for example, in ursids and mustelids. Thus theoretically, the roe deer and its ancestors, the miocene procapreolids and pliocene capreolids, might produce two offspring annually (see Sempéré, Chapter 14).

Among the plesiometacarpal deer of pliocene origin thriving far north of the Tropic of Cancer were the elaphuroid deer, which appeared in the Late Pliocene of Japan (Otsuka & Hasegawa 1976). Their subgenus *Elaphurus davidianus*, the Père David's deer appeared for the first time in Eastern China and Usuri in Mid-Pleistocene (Keqing 1987). It is the only representative of elaphurins. It can have two antler cycles, up to 19 recurrent estrus cycles (Curlewis et al. 1988), and two breeding seasons annually (Mohr 1962). The two annual antlers and breeding cycles point to an aseasonal plio-cervid which could adapt to the boreal climate and breed at the beginning of the summer due to an 8.5-months-long gestational period.

The Plio-Cervoids

The extant *Antilocapra americana* or pronghorn "antelope" is a pliocene offshoot of antilocaprids which emerged at the end of the Miocene. Due to its breeding strategy, it was capable of thriving in a strongly boreal climate of the savannah of western United States and Canada. The females ovulate and conceive up to three times subsequently, with each egg fertilized and developing into an embryo. Thus, up to three embryos in different stages of development can be found in the fallopian tubes. However, to secure the first embryo's survival and only one parturition period in the spring, all ova conceived later and subsequently descending into the uterus are killed by the sharp amnion end of the oldest one (O'Gara 1969a,b; O'Gara, Chapter 7). Theoretically, this gestational strategy looks like inaccomplished superfetation.

The pronghorn bucks also did not adapt their pronghorn cycle to the short-day breeding period. The pronghorn sheaths are sequestered late in the autumn, and regeneration proceeds through the late-winter and spring, similar to the antler cycle of the roe deer.

The Pleisto-Cervids

All cervids which have evolved since the Late Villafranchian, about 1.8 MYA (i.e., with advance of the main glacial periods) are obligate short-day breeders. The number of new pleisto-cervids, many of them giants, may again be an epigenetic response to "up-to-now" inexperienced stress of the winter severity.

All pleisto-cervids have antler construction that can be derived from the adaptive three-pointed rusin deer. Their explosive diversification happened in Eurasia and South America. Surprisingly enough, the diversification in North America remained far behind that in Eurasia (Kurtén & Anderson 1980).

During this epoch, deer with all imaginable antler constructions and shapes emerged and perished. The vast steppes with riparian forests, and the taiga-forests and tundras (Hopkins 1959; Frenzel 1967) produced giant deer, of which only the moose *Alces* spp. survived. Most of these giants perished when the dense forest began to reoccupy the tundra during each interglacial. Although it is difficult to assess the true reasons for their extinction, it is possible that the dense forests and the difficulty escaping predation might be either the main contributors or the ultimate causes of their disappearance. In Eurasia the most notable are the megacerins (Azzaroli 1979; Kahlke 1951; Lister 1987) and alcins (Bubenik, A. 1986b) with the broad-fronted moose, *Libralces* (= *Alces latifrons* (Kahlke 1951; Azzaroli 1979; Lister 1987). In North America the most notable was the tripalmated moose *Cervalces scotti* (Scott 1885; Kurtén & Anderson 1980; Sher 1986; Churcher & Pinsof 1987). All giant deer of the verticornis-megaloceros ring and the large alcins (Fig. 48) fit well into Goldschmidt's (1940) concept of "hopeful monsters" whose epigenetic development seems most probable.

The large antlers of Eurasiatic pleisto-cervids may be ascribed to the annually recurrent puberty with a long period of testicular quiescence, and the 24-hour light and food abundance during antler growth in the northern latitudes. The long-lasting testosterone absence allowed a long panty growth period for up to 3 months before the plasma testosterone begun to increase and hamper their growth. In this way a short-day breeder like the moose can produce antlers with an antler-to-body ratio between 4.5%–7% (Huxley 1931; Vogt 1947; Bubenik, A. 1966, 1984; Gould 1974).

However, the cortex of pleisto-cervids is relatively thin, and seems to be thinner the larger the antlers are. The core is homogenous, without any large vessels. In extant species with maximum antler mass (wapiti, moose), the panty may achieve, during the phase of dynamic growth (as interpolated from growing curves and growing radii), up to a longitudinal growth of 20mm/day and 200 cc/day, with a maximum weight of 24 kg. In the extinct giant elk, *Megaloceros giganteus*, the growth rates may reach up to 300 cc/day to produce antlers of total surface areas up to 13,000 cm^2 (Bubenik, A., unpublished) during a growing period certainly not longer than 150 days.

►

FIGURE 48. Goldschmid's "hopeful monsters" of the Pliocene/Pleistocene epoch: *1 Megaloceros giganteus*, the Giant Irish elk: spread up to 4 m. *2 Eucladoceros senesenzis* with beams about 1.6 m long. *3 Eucladoceros sedgwicki*, the most pointed cervin deer. *4 Cervalces scotti*, the tri-palmated moose. *5 Megaceros solilhacus*, the Giant fallow deer: spread around 3.5 m.

1
2
3
4
5

Surprisingly, the American telemetacarpal pleisto-cervids did not develop any true giants. The large-antlered pleisto-cervids of South America, like morenelaphins and deer of the antifer alliance, were versatile in antler form but cannot be called giants (Kraglievich 1932; Hofstetter 1952). Among extant North American odocoileins, only the mule deer *Odocoileus hemionus hemionus* seems to be a true endemic pleisto-cervid, as long as we do not have proof that it is not an invariably programmed seasonal breeder. The possibility of its pleistocene origin is corroborated by comparison of the mitochondrial DNA (mtDNA) analysis (Cronin et al. 1988) which provided unexpected dissimilarity with its closest relative, the black-tailed deer, *Od. h. columbianus*. This subspecies show, according to the mtDNA, greater similarity to white-tailed deer than to mule deer. Therefore Carr et al. (1986) hypothesize that if hybridization resulted by the introgression of mtDNA from white-tailed deer into mule deer, then it occurred in the past. However, because the mule deer is a paraphyletic relative to white-tailed deer, it seems likely that its speciation was fairly recent (Avise 1986).

Despite all known pleisto-cervids developed north or south of the Tropics of Cancer and Capricorn, at least one of the pleisto-cervids, the Java-rusa *R. timorensis*, is from China. There, it has survived long in the Pleistocene (Dubois 1891), apparently still as a seasonal breeder side by side with the aseasonal relative, the plio-cervid sambar (van Bemmel 1951; Stewart 1985). The migration route from China over Taiwan (C. Keqing, personal communication, 1985) and Malaysia, and eventually to the Java archipelago over the Malay shelf (Thenius 1980) is well documented (Keqing, personal communication, 1985).

Surprisingly enough, the Java-rusa exhibits a seasonal tuning of the breeding period and antler cycle when brought south of the Tropic of Capricorn. In South Australia it is capable of tuning the breeding period opportunistically to the dry and wet seasons (van Mourik & Stelmasiak 1986), whereas in the latitude of Beijing it is a typical short-day breeder (Tan Bangjie, personal communication, 1987).

The mentioned specific characters of mio-, plio-, and pleisto-eupecoran point to some discrepancies in the classification of bovoids and cervoids and the necessity to respect also the genetic flexibility of some lineages. In this regard the relative simplicity in diversification of bovoids allows a classification more correct than that of the cervids. Therefore, we consider it necessary to add the following remarks to the cervid taxonomy.

Part E. Concluding Remarks

Classification of Eupecoran in General and Cervoids in Particular

The taxonomy of eupecoran relies primarily on skeletal and dental characters, secondarily but incompletely on visceral anatomy of internal organs and skin glands (Brooke 1878). It ignores behavioral traits, despite their evolutionary importance, in order to eliminate convergences, as was shown, e.g., in anatides

(Lorenz 1941). The flexibility of antlerogenesis, tooth morphology, and skeletal anatomy may be sources of erroneous conclusions about nonexisting relationships, as is evident from papers of Stehlin (1928, 1939), Simpson (1945), Flerov (1952), Scott & Janis (1987), and others.

In this regard taxonomy of bovoids is not as confusing (Gentry, Chapter 6) as in cervoids, where at least 11 different cladograms exist (Janis & Scott 1987). They corroborate the fact that either we still do not have enough reliable characters or we do not try a multidirectional approach as the only way to avoid confusing relationships (Hillis 1987). In our view, a basic hurdle from the start is the question of whether giraffoids and/or paleomerycids were the one ancestral stock or whether both represent different ancestral groups with similar characters. Presently we do not have any clue that points to the possibility that there might be a hypothetical eupecoran stock with a dual mechanism by which epiphyseal ossicone and apophyseal apophyses could be simultaneously activated and controlled.

Because the known appendages of paleomerycids (e.g., *P. kaupi* or *P. tricornis*) have ossicones, we are reluctant to assume that they might also be ancestors for cervoids. In contrast, on the one hand we have giraffoids (e.g., giraffes) and probably giraffokerins (Churcher, Chapter 5) which have ossicones, and climacocerins and lagomerycins with cranial apophyses. The nature of cranial appendages in sivatherins is unknown. When we eliminate the paleomerycids as related to cervoids, we have to decide whether the giraffoids develop simultaneously both epiphyseal velericorn or cavicorn ossicones and apophyseal cervicorn pedicles. Presently, we do not have such proof and do not have any indication whether a hypothetical eupecoran with anlagen for both ossicones and pedicles existed.

In view that both ossicones and pedicles need quite different physiological and morphological mechanisms for their development and growth control, we have to ask whether the characters of giraffoids are absolutely reliable, or whether other lineages could develop parallel characters that may mimic false relationships. Therefore, and as a remote probability, we put the climacocerids and lagomerycids away from giraffoids and paleomerycids. We deliberately left out the hoplitomerycids because A. Bubenik (unpublished) showed to Leinders in 1986 that the cranial appendages are ossicones. Thus, despite the lacrimal fossae, the hoplitomerycids cannot be considered as cervoids.

The lack of histological evidences in cervid classification limits the accuracy of any cladogram. Therefore, we decided to focus the attention of taxonomists also on characters which should on the one hand be respected, and on the other hand be eliminated as possible convergences.

1. The evolutionary stages of cranial appendages in the lagomerycid-procervulus group show that *Procervulus aurelianensis* and *P. dichotomus* differ fundamentally in branching pattern; the former uses only sprouting like lagomerycids, the latter splits the shaft in a fork like dicrocerids or euproxins. Thus their phyletic relationships might be questioned.

2. The difference between the pristine centrifugal mineralization of the osteon of pseudo- and protoantlers, and the centripetal mineralization of the antler cortex, cannot be neglected due to their different morphological and physiological effects, i.e., perennial or deciduous nature of the appendage.
3. In context with this, the importance of the appearance of a true coronet is underestimated, possibly due to a false interpretation of the term coronet.
4. The cladograms ignore the fact that during cervoid evolution the holometacarpal metapodia were transformed by distal or proximal regression of the metacarpi that was independent of space, time, and antlerogenetic stage. Thus the pliocervins (Khomenko 1913) with deciduous antlers and coronet retained holometacarpal metapodia despite the fact that the telemetacarpal type was developed just in early Mid-Miocene by merycodontins, and in Upper Miocene by capreolids, whereas the cervulins of the upper Mid-Miocene were plesiometacarpal. We do not have any studies that relate whether both tele- and plesiometacarpal metapodia could influence the gait in general or give any advantage in respect to the antler design and symmetric or asymmetric contests in particular.
5. Cautiousness also seems necessary in overemphasizing the importance of tooth characters in view of the possibly epigenetically affected adaptability to the texture of the forage (Obergfell 1957; Sondaar 1977; Slavkin 1988).
6. It is erroneous to believe that the number of points, which is not equal to the number of tines in antler construction, indicates the phyletic position of the genus or subspecies (Geist 1987a,b), because the speed of construction improvement was different from lineage to lineage and may change with latitude.
7. The reliability of karyotype characters in distant phyletic lineages (Koulischer et al. 1972; Spotorno & Fernandez-Donoso 1975; Spotorno et al. 1988; Harrington 1985; Groves & Grubb 1987; Mayr et al. 1987) can lead to false phyletic conclusions (Groves & Grubb 1987).
8. The use of mitochondrial DNA (Carr et al. 1986; Moritz et al. 1987; Cronin et al. 1988) may lead to false phyletic conclusions, because we do not know yet how conservative it may be in cervoid lineages so distant as, e.g., miocervids and pleistocervids.
9. Under any circumstance we should not leave out of consideration the regenerative potential of the velvet, and its possible precursors.
10. Our knowledge of the role of androgens and estrogens in the growth and mineralization of the pedicle apices needs deeper study. We need to know how the pedicle and the regeneration of its apex behave under different plasma levels of both hormones.

Using the above characters, we designed a cladogram of eupecoran, which again shows phyletic relationships from quite a different angle (Fig. 49). Obviously, we also could make erroneous conclusions; but the 11 cladograms presented by Janis & Scott (1987) prove that other scholars were also more or less biased.

For the same reason, our cladograms of living cervoids will differ from all others because we emphasized the morphological and physiological characters. Therefore, before we put them into a cladogram, we listed them for each taxon, as follows:

Muntiacidae, Fam. Nov. (Bubenik, A. 1982c) (= Muntiacinae Pocock 1923)

The relationships of the extinct and extant muntjac-linked lineages with the new family Muntiacidae, syn. Muntiacinae (Pocock 1923 = Cervulinae Sclater 1870), is corroborated by many characters extracted from the papers of Koenigswald (1933), Dehm (1944), van Bemmel (1952), Colbert (1958), Todd (1975), Dansie (1983) and personal communication (1986), A. Bubenik, (1982b, 1989), Groves & Grubb (1987), and generous advice of N. Chapman (marked with *), and from the author's own records. (See also Groves & Grubb, Chapter 3.)

All morphophysiological peculiarities of extant cervulins, i.e., *Elaphodus*, and muntjacs from *M. atherodes* to *M. zimmermanni* contradict the notion that the antlers become secondarily small (Groves & Grubb 1987 vs. Groves & Grubb, Chapter 3). Their pristine character is supported by the following long-known and unknown features:

1. Muntjacs are plesiometacarpal deer, which have two important characters of telemetacarpal deer: lack of metatarsal glands, and a rumen with only two blind chambers.
2. The long and divergent supraorbital pedicles develop slightly posteriorly and the latter extends anteriorly toward the naso-orbital ridges, which are developed before the pedicles' growth is induced. This, of course, may corroborate quite another function than reinforcement of pedicle base. In contrast to the supraorbital, more-or-less perpendicular and parallel pedicles of protocervids, they are streamlined with profile and are divergent. Their tips are generally convergent, but can be forked. In some specimens numerous points are developed, probably by sprouting, around the coronets. In subsequent antler sets a coronet is developed, but the diameter of the shaft may be almost twice as large as that of the pedicles, which points to long lasting peripheral apposition of bone in contrast to eucervids.
3. In the tropics the pedicles begin to develop at 4 weeks of age, in captive *M. reevesi*; in England in 5–8 weeks. They need another 4–12 months to attain their full length and start to grow into first antlers (in *M. reevesi* they are 6–50 mm. long). Both the antler length and the timing of first antler growth are independent on latitude. The velvet shedding may last several months (van Bemmel 1952; N. Chapman*) which may be due to the very slow progress of mineralization of antler cores.

   This behavior may be an atavism linked with protoantlers, and/or a response to the undulating testosterone level which does not reach a nadir below 1.2 ng, considered as crucial for testicular quiescence. When the petrification is accomplished the whole antler core becomes petrified. This also is a stage common in the telemetacarpal deer, such as roe deer, pudu, and mazama deer.

4. Another ancestral feature is the asynchronous cast and an irregular antler cycle which can be repeated at intervals of 60 to 112 weeks*; also an unilateral delay in antler cycle is frequent: one hard antler can survive the regeneration cycle of that on the heterolateral side. The concave seal can be biased and may lie up to 10 mm below the coronet. The core of the first antlers survives some months after velvet shedding.
5. Similarily to roe deer (Bubenik, A. 1966), the control of antler cycle and shape is totally dependent on testicular testosterone. Hence, castrates develop fast proliferating perukes.
6. The cores of the first, and perhaps the subsequent antlers as well, survive after velvet shedding for a yet unknown period of time. The radiographs of *M. atherodes* skulls, provided us gratefully by the British Museum of Natural History and New York Museum of Natural History show, contrary to Groves & Grubb (1987) that this most primitive muntjac can cast their antlers, because specimens with completely petrified core have coronets.
7. Muntjacs karyotypes have the following peculiarities: lowest chromosome number, a single X-chromosome in the female of *Elaphodus cephalophus*, and an X-autosome complex with an additional Y-chromosome in the males of [*Muntiacus muntiac*].
8. Tusks, as weaponry, have a dominant role over antlers in intraspecific combat (Barrette 1977), but not toward predators.

►

FIGURE 49. Cladogram of eupecoran: as leading characters were used the evolutionary stages of cranial appendages of both bovids and cervoids. The term *Protobovidae* may be questioned because of lack of evidence that they existed. But they might develop from paleomerycids. *Procervulus aurelianensis* and *Triceromeryx* show the same sprouting process which may or may not point to a close, remote or no relationship of both. Climacocerids and lagomerycids figure among eupecoran with cranial apophyses because of lack of evidence that they might be an offshoot of giraffoids, which were equipped with ossicones. The early dicrocerids encompass also the heteroproxins. We separated these early dicrocerids from the euproxins represented by *Euprox furcatus*; his antlers were deciduous, have mineralized centrifugally, antler casting occurred below well-developed coronet, but pedicles were located supraorbitally. The *E. furcatus* cannot be an offshoot of dicrocerids, and the primitive spiker *Euprox minimus* (Toula 1884) with pedicles located on the frontal cristae cannot be relative of *E. furcatus*; otherwise we must accept the improbable reversal that the more advanced, i.e., bifurcated *E. furcatus* relocated the pedicles again back on the upper orbits. Therefore we presume that *Euprox minimus* was a member of a special lineage, we call *Proxini*. We derived the term from the preoccupied *Prox furcatus* (Hensel 1859) = *E. furcatus*. In our view the euproxins might be closer to the cervulins or muntjacs than to proxins and eucervids.

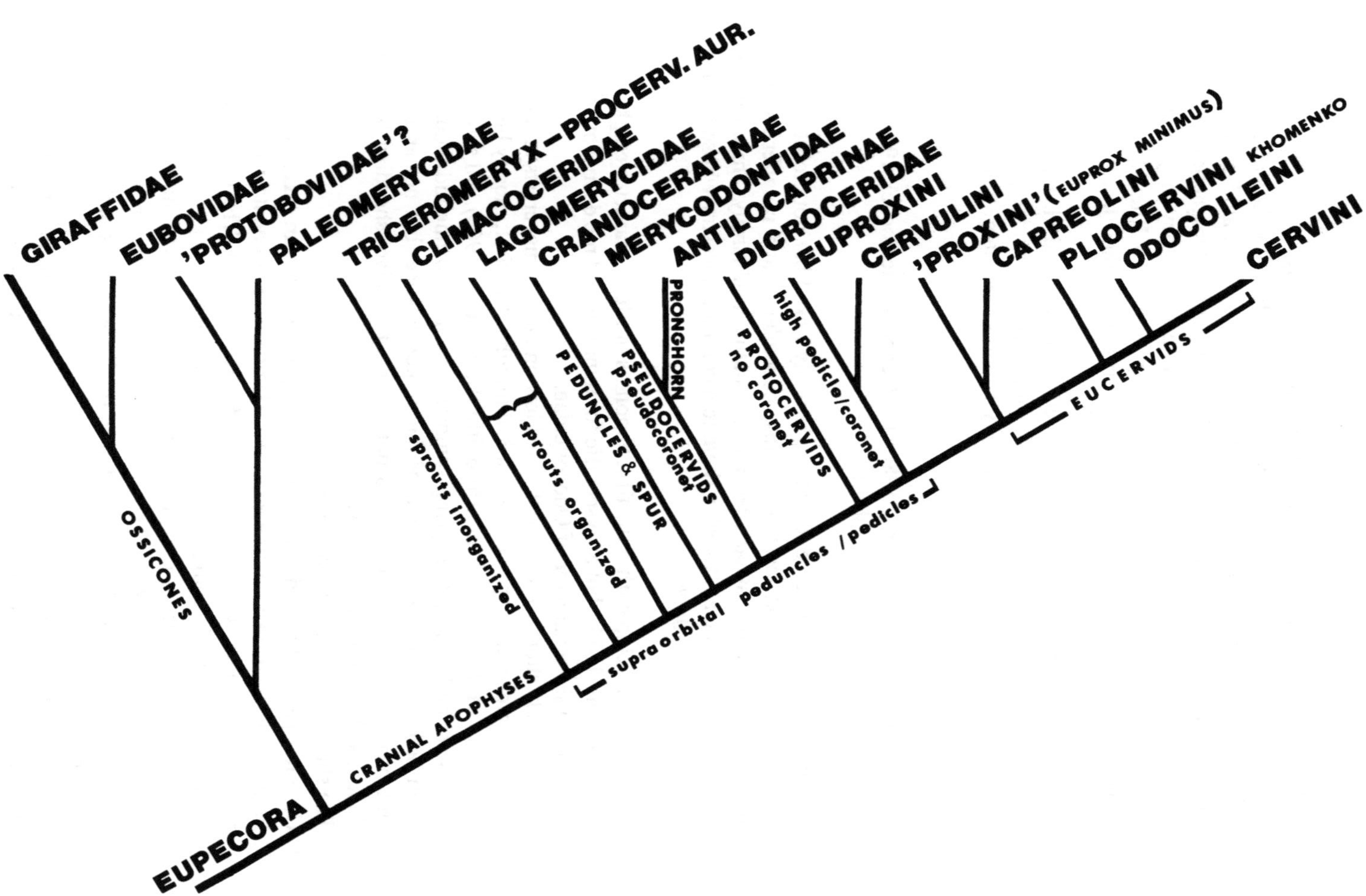
GIRAFFIDAE
EUBOVIDAE
'PROTOBOVIDAE'?
PALEOMERYCIDAE
TRICEROMERYX–PROCERV. AUR.
CLIMACOCERIDAE
LAGOMERYCIDAE
CRANIOCERATINAE
MERYCODONTIDAE
ANTILOCAPRINAE
DICROCERIDAE
EUPROXINI
CERVULINI
'PROXINI' (EUPROX MINIMUS)
CAPREOLINI
PLIOCERVINI KHOMENKO
ODOCOILEINI
CERVINI
PRONGHORN
OSSICONES
sprouts inorganized
sprouts organized
PEDUNCLES & SPUR
PSEUDOCERVIDS
pseudocoronet
PROTOCERVIDS
no coronet
high pedicle/coronet
EUCERVIDS
supraorbital peduncles / pedicles
CRANIAL APOPHYSES
EUPECORA

9. Muntjacs are the only deer without metatarsal glands.
10. Muntjacs are polyestrous; overt estrus follows only a few days after parturition (Dansie 1983). The extant representatives of Muntiacidae, the genera *Elaphodus* and *Muntiacus* (Groves & Grubb, Chapter 3) might be close to the protocervids of the Tortonian and Helvetian periods; they may be close to the miocene genera *Stephanocemas*, *Procapreolus*, and *Paleoplatyceros*.
11. The karyotypes hide many obscure problems due to their low number.

Thus the muntiacids should be considered as an ancestral clade, not directly linked to other cervids and certainly not a subfamily of Cervidae (Flerov 1952). Therefore they deserve the status of family Muntiacidae, fam. nova (Groves & Grubb, Chapter 3).

Odocoileidae, Fam. Nov. Bubenik 1986c (= Odocoileinae Pocock 1923, Without Hydropotini and Capreolini)

A. Bubenik (1986c) pointed out that phyletically linked but not closely related telemetacarpal clades—at least the alcins, odocoileins, and rangiferins—have such distinct characters that each deserves the status of subfamily. Thus, we need a new family Odocoileidae, to bring them to a common ancestry whose features can be described as follows:

1. First antlers develop within the first 5 months of life, a character emphasized first by Caton (1874), but also specific for muntjacs and roe deer. Thus this is not an exclusive character of Odocoileidae.
2. The relatively short pedicles are located low on the external crests of the frontals and begin to develop at the age of 3–4 months. The first, mostly bud-sized antlers are hard within 4–5 months of life (Bubenik, A. 1966). All subsequent antlers are built during the spring-summer season and prime individuals cast them before or around the time of the winter solstice, but not around the autumnal equinox. The antlers do not survive the velvet shedding by more than few days (Waldo & Wislocki (1951).
3. The shaft can ramify by sprouting.
4. Castrates generally build velericorn antlers; perukes are an exception.
5. Permanent dentition is completed within 18 months. Rudimentary canines are relatively frequent, i.e., between 2%-20% (Krieg 1948). The early replacement of milk teeth is common in all primitive lineages.
6. All "true" Odocoileidae, except of some tropical relatives, have metatarsal glands (Hershkowitz 1969). Those that have them urinate on the heels, rubbing the tarsal brushes together. All have vestibular nasal glands either active or regressed.
7. The rumen has two blind chambers, again an indicator of a primitive stage (present as mentioned above also in *Capreolus* and *Muntjacus*).
8. All genera have active or regressed nasal glands (Atkeson et al. 1988).

Odocoileinae (= Odocoileini Pocock 1923, Without Hydropotini, Capreolini, Rangiferini and Alcini, Simpson 1945)

This primitive (but in comparison to muntjacs more advanced) deer group is highly polymorphic (Groves & Grubb 1987) and includes genera *Pudu*, *Mazama*, *Blastocerus*, *Dorcelaphus*, *Hippocamelus*, and *Odocoileus*.

1. In *Odocoileus*, *Blastoceros* = (*Ozotoceras*), and *Dorcelaphus* it seems that it is the posterior half of the beam which carries the potential for ramification by tines or dichotomous branching. In *Odocoileus* a prong is regular.
2. The karyotypes point to relatively primitive or advanced stages.
3. Most present-day genera are polyestrous aseasonal breeders. *Od. virginianus* has, and *Pudu* may have, the flexibility to be both polyestrous aseasonal or monoestrous short-day breeders. *Od. hemionus* seems invariably to be a short-day breeder, i.e., a pleisto-cervid. Whether the *Huemul* is photoperiodically flexible or invariably a short-day breeder must be investigated.

Alcinae Jerdon 1874

The anatomical and morphological characters of living moose point to the taxon of subfamily, a view shared by many students (for references, see Bubenik, A. 1986c). The different morphology of the premaxillary bones between *Alces a. caucasicus*, *Alces a. andersoni*, *A. a. gigas*, and probably the Fennoscandian *Alces a. alces* may be also of taxonomical value (Bubenik, A. 1986b). The special characters of Alcinae are:

1. Skull anatomy with the progressive elongation of the premaxillary bones which finally may split into an upper and lower part (opposed by Tikhonov, personal communication, 1988), or with the upper part regressed.
2. Horizontally oriented pedicles and capability to develop a relatively long, downward pointing prong in a second antler set (Bubenik, A. 1973, 1986c).
3. The tendency for palmation as a hypertelic product of an originally cervicorn design.
4. Invariably a short-day breeder due to its pleistocene origin.

Rangiferinae New Subf. Bubenik 1986c (= Rangiferidae Brookes, 1828)

Although the only living genus *Rangifer* has the morphological characters of other Odocoileidae, except permanent upper canines and antlers in both sexes, the morphophysiological character of antlerogenesis is unparalleled in any other deer lineage.

1. The antlerogenesis is the most peculiar we know. Pedicle development and annual antler cycles are independent on testicular androgens (Berthold 1831; Tandler 1910; Wika et al. 1975). The antler core of females survives the velvet shedding almost to winter's end (Bouchud 1966; Thing et al. 1986).

2. Also peculiar is the branching of rangiferins antlers. They usually ramify by producing tines. However, they can develop sprouts alongside the whole beam, even below the tines. An amputated tine will be replaced in the same fashion by a sprout (Bubenik, A. 1956, 1959b). The pedicles are very short; coronets are poor or almost absent.
3. The origin of rangiferins is unknown. The possibility that their propagule could be *Algamaceros blicki* (Hoffstetter 1952), cannot be accepted. Algamacerins were not blastocerins (Webb, personal communication, 1988), as presumed by Frick (1937). Algamacerins are probably one of tribes of Odocoileinae, similar to the morenelaphins (Kraglievich 1932). The hypothesis of Kurtén & Anderson (1980), and Hershkowitz (1982)—that morenelaphins may have crossed the Panamanian isthmus and established themselves in the northern hemisphere as ancestor of rangiferins—is questionable unless it can be proved that morenelaphins could have ramified the antlers by sprouting and that both sexes could carry antlers.

Capreolidae, Fam. Nov., Brookes 1828

The only common characters with odocoileids are the telemetacarpal metapodia, early exchange of permanent dentition, two blind chambers of the rumen, and first antler sets which develop within 6 months of life. The following characteristics are too numerous to substantiate the status of a tribe Capreolini of Odocoileidae (Simpson 1945).

1. The vomer is long and low as in alcins or cervins, which was the reason why Flerov (1952) classified *Capreolus* as cervin deer. The embryogenesis of metapodia from the holometacarpal into telemetacarpal stage goes faster than in the plesiometacarpal *C. elaphus* (Slabý 1962).
2. The pedicles are long. The shaft does not produce a prong, but rather numerous, mostly medially protruding pearls. The coronets are well developed.
3. The antler cycle cannot be maintained without testicular testosterone, as in muntjac.
4. The winter antler growth period and the extraordinarily long testicular activity do not fit the odocoileid scheme. The finding of Brockstedt-Rasmusen et al. (1987) that roe deer antlers also survive the shedding period is questionable. However, if this should be verified, it would be just another character pointing to an origin far from that of Odocoileinae.
5. It is the only genus with high frequency of intersexes and females carrying velericorn antlers.
6. Neither species of *Capreolus* (Sokolov & Gromov 1985) has tarsal glands. Both sexes do not urinate on the heels or show behavioral residue of this pattern.
7. There is no trace of the vestibular nasal glands, typical for Odocoileinae (Atkeson et al. 1988).
8. The penis is not of odocoileid type (Garrod 1877).

9. The roe deer is the only typical long-day breeder in northern latitudes with delayed ovum implantation (Stieve 1950b). Among living mammals we do not know of one example where only one genus or tribe has a diapause (Stieve 1950b; Short 1967). Due to these strange characteristics, the genus *Capreolus* should be considered as the only extant survivor of the family Capreolidae. The ancestral origin of the genus is well documented in its karyotype (Todd 1975; Groves & Grubb 1987). In addition, and according to Schlosser (1924), the *Capreolus* group may have developed together with *Dicroceros* and *Cervulus* from paleomerycids. However, procapreolids like *P. lóczyi* or *P. rüttimeyeri* were true antlered deer unknown in the Mid Miocene. The fact that some characters of muntjacids and capreolids are similar may point to a common, apparently holometacarpal clade.

Thus, we should respect Garrod's (1877, p. 18) view that *Capreolus* is "one of the most difficult of the deer tribe to localize [and] I have placed it not far from *Cervulus* on account of the configuration of its glans penis."

Eucervidae Fam. Nov.

The family of Cervidae Gray, 1821 encompasses all known plesiometacarpi and some holometacarpi as well (Simpson 1945). In the view of the facts presented here and earlier (Bubenik, A. 1986c) their subfamilies and tribes represent not one phyletically related group. Therefore, we propose a new family Eucervidae for all plesiometacarpal clades which developed from the end of the Miocene and are characterized by following monophyletic characters.

1. Long and low vomer, and rumen with three blind chambers (Bubenik, A. 1959a).
2. Permanent dentition completed within 24–28 months.
3. Medium long pedicles which begin to grow at the age of 5–6 months and need another 6 months before they can grow into the first long antlers. Pedicle bases are relatively high on the frontal cristae and are longer than in odocoileids. Coronets are well developed.
4. In pleistocene eucervids the antler growth period is during spring and summer. Antler casting occurs after the December solstice until the March equinox. In plio-cervids the antler cycle is aseasonal.
5. Sprouting appears only in the first antlers of elaphins (*Cervus elaphus* spp.); however, the bez-tine still carries the character of a former sprout, protruding from the beam out of the entire antler plane as formulated by Thompson d'Arcy (1940).
6. The antler cycle is only partially dependent on testicular androgens. Cryptorchids and castrates produce velericorn antlers, whose growth is better controlled than in hypogonadal or castrated males of odocoileids.
7. The ramification pattern evolved the greatest known variance and size in the whole history of all antlered deer.

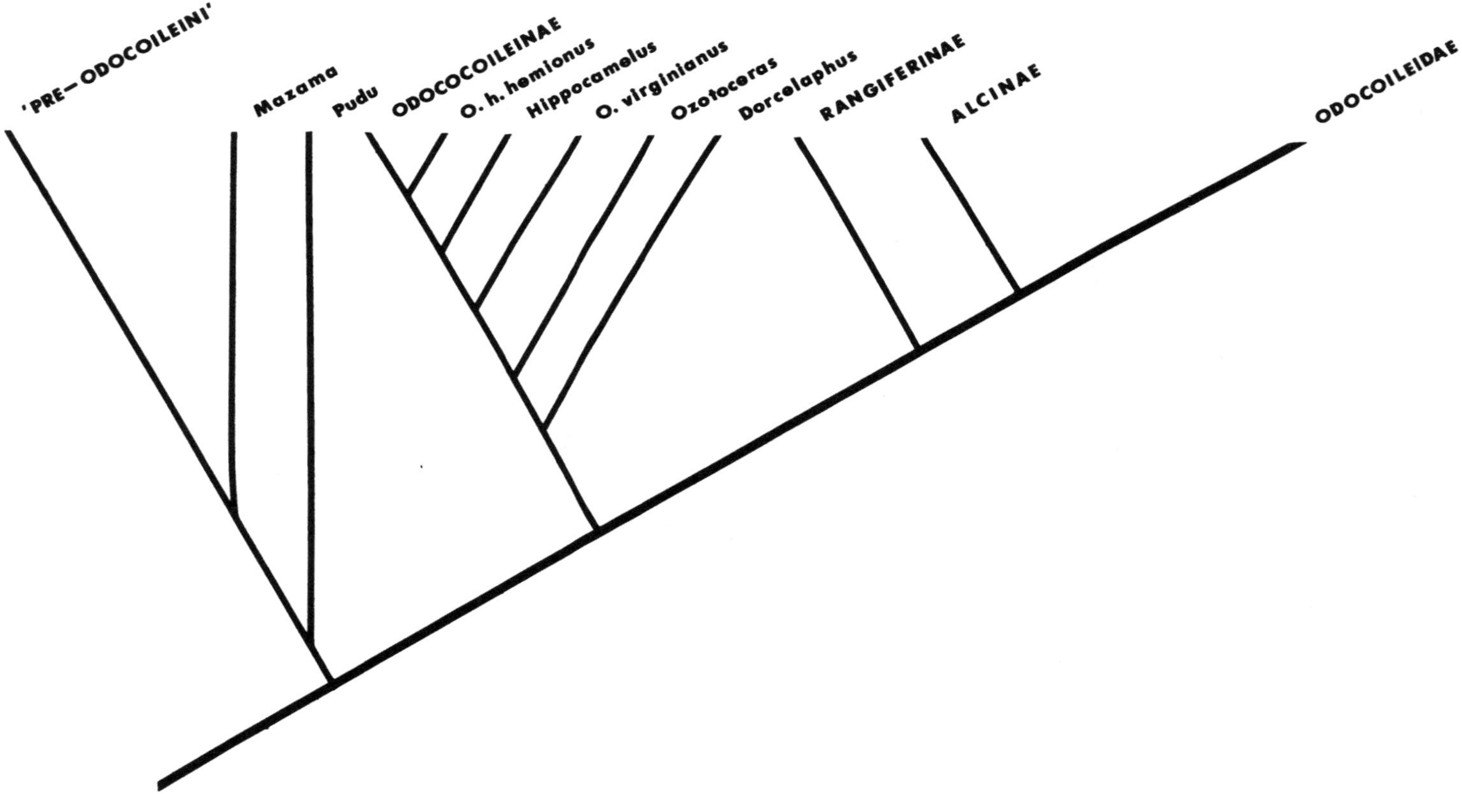

FIGURE 50. Cladogram of American (New World) eucervids: Due to characters described in text we separated the pudu and mazama deer from the main lineage of Odocoileidae (Bubenik, A. 1986b) and termed them Pre-odocoileinae. However, both pudu and mazama may be mio- or pliocervid offshoots of just older eucervids which ancestors could share some characters with proxins.

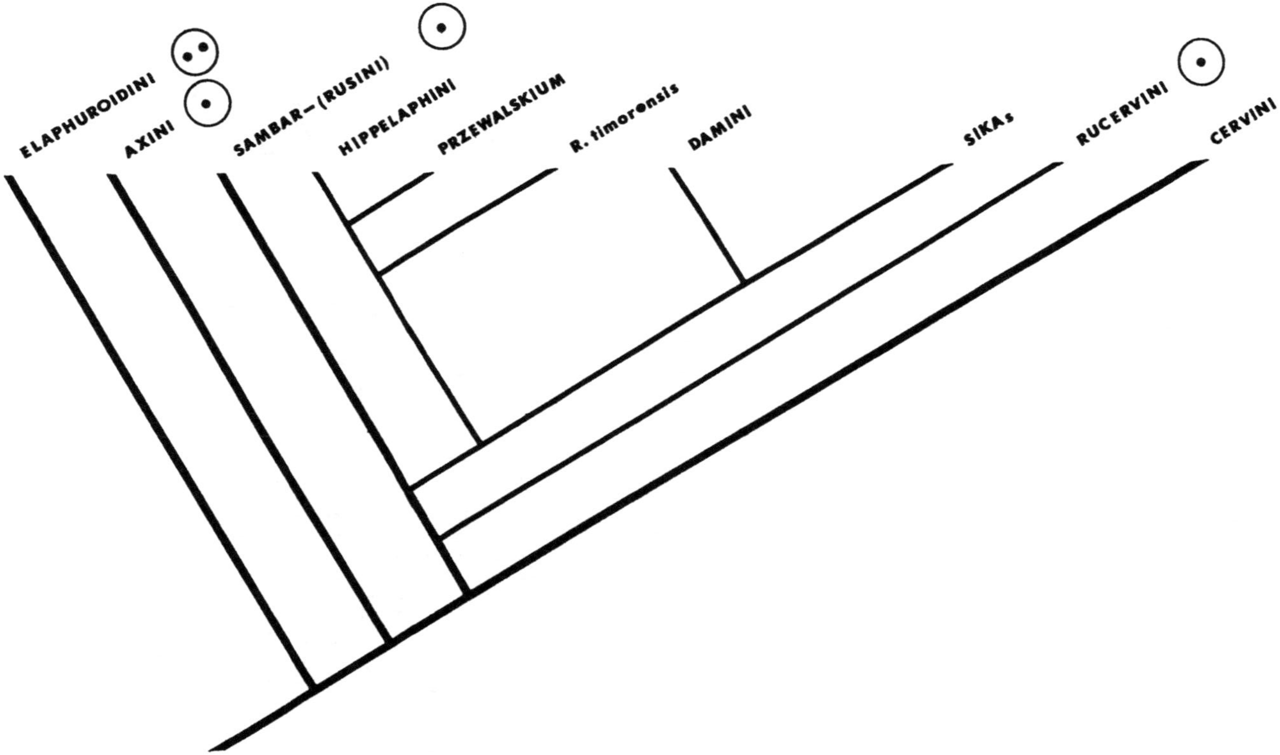

FIGURE 51. Cladogram of Eurasiatic (Old World) eucervids: The circles with one or two black points indicate that the core has two or one large vessel. They indicate the earliest stages in contrast to modern plio-, and pleistocervids with more or less homogenous and spongious core. Due to this evolutionary difference we separated from the sambars *R. equina*-ring the *R. timorensis*, using for its lineage the term *Hippelaphini* according to *Hippelaphus timorensis* Heude (1896) = *C. timorensis* de Blainville (1822) (Mohr 1918). In accord with Teilhard and Trassaert's (1937) theory of adaptive terminal fork we incline to the assumption of Flerov (1952) that Thorold's deer *Przewalskium* = *C. albirostris* may develop as an offshoot of hippelaphins. In our collection is a picture of a three-tined *R. timorensis* (Museum of Natural History, Leyden). It has three tines from which the longest is the third or dagger. It is flattened in a fashion similar to that of *Przewalskium*.

8. All recent pleisto-cervids seem to be descendants of plio-cervid lineages, however many of the propaguls are still unknown. Both plio- and pleisto-cervids are closely related rings with very dynamic gradual evolution, which may be ascribed to epigenetic responses to very rapid changes of habitat conditions.

Using the morphological and physiological characters of living cervids, we designed cladograms for both tele- and metacarpi (Fig. 50, 51). They differ from all previous reasons we gave for the cladogram of eupecora. We separated the pudu and mazamas from the main stock of odocoileids, and put some hypothetical dichotomous elaphuroid as the ancestral clade from which, according to the rule of adaptive and inadaptive fork, any other cervin form could result. A detailed explanation of the reasons for diversification are in the corresponding legends.

Conclusion

The morphophysiological characters presented here show clearly that the phyletic tree of deer needs revision. To do it now will be premature until we have more knowledge about the flexibility of skeletal characters and will find how far all known discrepancies among leading characters can be eliminated or brought along with each other. The hypotheses offered in this paper are merely a challenge to paleontologists, neuroendocrinologists, and taxonomists. The justification of families Muntiacidae and Capreolidae, need a new joint taxon for all cervidae. Due to the different meaning of the superfamily Cervoidea (Simpson 1945), a new subgroup within the eupecoran seems necessary.

Using these and other characters which we consider as specific, we designed the cladograms for extant deer. In order to characterize some of the lineages we chose in few cases terms, whose roots (at least) are preoccupied. The meanings and reasons why we used them are explained in the legends. We are aware that the phyletic age or close relationship of some taxa or lineages is questionable. However, this is also the case in other published cladograms.

We deliberately omitted a conclusive summary, partly because of "heretic" views we have developed, partly to avoid repetition in view of the limited space. We decided to close this chapter with two citations suited well to our paper:

"All scientific theories need to be continuously challenged." Popper, K.R. 1984: Evolutionary Epistemology. In: Evolutionary Theory: Paths into the future. Pollard J.W. (ed.). John Wiley & Sons, London.

"Those who lack all idea that it is possible to be wrong can learn nothing except know-how." G. Bateson, in: Mind and Nature, 1979. E.P. Dutton, New York.

Acknowledgments.. Working over 40 years in antlerogenesis and a little bit in cornuogenesis it is impossible to thank all my colleagues to whom I am obliged for their ideas, suggestions and comments. They came from all fields of biology, physiology and paleontology. Looking back, I cannot omit my great teachers and personal friends, the late and unforgettable university professors at Prague and Brno: K. Absolon, J. Augusta, Z. Frankenberger, R. Pavlansky and R. Tachezy. Furthermore, I have to thank N.K. Vereshchagin and the late K.K. Flerov for the encouragement to search for paleontological relationships. From the 'inner circle' of my acquaintances and friends I would like to acknowledge the criticism of my son George Bubenik and his colleague E. Balon, further the amiable exchange of ideas with R.J. Goss and H. Hartwig. Among other colleagues who gave me every support special gratitude goes to A. Azzaroli, N. Chapman, O. Dansie, M.D. Dermitzakis, J. De Vos, B. Engesser, A.W. Gentry, L. Ginsburg, C.P. Groves, P. Grubb, E. Heintz, E.P.J. Heizmann, W. Hügin, Z. Jaczewski, C.S. Keqing, J.J.M. Leinders, J. Morales, S.S. Moyà, P.Y. Sondaar, N.K. Symeonidis, C. Sudre, E. Thenius, H. Thomas, and M.R. Voorhies. I also appreciate the critical comments of N.J. Boggard, C.S. Churcher and S.D. Webb. Without their help and professional experience, I would not be able to upgrade the paper in this form. Special thanks go to the librarians H. Arro and A. Chalk for their help in collecting the literature. Last but not least, I express thanks and admiration for the untiring help of my secretary and wife Mary, for checking and rechecking the many versions of this paper up to the final edition.

2 Correlation of Reproductive and Digestive Strategies in the Evolution of Cranial Appendages

CHRISTINE M. JANIS

Introduction

Cranial appendages are characteristic of most living and fossil pecoran genera, where they consist of bony outgrowths from the skull that may or may not be covered by skin or keratin. Cranial appendages have also evolved in other ungulate lineages: as bony protruberances from the head [in the extinct families Uintatheriidae (order Dinocerata) Brontotheriidae (Perrisodactyla), and Protoceratidae (Artiodactyla], or as dermal structures (in the Rhinocerotidae). Fossil record evidence shows that bony cranial appendages have also arisen in certain taxa in other families; for example, in the Suidae (*Kubanochoerus*), in the Rhinocerotidae (*Diceratherium* and *Menoceras*), in the Merycoidondontidae (*Cyclopideus*), and even in the Rodentia (*Mylagaulus*). Bony cranial appendages are also found in certain dinosaurs, most notably in the Ceratopsia. In addition, other ungulate lineages have modified canine or incisor teeth into enlarged tusks, which may function in an analogous fashion to outgrowths from the skull; this is apparent in the Elephantidae (and also in extinct proboscidean families), the Suidae, the Tragulidae (and in other extinct traguloid families), and to a less pronounced extent in the Equidae, Tapiriidae, Camelidae, Procaviidae (Hyracoidea), Tayassuidae, and Hippopotamidae (Kiltie 1985).

All living ungulate taxa that possess cranial appendages or enlarged tusks show evidence of sexual dimorphism in this character, with the condition being less developed in the females. Yet pecorans are unusual in that the initial evolutionary condition, and the usual condition in living animals, is for cranial appendages to be found in the males alone. The only non-pecoran lineages in which cranial appendages are lacking in the females are the extinct camelid-related Protoceratidae and the extinct diceratherine rhinos (Janis 1982). Preliminary studies of the extinct rhino-like Brontotheriidae and Uintatheriidae suggest that bony horns were present in both sexes in some genera, but may have been larger in the males. Thus, it appears that ungulate taxa in which the males alone possess cranial appendages are usually limited to ruminating artiodactyls: the pecorans and the pecoran-mimicking protoceratids.

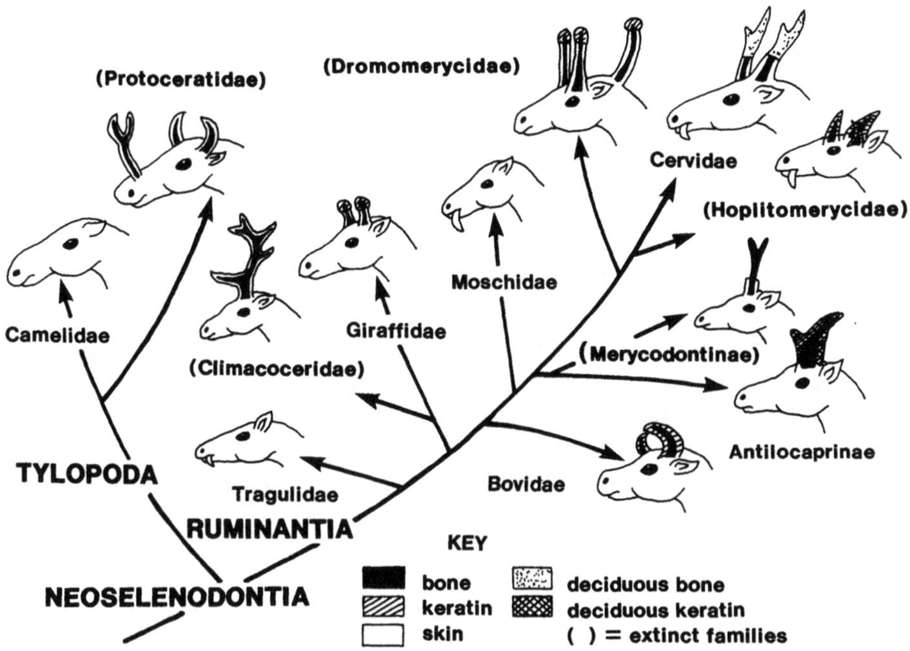

FIGURE 1. Phylogeny of the Neoselenodontia, showing the independent acquisition of cranial appendages in the different families. (Modified from Janis & Scott, 1987.) Originally appeared in American Museum of Natural History Novitates, Issue no. 2893 (p. 22).

Notes:(a) The relative proportions of skin, bone and keratin of the cranial appendages are not shown to scale. (b) Cranial appendages may have evolved only once within the Antilocapridae. (c) The Dromomerycidae may be better considered as a subfamily of the Palaeomerycidae (Janis & Scott 1987). (d) The keratin tips of the dromomerycid cranial appendages are based on an analogy with the form of the underlying bone of giraffid ossicones: there is no direct fossil record evidence for such structures. [However, there is no evidence at all for *deciduous* and forked keratin tips to the horns in this family, as depicted by Frick (1937).]

It is apparent from these comparative data that the evolution of cranial appendages is something that has happened many times in parallel in vertebrates. Even within the pecoran artiodactyls, evidence from developmental studies and the fossil record shows that cranial appendages evolved in parallel in every horned family (Janis & Scott 1987) (Fig. 1). (In this paper I will use the term horned to describe the possession of cranial appendages in ungulates, as there is no other appropriate concise adjective, but it should be appreciated that this does not imply an homology of all cranial appendages with the true horns of bovids.) A central question in the evolution of cranial appendages in the evolution of the Pecora is this: if such appendages are apparently so "easy" to develop in evolution, why are they absent in most other ungulate lineages which paralleled the Pecora in aspects of their evolution, such as the Equidae and the Camelidae? And why is the typical condition (certainly the initial evolutionary condition) in the Pecora the possession of cranial appendages in the males alone?

Correlation of Cranial Appendages with Ecology in Living Ungulates

Aspects of the biology of living herbivorous mammals such as body size, diet, feeding, and digestive strategy and habitat preference can be shown to be interrelated (Janis 1982; Demment & Van Soest 1985 for review). For example, small herbivores tend to be selective feeders on succulent vegetation, living in closed habitats such as forests, whereas larger herbivores tend to feed less selectively on more fibrous vegetation and may be found in more open grassland habitats. Many authors (e.g., Estes 1974; Jarman 1974; Packer 1983; Kiltie 1985) have noted that these aspects of pecoran biology are also correlated with the extent of sexual dimorphism in the possession or development of cranial appendages, which they interpret as reflecting various habitat appropriate aspects of their socioecology. These include aspects of social behavior (e.g., group-forming or solitary, length of association of members of a group, size and membership of typical groups and patterns of seasonal variation of these behaviors); reproductive behavior (e.g., territory maintenance and defense by males, lie-out or follower types of young); and antipredator strategies (crypsis, flight, or active predator defense).

The classic paper by Jarman (1974) divided the African Bovidae into five categories, based on the relationship of body size and group size to the distribution of food resources in the habitat, and showed that the type of social behavior and morphology of the horns was characteristic for each category. To summarize Jarman's findings, in progressing from his Category A to his Category E there is a general trend for increasing body size (Category A containing the dwarf antelope species, and Category E containing the buffalo and the eland); for increasing preference for open habitat (ranging from tropical forest to open savanna); and for increase in both group size and the permanence of the associations of the group members (ranging from solitary individuals or monogamously bonded pairs in Category A, to large, year-round, mixed-sex herds in Category E). However, aspects of reproductive biology and patterns sexual dimorphism do not follow a simple progression through the categories.

The bovids in Category A are small, mostly ranging from 5–15 kg. Ruminant artiodactyls at this body size have no dietary option but to select items of herbage containing little fiber, such as the reproductive or young growth parts of plants (Janis 1976; Demment & Van Soest 1985). These small ruminants usually have a distinct home range which they mark and defend to a certain extent, but the individuals are usually widely spaced (due to the patchy nature of the distribution of their food items). There is no pronounced male/male antagonism in relation to defense of resources or females, and the resulting reproductive system is one of facultative monogamy. There is little evidence of sexual dimorphism in bovids in this category. Small horns may be possessed by one or both sexes, and the females are the same size as the males, or a little larger, depending on the species. The duikers (genus *Cephalophus*) are an example of a Category A type of antelope.

A critical difference between bovids in Category A and in the other categories appears to be in the increase of body size to a value of at least 15 kg. At this body size it is possible for a ruminant to obtain most of its dietary requirements from more fibrous sources of vegetation, such as the structural plant parts (leaves and stems) of both grasses and dicotyledonous trees and shrubs. As a result of this increase in body size, a ruminant finds a greater number of available food items within any given area and may exist in a habitat that lacks a year-round supply of succulent, nonfibrous items. Additionally, it may benefit from the fact that the food items are more uniformly distributed within the habitat, and hence encounter less competition with other individuals if feeding in a group situation. Bovids in Categories B-E tend to forage together in groups (although the group size, membership, and patterns of association vary with the category), a strategy that may be related both to maximizing the available food resources and to antipredator tactics. Of course, it would not be impossible for a ruminant larger than 15 kg to display a Category A type of social and reproductive behavior in a dense forest habitat. This is certainly the case for the larger species of duikers. However, animals in Category A are usually small, probably because of the search-time constraints on foraging for small succulent items and the difficulty of locomotion for a large animal in a dense forest habitat.

The salient point is this: below a certain critical body size, the constraints of the digestive physiology of a ruminant artiodactyl dictate that the diet must consist of herbage with a low proportion of cellulose (i.e., dietary fiber). As a consequence of this, ruminants below about 15 kg encounter a patchy distribution of available food items and hence tend to lead a solitary life. Only at a larger body size is it possible for a ruminant to forage in a less selective fashion in the company of conspecifics, and at this juncture their reproductive strategy may change from one of facultative monogamy to any one of a variety of polygamous tactics. The change from a monogamous to a polygamous mating system results in a change in the types of social interaction within and between the sexes of any given species, resulting in a difference in the pattern of sexual dimorphism (most notably expressed in the morphology of the cranial appendages).

The bovids in Jarman's Category B have the greatest amount of sexual dimorphism in terms of their horns. Category B mainly contains medium-sized antelope (15–100 kg) that live in woodland or ecotonal habitats and feed selectively on dicotyledonous leaf material or on fresh grass. Horns are found in the males only. Adult females are usually found together in groups of three to six, and the males hold year-round feeding and reproductive territories, defending the boundaries against other males and mating with the females that pass through their territory. The bushbuck *Tragelaphus scriptus* is an example of a Category B type of antelope. In Categories C through E there is a progressive tendency for the animals to live in more open habitat areas, for the mean group size to be larger, with females and males living in mixed-sex herds for at least part of the year, and for male territoriality to be less pronounced. Bovid species in Category C are of a similar size range to the species in Category B. They live in woodland to open habitats, are fairly selective feeders taking a mixture of grass and dicotyledonous browse material, and the males hold seasonal territories or form lek systems. The males of these species tend to have elaborate horns and are larger than the females, as is also characteristic of the males in Category B, but the females of these species may also possess small horns. The impala, *Aepyceros melampus*, is an example of a Category C type of antelope.

The bovid species in Category D are less selective feeders, their diet consisting predominantly of grasses, and they are of a larger mean body size than the species in Categories B and C (120–300 kg). They live in mixed-sex herds in open savanna habitats, and some species are migratory. Although they do not hold year-round feeding or reproductive territories, the males may set up small temporary territories during the breeding season. Species in this category have males and females of equal or near equal body size, and the females have horns that are similar to those of the males, although the horns are usually somewhat smaller. The wildebeest, *Connochaetes taurinus*, is an example of a Category D type of antelope. Category E includes large-bodied species (300–1000 kg), which rove in large mixed-sex herds over a large home range, feeding unselectively on both grass and dicotyledonous browse. Both males and females have horns that are similar in appearance, although again the horns of the female are usually smaller and lighter. But in Category E there is a marked dimorphism in body size, with the males weighing up to half as much again as the females, and male growth continues throughout most of adult life, so that the older males are considerably larger than the younger ones. The males do not set up any type of territory, but form dominance hierarchies within the mixed-sex herds. Body size is an important factor in determining position in the hierarchy, which in turn determines the access to females. The buffalo, *Syncerus caffer*, is an example of a Category E type of antelope.

Jarman's categories have been criticized as overly simplistic, but they do seem to have withstood the test of time in their description of bovid socioecology, and no other worker has devised a better unifying theme. They also appear to hold when applied to other pecoran taxa besides African antelope, although no other family possesses the diversity of habitats and morphologies that characterize the Bovidae. For example, the Eurasian and New World bovids in the tribes Caprini and Rupicaprini are primarily mixed feeders (i.e., taking both grass and browse) living in mixed-sex herds for part of the year in fairly open habitats (Schaller 1977), and based on diet and socioecology would presumably fit into Category C. In accordance with this, it can be seen that the females of most species also possess horns, but these horns are smaller than those seen in the males, and the females are usually somewhat smaller than the males. This pattern of sexual dimorphism also accords with Category C. In contrast, the more woodland-dwelling Asian bovids in the tribe Boselaphini would fit into Category B in terms of diet and socioecology, and the species in this tribe lack horns in the females.

Most cervids are woodland or ecotonal browsers, with the males living separately from the females for most of the year, and again in accordance with this type of Category B diet and socioecology, the females lack cranial appendages. The one exception among the Cervidae is the reindeer, genus *Rangifer*, which is the only genus to live in open habitats in permanent mixed-sex associations (thus presumably categorizable as Category D). In addition the males and females are of similar body size, although living in the tundra they have a diet of forbes and lichens rather than grass. Correspondingly *Rangifer* is the only genus in which antlers are seen in the females of the species. The pronghorn, *Antilocapra americana*, is a mixed feeder living in seasonably variable mixed-sex herds in open habitats, presumably fitting into Category C, and small cranial appendages are found in the females.

Finally, in the Giraffidae, the giraffe, *Giraffa camelopardalis*, is an open-habitat browser, living in variable mixed-sex herds (but with no fixed herd membership). The males are somewhat larger than the females, and the females possess smaller horns than the males. Thus giraffes appear to fit a Category C type of socioecology and pattern of sexual dimorphism. In contrast, the okapi, *Okapia johnstoni*, is a forest-living, more-or-less solitary species. Small cranial appendages are present in the males only, and the females are slightly larger than the males. The pattern of socioecology and sexual dimorphism here appears to resemble the small forest bovids in Category A, despite the much larger body size of the okapi, and its more folivorous diet. (See Janis, 1982, and Kiltie, 1985, for further discussion of classification of other living pecorans into Jarman's scheme for the African Bovidae.) Jarman (1983) has also shown that the same broad correlation of habitat, body size, social behavior, and sexual dimorphism can be applied to the macropod marsupials, which can be considered as the Australian ungulate equivalents.

Correlation of Cranial Appendages with Ecology in Fossil Ungulates and Evolutionary Problems

Work in recent years on the correlation between morphology and ecology has enabled the use of morphological proportions of extinct mammals in the inference of paleohabitat (Andrews et al. 1979; Janis 1984, in press a; Scott 1979, 1983, 1987; van Valkenburgh 1985, 1988). Diet in fossil ruminants can be determined from the correlation between craniodental morphology and dietary type in living ungulates (Janis 1988, press a; Janis & Ehrhardt 1988), and habitat preference can be determined from limb proportions (Scott 1979, 1985, 1987). Additionally, body weights of fossil ungulates can be estimated from both craniodental and postcranial proportions (Janis, in press b; Scott 1979, in press). The body weights of ungulates in this chapter have been estimated from craniodental proportions in the case of North

American taxa (Janis, in press b), and approximated for other fossil taxa from a comparison of the skull length with living ungulates and North American fossil taxa for which accurate estimates are available.

I have shown (Janis 1982) that Jarman's categories apparently hold true for the extinct pecoran family Dromomerycidae and the extinct tylopod family Protoceratidae. Almost all dromomerycid and protoceratid taxa have cranial appendages in the (presumed) males only and appear to have been woodland browsers, classifiable in Category B. The only dromomerycid which possesses small cranial appendages in both sexes, *Aletomeryx*, is also the genus with the most fibrous diet (as evidenced from the dentition), and the most open habitat preference (as evidenced from the postcranial proportions). *Aletomeryx* was probably a gazelle analogue, a mixed feeder in a more open habitat woodland savanna environment, classifiable in Category C. Thus *Aletomeryx* was probably more gregarious than the other dromomerycid taxa, which may be the reason for the relative abundance of the remains of this taxon in comparison with other members of the family. The extinct species of the family Antilocapriadae all appear to have been fairly open-habitat mixed feeders. However, the limb proportions of the earlier and smaller merycodontines suggest a more ecotonal habitat, resembling present-day small bovids like the orebi, *Ourebia ourebi*, and cranial appendages are never seen in the females. In contrast, some taxa contained within the larger, and apparently more open-habitat, antilocaprines had small cranial appendages in the females. (The single living species of the family, the pronghorn "antelope," is a surviving example of this condition.) (These data are taken from work in progress by myself and by Kathleen Scott.)

There remain some anomalies in Jarman's scheme. For example, many browsers in Category B (principally the members of the bovid tribe Tragelaphini) are not territorial and have large home ranges, as is also true of most cervids. (This issue will be discussed further in a later section.) The bongo, *Tragelaphus euryceras*, is the most forest-dwelling species of the tribe Tragelaphini, yet is the only species in which the female has horns (although one might argue that, living in a Category A type of dense forest environment, the acquisition of horns by the female is related to a reversal to a Category A type of sexual dimorphism). Alternatively, the bongo may be more closely related to the eland (*Taurotragus oryx*) than to the other members of the tribe (E. Vrba, personal communication). The eland is an open-habitat antelope with horns in both sexes, and the bongo may have only recently reentered a forest type of environment, retaining the horned condition of the females.

Recent workers (e.g., Packer 1983; Kiltie 1985) have further addressed the question of the evolutionary and ecological reasons for the acquisition of cranial appendages in female pecorans. Previous suggestions have included male mimicry to avoid female-oriented predation, or defense against conspecific males in mixed-sex herds (see discussions in Janis 1982; Kiltie 1985). Packer (1983) performed a quantitative correlation of a variety of socioecological variables with the incidence of horns in female African antelope, and showed that body size was the only significant determining variable. Once the correlation of body size with other variables was removed from the analysis, group size and habitat type had no correlation with the incidence of horns in the females of a species. He concluded that the evolution of horns in female bovids is related to strategies of predator defense: horns are more likely to occur in the females of large species than they are relatively larger than their predators, and hence more likely to rely on aggressive defense than on crypsis or flight. Kiltie (1985), in a similar analysis including all living pecoran species, showed that both body size and group size were independently significantly correlated with the incidence of cranial appendages in females, but that no variable, alone or in combination, was a reliable predictor of the presence of cranial appendages in the females of any given species.

Data from the fossil record throws doubt on Packer's (1983) assertion that cranial appendages in females is correlated only with body size. Cranial appendages in female dromomerycids are seen only in *Aletomeryx*, which was the smallest genus (body weight around 20 kg) and probably also the most open-habitat one. As previously noted, in the Cervidae, cranial appendages are found only in the reindeer, genus *Rangifer*, which is not of especially large body size in comparison with other members of the family. Yet this genus did not appear in the fossil record until the Middle Pleistocene (Lister 1984), which was some time after first appearance (Late Pliocene) of their present day preferred open-habitat type of tundra vegetation in the Northern hemisphere (Wolfe 1985). The fossil record also fails to support Kiltie's (1985) assertion that reversals can occur from presence to absence of cranial appendages in the females in pecoran lineages. He bases this assumption on the fact that the bovid tribes Antilopini and Alcelaphini, which now contain hornless females in some species, were derived from the tribe Caprini (Gentry 1978) in which the females possess horns in all living species. However, there is no firm evidence to suggest that horns were present in the females in all fossil members of the tribe Caprini, and there is clear evidence from the fossil record that horns in the females were evolved within the genus *Gazella* in the tribe Antilopini (Heintz 1969).

The precise socioecological reasons why cranial appendages should evolve in female pecorans remains unclear, although a multivariate analysis which includes a range of fossil taxa as well as living species may yet yield some interesting results. None of these quantitative analyses address the issue of why the females of the small, forest-dwelling duiker species should possess horns: they remain as an "anomolous" case that does not fit with the correlations with body size or habitat. Yet duikers are unusual in the close pair-bonding of the males and females, in contrast to the more solitary behavior typical of most small ruminants, such as the forest-habitat neotragine antelope (Kingdon 1982). In this year-round association of the sexes they resemble the more gregarious, open-habitat ruminants of larger body size. I suggest that the critical socioecological variable of horns in female ruminants may be this close association of the sexes. Perhaps horns in female ruminants serve as a means of defense in intersexual intraspecific agonistic encounters, for example over foraging areas (as noted in Chillingham cattle, S.J. Hall, personal communication).

Climatic Changes During the Tertiary

In trying to determine the evolutionary events that led to the evolution of cranial appendages in pecoran ruminants, it is first necessary to review some aspects of mammalian evolution during the Cenozoic Era. Although small, insectivorous types of mammals first appeared at the end of the Triassic Period of the Mesozoic Era, some 170 million years ago, mammals did not start to diversify into forms larger than approximately rabbit-sized until the beginning of the Cenozoic Era, after the demise of the dinosaurs, some 65 million years ago. The Cenozoic is divided into two periods: the Quaternary, stretching back from the Recent to include the Pleistocene Epoch, which commenced about 2 million years ago; and the Tertiary, which encompassed the first 63 million years of the Cenozoic. The Tertiary is further subdivided into 5 epochs of unequal length. In order, there were the Paleocene, the Eocene, the Oligocene (bracketed together as the Paleogene), the Miocene, and the Pliocene (bracketed together as the Neogene). These epochs may be further subdivided into "ages," which (unlike other episodes of geological time) are based on local land mammal faunal assemblages, and hence vary in name and duration between the different continents. In particular, the European land mammal ages have undergone much revision in terms of dating and naming in the past decade, and it is easy for the nonspecialist to get confused when tackling the literature on pecoran evolution. (Savage and Russell, 1983, provides an excellent guide to the current terminology and dating of the land mammal ages on different continents.)

The radiation and diversification of mammalian orders during the Cenozoic was accompanied by changes in climate and vegetation throughout the period, affecting the higher latitudes in particular. Of comparatively recent (Plio-Pleistocene) origin is the type of climatic zonation seen today, with polar ice caps devoid of vegetation, tropical evergreen forests in the equatorial zones, tundra and coniferous forests in the higher latitudes, and deciduous woodland or steppe characterizing the middle latitudes. The present-day patterns of vegetation and climatic regimes, and the changes in these patterns during the Cenozoic, have their origins in the changing geography of the earth. In the mid Mesozoic, all the continents were clustered together around the equator in the supercontinent Pangea, and since that time they have been gradually moving apart. Thus geographic conditions were very different in the early Tertiary from the situation seen today, where much of the world's land mass is in the higher latitudes (especially in the Northern Hemisphere), and the existence of land mass over the polar regions has resulted in the formation of polar ice caps. It is not my intention in this paper to attempt to provide a precise geological interpretation for the details of local and global Cenozoic climatic changes, nor of the nature of the evidence by which changes in climate and vegetation have been deduced. However, I shall present a brief review of major world vegetation changes during the Cenozoic, especially as this affected the evolution of ungulate mammals in the Northern Hemisphere and Africa, as an understanding of these changes is essential to understanding the possible evolutionary reasons for the evolution of cranial appendages in pecoran ruminants. (This review is taken mainly from Wolfe 1985.)

The world in the Cretaceous appears to have been warmer than at the start of the Tertiary, but in the generally cooler world of the Paleocene there was still little evidence of climatic zonation. Broad-leaved evergreen vegetation extended to well within the confines of the Arctic Circle, and multicanopied evergreen forests were found as high as latitude 60. Most Paleocene floras (paratropical rain forest) were represented by a low-diversity type of vegetation that was not temperature sensitive, suggesting a world with little seasonal fluctuation in either temperature or rainfall. The global climate for the Eocene was warmer, reaching a maximum for the Tertiary in the early middle part of the epoch (Burchardt 1978). Coniferous forests appeared for the first time above latitude 50, with semideciduous paratropical rain forest typical of latitudes 30–45. A change in mammal faunas during the Eocene, with the first appearance of a diversity of small terrestrial folivores (mainly perissodactyls) in the the Northern Hemisphere, suggests a difference in habitat from the midlatitude paratropical forests of the Paleocene (Collinson & Hooker 1987). The most profound climatic event in the Tertiary occurred during the Late Eocene, with a resulting plummet in mean annual temperature and a rise in the mean seasonal temperature variation in higher latitudes. At this time the climate regime changed from one with little seasonal variation to one that was marked by cold winters and frost (Wolfe 1978; Collinson et al. 1981; Collinson & Hooker 1987). Following this event, dense coniferous forests predominated above latitude 50, and multistrated tropical forests were restricted to 20 degrees either side of the equator. In the Northern Hemisphere, the predominant vegetation was broad-leaved evergreen forests between 20–35 degrees, and a new type of vegetation, small-leaved deciduous forest, was found between latitudes 35–50.

The global climate was somewhat ameliorated during the Late Oligocene and Early Miocene, with the thermal maximum of the Neogene occurring at the start of the Middle Miocene (16 million years ago). However, this was accompanied by the formation of the Antarctic ice cap, with a resultant build-up of high-pressure systems over the tropical latitudes, and a change in rainfall patterns resulting in seasonal periods of drought, especially along the western edges of continents. The deciduous forests across the middle latitudes in North America and Eurasia were replaced by woodland, and by the Late Miocene there was vegetational evidence of intense winter Arctic fronts. Despite the first appearance of grass pollen in the Eocene (Truswell & Harris 1982), there is no evidence of open savanna or grasslands anywhere in the world until the Late Miocene, although savanna woodland and scrub may have been present in the Early Miocene of the Southwest United States (Axelrod 1939). The rise of the Himalayas resulted in the appearance of steppe vegetation in Central Asia during the Miocene (via a rain shadow effect), but steppe vegetation was not apparent in North America until the Pliocene. The development of the East African savanna in the late Neogene was probably correlated with the rain shadow effect caused by the uplift of the East African rift system. Desert was unknown during the Tertiary, and the Arctic types of tundra and taiga vegetation were not seen until the Pleistocene.

In summary, the evolution of artiodactyls in the middle latitudes of the Northern Hemisphere took place against a background of progressive climatic and vegetational change—from the paratropical, nonseasonal forests of the Eocene, to the seasonal, deciduous forests of the Oligocene, to the development of deciduous woodlands giving way to more open savanna during the course of the Miocene. Finally, in the Pliocene, a vegetational pattern broadly similar to that of the present day was established, with steppe or deciduous woodland predominating. Meanwhile, in Africa the open savanna habitats were not in evidence until the Late Miocene.

Evolution of Ruminant Artiodactyls

All artiodactyls can be united taxonomically by the possession of a unique type of astragalus bone in the ankle joint, which has a "double pulley" type of articulation (Schaeffer 1947). The order first appeared at the start of the Eocene in North America and Europe. They were small (none much above 1 kg in body weight) and had low-crowned, bunodont cheek teeth suggesting an omnivorous rather than herbivorous diet. Yet many taxa had comparatively advanced postcranial skeletons (Rose 1985), suggesting a greater capacity for rapid locomotion than evidenced in living "primitive" artiodactyls such as pigs. The division of the Artiodactyla into the three suborders—Suina, Tylopoda, and Ruminantia (including infraorders Tragulina and Pecora)—was not apparent until the end of the Eocene. Of these suborders, the Ruminantia is the only one that can be uniquely taxonomically defined in a cladistic sense: all members of the suborder are characterized by the fusion of cuboid and navicular bones in the tarsus.

The taxonomic term Ruminantia is of course rather misleading in this sense, as the tylopod camelids (and presumably also their extinct close relatives, the protoceratids) also ruminate or chew the cud. The Tylopoda, as this term is currently used (e.g., Romer 1966) is a rather ill-defined assemblage of artiodactyl families that resemble ruminants (and differ from suines) in the tendency to develop selenodont cheek teeth, diastemas between the cheek teeth and the incisors, and elongated metapodials with the reduction of the lateral digits. Webb and Taylor (1980) have identified a clade within the Tylopoda of the Camelidae plus the Protoceratidae and place this as the sister group to the Ruminantia, terming the entire group the Neoselenodontia. Other families included in the Tylopoda are the North American endemic oreodonts (families Agriochoeridae and Merycoidodontidae) and a variety of small, rather traguloid-like animals known from the Late Eocene to the Early Miocene of Europe.

During the Oligocene a variety of small, hornless tylopods and traguloid ruminants were found in both North America and the Old World (tylopods in Europe and traguloids in Asia). The living family Tragulidae was not known from the fossil record until the Early Miocene of the Old World, although its relatively primitive position among the Tragulina suggests that it must have evolved considerably earlier (Webb & Taylor 1980). Traguloids migrated into Europe later in the Oligocene, presumably resulting in the demise of the endemic European tylopods, which were all extinct by the end of the Early Miocene. Most of the North American traguloids had also disappeared by this time, probably as the result of climatic changes. The earliest pecoran ruminants, the gelocids, were small, hornless animals (about the size of a present-day tragulid), known from the Early to Middle Oligocene of Europe and Asia. Gelocids can be distinguished as pecorans by details of the dentition, braincase, and astragalus (Webb & Taylor 1980). However, gelocids represent an assemblage of primitive pecorans from which the higher families of ruminants were derived, rather than a distinct family (Janis 1987). The earliest higher pecorans were the cervoid genera *Dremotherium* (Moschidae) and *Amphitragulus* (Palaeomerycidae), which appeared in the later Oligocene of Europe, probably derived from an Asian gelocid taxon such as *Eumeryx* (Janis & Scott 1987). They were somewhat larger than the gelocids (approximately 8–10 kg), and were also hornless, but they possessed (presumably sexually dimorphic) large saber-like canines resembling those of the present-day musk deer (genus *Moschus*), in contrast to the shorter, more traguloid-like canines of the gelocids (Janis & Scott 1987).

The earliest horned pecorans in the Old World were known from the earliest part of the Miocene in Eurasia, with the appearance of *Palaeomeryx* and the cervid genera *Dicroceros* and *Stephanocemas*. The latter part of the Early Miocene saw the appearance of the earliest horned bovids and giraffoids (families Giraffidae and Climacoceridae). The bovid *Eotragus* was known from both Africa and Europe, the date in Europe slightly preceding the one in Africa. [Although some supposed bovids, such as the taxon *Palaeohypsodontus*, were known from the Oligocene and Miocene of Asia, they consist of dental remains only, and there is no firm basis for the inclusion of these animals in the family Bovidae (Janis & Scott 1987).] The giraffid *Giraffokeryx* was known from Europe, and the genera *Climacoceras* (Climacoceridae), and *Palaeotragus* (Giraffidae) were known from East Africa [although the form of the cranial appendages is not homologous between the climacocerid giraffoids and the true giraffoids (Janis & Scott 1987).] Giraffoids may have evolved ossicones in Africa, and the earliest and most primitive forms have been found on this continent. However, the place of origin of the Giraffoidea is unknown, and recent speculation suggests an early Asian appearance of the most primitive horned giraffoids (Hamilton, quoted in Patterson 1981). Late Early Miocene faunas from North Africa also contained the horned bovids *Eotragus*, *Gazella* and *Protragocerus*, the peculiar giraffoid *Zarafa* with horizontally projecting ossicones, and the horned ruminant *Prolibytherium*, of uncertain taxonomic affinities (Hamilton 1973).

Thus, the evolutionary history of the camelids and protoceratids was based in North America, while the evolutionary history of the Pecora was based primarily in the Old World [with the exception of the migration of the horned cervoid families Dromomerycidae and Antilocapridae, and the hornless Blastomerycinae (family Moschidae) from Eurasia to North America in the Early Miocene (Webb

& Taylor 1980).] Camelids did not appear in the Old World, nor cervids in the New World, until the Pliocene, and bovids did not arrive in North America until the Pleistocene. (The diversification of the South American cervids was extremely rapid. The small size of the forest-dwelling forms, such as the pudu, must surely represent an evolutionary decrease, as all the fossil North American odocoilines were considerably larger.) Ruminant artiodactyls were absent from Africa until the Early Miocene, and the evolution of the horned pecoran and tylopod families appears to have had its center in the higher latitudes. However, the earliest incidence of endemic horned artiodactyls in North America was in the Late Oligocene, with the appearance of the protoceratid genus *Protoceras* predating the appearance of the horned pecora of the Old World in the Early Miocene by a good 10 million years.

All these taxa that possessed cranial appendages were larger than their preceding hornless relatives. The European genera *Dicroceros* (Cervidae), *Eotragus* (Bovidae), *Hoplitomeryx* (Hoplitomerycidae, Cervoidea) and *Stephanocemas* ('Lagomerycidae') were approximately 18–20 kg, as were the North American genera *Aletomeryx*, *Barbouromeryx* (Dromomerycidae), *Paracosoryx* (Merycodontinae, Antilocaprideae) and *Protoceras* (Protoceratidae, Tylopoda). The early species of *Palaeomeryx* [e.g., *P. tricornatus* from China (Qiu et al. 1985)] probably also fell within this size range. Other early bovid and giraffoid genera were slightly larger, with body weights probably ranging between 20–50 kg. Some smaller horned lagomerycids, of a body weight of around 10–12 kg, were known from the Miocene of Eurasia. However, as lagomerycids have now been shown to belong within the muntiacine cervids (see discussion in Janis & Scott 1987), it is possible that they represent a decrease in body size from the earliest and most primitive cervid, *Dicroceros*. Whether or not this precise "critical weight" of 18 kg can be shown to hold true for all first appearances of horned pecoran lineages, it remains true that horned pecorans are in general larger than their hornless relatives, and that evolutionary decreases in size can occur once horns have been acquired.

Discussion

Piecing together the fossil record evidence for the evolution of cranial appendages in ruminant artiodactyls, two facts become apparent. Cranial appendages were evolved independently in a number of lineages in the Early Neogene in the Northern Hemisphere, at a point when the vegetation changed from closed forest to more open woodland, and the body size of the animals increased to a threshold level of about 18 kg. This was seen in North America in the Late Oligocene, with the appearance of the horned tylopod *Protoceras*, and in the Early Miocene of Eurasia, with the appearance of several independently horned lineages of pecoran ruminants (Janis 1982). These changes in global habitat and in body size over evolutionary time suggest a parallel with the changes in habitat preferences and increases in body size seen today in African bovids (Jarman 1974), in the change from a Category A type of nonterritorial, forest-dwelling selective browser, to a larger Category B, territorial, woodland-dwelling browser of more purely folivorous habits. I would suggest, by analogy with a comparison of spatial variation seen in aspects of socioecology in living pecorans today, that the evolutionary pressures that led to the evolution of cranial appendages within the various pecoran lineages were as follows.

In the Oligocene of the Old World, the gelocids and early cervoids were woodland-dwelling browsers (as suggested by their small size and the unspecialized morphology of their cheek teeth, and also by paleobotanical evidence of available habitat type). They would probably have been solitary or pair-forming in their reproductive behavior, as there is little evidence of sexual dimorphism. It is possible that, as in living small ruminants, the females were slightly larger than the males (Ralls 1976). In the Oligocene cervoids *Dremotherium* and *Amphitragulus*, the saber-like canines were probably larger in the males, as in present-day musk deer and tragulids. These animals may have had distinctly marked home ranges, but probably would have been largely solitary in their habits with few incidents of boundary disputes between males, as the patchy distribution of the food in the large home ranges would not have been conducive to a male's territory attracting bands of foraging females. Thus there would be little reproductive incentive for intense defense of a resource area. (This type of behavioral ecology is seen in present-day cervoids of similar body size, habitat, and degree of sexual dimorphism, such as the musk deer, *Moschus*, and the muntjac, *Muntiacus*.)

As the climate in the Early Miocene became drier, the woodlands of the Northern latitudes became more open. The increased seasonality of rainfall in this time period would have resulted in parts of the year when there was little availability of succulent items of vegetation such as buds, berries, and young leaves. There would have been selection pressure for ruminant species to increase in body size to reduce their relative requirements for food, and at a larger body size they would also benefit from an increased fat storage capacity to last out periods of reduced food availability. As the average body size of the various pecoran lineages increased to about 18 kg, it would then have been within the physiological capacity of the animals to select more fibrous leaves as the main component of their diet (Demment & Van Soest 1985). In addition, at this body size the acquisition of a more fully developed selenodont dentition (the specialized ruminant form of molar pattern characterizing animals with a folivorous, rather than omnivorous, diet) was seen in fossil taxa. Note that at this point there would not necessarily be an evolutionary compulsion to adopt a more folivorous diet. Tragulids with the more primitive type of bunoselenodont cheek teeth (suggesting a more selective folivorous/frugivorous type of diet seen in present day tragulids) were common in the Early and Middle Miocene in Northern latitudes, and some species of the tragulid genus *Dorcatherium* were fairly large. The range of sizes for species of this genus was about 5–10 kg, a similar size range to the selective browsing species of the bovid genus *Cephalophus*.

However, as the availability of a selective browsing type of diet decreased in the Northern latitudes, ruminants of larger body size would at least have had the physiological capacity to adopt a more fibrous folivorous diet. As the distribution of fibrous leafy food items is more uniform than the distribution of succulent browse items, animals adopting this type of diet would face less competition from neighboring individuals for food. It thus would not have been disadvantageous, in terms of food selection by an individual, for animals to forage together in small groups, and there may have been advantages to this strategy either in terms of habitat utilization or predator defense (or both). The adoption of a folivorous diet would also mean that, for any individual, there would be more actual food items available for selection in any given area, and it would have been possible to decrease the foraging radius. At some critical point, the effective home range of an individual ruminant would be small enough to make defense of a discrete feeding and reproductive territory a feasible male reproductive strategy, in terms of energy budgets and costs of boundary patrol.

Thus at this threshold body weight of about 18 kg, it would become possible for female pecorans to forage together in small groups, moving through a more open woodland habitat than seen earlier in the Miocene. It was then not only possible, but also advantageous, for the males to maintain and defend resource territories, to maximize their chances of impregnating a number of females that would be attracted to the territory to feed. At this point in time, male/male competition for the best territories to attract the maximum number of females would have led to the evolution of cranial appendages to engage in more ritualized fighting and display, as opposed to the fighting by canine-slashing seen in present day tragulids (Ralls et al. 1975), which was also probably the mode of combat employed by the gelocids and hornless cervoids. The initial development of small cranial appendages was probably to ward off canine slashes by the opponent, as seen today in the muntjac (Barrette 1977). With a higher density of animals in the habitat (as could now be supported by the folivorous habits of these early horned pecorans) and a territorial, polygamous life-style in the males of these species, the number of agonistic encounters an individual would encounter in a lifetime would be greatly increased, and there would be selection for a more ritualized more of male/male interaction which resulted in less danger of serious injury to the individual (Geist 1966a).

Although it is true that many present-day woodland browsing bovids have a large home range and are not territorial (Owen-Smith 1977), it must be remembered that the drying trend that started in the Early Miocene has continued through to the present day. The woodland habitats of the Early Miocene were likely to have supported a much higher density and diversity of plants than seen today even in tropical woodlands (Wolfe 1985), and perhaps it was possible at that time for a browsing ruminant of 18–30 kg to have a feeding range small enough to defend as discrete territory. Thus, even though many species in Category B are not territorial today, this type of ruminant may have held territories in the Miocene woodland habitats. As the global climate became progressively cooler and drier in the Late Miocene, many pecoran lineages in the higher latitudes would have encountered a still more open type of habitat, making the holding of discrete territories by the males difficult, and encouraing the formation of more permanent, mixed-sex herds for part or all of the year. It appears that, once cranial appendages had evolved in pecoran lineages, they were not secondarily lost, even if their original function was later somewhat modified. When certain pecoran species adopted this behavioral strategy of both sexes living together in herds, there was obviously some advantage (for whatever evolutionary reason) in the reduction of the degree of sexual dimorphism in the species, and this was accomplished by the evolution of cranial appendages in the females. It was in this Late Miocene time of increasing drying of global habitats that the possession of horns in the females of bovid species became apparent (Thomas 1985).

The history of bovid evolution has been primarily in Africa and southern Eurasia (Thomas 1985). It is interesting to note that, while all pecoran lineages saw the evolution of taxa of larger body size during the Neogene (ranging from 20–1000 kg), cranial appendages were never apparent in the females of cervid species before the Pleistocene appearance of the genus *Rangifer*. In contrast to the bovids, cervids maintained a more northern distribution during the Neogene in hardwood woodland habitats (A. Lister, personal communication). Families that contain mainly browsing taxa (such as the Dromomerycidae and the Giraffidae), rarely have species where horns are present in the females. However, horned females are more common in species in the Antilocapridae, where the early evolution of long limbs and hypsodont cheek teeth suggests an initial commitment to an open-habitat, mixed-feeding type of life-style, despite the fact that the average body size of antilocaprid taxa is fairly small (approximately 20–80 kg). It is my contention, from an examination of the fossil record evidence, that the evolution of cranial appendages in the females of pecoran species is more closely tied to habitat type and the year-round association of the sexes than to absolute body size, in contrast to the recent suggestions by Packer (1983) and Kiltie (1985).

An evolutionary theory which attempts to explain why cranial appendages should appear in pecoran lineages at a critical junction of changing habitats and increase in body size must also explain why cranial appendages were never developed in those ungulates which paralleled the Pecora in various aspects of their evolution. That is, the tylopod camelid artiodactyls and the equid perissodactyls, which also evolved into gregarious, open-habitat, large-bodied animals subsisting on fibrous diets.

As previously discussed, the early evolution of camelids was restricted to North America, but the related protoceratids on the same continent paralleled (but predated) the Old World Pecorans in evolution of cranial appendages in the males alone. The patterns of ungulate diversity in the North American Tertiary faunas suggest that an open-woodland type of habitat appeared earlier in the New World than in the Old World (Webb 1977, 1978; Janis 1982), and that during the Miocene, the drying effect on the habitat was more pronounced in North America than in Eurasia (Janis 1984). The protoceratids were all animals with short legs, and low to moderately high-crowned cheek teeth, suggesting a woodland, browsing (possibly riverine, as suggested by the moose-like snouts) type of life-style for the entire family. In contrast, camelids showed a tendency from the start of their evolutionary history to have longer legs and more hypsodont cheek teeth, suggesting a preference for a more open type of habitat, with an antilocaprid-like low browsing type of foraging behavior, or a giraffe-like high browsing habit (Janis 1982).

I suggest that the change from woodland to savanna in the mid Tertiary not only occurred slightly earlier in North America than in the Old World, but also that the change occurred in a more rapid fashion, and that in the Late Oligocene there was a split in habitat preference in the endemic tylopod artiodactyls; the protoceratids remained in the dwindling areas of closed woodland habitat, and the camelids radiated into the newly formed areas of more open habitat. (Some paleoecological evidence also exists for this Late Oligocene scenario of camelid habitat preference, see Clark et al. 1967). In this type of open-bushland habitat, territorial defense by the males probably would not have been a viable reproductive strategy. The feeding ranges would have been relatively large, due to the patchy dispersal of quality food items, and the more two-dimensional nature of the distribution of food in the habitat. Thus camelids probably moved over evolutionary time from a Category A type of socioecology (solitary), to a Category C type of socioecology (herd-forming), bypassing the Category B territorial stage which appears to have been critical in the evolution of cranial appendages in pecoran lineages.

Present-day camelids are all harem-forming, and the males exhibit defense of female groups rather than resource defense (Koford 1957; Franklin 1974). Males in harem systems appear to have fewer agonistic encounters with other males than do those which defend food-resource territories (Owen-Smith 1977). I suggest that camelids never passed through a critical evolutionary stage of defending food-resource territories in a woodland habitat, and hence were never subjected to intense evolutionary pressures to develop sexually dimorphic cranial appendages for ritualized male/male combat. In contrast, the related protoceratids, which remained in such woodland habitats that were available during the Oligocene, paralleled the evolution of the Pecora in their morphology, and presumably also in their socioecology.

Equids also evolved on the North American continent, and the available morphological (Janis 1979) and paleoecological (Clark et al 1967) evidence suggests that they remained as woodland browsers during the Oligocene. However, the difference in digestive physiology between equids and neoselenodont ruminants (the Ruminantia plus the tylopods and protoceratids) is of profound importance in this case. Equids are hind-gut fermenters, and hence are not restricted to a nonfibrous diet at small body sizes, in contrast to ruminating artiodactyls (Janis 1976). An additional consequence of a hind-gut system of fermentation is that equids must consume more food per day than a ruminant of similar body size, and thus would require a relatively larger home range to forage for their daily requirements. In the woodlands of the Oligocene of North America, while the male protoceratids were presumably maintaining resource territories (as evidenced by the evolution of cranial appendages), equids of similar body size would have required a larger home range for daily maintenance, which may have been too large for a male to defend as a feeding and reproductive territory.

There is a lower size limit for a nonruminant folivore of about 1 kg (Kay & Covert 1983), as opposed to a limit of about 10 kg in ruminants (Demment & Van

Soest 1985). However, the earliest equid (*Hyracotherium*) was considerably larger than this lower limit, with a body weight of around 10 kg. Equids would have been able to steadily progress, in an evolutionary sense, from the type of selective browsing diet seen in *Hyracotherium* to the more folivorous diet of the larger Oligocene genus *Mesohippus* (diets determined from dental morphology and dental wear patterns, see Janis 1979). Equids have always shown little evidence of sexual dimorphism, although males tend to possess larger canines than females, especially in the small Eocene taxa (Gingerich 1981). Thus, because of their digestive physiology, there was probably never a "critical moment" in equid evolution, as postulated for ruminants, where a radical change in diet became possible as a consequence of both changes in local vegetational habitat and an increase in body size. Hence they were never subjected to evolutionary pressure to develop sexually dimorphic morphological changes in correlation with profound changes in socioecology. I suggest that Oligocene woodland browsing equids foraged in a solitary fashion, or in small groups, as is typical of present-day browsing perissodactyls such as tapirs and small rhinos. The lineages which adopted an open-habitat grazing habit in the Middle to Late Miocene (as evidenced by the evolution of longer legs and hypsodont cheek teeth), probably adopted the harem system typical of most present-day equids.

It is interesting to note that those perissodactyls which do display territorial behavior today are the ones in the most open habitats, with the most sparse distribution of available food, in contrast to the pattern of territoriality seen in ruminants. Such animals include the white rhino, *Ceratotherium simum* (Owen-Smith 1974), Grevy's zebra, *Equus grevyi*, and the African and Asian wild asses, *Equus asinus*, *Equus kiang*, and *Equus hemionus* (Klingel 1974, 1977). It seems likely that this difference is the reflection of differences in digestive physiology. There is evidence that browsing rhinos have larger home ranges than grazing ones (Laurie 1982), whereas the reverse is true for antelope (Owen-Smith 1977). A perissodactyl is limited by the quantity, rather than by the quality, of the available food, in contrast to a ruminant artiodactyl (Janis 1976). Hence perissodactyls may be better able to form discrete territories in open areas where there is a large quantity of low-quality food, when a ruminant of similar body size would require a larger home range to obtain enough high quality food items. In contrast, the actual quantity of available food in a similar area of woodland habitat might be too small to sustain a browsing perissodactyl, although it might be sufficient for a ruminant of similar body size, with its lower daily intake requirement. (See Janis, 1982, for an extended discussion of this argument.)

Conclusion

The evolution of cranial appendages in pecoran lineages appears to have been the result of an interaction of the requirements of their digestive physiology with the physionomic features of changing climate and vegetation seen during the Miocene in Eurasia, which was the center of their evolution. As the climate became cooler and drier, and the various pecoran lineages experienced a compensatory increase in body size, they experienced a threshold level at the body weight of about 18 kg. At this point, a radical change in diet was physiologically possible (shifting from a habit of selective browsing on nonfibrous items of vegetation to one of consuming more fibrous leaves). This would have resulted in a vastly different perception of the environment for these ruminants, in terms of the distribution of the available food resources, making it possible for them to forage in social groups and decreasing the effective size of the home ranges. It is at this point in pecoran evolution, when the fossil record shows an increase in body size together with a change from forest to woodland habitats in Eurasia, that cranial appendages are first apparent in the males of the various lineages.

I make the hypothesis that the emergence of sexual dimorphism was correlated with a change in socioecology, with the males of the species now competing for feeding resource territories to attract roving groups of foraging females, and with a change in the mating system from facultative monogamy to polygamy. The increased incidence of male/male combat that this change in socioecology would presumably entail would have resulted in a strong selective pressure for organs of ritualized combat and display, to reduce the possibility of severe injury or death to individuals. The evidence from the fossil record suggests that cranial appendages in female ungulates are developed primarily in those lineages which adopt a preference for more open habitats, and which spend all or part of the year in mixed-sex herds.

In contrast to the evolution of the pecoran artiodactyls, cranial appendages were never developed in the camelids or the equids, ungulate lineages which in many other ways paralleled the evolutionary patterns seen in the Pecora. The absence of cranial appendages in camelids can be explained by their evolutionary origin in North America, with a rapid adoption of an open habitat lifestyle during their evolutionary history. I suggest that camelid socioecology changed directly from one of solitary forest dwelling to living in mixed-sex herds in open habitats. The absence of cranial appendages in the Equidae can be explained by the differences in digestive physiology between perissodactyls and ruminant artiodactyls. Equids are hind-gut fermenters, and hence require a greater absolute amount of food per day than a ruminant of similar body size. In the Tertiary woodland habitats, a browsing equid would always have required a larger home range than a similar-sized ruminant, and hence may never have been able to establish a resource territory for energetic reasons. Thus neither camelids nor equids ever passed through an evolutionary period where there was defense of resource territories by the males in a woodland habitat, with the result that neither family was ever under selection pressure for the evolution of cranial appendages.

Acknowledgments. I would like to thank A. Gentry, A. Lister, and S.J. Hall for discussion of some of the ideas presented in this paper, and A. Bubenik and an anonymous reviewer for comments on the original version of the manuscript. I also thank Eileen Gibbs, whose scholarship funds enabled me to travel to Africa in 1982, giving me an unparalleled opportunity to tie together my knowledge of the fossil ungulate faunas with the living diversity of African ungulates.

3
Muntiacidae

COLIN P. GROVES AND PETER GRUBB

Introduction

Muntjac and tufted deer are small-sized deer with small antlers, raised up on long pedicles which are prolonged forward onto the face as ridges and with upper canines in both sexes (much longer in males) (see Figs. 1, 2, 3). Their taxonomic status is debatable (Bubenik, G. 1983).

One of us (Groves, 1974) has argued that the muntjac and tufted deer are true cervids, related to the Cervinae. We have followed this proposal more recently (Groves & Grubb 1987) and shown that if one assumes the premise that two characters (antlers and the plesiometacarpal condition) each evolved only once, then the muntjacs must be the sister group of the Cervinae and the two taxa can be treated as tribes.

However, the premise remains speculative, especially as cranial appendages generally, and the alternative (telemetacarpal) condition of the lateral metacarpals, have undoubtedly evolved on more than one occasion (reviewed by Groves & Grubb 1987).

Apart from the plesiometacarpal state, we know of no putative yet trenchant synapomorphies shared uniquely by the muntjacs and the Cervinae. The elongated pedicles could be an exaggeration of characters already present in some Cervinae, but it is also possible that they have been developed independently. Indeed the strong differences between muntjac and Cervinae could reflect a long independent history following descent from an antlerless, holometacarpal morphotype. The muntjacs differ, particularly in their anomalous chromosome complement. The absence of metatarsal glands is surprising, since they are present in telemetacarpal as well as plesiometacarpal branch-antlered deer. Again, a long separate history could account for this.

The idea that the muntjacs are primitive is supported by general observations on skull shape. Facial elongation is very marked, such that the posterior sutures of the nasals are forward of a line joining the anterior orbital margins, and the ethmoid fissure is carried well forward of the nasolacrimofrontal angle. In these respects even the antlerless Hydropotes resembles other deer, as indeed (perhaps independently?) do the Moschidae.

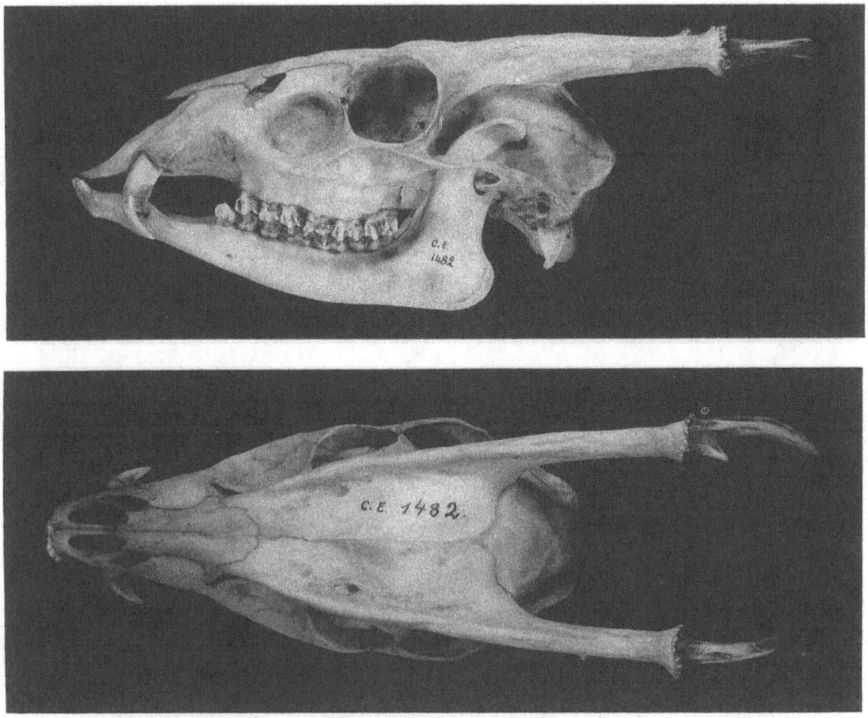

a

b

FIGURE 1. Skull of male holotype of *Muntiacus feai*: *a* lateral view; *b* dorsal view. Greatest skull length, 201 mm. (Photographs courtesy of Professor Enrico Tortonese, Genoa.)

A. Bubenik (1982) separated muntjacs into their own family Muntiacidae, distinguished by having short antlers on long supraorbital pedicles, which are supported by bony thickening along the supraorbital margins; whereas true Cervidae have long antlers borne on short, frontal (not supraorbital) pedicles, and in addition have reduced canines. Later, A. & G. Bubenik (1986) separated the Odocoileidae as well. In the Cervidae, the antler pedicles reach their full size only after 14–16 months. In the Odocoileidae, they are full-sized in 4–5 weeks, and the first antlers are long and ramified. In the Odocoileidae, mineralization takes a mere 2 weeks, and the first set of antlers are short and sometimes remain unshed. The antler branching occurs to a different plan in the two families, suggesting that branching originated independently in the two lineages. The Muntiacidae resemble the Odocoileidae in most of these ontogenetic parameters: pedicles are full-sized by 6 months of age; the first set of antlers is simple, unbranched, and is fully developed shortly after (van Bemmel 1952). On the other hand, the branching pattern could be said to more resemble that of true Cervidae, in that there is (in those species whose antlers do branch) a brow-tine.

In Muntiacidae, as in Odocoileidae, the rumen has only two blind sacs; in the Cervidae there are three. Tooth replacement takes place at 10–12 months of age

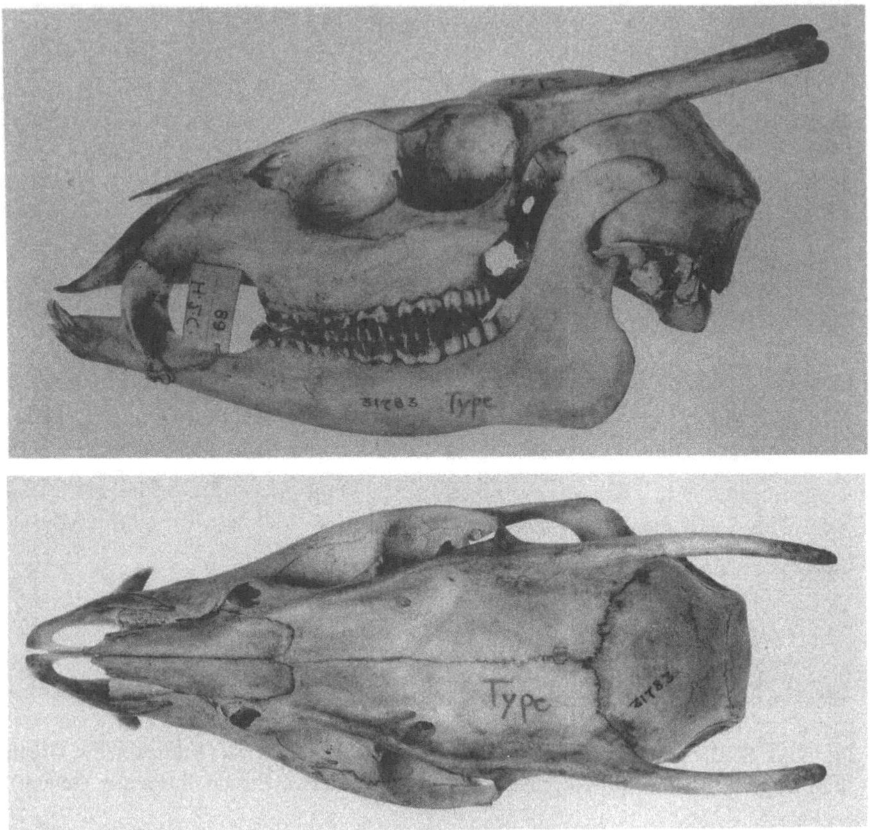

a

b

FIGURE 2. Skull of male holotype of *Muntiacus rooseveltorum* (= *M. feai*): *a* lateral view; *b* dorsal view. Greatest skull length, 188 mm. (Photographs courtesy of Field Museum of Natural History, Chicago, through Robert Izor.)

(van Bemmel, 1952); this resembles the Odocoileidae but contrasts with the Cervidae, in which replacement occurs at 18 months at the earliest. In Muntiacidae, alone among deer with the exception of the enigmatic and (perhaps primitively) antlerless *Hydropotes*, the permanent upper canines are enamel-coated; in other deer, only the milk teeth are enameled.

The hormonal control of antler development in the Muntiacidae differs from both Cervidae and Odocoileidae. In castrated males of the two latter families, adrenal steroids ensure the continued development of normal antlers; in the Muntiacidae, there is no such hormonal compensation, and the formation of a perruque, uncontrolled proliferation of unmineralized tissue, succeeds castration as in the taxonomically isolated genus *Capreolus*.

It is clear, at any rate, that the Muntiacidae are much more different from other deer than was appreciated, for example, by Groves (1974); and that they are

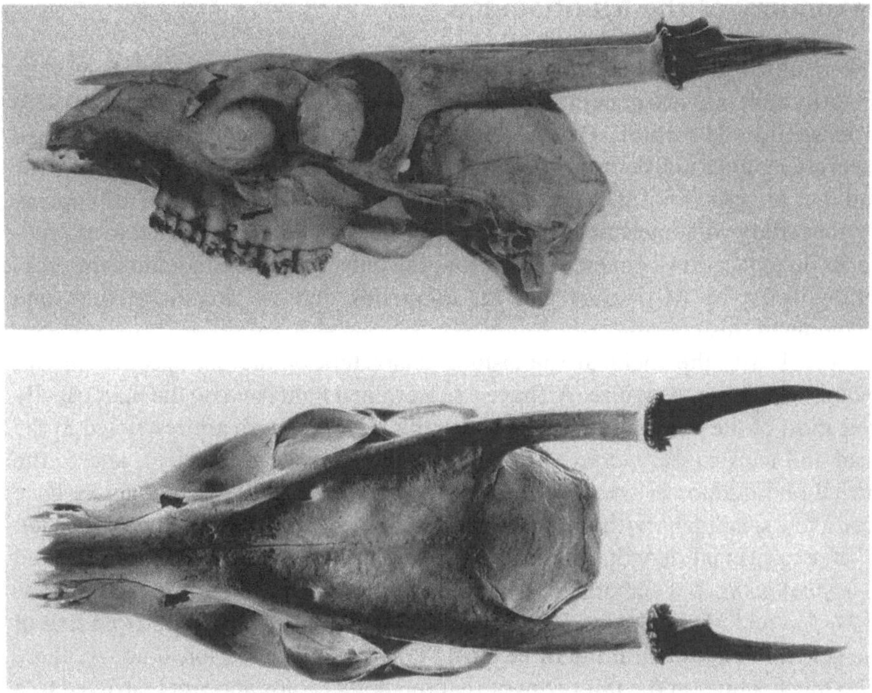

a

b

FIGURE 3. Skull of male *Muntiacus crinifrons*: *a* lateral view; *b* dorsal view. The premaxillae are missing. Skull length from occipital crest to base of nasals 127 mm. (Photographs courtesy of Karl Koopman from American Museum of Natural History.)

primitive in a way that no other living deer are (*Hydropotes* always excepted). Muntjac-like deer occur as long ago as the Miocene. The implication that antlers have been evolved in parallel two or three times is irresistible.

The form of the preorbital fossa in the muntjac group is characteristic. While in other deer, as in the Moschidae, it forms a simple depression, sloping gradually into the facial surface, in the muntjac group it is clearly and sharply demarcated and pit-like, especially superiorly.

It will be argued below that the muntjac group forms a graded series as far as antler development is concerned. It is not immediately apparent whether these appendages have been elaborated or have degenerated during the evolution of the group; but when other characters are considered, particularly karyotypic variation, the series is seen to conform with the elaboration hypothesis. This in turn supports the hypothesis (above) that the muntjacs acquired antlers independently of other deer and should be segregated in a family of their own – the Muntiacidae, a remarkable group in which important stages of antler development and deciduousness may be observed among contemporary species.

Taxonomy of the Muntiacidae

The muntjac group—henceforth referred to as Muntiacidae—have been customarily split into two genera, *Elaphodus* and *Muntiacus*. To these Pocock (1923) proposed to add a third, *Procops*, for Fea's Muntjac which he thought lacked preorbital glands unlike other species; but it seems clear that Pocock was in error and that *M. feai* is not the most distinct of the species remaining in *Muntiacus*. Further study will determine whether any of the species here referred to as *Muntiacus* do in fact merit generic separation; the only plausible candidates for separation would be *M. reevesi* and *M. atherodes*, but we do not at this time recommend this course.

Table 1 lists the main distinguishing characters of the six species (in two genera) which we recognize. All have a subterminal light band on the hair, usually over most of the body surface; hairs with this agouti pattern are restricted to the head and neck in *Elaphodus* (which hence falls outside the series), and to the nuchal and middorsal areas, at the very most, in *M. muntjak*, though we have seen a few specimens with the agouti pattern over most of the upperparts. The tail is of exceptional length in *M. crinifrons*, and is short in *M. reevesi* and *E. cephalophus*, but elaborately fringed and tufted with white fur in the latter, as in *M. feai* and *M. crinifrons*. The dorsal hue of the tail often differs from that of the general body tone; in the three bushy-tailed species mentioned above, there is so much white on the tail's fringes that the dorsal color appears restricted to a median line. The general body tone differs among the species, as well as (to a lesser degree) among subspecies within *M. muntjak* and *E. cephalophus*. Lateral hoofs are relatively long—even though unsupported by bony elements—in *E. cephalophus*, *M. feai* and *M. crinifrons*. Ears are narrow and pointed in *M. muntjak* and *M. crinifrons*, more rounded in other species of Muntiacus, and broad and rounded in *Elaphodus*.

The frontal pattern is primitive in both sexes in *Elaphodus* and *M. atherodes*; primitive in female, derived in male, in *M. reevesi*; derived in both sexes in *M. feai* and *M. muntjak*; and still further derived, in both sexes, in *M. crinifrons*.

In the skull, the premaxilla is broad along the whole length of its ascending process in *E. cephalophus, M. crinifrons, M. reevesi*, and *M. feai*, but usually (almost invariably in *M. reevesi*) is separated from the nasal by a thin strip of maxilla. In *M. muntjak* and *M. atherodes*, however, the premaxilla tends to narrow superiorly, but always contacts the nasal.

In other skull characters, the development of strong ridges, a downward continuation of the antler pedicles, in *Muntiacus*, but only weak ones in *Elaphodus*, gives the skulls of these two genera a distinctly different appearance. The pedicles of *M. muntjak* are much more strongly developed, in both length and thickness, than in any other species. The pedicles slant backward in line with the facial profile in *E. cephalophus, M. reevesi, M. feai*, and *M. muntjak*. They may be somewhat depressed below this line in *M. atherodes* and *M. crinifrons*, in the latter curving up slightly towards the antler bases.

TABLE 1. Taxonomic characters in the *Muntiacidae*.

	Elaphodus cephalophus	*Muntiacus atherodes*	*M. reevesi*	*M. feai*	*M. crinifrons*	*M. muntjak*
Hair banding	yes (lower neck & face only)	yes	yes	yes	yes	little/none
Tail: length (mm)	90–120	140–200	65–120	175	180–240	140–180
dorsal hue	black	brown	chestnut	black	black	chestnut
Body hue	black-brown	yellow	red-yellow	brown	black	reddish
Tuft:						
size	short, thick	absent	absent	long	long, thick	short (along pedicles only)
color	black	–	–	red & black	red	red
Lateral hoofs (mm)	10–12	(small)	4–7	8–12	9	(small)
Ears	rounded	f.rounded	f.rounded	f.rounded	pointed	pointed
Karyotype:						
2n – female	48?	unknown	46	12/13/14	8	6 or 8
male	unknown	unknown	46	unknown	9	7 or 9
X-autosome translocation	no	unknown	no	yes	yes	yes
Premaxilla-nasal contact	sometimes	yes	no	sometimes	yes	yes
Sex size relationship	equal	male larger	male larger	female	female	equal
Antler:						
length (mm)	(virt.abs)	20–40	40–80	50	20–60	70–120
shedding	no	occasional	yes	yes	yes?	yes
pedicle l.	55–70	60–80	50–80	70–100	80–100	90–140
pedicle b.	(thin)	10–12	14–17	8–9	7–8	15–30
branching	no	no	yes	yes	yes	yes

The dorsal profile of the facial skeleton is convex in *E. cephalophus* and *M. crinifrons*, giving these two species a deep, ram-faced muzzle. The braincase is also more rounded and convex in these two species (and, somewhat less, so, in *M. atherodes*), with a distinct concavity, between the two convex surfaces, in the interorbital region. The preorbital fissure is very large and wide in *M. feai* and fairly so in *M. atherodes* and *M. muntjak*; longer and narrower in *M. reevesi*; very small and narrow in *M. crinifrons*; and small, or absent altogether, in *E. cephalophus*. The preorbital fossa is very large, hollowing out almost the whole of the lacrimal and extending into the maxilla, in *E. cephalophus* and in *M. reevesi*; occupying about half (the lower half) of the maxilla in *M. atherodes, M. feai*, and *M. crinifrons*; and generally smaller still in *M. muntjak*. The fossa in *M. muntjak* and *M. crinifrons* tends also to be less well defined than in other species.

In *M. muntjak* and *M. crinifrons* the nasals stand above the maxillae on a convex ridge, and they are narrow throughout their length. In the other species the nasals are flattened; this is most markedly the case in *M. feai*, in which they expand posteriorly. In *M. feai* the snout narrows evenly anteriorly, instead of being somewhat pinched in at the level of the nasal opening, with the premaxillae expanded in front, as in other species. The forehead is narrow between the frontal ridges in *M, crinifrons*, unlike other species. The orbits lie flat against the lateral facial profiles in *M. feai*, but their lower margins protrude in other species, especially *M. muntjak* and *M. crinifrons*.

In *M. muntjak* the antlers are well developed (in males), and cast annually (as far as is known). In *M. feai* they are slightly less well developed, but again subject to periodic casting (as indicated by the development of a coronet). In *E. cephalophus*, they are represented by tiny stumps, invisible externally except under close inspection. In the three species mentioned above, almost any antler, however small, will be branched, remaining unbranched only in some examples of *M. reevesi*.

In *M. atherodes* the picture is quite different: the antlers are short, slender, unbranched, and almost never shed. As the antler grows for the first time, it of course contains an extension of the frontal cavity, which extends up into the pedicle; when cast, the coronet seals off the cavity, so that subsequent antlers develop as solid structures. The difference between *M. atherodes* and *M. muntjak* in this respect is very marked: the extension of the frontal cavity into the antler in the former confirms the evidence of the absence of a burr: the antler is not cast, except in a few individuals. The frontal sinus extends further up into the pedicle in *M. atherodes*. (These features are illustrated in Figs. 4, 5.) Under what circumstances the antler was cast in the two specimens known out of 26 adult males studied (Groves & Grubb 1982) with a burr is obscure. In *M. reevesi*, casting appears to be more usual than in *M. atherodes*, but is, again, not a regular annual occurrence.

The antlers of *E. cephalophus*, as already noted, are quite tiny; nonetheless, occasional shedding may occur. Lydekker (1915) illustrated a specimen with a well-developed burr.

M. crinifrons is another species in which shedding appears to be irregular. We have seen six male specimens. In two (in the Kunming Institute of Zoology) the

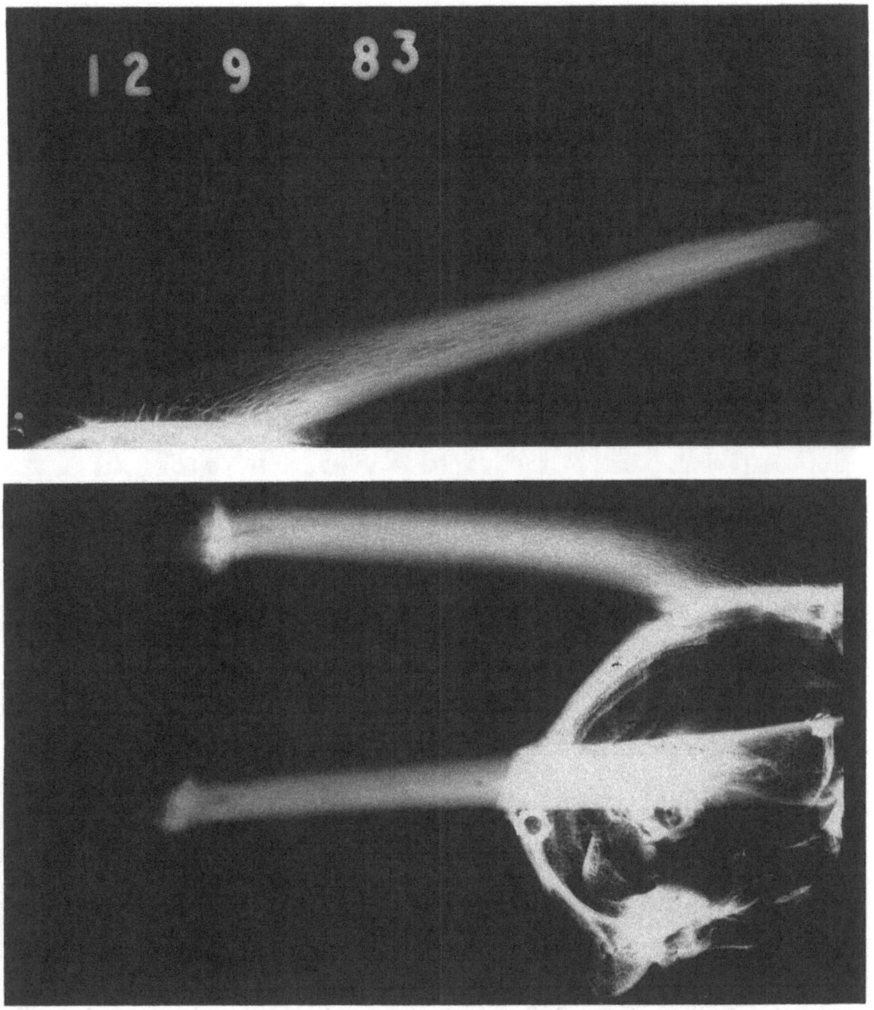

a

b

FIGURE 4. *a* Radiograph of the first antler of a subadult male *M. muntjak*. Note that the core of the pedicle proceeds up to the tip of the antler. *b* Radiograph of a subadult male *M. muntjak* with antlers which have a coronet and completely petrified antlers, a precondition for casting.

antlers are small and unbranched, without a burr; but they are not quite mature, and the antlers could be first-growth antlers. In the type skull (British Museum [Natural History]) and in skulls in the American Museum of Natural History and the Shanghai Natural History Museum, the antlers have clearly been shed, although they are very small. Finally, in another skull in the Shanghai Museum the antlers have not only a coronet but are quite large. Sheng & Lu (1980) do not specifically comment on this problem, but the antler measurements they give are

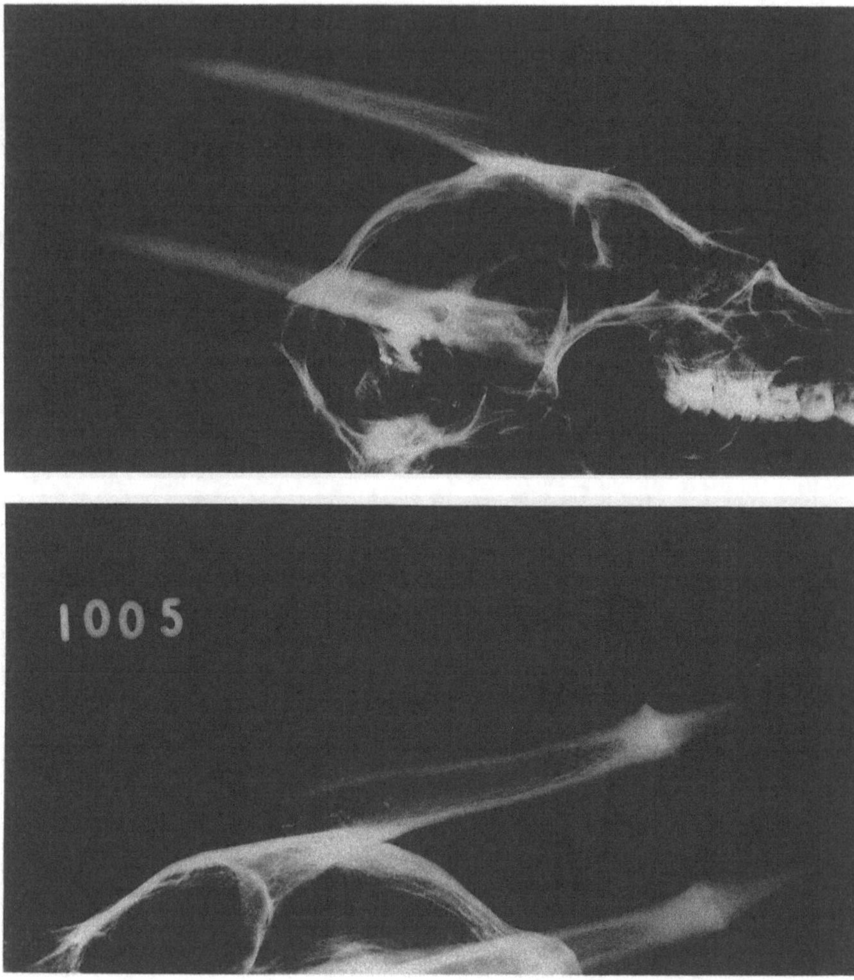

a

b

FIGURE 5. *a* Radiograph of a skull of a submature male *M. atherodes* mature but still in first antlers; the beginning of mineralization of the antler core. *b* Radiograph of a skull of a mature male *M. atherodes*. Subsequent antler with coronet and compact plug toward the pedicle.

extremely small, much less than those of the type of Andrews' specimen, raising the possibility that the antlers are not always shed even in adults; however, in a later paper (Sheng & Lu 1981), they do imply that shedding is the rule.

There is finally the matter of karyotypes to be discussed. That *M. muntjak* and *M. reevesi*, superficially so alike, are strongly different in their karyotypes, has been known for a long time. But the picture now appears more complicated than this.

(1) *M.muntjak*. Specimens of the species *M. muntjak* in the Catskill Game Farm, from stock collected on the northern slopes of the Garo Hills, Assam, have the lowest chromosome number known among mammals: 2n=6 in the female, 7 in the male. The X chromosome is fused to the centromeric end of the smallest acrocentric autosome, accounting for the odd number in the male. A similar chromosome number has been recorded several times in red muntjac from the northern part of mainland southeast Asia, whether they are identified as *M. m. vaginalis* (Wurster & Benirschke 1970; Shi et al. 1980; Neitzel 1982; Wurster-Hill et al. 1983) or as *M. rooseveltorum* (Wurster-Hill and Seidel 1985). The latter is an obvious misidentification of a specimen from Laos. Both the black-legged and the light-colored subspecies found in Yunnan have the same karyotype, as kindly communicated by Professor Shi (in letter to C.P.G.).

A female, identified as *M. m. muntjak*, collected at Tanda Baik, Pahang, West Malaysia, had a different karyotype: 2n=8 (Wurster & Aitkin 1972). The same autosome-X translocation was however present. White (1978) suggested that the fertility of hybrids between the two taxa would be severely reduced—a proposition which we will examine below.

(2) *M. reevesi*. Specimens of this species, of both mainland and Taiwanese origin, always have 2n=46 in both sexes (Neitzel 1982; Wurster-Hill et al. 1983). Hybrids between the northern form of *M. muntjak* and Chinese *M. reevesi* have intermediate chromosome numbers (26 in the female, 27 in the male), and enable the chromosome changes in the evolution from the higher number to the lower to be worked out (Shi et al., 1980).

(3) *M. feai*. A female of the species *M. feai* in Bangkok Zoo had a chromosome number of 13. The autosome-X translocation of *M. muntjak* occurred. The odd total number occurred because the animal was heterozygous for a Robertsonian translocation between chromosomes 2 and 7; hence, 2n=12 or 14 could also, theoretically, be expected in this species (Soma et al. 1983).

(4) *M. crinifrons*. Shi (1983) reports the karyotype of the species *M. crinifrons*: it is identical to the southern type of *M. muntjak*, i.e., 2n=8 in the female, 9 in the male, with the same autosome-X translocation. There was one so-far-unexplained anomaly: all males tested heterozygous for a unique pattern on chromosome 1, but none of the females had this pattern.

(5) *E. cephalophus*. In a female of the species *E. cephalophus*, from Yunnan, the chromosome number was 47 (Shi, 1981). The largest chromosome, a telocentric, was unpaired; Shi suggested this was the X, and that the specimen was aneuploid, or perhaps a gonosomic mosaic (where the somatic cells are XO even though the gonads are all XX).

It can be seen that there are plenty of unsolved problems in muntjac karyotypes. The species *M. atherodes* is still karyotypically unknown; we predict a high chromosome number for it.

The phylogeny of the Muntiacidae will be reconstructed in the last section. In the meantime, we are satisfied that many sources of evidence—craniology, external features, antler physiology, karyotype—all go to support the real existence of the six species here recognized; as well, of course, as the partial or wide sympatry exhibited between given pairs.

Genus Elaphodus Milne-Edwards *1871*

1871 *Elaphodus* Milne-Edwards, Arch.Mus.H.N.Paris, 7:93. *E. cephalophus* Milne-Edwards.

1874 *Lophotragus* Swinhoe, P.Z.S.1874:453. *L. michianus* Swinhoe.

Diagnosis: A genus of Muntiacidae in which antlers are tiny, stumplike, unbranched, irregularly or never shed, their pedicles relatively short (under 70 mm), very slender, extended onto face in the form of weak ridges which follow a curved course to upper orbital margins, then turn along margins and gradually fade out; preorbital fossa very large, well defined, occupying the whole of lacrimal bone and overlapping onto maxilla; preorbital fissure poorly developed or absent; ears broad and rounded; lateral hoofs very long, 10–12 mm; frontal tuft short but very thick; chromosomes 48(?), at least in female.

Species: The genus has been considered to contain but a single species, since the turn of the century. There are no indications that more than one species might exist.

Elaphodus cephalophus Milne-Edwards 1871 (Tufted Deer)

The question of subspeciation in this widespread, mainly mid-montane species (Sheng & Lu 1980) is a difficult one. Groves & Grubb (1986) were inclined to doubt whether the geographic variation in the species was other than clinal, but examination of further data indicates that *E. c. michianus* must be distinguished from the nominate subspecies. Two other subspecies are poorly known and their status remains provisional.

Elaphodus cephalophus cephalophus Milne-Edwards 1891

1871 *Elaphodus cephalophus* Milne-Edwards, Arch.Mus.H.N.Paris, 7:93. "Moupin" (Baoxing), Sichuan.

Diagnosis: A large subspecies, averaging larger than any other (Table 2).

Distribution: Specimens assigned to this subspecies come from Sichuan; Lichiang in Yunnan; and the Myitkyina district of northern Burma.

Notes: The range of variation in coloration encompasses the range of variation in the species as a whole, from uniform dark chocolate brown to streaked, light brown with more contrasting darker dorsal stripe and limbs. The frontal tuft may be all dark, or frosted with whitish hair bands demarcating marginal brown tracts (recalling the frontal stripes of some *Muntiacus* species). Similarly, the ear bases may be speckled with white, or uniform with the color of the cheeks.

The skulls of the two sexes are exactly the same size. The greatest length in males averages 197.1 mm (s.d. 5.91, n=11); in females, 197.6 (s.d. 4.83, n=10); other measurements are likewise nearly identical. This being the only large

TABLE 2. Range of skull measurements (in mm) in *Elaphodus cephalophus.*[a]

	Greatest length			Bizygomatic breadth			Length preorbital fossa		Premaxilla/ nasal contact (%)	
	mean	s.d.	n	mean	s.d.	n	range	n	%	n
Sichuan	196.6	6.00	21	84.6	3.98	22	29.3–35.5	5	37	27
Cheijang/N.Fujian	178.9	5.95	7	78.5	1.83	1	30.0–31.5	3	43	7

[a] Skulls from: Burma, Lichiang (Yunan), Puchi (Hubei), Ichang, and Fingling have parameters between those from Sichuan and from Cheijang/N.Fujian.

sample available to us we cannot guarantee that the sexes are identical in size in other subspecies, but there is at any rate no marked disparity and we have felt at liberty to combine the sexes in Table 2.

Elaphodus cephalophus michianus (Swinhoe 1874)

1874 *Lophotragus michianus* Swinhoe, Proc.Zool.Soc. London 1874:453. Ningpo, Chejiang.

Diagnosis: Of six skins seen, all were relatively dark especially in the mid dorsal region, but none was as dark as the darkest of the nominate subspecies. Much smaller than the latter (Table 2).

Distribution: Chejiang (Ningpo) and north Fujian (Chungan Hsien) Provinces.

Elaphodus cephalophus ichangensis Lydekker 1904

1904 *Elaphodus ichangensis* Lydekker, Proc.Zool.Soc. London 1904 (2):169. Ichang, Hubei.

Diagnosis: Resembles *michianus* and similarly small, but with very broad skull (zygomatic breadth 85mm against 78.5).

Distribution: Hubei Province (Ichange and Puchi).

Notes: Given the very small available sample sizes, the validity of this subspecies must be considered provisional.

Elaphodus cephalophus fociensis Lydekker 1904

1904 *Elaphodus michianus fociensis* Lydekker, Proc.Zool.Soc. London 1904 (4) 169. "Fingling," Fujian (actually, from the specimen label and Rickett's correspondence in the British Museum, "Kohwang near Ching Feng Ling, about 100 miles northwest of Foochow").

Diagnosis: Not separable from the previous two subspecies on pelage characters, but the single available complete skull is much larger.

Distribution: Fujian Province (the type locality, and Yenping).

Notes: Allen (1930) synonymised this form with the closely adjacent race *michianus*, but typologically it falls into the range of variation of the nominate subspecies. However, to unite them would be to recognize a polytopic subspecies and to imply a very elaborate pattern of semispeciation on very little evidence.

Genus Muntiacus Rafinesque 1815

1815 *Muntiacus* Rafinesque, Analyse de la Nature, 56. *Cervus muntjak* Zimmermann.
1816 *Cervulus* de Blainville, Bull.Soc.Philom. 74. *Cervus muntjak* Zimmermann.
1825 *Muntjaccus* Gray, Thomson's Ann.Philos., 26:432. nomen nudum.
1827 *Stylocerus* Hamilton Smith, Griffith's Cuvier Anim.Kingd., 5:319. (As a subgenus of Cervus). *Cervus muntjak* Zimmermann.
1836 *Prox* Ogilby, P.Z.S. 1836:135. *Prox moschatus* Ogilby.
1843 *Muntjacus* Gray, List.Spec.Mamm.B.M., 173. *Muntjacus vaginalis* Gray.
1923 *Procops* Pocok, P.Z.S. 1923:207. *Cervulus feae* Thomas & Doria.

Diagnosis: A genus of *Muntiacidae* in which antlers are always at least 20mm long, and may be cast at least sometimes; their pedicles long, more than 70mm (except in the dimunitive species *M. reevesi* and *M. atherodes*), more than 7mm thick, extended strongly onto face as thick "ribs" which do not deviate much, if at all, towards orbital rims but end halfway along rims and end in a thickening of orbital rims; preorbital fossa smaller and less well-defined than *Elaphodus*, not occupying whole of lacrimal bone and only minimally extending onto maxilla; preorbital fissure always well-developed; ears narrower and more pointed than *Elaphodus*; lateral hoofs shorter, but occasionally reaching 12 mm; frontal tuft present or absent, but when present less thick, mat-like than *Elaphodus*; chromosomes at most 46.

Species: Since the turn of the century, the four species *M. muntjak, reevesi, feae* (correctly *feai*) and *crinifrons* have invariably been recognized. To these Lydekker (1915) added *M. lachrymans* and *M. sinensis*, but Allen (1939) showed these to be no more than individual variants of *M. reevesi*. Osgood (1932) added *M. rooseveltorum* to the list, and this species has been recognized by all subsequent authors up to the present, if with some embarrassment caused by the stubborn refusal of any further specimens to turn up; a specimen identified as *rooseveltorum* was karyotyped by Wurster-Hill & Siedel (1985), but no description was given of the animal and there are grounds for doubting whether it was anything more than *M. muntjak*. We will show below that *M. rooseveltorum* is a synonym – or, at most, a subspecies – of *M. feai*; but a further species does await description in the China/Burma border region. Finally there is the intermittent recognition of a second species, sympatric with *M. muntjak*, on Borneo, generally referred to as *M. pleiharicus*, although Montulet (1984) allocated it to *M. rooseveltorum*. Groves & Grubb (1982) showed that this second species does indeed exist; that it occupies a special place in the genus; and that it cannot be

TABLE 3. Range of skull measurements (in mm) in *Muntiacus reevesi*.[a]

		Greatest length			Toothrow length			Bizygomatic breadth			Preorbital length		
		mean	s.d.	n	mean	s.d.	n	mean	s.d.	n	mean	s.d.	n
Gwangdong	(M)	170.5	3.0	4	49.3	2.75	4	74.3	2.63	4	87.5	3.11	4
	(F)	165.5	–	2	50.0	–	1	72.0	–	1	84.0	–	1
Taiwan	(M)	155.0	1.90	6	47.5	1.87	6	71.3	2.50	6	79.9	1.63	6
	(F)	150.2	6.34	6	47.2	2.32	6	64.7	3.15	7	73.7	4.76	6

[a]Male (M) and female (F) skulls from: Sichuan, Anhui/Cheijang, Ichang/Hubei, Hunan, Fujian, Yenping have parameters between those from Gwangdong and from Taiwan.

called *pleiharicus*, this latter name being referable to *M. muntjak*, but had to be given a new name, *M. atherodes*.

Notes: the difficulty (above) in firmly diagnosing this genus is due in part to the graded relations between the species. *M. atherodes* is marked by the retention of many traits which are symplesiomorhpic for the genus as a whole, and *M. muntjak* is highly autapomorphic, the other species being serially intermediate.

Muntiacus atherodes Groves & Grubb 1982 (Bornean Yellow Muntjac)

1982 *Muntiacus atherodes* Groves & Grubb, Zool.Meded.Leiden, 56:203.

Diagnosis: A species of *Muntiacus* of small size; antlers very short, 20–40mm long, rarely cast; pedicles short and slender, bent down below line of face; frontal tuft absent; predominately yellow, with agouti-banded hairs; dorsum of tail brown; a dark median dorsal stripe; frontal region blackish, extending onto inner halves of pedicles in males, and continuous with nuchal-dorsal stripe, not differentiated into separate dark stripes as in other species; occiput not contrastingly orange colored as in all other species. Male larger than female.

Distribution: the whole of Borneo. Payne et al. (1985) consider it a more lowland-living, coastal species than sympatric *M. muntjak*.

Notes: This species has been extensively described and figured by Groves & Grubb (1982), who survey the history of its nomenclature and taxonomic treatment, and list measurements. Ma et al. (1986) state that the type specimen is actually a subadult, and that this accounts for the characters ascribed to the species; but this is not the case.

Muntiacus reevesi (Ogilby 1839) (Chinese or Reeves' Muntjac) (Table 3)

Diagnosis: a species of *Muntiacus* of very small size; antlers fairly short, 40–80 mm long, often unbranched but irregularly cast; pedicles long and broad for the skull size; preorbital fossa very large, often abutting posteriorly against orbital margin for a substantial distance; preorbital fissure very long and narrow; almost never a premaxilla-nasal contact; frontal tuft absent; predominately reddish-yellow, but with varying amounts of grey overtones; head more reddish toned than body, and seeming large for body size. Male larger than female. 46 chromosomes.

Unique in the genus in having strong sexual dimorphism in color pattern. In the female, the frontal region is black, narrowing between the ears and continuing as a nuchal stripe, and the occipital region is orange. This pattern recalls that of *M. atherodes* (both sexes) where however the black markings are less well demarcated and the color of the occiput is not contrasting. The male on the other hand resembles both sexes of *M. muntjak* and *M. feai* in that the frontal region is bright orange, edged with black stripes (which continue onto the pedicles), and the orange color extends onto the occiput and ears, uninterrupted by a median stripe. The nuchal stripe commences at the top of the neck and has no connection with markings on the head.

Distribution: China, from about the Yangtze River south to about Guangdong, and including Taiwan.

Muntiacus reevesi reevesi (Ogilby 1839)

1839 *Cervus reevesi* Ogilby, P.Z.S. 1838:105. Canton.
1871 *Cervulus lachrymans* Milne-Edwards, Nouv.Arch.Mus.H.N.Paris, 7, Bull., 93. Moupin, Sichuan.
1872 *Cervulus sclateri* Swinhoe, P.Z.S. 1872:814. Ningpo.
1905 *Cervulus sinensis* Hilzheimer, Zool.Anz. 29:297. Hwai Mountains, Anhui province (selected by Lydekker, 1915:31).
1906 *Cervulus reevesi pingshiangicus* Hilzheimer, Abh.Mus.Naturk.Magdeburg, 1:169. Pingshiang, central China.
1910 *Cervulus bridgemani* Lydekker, P.Z.S. 1910:989. Hwai Mountains.
1914 *Muntiacus lachrymans teesdalei* Wroughton, in Lydekker, Cat.Ung.Mamm.B.M., 4:27. Tatung, Yangtze valley.

Diagnosis: Skull larger, but relatively narrow and small-toothed; general color relatively dull; tail chestnut on dorsal surface.

Distribution: Mainland range of species; introduced to southern England in the early 20th century, and now widely distributed there.

Notes: The taxonomy of Chinese muntjac as given by such authors as Hilzheimer (1985) and Lydekker (1915) was complex and elaborate. Hilzheimer, following Brooke, distinguished a highland species *M. lachrymans* from the coastal *M. reevesi* by its more parallel, less divergent pedicles, the yellow rather than red forehead, occiput and ear-backs, and the smaller preorbital fossae (smaller than the orbits, and not abutting on them, unlike *reevesi*). To this pair he added *M. sinensis*, in which the antlers were as in *reevesi* but the forehead-occiput-ear color as in *lachrymans*, the preorbital pit was equal in size to, or slightly larger than, the orbit, and the color was more yellow-brown than either (being blue-brown in *reevesi*, red-mixed in *lachrymans*), with more white on the underside. Lydekker (1915) added that a neck-stripe was present in *M. reevesi* and *M. sinensis* but not in *M. lachrymans*, and that there is a sexual difference in the color of the ears and forehead (these being darker in females), the difference being most marked in *M. sinensis*; but confusion concerning the nature and extent of the sexual dimorphism led him to identify three males from Ningpo as

M. lachrymans sclateri and four females from the same locality as *M. reevesi reevesi*! Subspecies were recognized by both Hilzheimer and Lydekker in *M. reevesi* and *M. lachrymans*, differing in color details.

The distribution of these three species was supposed to involve some sympatry. The type localities of both *M. lachrymans teesdalei* and *M. reevesi pingshiangicus* were in Anhui province, but from lowland and highland areas respectively: an unexpected reversal of the altitudinal difference between the nominotypical subspecies of the respective species. Such anomalies, not to mention the peculiar mosaic nature of the differentiating traits, might have been expected to sound a warning as to the reality of the taxonomic scheme.

It was Allen (1939) who, on the basis of the major variations in a large series from Fujian province, finally synonymized all these forms. On the evidence of the additional specimens in the British Museum, we agree with Allen. Color is variable, with greater or lesser amounts of blue tint (due to the length of the dark tips); and preorbital fossae which, though they do vary in size, are always much larger than in any other species of muntjac (except, on occasion, *M. feai*). Coastal skins do tend to be less reddish than others; those from southeast China tend to have more white on the chin and throat, and to have a better developed neck-stripe; the preorbital fossae do indeed tend to be larger than the orbits in Sichuan, slightly larger (or equal in size) in Hubei and southeastern specimens, and slightly smaller than the orbits on the northeast coast and in Anhui province.

Size varies in a mosaic fashion, not clinally. The largest are those from Guangdong, closely followed by Sichuan. Among the males, Ichang/Hubei specimens average smallest, followed by those from Hunan; but among females, Ichang specimens are rather large, those from Fujian province and Yenping being the smallest. Antlers are largest in Guangdong and, curiously, smallest in Sichuan, but the length of the pedicles does not reflect this. In no case do these differences, even between largest and smallest, even approach the conventional level of subspecific differentiation.

Ma et al. (1986) refer four skins, purchased in western Yunnan, to *Muntiacus rooseveltorum*. Examination of these specimens by one of us (C.P.G), in company with N. Chapman, revealed that they are actually referable to *M. reevesi*, and so perhaps constitute a range extension for the species.

Muntiacus reevesi micrurus (Sclater 1875)

1875 *Cervulus micrurus* Sclater, P.Z.S. 1875:421. Taiwan.

Diagnosis: Smaller in size, though only slightly smaller than some mainland populations; with relatively large teeth and broad skull for its size; averaging darker than mainland samples, with distinctly darker dorsal surface to tail.

Distribution: Taiwan.

Notes: For an insular form this subspecies is surprisingly poorly differentiated. Of the skull measurements, only preorbital length and biorbital breadth of females come close to 75% differentiation. The chin and throat are never as white as in some mainland skins; the neck-stripe is never more than vaguely developed and the ear-backs are never very dark, except at the tips; and the preorbital fossae may be somewhat larger or somewhat smaller than the orbits.

Muntiacus feai (Thomas & Doria 1889) (Fea's Muntjac) (Table 4, Fig. 1a,b)

1889 *Cervulus feae* Thomas & Doria, Ann.Mus.Genova, (2) 7:92. Thagata Juva, southeast of Mt. Muleiyit, Tenasserim. (On the correct spelling of this name see Grubb 1977) (Fig. 2a,b)

1932 *Muntiacus rooseveltorum* Osgood, Field Mus.N.H.Zool.Ser., 18:332. Muong Yo, Laos, 2300 ft.

Diagnosis: A species of *Muntiacus* of medium size, with short but fairly stout, generally branched, and regularly cast antlers; pedicles long but very slender, only 8–9 mm thick; preorbital fossa large, overlapping some way into maxilla; ethmoid fissure very big; nasals very flat, their posterior ends expanded; tail rather long with long white underside fur extending onto upper side as a long fringe; lateral hoofs long, 8–12 mm; body color dull olive-brown, agouti-banded; limbs black-brown, contrasting with body; a white ring round hoofs; dorsum of tail usually black; female with a long scruffy yellowish frontal tuft with black stripes; male similar in pattern but with no tuft. Female larger than male. Hair reversed on midline of neck. Chromosome number in female apparently 12, 13, or 14, with X chromosome translocated onto an autosome. Skull convex, broad, with maxillae bulging outward; rostrum short, deep; nasal process of premaxilla narrow, never reaching nasal bone.

Distribution: Known from the following localities: Dawna Range, Burma-Thailand border (Peacock 1933); east of Moulmein (Thomas & Doria 1889); Tenasserim boundary at 14.23N (Gairdner 1914); Hue Sut Yhaw, Burma, 12.40N (Gairdner 1915); Raheng, Pangna Province, northeast of Phuket island (Frädrich 1981); Muang district Surathani Province (9.08N, 99.14E) (Bangkok, National Zoological Institute); Muong Yo, Laos (Osgood 1932: type of *rooseveltorum*). Probably the species is sporadically distributed, wherever there are suitable habitats, from the Isthmus of Kra north and east through Burma and Laos into southernmost China.

Notes: There is some variation in this species, such that a number of specimens have not been recognized in the past. Of the British Museum material, the Raheng skin is brown, the limbs colored more or less like the body; the forehead is also like the body; the ear-backs yellow at the base, but darker over most of their surface; the chin and throat lighter (but not white). A skin (B.M. 15.12.1.24), labeled just "South-west Siam", is more grey-brown. Frädrich (1981) would not definitely associate the muntjac he saw in Bangkok zoo with this species, especially because it had large face-glands, differing from Pocock's (1923) description. There is no doubt, however, but that all these descriptions really do refer to the present species, which is somewhat variable in its external features (in fact, about as variable as *M. muntjak* subspp. or *M. reevesi*: no more, no less), while Pocock's characterization of it as lacking face-glands would seem simply to be in error.

TABLE 4. Range of individual skull measurements (in mm) in *Muntiacus feai*.

	Greatest length	Basal length	Mastoid breadth	Bizygomatic breadth	Pedicle length	Antler length	Biorbital breadth
Yunnan	198	173	58	87	74	51	84
Raheng (type)	201	173	62	91	103	50	84
Adung valley	196.5	174	55	83	–	–	82
rooseveltorum type	188	162.5	55	73	82	–	78

It is undoubtedly this unfamiliarity with the real nature of this species which led to the description of Osgood's *M. rooseveltorum*. Osgood had clearly never seen the type of *M. feai*, and did not even mention this taxon in his monograph. His description recalls the British Museum specimens of this species in every respect, and inspection of the type by one of us (P.G.) confirms this suspicion. We must say, however, that we have not observed such large mental glands in any other specimen, and can only assume that their enlargement is a periodic one connected, perhaps, with sexual activity although it is not fully mature. The skull (Table 4) is the smallest of this species that has been available to us; this, and its relative narrowness, might be due in part to its subadult status, but it is also possible that there might be subspecific differentiation within this species.

There are also specimens in the British Museum and in the Kunming Institute of Zoology, from northern Burma and southwestern China (western Yunnan, southeastern Tibet) which most closely resemble this species, but show certain characteristic differences: the color is brighter, more reddish; the neck hair is not reversed; the pelage is long and thick; the hoof-ring may be absent; the forehead and crown are reddish, not yellow, and the hair in this region is long in males, and strongly blackened in females; the skull is narrower, with a long narrow rostrum; the orbital rims are well-defined; there is a constriction between orbit and lower margin of preorbital fossa; the zygomatic arched in angled; and the nasal process of the premaxilla is broad, making a short contact with the nasal bone. Specimens are from Wadan (26.46N, 98.08E), Nam Tamai (27.42N, 97.54E), Adung (28.10N, 94.40E), all, British Museum specimens; Yunnan, west of the Salween (26N, 98.40E) and Gongshan (28N, 98.20E) (specimens in Kunming Institute of Zoology); southern Tibet (Zhang et al. 1984); and possibly Meiktila, Hlaingdet, Mansi-Katha, Thaungyin valley, and the Darjeeling district (U Tun Yin 1968); and Jingbin and Binbien, Yunnan (Sokolov 1957). These localities are, thus, at high altitudes: Wadan, 4,000 ft.; Nam Tamai, 3,000 ft.; Adung, 6,000 ft.

We had been inclined to include all these specimens in *M. feai* (perhaps as a northern, highland subspecies) until recently, following Ma et al. (1986), but the karyotype recently obtained from a specimen from Gongshan is uniquely different from all other muntjacs (Shi Liming, personal communication), and a new species is shortly to be described.

TABLE 5. Range of male skull measurements (in mm) in *Muntiacus crinifrons*.

	Greatest length	Basal length	Bizygomatic breadth	Pedicle length	Antler length
(n=23)	213 (205–224)	200 (193–212)	88 (80–93)	55 (40–70)	11 (1–52)

From Sheng & Lu (1980).

Muntiacus crinifrons (Sclater 1885) (Black, or Hairy-fronted, Muntjac) Table 5, Fig. 3a,b

1885 *Cervulus crinifrons* Sclater, P.Z.S. 1885:1. Ningpo

Diagnosis: A species of Muntiacus of large size, with short, often unbranched antlers, on long but slender pedicles; skull distinctive, with convex facial and calvarial profiles, and a distinct depression between, in anterior interorbital region; ethmoid fissure very small, narrow; antler pedicles may be depressed below line of facial profile; forehead very narrow between frontal ridges; nasals very convex from side to side, and somewhat compressed, their posterior ends narrow; nasal process of premaxilla always fails to reach nasal bone by a couple of millimeters, with a round superior end; preorbital fossa restricted to lower half of lacrimal bone; lateral hoofs rather long; tail strikingly long, with long white fringe restricting black dorsal hue to a narrow median area; color of body black, with some red banding of hairs; frontal tuft very long, thick, hiding antler pedicles, and bright red in color, as are ear-backs and occiput, this color fading into the reddish-grey of face and reddish-black of neck; underside only slightly lighter than upper side, only groin and inner aspect of thighs being white, like underside and fringe of tail. Female larger than male. Chromosomes 8 in female, 9 in male, with characteristic X-autosome translocation.

Distribution: Long known only from the environs of Ningpo, this species was shown by Sheng & Lu (1980) and Lu & Sheng (1984) to be quite common in western Zhejiang, on the border with Anhui, in southeastern Anhui, and extending to Pucheng county, Fujian, between about 28–31N, 117–120E. It lives in mountain forest at some 1000 m; in habitat that is steep, even precipitous, but warm, humid, monsoonal; the vegetation is deciduous, evergreen and bamboo in patches. At low altitude in the same region lives *M. reevesi*, with *Elaphodus cephalophus* at intermediate altitudes.

Notes: As a species which has been recently rediscovered after a silence, broken only three times, since its initial description a century ago, there have been no real problems with misidentifications and dubious attributions, unlike *M. feai*. Sheng & Lu (1980) describe its general range of variation: the coat is blacker, denser in winter, and sparser, more grey-brown in summer. A specimen in the Kunming Institute of Zoology has more reddish tones on the foreparts than seems to be implied by this description. From Table 5, it would seem that the highland population, described by Sheng & Lu, have shorter antlers and pedicles than the Ningpo specimen. There may thus be some geographic variation within the species.

Muntiacus muntjak (Zimmermann 1780) (Red Muntjac) (Tables 6, 7)

Diagnosis: A species of *Muntiacus* of medium to large size, with stout, well-developed antlers which are always branched and regularly cast, i.e., provided with a coronet after the first set; pedicles long and stout, extending down onto face as high, thick ridges; skull with relatively protruding orbits, fairly small preorbital fossa, large preorbital fissure, straight dorsal profile, broad forehead, transversely convex nasals (but not as strongly so as *M. crinifrons*), and premaxillae that almost always contact nasals; lateral hoofs very small; tail short, its dorsal color not greatly different from that of the body, and with only a short white fringe; frontal tuft short, but noticeable, colored as body, but with strongly marked black lines running up pedicles; general tone of body reddish, with no agouti-banding as such, but long black tips in dorsal region, especially on withers and nape. Sexes similar in size. Chromosomes 6 in female, 7 in male; or 8 in female, 9 in male; apparently varying geographically.

Distribution: Sri Lanka, India, Nepal, Bhutan, Bangladesh; from southern China (Yunnan, Guangxi) south through mainland southeast Asia to the islands of Sumatra, Borneo, Java, Bali, and Lombok. There is some doubt whether animals from all this range ought to be referred to one species, or whether two (a mainland, and a Malaysia/Indonesia species) might be involved.

Notes: Geographic variation in this species is very marked, with cases of fairly sharp transition between one distinctive form and another over narrow boundaries. What makes revision difficult is the tendency for characteristic phenotypes to crop up in unexpected places, so that polytopic subspecies must be recognized. The most remarkable example of this is *M. m. aureus*, which is widespread in northwestern India and recurs in central Burma; were simple size shifts or a single color pattern feature involved, we would be able to dismiss it as a case of independent evolution of the particular morphology, but as a number of features occur in the same combination in the two places, we are tempted to postulate a rather complex range reshuffle in the not-too-distant past.

We here recognize ten described subspecies as distinct.

Muntiacus muntjak muntjak (Zimmermann 1780)

1780 *Cervus muntjak* Zimmermann, Geogr.Geschichte, 2:131. Java.
1788 *Cervus muntjac* Gmelin, Linn.'s Syst.Nat.ed.13, 1:180.
1816 *Cervulus moschatus* de Blainville, Bull.Soc.Philom., 77. Bencoolen (Bengkulu), southern Sumatra.
1816 *Cervulus subcornutus* de Blainville, loc.cit. Java.
1824 *Muntjacus vaginalis* Horsfield, Zool.Res.Java, a and b. Java. (Not of Boddaert, 1785).
1896 *Cervulus pleiharicus* Kohlbrugge, Natuurk.Tijdschr.Ned.-Ind., 55:192. Pleihari, southeast Kalimantan. (See Groves & Grubb, 1982).
1907 *Muntiacus bancanus* Lyon, Proc.U.S.Nat.Mus., 31:582. Tanjung Bedaan, Bangka.
1910 *Cervulus muntjac typicus* Ward, Records of Big Game, ed. 6:80.

TABLE 6. Skull measurements in *Muntiacus muntjak*.

Males	Greatest length			Basal length			Toothrow length			Mastoid breadth			Bizygomatic breadth			Nasal length			Preorbital length			Antler pedicle length		
	mean	s.d.	n	mean	s.d.	n	mean	s.d.	n	mean	s.d.	n	mean	s.d.	n	mean	s.d.	n	mean	s.d.	n	mean	s.d.	n
M. m. malabaricus																								
Sri Lanka	178.1	6.34	9	158.3	4.71	8	53.0	3.84	9	57.8	2.20	8	77.8	1.60	6	44.8	3.04	9	91.0	4.07	8	102.2	14.89	7
S. W. India	175.7	1.53	3	153.8	3.75	3	55.7	1.15	3	58.3	3.21	3	76.7	3.51	3	43.2	1.89	3	89.0	1.00	3	95.0	–	2
M. m. aureus																								
N. W. India	188.8	6.53	5	161.5	2.38	4	55.3	3.09	7	59.7	1.86	6	83.7	1.86	6	51.5	3.51	6	96.8	3.63	5	110.3	9.07	3
Central Burma	185.5	–	2	159.0	–	1	57.0	3.61	3	59.3	4.04	3	80.0	–	2	49.5	5.74	4	96.0	–	2	118.3	8.47	4
M. m. vaginalis																								
Nepal	199.1	8.88	7	173.7	7.65	7	59.4	3.01	12	66.3	3.32	11	87.3	4.31	11	54.8	4.49	13	103.6	6.71	9	125.8	18.63	4
N. E. India	207.6	3.78	5	181.6	3.05	5	61.6	1.51	7	68.0	4.51	7	90.0	4.69	6	53.3	2.07	6	108.8	2.49	5	124.3	15.71	4
Arakan	204.0	–	2	179.0	–	2	59.3	–	2	68.5	–	2	89.0	–	2	53.0	–	2	106.8	–	2	124.0	–	1
M. m. yunnanensis × *curvostylis* intermediates																								
North Burma	215.7	5.60	9	191.0	4.69	6	61.9	3.38	9	69.0	1.58	9	92.8	2.17	9	58.7	3.82	9	115.2	4.49	9	116.6	13.71	7
M. m. yunnanensis																								
Central Yunnan	206.8	6.27	15	18.41	6.10	15	61.1	2.08	16	65.6	2.70	16	90.9	3.90	16	55.3	6.42	15	109.7	5.08	14	113.0	8.43	15
M. m. nigripes																								
Hainan	196.0	–	2	166.7	6.66	3	60.0	3.61	3	60.8	2.36	3	85.8	4.54	3	57.0	–	2	100.0	5.57	3	122.0	–	2
M. m. yunnanensis × *annamensis* intermediates																								
Chapa, N. Laos	208.8	–	2	185.5	–	2	65.3	2.08	3	64.7	1.53	3	88.7	0.58	3	52.0	–	2	111.0	4.00	3	119.0	–	1
Hoi Xuan	213.5	–	2	188.5	–	2	59.0	–	2	69.5	–	2	95.0	–	2	57.5	–	2	113.8	–	2	97.0	–	1

M. m. ammamensis																								
S. Vietnam	208.1	5.18	5	183.5	3.04	3	60.8	2.27	6	66.4	3.96	7	89.2	3.27	5	53.3	5.12	6	110.5	3.12	5	124.6	8.32	7
Thai/Kampuchea	196.3	4.93	3	174.0	–	2	55.3	–	2	61.7	2.08	3	85.7	4.04	3	53.7	1.53	3	105.8	–	2	127.5	16.22	3
M. m. curvostylis																								
Xishuanbanna	207.0	–	1	181.0	–	1	57.0	–	1	70.1	–	1	91.0	–	1	54.0	–	1	111.0	–	1	89.0	–	1
N. Thailand	212.8	3.40	3	187.0	–	2	58.9	3.94	5	66.9	4.48	5	91.3	3.50	4	55.4	2.41	5	114.0	1.35	4	117.9	15.89	4
M. m. curvostylis																								
Victoria Point	211.8	1.61	3	186.0	2.60	3	62.6	3.04	4	65.8	1.50	4	92.1	3.33	4	57.0	1.83	4	110.7	1.15	3	130.7	5.51	3
Trang/Perlis	236.0	–	1	207.0	–	1	62.1	1.31	4	68.5	3.00	4	97.9	2.25	4	61.3	6.45	3	128.0	–	1	117.3	7.64	3
Subspecies?																								
W. Thailand	196.3	4.62	3	175.7	7.51	3	58.8	4.27	4	65.2	2.68	5	89.7	1.53	3	56.7	4.04	3	105.7	8.02	3	118.1	16.99	4
M. m. muntjak?																								
Malaya	226.0		1	198.0	–	1	64.5	–	2	66.0	–	2	90.0	–	2	59.0	–	2	123.0	–	1	–	–	–
M. m. montanus																								
Aceh	175.0	–	1	153.0	–	1	54.0	–	1	55.0	–	1	78.5	–	4	39.5	–	1	86.5	–	–	82.0	–	1
Kerinci	193.0	6.56	7	–	–	–	55.3	2.04	7	–	–	–	85.7	3.46	6	–	–	–	–	–	–	–	–	–
M. m. muntjak																								
Borneo	192.8	5.76	5	168.3	5.95	5	55.3	3.84	9	61.2	4.60	7	85.9	5.75	7	54.5	3.28	5	99.7	4.15	5	104.0	16.23	26
Java	207.5	6.90	17	181.5	6.33	15	60.5	2.93	21	61.3	3.21	23	88.6	4.93	22	49.9	4.43	25	112.7	5.38	17	110.6	9.39	28
Bali/Lombok	188.4	4.40	7	165.4	4.76	7	55.4	2.14	7	56.0	2.24	7	80.0	4.65	6	45.6	2.91	8	99.5	2.88	6	85.2	8.50	12

(*Continued*)

TABLE 6. (*Continued*).

Females	Greatest length			Basal length			Toothrow length			Mastoid breadth			Bizygomatic breadth			Nasal length			Preorbital length			Antler pedicle length		
	mean	s.d.	n	mean	s.d.	n	mean	s.d.	n	mean	s.d.	n	mean	s.d.	n	mean	s.d.	n	mean	s.d.	n	mean	s.d.	n
M.m malabaricus																								
Sri Lanka	184		1	159		1	52		1	52.5		1	77		1	45		1	93		1			
Pune	177		1	153		1	52		1	52		1	73		1	37		1	89		1			
M.m aureus																								
Kumaun	194.5		2	170.5		2	52.7	3.06	3	58.7	4.93	3	82.5		2	54.0		2	102.5		2			
M.m vaginalis																								
Nepal/Sikkim	191.0	4.58	3	167.3	5.51	3	58.3	4.51	3	58.0	3.61	3	84.0		2	56.0	3.46	3	101.0	5.29	3			
N.E. India	200.0	6.22	4	176.3	4.19	4	61.2	1.30	5	61.8	0.50	4	86.0	2.58	4	56.0	3.81	5	107.4	4.16	5			
Arakan	202.5		2	178.5		2	58.8		2	64.0		2	87.0		2	54.5		2	105.0		2			
M. m. yunnanensis × *curvostylis*																								
N. Burma	216.7	6.75	7	188.6	6.04	6	63.5	3.42	8	61.8	1.83	8	89.6	2.53	8	60.1	2.61	7	115.3	4.23	7			
M. m. yunnanensis																								
C. Yunnan	204.2	4.25	3	179.0	3.74	4	60.5	2.68	4	58.9	3.01	4	88.3	7.54	4	52.8	2.75	3	109.0	2.58	4			
M. m. nigripes																								
Hainan	–		–	166		1	63		1	51		1	79		1	–		–	97		1			

M. m. yunnanensis × *annamensis*																					
Can Ho/Muong Yo	216.5		2	188.5		2	62.5		2	64.0		2	91.0		2	56.8		2	115.5		2
M. m. annamensis																					
S. Vietnam	196		1	172.5		2	56.0		2	58.3		2	83.5		1	51.5		2	106.5		2
M. mm. curvostylis																					
N. Thailand	208.3	5.19	4	181.0	6.08	3	61.4	3.59	4	59.3	1.53	3	85.0	3.46	3	58.5	5.80	4	113.0	4.76	4
Victoria Point	212.7	8.50	3	189.0	10.54	3	63.0	1.00	3	64.0	5.29	3	89.3	3.06	3	56.3	4.73	3	114.0	8.72	3
Subsp.nov.? cf.annamensis																					
W. Thailand	197.0	6.78	4	172.8	7.18	4	61.0	4.18	5	61.8	3.30	4	84.3	2.87	4	49.0	3.92	4	104.7	2.52	4
M. m. muntjak and cf.muntjak																					
P. Pangkor	211		1	189		1	63		1	61		1	87		1	52		1	114		1
Java	208.3	6.80	7	182.9	5.84	7	62.1	5.67	7	–			84.3	3.67	7	50.9	4.31	7	112.4	4.29	7
Belitung	187		1	164		1	51		1	–			82		1	49		1	100		1
Banka	188		2	166		2	55.8		2	51.8		2	82.5		2	56.5		2	103		2
P. Matasiri	195		1	174		1	55		1	59		1	87		1	53		1	103		1
Borneo	195.2	8.81	5	170.5	7.03	5	55.0	2.65	5	56.3	2.50	5	82.3	2.22	5	48.2	4.21	5	102.0	4.41	5
M. m. montanus																					
Sumatra	194.7	3.06	3				55.3	2.58	3				84.0	0	3						

TABLE 7. Skull measurements in *Muntiacus muntjak*.

Males and females	Greatest skull length			Basal length			Toothrow length			Nasal length			Preorbital length		
combined	mean	s.d.	n	mean	s.d.	n	mean	s.d.	n	mean	s.d.	n	mean	s.d.	n
M. m. malabricus															
Sri Lanka	178.7	6.26	10	158.3	4.40	9	52.9	3.63	10	44.9	2.87	10	91.2	3.87	9
S. W. India	176.0	1.41	4	153.6	3.09	4	52.8	2.87	4	41.6	3.45	4	89.0	0.82	4
M. m. aureus															
N. W. India	191.8	5.19	7	164.5	5.24	6	54.6	3.00	12	52.9	2.10	8	100.4	3.82	7
C. Burma	184.5	–	2	159.0	–	1	57.0	3.61	3	49.5	5.74	4	96.0	–	2
M. m. vaginalis															
Nepal/Sikkim	195.9	7.80	14	172.1	6.79	12	59.3	3.00	20	55.0	4.07	21	103.8	5.55	16
N. E. India	206.6	4.58	7	180.9	3.29	7	61.8	1.39	9	54.3	2.45	9	108.8	2.34	7
Arakan	203.3	3.10	4	178.8	3.77	4	59.0	2.48	4	55.9	3.57	4	105.9	2.84	4
M. m. yunnanensis × *curvostylis*															
N. Burma	216.2	5.93	16	189.0	6.86	12	62.7	3.39	17	59.3	3.32	16	115.3	4.23	16
M. m. yunnanensis															
C. Yunnan	206.4	5.96	18	180.9	5.69	19	61.5	2.75	20	54.9	5.98	18	109.6	4.58	18
M. m. nigripes															
Hainan	196.0	–	2	166.5	5.45	4	60.8	3.30	4	57.0	–	2	99.3	4.79	4
M. m. yunnanensis × *annamensis*															
N. Vietnam	212.9	4.96	6	187.5	3.15	6	62.7	3.13	7	55.4	5.22	6	113.1	3.47	7
M. m. annamensis															
S. Vietnam	204.3	7.33	8	179.3	5.74	6	60.5	3.95	11	54.2	5.23	11	109.1	3.10	8

M. m. curvostylis															
Xishuanbanna	207.0	–	1	181.0	–	1	57.0	–	1	54.0	–	1	111.0	–	1
N. Thailand	210.2	4.83	7	183.4	5.46	5	60.0	3.78	9	56.8	4.27	9	113.5	3.28	8
Victoria Point	212.3	5.49	6	187.5	7.06	6	62.8	2.23	7	56.7	3.04	7	112.3	5.85	6
M. m. yunnanensis? cannamensis															
Thai/Kampuchea	196.3	4.93	3	174.0	–	2	55.3	–	2	53.7	1.53	3	105.8	–	2
M. m. yunnanensis? cf.annamensis (cont.)															
W. Thailand	196.7	5.50	7	174.0	6.86	7	60.0	4.12	9	52.3	5.47	7	104.4	5.26	7
Subspp.?															
Trang/Perlis	236.0	–	1	207.0	–	1	62.1	1.31	4	61.3	6.45	3	128.0	–	1
Pahang	226.0	–	1	198.0	–	1	66.0	–	1	64.0	–	1	123.0	–	1
M. m. muntjak and *cf.muntjak*															
W. Malay coast	211.0	–	1	189.0	–	1	63.0	–	2	53.0	–	2	114.0	–	1
Deli	211.5	–	2	186.0	–	2	65.5	–	2	56.5	–	2	112.5	–	2
Bengkulu	207.0	–	2	180.0	–	2	61.6	–	2	56.3	–	2	112.0	–	2
Java	207.9	6.84	23	182.1	6.16	21	60.4	2.72	27	50.3	4.35	31	112.8	5.05	23
Borneo	193.2	7.15	12	168.3	5.95	5	55.3	3.84	9	54.5	3.28	5	99.7	4.15	5
Bangka/Belitung	187.7	2.08	3	165.3	1.53	3	54.3	4.25	3	50.0	2.65	3	98.0	2.12	3
P. Bintan	195.0	–	1	168.0	–	1	57.0	–	1	50.0	–	1	100.0	–	1
P. Matasiri	195.0	–	1	174.0	–	1	55.0	–	1	53.0	–	1	103.0	–	1
Bali/Lombok	188.4	4.40	7	165.4	4.76	7	55.4	2.14	7	45.6	2.91	9	99.5	2.88	6
M. m. montanus															
Kerinci	193.5	5.60	10	–	–	–	55.3	2.06	10	–	–	–	–	–	–
Aceh	175.0	–	1	153.0	–	1	54.0	–	1	39.5	–	1	86.5	–	1

1911 *Muntiacus rubidus* Lyon, Proc. U.S.Nat.Mus.,40:73. Pamukang Bay, southeast Kalimantan.
1915 *Muntiacus muntjak robinsoni* Lydekker, Cat.Ung.Mamm.B.M., 4:18. P.Bintang, Riau archipelago.
1915 *Muntiacus muntjak peninsulae* Lydekker, loc.cit. P.Pangkor, W. Malaysia.
1932 *Muntiacus muntjak nainggolani* Sody, Natuurk. Tijdschr. Ned.-Ind., 92:237. Sendang, West Bali.

Diagnosis: Deep rufous in color, the nape dusky (markedly different from the general body tone); forehead light cinnamon-rufous, well-set-off from the rest of the (dusky-toned) face, except for the reddish cheeks; ear-backs usually dark, except at the bases; thighs and shoulders dark chestnut-brown, contrasting somewhat with body color, becoming darker and greyer down the shanks; midregion of back dark chestnut, quite sharply distinct from rufous flanks, almost forming a broad dorsal stripe. Throat creamy-white, grading to light cinnamon-rufous chest and belly; pubic region creamy white, continuing down front of thighs as a creamy white stripe. Antler long, 80–200 mm. Skull relatively narrow, especially across mastoids; nasals relatively short in male (but those of female longer).

Distribution: Broadly speaking, Malaysian and Indonesian parts of species' range. In West Malaysia, the following localities are known for this subspecies: Biserat; Tanjong Antu; P. Pangkor; probably, Kuala Tahan, Pahang; Bukit Tangga, Ulu Gombak, and Air Kung, Negri Sembilan; Taiping, Dindings, and Ulu Temergoh, Perak. Specimens from Perlis and Trang are also included here, with some hesitation (see below).

Notes: A number of different subspecies have been described within this region, and are surveyed by van Bemmel (1952), who finds that they are not in fact very distinctive. Skins from all insular segments of the range can be matched up very well; while there are in some cases average differences—especially, Javan skins are in the main the reddest, those from Borneo being on average darker—the overlaps are such that no clear boundaries can be drawn. The only other geographically varying feature is size: Javan specimens are among the largest of the species, those from Borneo much smaller, those from Bali and Lombok smaller still. The standard deviations are, however, large enough to render these differences below the conventional level of subspecific difference, the Bornean sample intervening between the ranges of the Japanese and the Bali/Lombok samples which would otherwise be well differentiated.

The only exceptions to this uniformity are on Sumatra and in Malaysia. Muntjac from the central and northern Sumatran highlands are clearly differentiated from other Indonesian ones, but typical examples of *M. m. muntjak* do occur in the lowlands (Deli: though we have not seen the specimen from Tebingtinggi referred by van Bemmel to an intermediate between this race and *montanus*, a skull from Deli fits well into the present race; Siak River; Siolak Darus; Bengkulu; also Palembang according to van Bemmel (1952).)

In general, it is the contrast between the sharply darker midback and the red flanks which forms the best distinguishing character of this race. The black pedicle stripes are also thicker and more intensely black than in any of the noninsular forms. Unfortunately, the bimastoid narrowness and the short nasals are not absolutely distinctive, though most specimens could probably be separated out by these characters; the nasal character is also less marked in the females than in the males.

Specimens from Perlis and Trang differ from other examples of this subspecies in their much larger average size, although a skull from Pahang is also very large. They do not however show any real approach to *M. m. curvostylis* as would be expected from their geographical position; it is possible that the species should in fact be divided into two at this boundary.

Muntiacus muntjak montanus Robinson & Kloss 1918

1918 *Muntiacus muntjak montanus* Robinson & Kloss, J.Fed.Mal.States Mus., 8, 2:69. S. Kering, G. Kerinci, 7300 ft.

Diagnosis: Much smaller than most samples of nominotypical *M. m. muntjak*, and much darker in color; overall, dark chestnut, speckled with blackish; midback darkened; forehead, occiput and ear-bases, also pedicles, orange-brown; rest of face brown-grey; ear-backs dark; thighs and shoulders brown, becoming dark brown on shanks; throat olive-buff, rest of underside, and lower flanks, reddish-white, changing to whitish on groin, but front of thighs (to hocks) is only ochery, not white; feet white, this tone going up front of lower half of limbs; tail blackish brown above; pedicle stripes thick, black; skull narrow, nasals short in male; antlers short, under 100 mm, often unbranched; fully mature males make lack coronets.

Distribution: Sumatran highlands, from Kerinci north into Aceh; northerly specimens may be smaller than southerly.

Notes: van Bemmel (1952) discusses, and rejects, the idea that this may be a full species, noting the existence of an apparently intermediate specimen from Tebingtinggi. We note, in addition, the occasional tendency of skins from Borneo, as well as the Pahang skin mentioned above, to vary in the direction of this race. It also shares the skull narrowness, short nasals of the male, and strongly developed pedicle stripes, with *M. m. muntjak*, and may therefore be a local derivative of the latter.

Muntiacus muntjak curvostylis (Gray 1872)

1872 *Cervulus curvostylis* Gray, Cat.Rum.Mann.B.M. 94. Pachebon, Thailand.

1904 *Cervulus muntjac grandicornis* Lydekker, Field, 104:780. Thouagyen forest, Amherst district, Burma.

Diagnosis: Body color light reddish-yellow, midback tending to be darker (but the two tones grading, not sharply distinct); nape greyer; forehead, occiput and ear-backs as body, rest of face paler; limbs as body, or shanks greyer; throat buffy-white, becoming buffy on rest of venter; groin and thigh-stripe strikingly white, this color extending beyond hocks onto inner side of hind shank. Size large; antlers large, 110–150 mm.

Distribution: From Victoria Point, Burma, and Koh Lak, peninsular Thailand and Burma to the Mae Wong valley, northern Thailand (Hinlaem, Kanburi), and central Burma (Amherst district; lower Chindwin). All these localities appear to be in rainforest at low altitudes.

Notes: As in all subsequent (i.e., mainland) subspecies, this form has broad mastoid width and long nasal bones in both sexes (the next form, however, has rather short nasals). Two southerly skins – one from Koh Lak, one from Banlaw, Great Tenasserim River (both in the Singapore collection) – have a rather deeper rufous-toned midback than most, and the ear-backs tend to be grey-infused; this is the only possible indication of intermediacy between this subspecies and nominotypical *muntjak*, which otherwise might be ranked as specifically distinct.

Muntiacus muntjak menglalis (Wang & Groves 1988)

1988 *Muntiacus muntjak menglalis* Wang & Groves, in Ma et al., Acta Theriol, Sin. 8:96. Pujiao, Mengla County, Xishuanbana, southern Yunnan.

Diagnosis: Close to previous subspecies, but with remarkably short pedicles; color of upper parts similar, though limbs always colored as body, but underparts paler; throat and interramal region white. Nasal bones comparatively short and broad. Canines short.

Distribution: Xishuanbana district of Yunnan, and presumably neighboring areas of Laos, Vietnam and Burma.

Notes: Only a brief characterization of this newly described subspecies is given here. It is excluded from the table of measurements because only a single skull has been examined by C.P.G.

Muntiacus muntjak annamensis Kloss 1928

1928 *Muntiacus muntjak annamensis* Kloss, Ann.Mag.N.H.(10) 1:399. Langbian.

Diagnosis: Color similar to *M. m. curvostylis*, or paler but nape less grey as a rule; forehead, occiput and pedicles light brown, slightly redder than body; rest of face light orange-grey; ear-backs becoming dark (grey) beyond base; shanks as body, or greyer in midline; underside paler than upperside, tending towards white on belly and white on interramal region; no white line on hindlegs or a very short one hardly reaching hock, but usually, only pasterns having white marks; tail very red on upper side. Slightly smaller than previous race, but varying in size. Antler pedicles very long.

Distribution: Southern half of Vietnam; southern Laos; as far southwest as the Thai/Cambodia border region.

Notes: This race differs from the last by slight but consistent color details. There is, again, strong size variation, those from the Thai-Cambodia border region being rather small, but with the same relatively long pedicles as the much larger Laos and Vietnam samples.

Specimens from certain localities in southwestern Thailand (Hat Sanuk; "Southwest Siam," about 13.45N; Sai Yoke; Maa Wang; Pak Jong) and neighboring parts of Burma (Thaget) could belong to this race but are more like *curvostylis* in color, tending towards *M. m. aureus*—although there appears to be no geographic connection to the latter! They are similar in size to the Thai-Cambodia borderland specimens of the present subspecies, although quite without the long pedicles. Such a picture is a part of the almost inexplicable mosaic of characters within the present species.

Muntiacus muntjak nigripes G.M. Allen 1930

1930 *Muntiacus muntjak nigripes* G.M.Allen, Amer. Mus. Novit. 430:11. Nodoa, Hainan.

Diagnosis: Body color bright orange, with no darkening on midback; forehead and occiput deeper in color than any other races; rest of face slightly greyer; ear-backs nearly black, except at bases; shanks darkened, usually blackish-grey, sometimes only medium grey, this color extending in most cases as far as shoulders and thighs; throat whitish, rest of underside merely a paler version of upper side; groin light grey, not white; white on hindlegs confined to pasterns. Antlers short, 70–110 mm; size rather small, but very long pedicles.

Distribution: Hainan island.

Notes: Osgood (1932) and others have attributed mainland black-legged muntjac to this subspecies, but this is in error (see next subspecies).

Muntiacus muntjak yunnanensis Ma & Wang 1988

1988 *Muntiacus muntjak yunnanensis* Ma & Wang, in Ma et al., Acta Theriol.Sin.8:101. Wokang Dashan, Meglai, Cangyuan County, West Yunnan, 2200 m.

Diagnosis: Deep red-brown in color; forehead and occiput dark brown, not quite as dark as previous form; whole of limbs as far as shoulders from dark chestnut to blackish brown; chin and throat, axillae and groin region pure white, with a white line running down inside of hindlimbs to beyond hock; a whitish patch on pasterns. Antler pedicles extremely long; nasals relatively short.

Distribution: Yunnan, north of about 23.10N.

Notes: This is essentially the mainland version of the previous subspecies, from which it differs by its much larger size, shorter nasals, pure white throat and groin regions, and somewhat lesser contrast but greater extension of the dark color of the limbs. Skins intermediate between this new form and *M. m. annamensis* are from northernmost Vietnam (Chapa; Hoi Xuan) and Phong Saly in northern Laos, and between it and *M. m. curvostylis* are from northern Burma (Nanyaseik; Hpawshi; Myitkyina; Sumprabum; Tanghku; Hkamti) and northern Thailand (Me Pooan); but the boundary between it and *M. m. menglalis* appears to be very sharp in Yunnan where it corresponds to the boundary between the deciduous and evergreen forest.

Muntiacus muntjak vaginalis (Boddaert 1785)

1785 *Cervus vaginalis* Boddaert, Elenchus Anim., 1:136. Bengal.
1827 *Cervus moschatus* H. Smith, Griffith's Cuvier, 4:147. (Not of de Blainville, 1816). Nepal.
1833 *Cervus ratwa* Hodgson, Asiat.Researches, 18, 2:139. Nepal.
1839 *Cervus melas* Ogilby, Royle's Illust.Bot.Himalaya, 73. Himalayas.
1844 *Cervus stylocerus* Schinz, Synop. Mamm., 2:549. Renaming of *melas*.
1846 *Prox ratva* Sundevall, K. Svenska Vet.-Ak.Handl., 1844:85 1846 *Prox albipes* Sundevall, loc.cit.
1846 *Stylocerus muntjac* Cantor, J.Asiat.Soc.Bengal, 15:269.
1852 *Stylocerus muntjacus* Kelaart, Prodr.Faun.Zeylan., 85. Renaming of *vaginalis*.

Diagnosis: Body dark reddish, darker on midback than flanks; nape slightly greyer; forehead and occiput light orange-brown, rest of face greyish; earbacks reddish at base, remaining two-thirds dark grey; limbs dark brown to grey; underside paler; groin and line on front of hindlegs to hocks white. Antlers short, 80–120 mm; size medium; pedicles long.

Distribution: Nepal, eastern half of India south to Shevaroy Hills; Bhutan; western Burma (Arakan), south to Mt Victoria and Nkamat, Matanga River.

Muntiacus muntjak aureus (H. Smith 1826)

1826 *Cervus aureus* H. Smith, Griffith's Cuvier, Anim.Kingd. 4:148. "Perhaps Malacca"; but "some part of southern India" (Lydekker, 1915).
1844 *Cervus albipes* Wagner, Schreber's Saügethiere, Suppl.4:394. "Bombay and Poona".
1872 *Cervus tamulicus* Gray, Cat.Rum.Mamm. B.M.94. "Daccan".

Diagnosis: Body pale yellowish; nape greyer; forehead and occiput pale orange-brown; rest of face light grey orange; ear-backs orange at base, becoming grey, and tips and rims may be dark grey; limbs colored more or less same as body; underside somewhat paler than upper side; a line down front of thighs, to hocks, white; antlers short, 70–100 mm; size small.

Distribution: (1) Northwestern India (Kumaun and Kheri, on the Nepal border) as far southeast as the Deccan; (2) central Burma (Mongwa, Pyawbwe and Yin, all on the lower Chindwin).

Notes: The peculiar polytopic distribution of this form has already been mentioned. Involving, as it does, several apparently unlinked aspects of coloration and size, it is difficult to see the distribution as due to a random reassortment of characters in unrelated populations.

Muntiacus muntjak malabaricus Lydekker 1915

1915 *Muntiacus muntjak malabaricus* Lydekker, Cat.Ung.Mamm.B.M.,4:24. Nagarhole.

Diagnosis: The smallest mainland subspecies; general color washed-out reddish, with much greying on nape and back; limbs as body; underside drab; white area on lower limbs prominent and extending round to front of pasterns, limiting reddish coloration to a narrow band down the limb; antlers short, 60–100 mm, with short pedicles.

Distribution: Sri Lanka; and southwestern India (Nilgiri Hills; Sampaje, Coorg; Pune).

Notes: The nomenclature of this form is not quite clear. The type locality of *albipes* Wagner is stated to be "Bombay and Poona," but it is unclear how authoritative this localization is. The present race evidently does extend as far north as Pune (Poona); moreover the name *albipes* (white-footed) does suggest the present race rather than the last. But the previous race, *aureus*, extends to the Deccan (type of *tamulicus*, collected by Sykes, hence likely to be from the Hyderabad district). Under the present circumstances, we continue to use the name *malabaricus* for this race, while realizing that the name may in fact be *albipes* should examination of the type of the latter prove it identical to the southwestern form.

Phylogeny

With the aid of the chromosome data, in particular, the outlines of muntjac phylogeny can be unravelled fairly clearly. The karyotype of *M. muntjak* is so unusual that it cannot be anything but highly derived.

The most primitive surviving muntiacid is *Elaphodus cephalophus*. The karyotype demands this, and confirms that the development of antlers in the Muntiacidae is independent of their development in the true Cervidae. Other primitive characters possessed by *Elaphodus* include the rudimentary antlers, never cast, the short slender pedicles, the agouti-banding of the hairs, and the equality of size of the two sexes. The broad rounded shape of the ears and the large size of the preorbital fossa (indicating a large face gland) are characters shared with some of the species of *Muntiacus*, and so are most parsimoniously considered as primitive. The presence of elongated frontal hairs is common to all Muntiacidae except for two *Muntiacus* species, so could be regarded as either a primitive condition which has been lost in these two or a potentiality which has been brought out in some taxa but not in others. In that the form of this frontal elongation, whether truly tuft-like or otherwise, differs in different taxa, and as the two species which lack a tuft are the two most primitive in *Muntiacus*, we prefer the second alternative.

Muntiacus species agree in possessing certain characters more derived than those of *Elaphodus*: longer pedicles and developed, if spike-like, antlers; smaller preorbital fossae; chromosome number reduced (if only slightly). The greater development of the preorbital fissure would also be a synapomorphy of the genus.

The most primitive species of *Muntiacus* is *M. atherodes* (Fig. 5a,b): tiny antlers, which are as a rule not cast, and are never branched. Other features support its primitive status, but are not exclusive to it: agouti-banding on the hair, somewhat rounded ears, unspecialized skull features. Unfortunately its karyotype is unknown; if we are correct about its phylogenetic status, its chromosome number should be high.

Next most primitive is *M. reevesi*. It shares all the primitive characters of *M. atherodes* except that its antlers seem to be more frequently cast and, if large enough, are generally (though not invariably) branched. In one respect, its large preorbital fossae, it is more primitive than *M. atherodes*: impressed with antler form, we suggest that face gland reduction is one of the few derived traits of *M. atherodes* (derived in parallel to more specialized species: an inference supported in that *M. feai* also retains a large one). Two characters shared by *M. reevesi* and *M. atherodes* would be considered derived by comparison to *Elaphodus*: sexual dimorphism in size (male larger than female), and lack of frontal tuft. As the evidence indicates (above) that these two species, though both primitive, are seriated and do not share a common stem, either they are convergent in these two respects or, more likely (as already indicated with respect to the frontal tuft character), these two traits are primitive for *Muntiacus* and hence reversed from the *Elaphodus* condition.

The other three species of *Muntiacus* share the derived traits of karyotypes with extensive Robertsonian changes (including the X translocated onto one of the autosomes); elongated antler pedicles; frontal tufts developed, of a type different from that of *Elaphodus* (concentrated on the pedicles); male not larger than female. The most divergent of the three, sister-group to the other two, is *M. feai*: not only by virtue of the less reduced chromosome number, but because of its retention of rather broad, rounded ears, its unprotruded orbits (despite its large size), its large preorbital fossa, and lack of the convexity of the nasals and the muzzle in general.

M. muntjak and *M. crinifrons* are together the most derived taxa in the family. The large frontal tuft and rostral convexity are more advanced in *M. crinifrons*; the antlers and color pattern, in *M. muntjak*. (There is in fact, as has been noted above, some suggestion that the antlers have been actually reduced in one population of *M. crinifrons*). Within the latter species the mainland group of subspecies have undergone a further karyotypic evolution, the insular race(s) retaining the less derived form seen in *M. crinifrons*.

Ma et al. (1986) have recently produced a further assessment of polarity in *Muntiacus*. To antler and pedicle characters they add longer, wider nasals, separation of premaxilla from nasal, and lacrimal pit apomorphic characters. We are unsure of these features, although at least the nasal character's proposed polarity seems plausible.

Conclusion

When examined geographically, the species and subspecies of *Muntiacus* show the "centrifugal" pattern first described by Brown (1957) and emphasized as a common pattern by Groves (1989). The two most primitive species are found at the edges of the range of the genus (China; Borneo), though sympatric, at least in part, with other species. The three more derived species may have differentiated as a habitat-segregated pair (*M. feai* in the evergreen forest, *M. muntjak/crinifrons* in monsoon forest) in the southeast Asian mainland region, from which the monsoon forest species dispersed to the north, south and west as this forest type spread, adapting in situ to rainforest conditions when the environment changed (in southwestern India/Sri Lanka, and southern Southeast Asia). *M. crinifrons*, more isolated than other populations of the most-derived pair, evolved to species status, while the remainder of the populations, in at least intermittent contact, remained conspecific (*M. muntjak*). Finally the centrally distributed population of the derived pair (mainland *M. muntjak*) acquired the still more highly derived karyotype. The essence of the centrifugal model is that evolutionary innovations arise in the middle of a species' range and, if successful, spread out and swallow up more primitive phenotypes (if conspecific) or overlie or replace them (if reproductively isolated) as far as environmental variables permit. The genus *Muntiacus* answers to this model extremely well.

It is also interesting that in the present case the phyletic series forms a zigzag pattern: *Elaphodus* in China, then south to *M. atherodes*, north to *M. reevesi*, south to *M. feai*, north to *M. crinifrons*, south to *M. m. muntjak*, north to mainland *M. muntjak*. This is imposed on the centrifugal pattern and possibly represents an illustration of the limits of sympatry: each stage in the series replaces the one before, up to the limits of the dispersal ability, but can coexist with a species that is two stages up or down the series (that is to say, different enough that the two would not compete).

Acknowledgments. We would like to thank the following for providing access to specimens in their charge: John Edwards Hill (London); Sydney Anderson and Karl Koopman (New York); Robert Izor (Chicago); Richard Estes (Philadelphia); Henry Setzer (Washington); Renate Angermann (Berlin); Ralf Angst (Karlsruhe); Fritz Dieterlin (Stuttgart); H. Schliemann (Hamburg); Peter Lups (Bern); X. Misonne (Brussels); Chris Smeenk (Leiden); Louis de Roguin (Geneva); P.K. Das (Calcutta); Niphan Ratanaworaphan (Bangkok); Wang Yingxiang (Kunming); Zhang Song Ling, (Shanghai); Yang Chang Man, (Singapore). E. Tortonese,

K. Koopman and R. Izor very kindly supplied photographs; Kim Dennis-Bryan kindly arranged the X-rays. One of both of us would like to record the pleasure we have taken in the fruitful discussions we have had with Wang Yingxiang, Shi Liming, and Tony Bubenik; and to thank Doris Wurster-Hill for useful discussion and information by correspondence.

4
Cervidae of Southeast Asia

PETER GRUBB

Introduction

Old World cervoids predominate in Southern and Eastern Asia: Of the 33 species (Groves & Grubb 1987), 13 are tropical Asiatic, 13 are Sino-Himalayan (as defined by Vaurie 1972) or are restricted to the flood plains of China's major rivers, and only seven are Palearctic in overall distribution. Of the "typical" plesiometacarpal deer (*Cervus* and allies), seven Southeast Asian genera or subgenera can be differentiated. Each is probably a natural clade (Groves & Grubb 1987). Indian and Indochinese species belong to flood-plain and deciduous woodland ungulate faunas, closely analogous to those of Africa but less diverse. These deer cannot be differentiated as a browsing guild, for some are as graminivorous as many of their African analogues, *Rucervus* species being particularly stenophagous (Kurt 1978; Martin 1977).

Species Groups of Asiatic Deer

The hog deer (*Hyelaphus* species) range from northern India to Yunnan, with some insular populations. They are small deer of tall grassland, ecological equivalents of the African reedbuck *Redunca redunca*. Their antlers are very like those of *Rusa* species but are smoother and thinner (Fig. 1b).

In contrast, the chital, *Axis axis*, of peninsular India has specialized antlers for a six-pointer (Fig. 1a). They are very long but with relatively short, forward-curved brow tines, elongated front-outer tines of the terminal fork (A2 in Pocock's 1933 terminology), and a tendency to form snags between brow tine and beam. Ecologically, the chital is the analogue of the African impala, *Aepyceros melampus*, and like it may feed communally with monkeys on fallen fruit (Kurt 1978).

The species of rusa and sambar (*Rusa*) are naturally distributed from Sri Lanka and India to Yunnan, the Philippines, and Java. Except for the most specialized species, *R. timorensis*, they are deer of relatively dense cover, paralleling the African antelopes of the genus *Tragelaphus*, but are nevertheless exceptionally

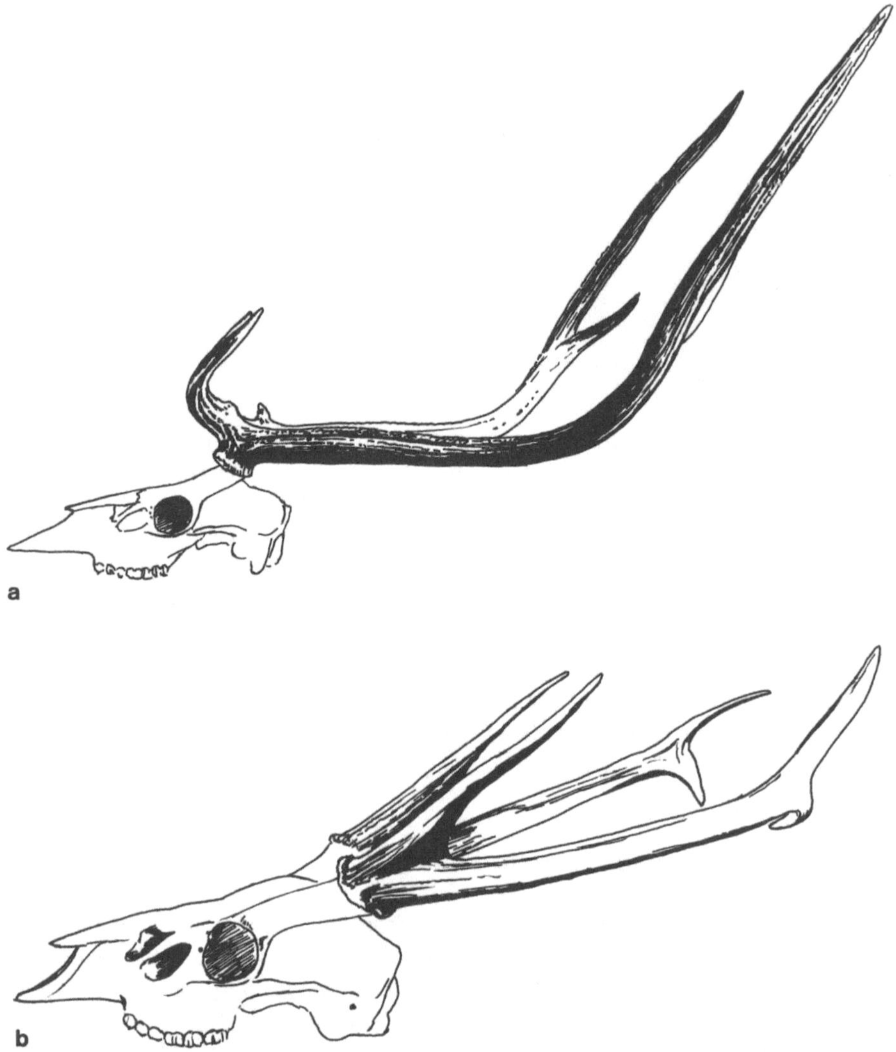

FIGURE 1. Skull and antlers of chital, *Axis axis* (*a*) and of hog deer, *Hyelaphus porcinus* (*b*).

widely dispersed ecologically (Engelmann 1938; Green 1985; Hoogerwerf 1970; Schaller 1967). Size range is from the diminutive *R. mariannus nigellus*, smaller than a hog deer, to the sambar, *R. unicolor unicolor*, largest Southeast Asian deer bar the shou, *Cervus elaphus affinis*. The small Philippine species have relatively short antlers, with a large brow tine and the second anterior tine (A2, Pocock) dominating the terminal fork. The Malay and Indonesian subspecies of *R. unicolor*, though larger, have similar antlers (Fig. 2b), but the continental races have proportionally longer ones (Fig. 3a). The Javan *R. timorensis* has relatively long antlers too, but of a different form: The back-inner tine of the terminal fork (P2, Pocock) is very much lengthened (Fig. 3b).

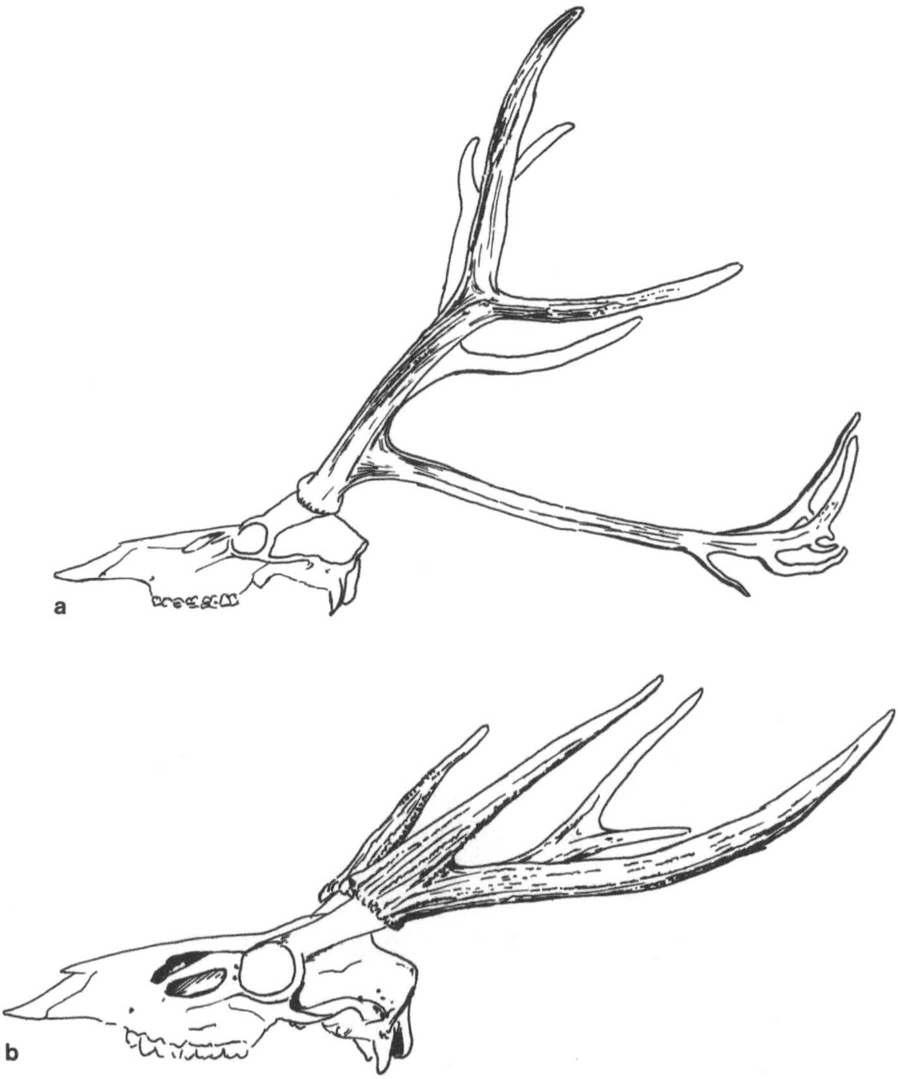

FIGURE 2. Skull and antlers of Père David's deer, *Elaphurus davidianus* (*a*) and of Malay rusa, *Rusa unicolor equinus* (*b*).

Rucervus species are deer of open woodland and the tall grass of flood plains, associating in large herds. They are specialized graminivores with uniquely folded cheek teeth, ecological parallels of the somewhat smaller African lechwe, *Kobus leche*. The rucervines have a more restricted distribution than the rusas, from Northern India to Guangxi. The second anterior antler branch (A2, Pocock) in these species is as thick as, or thicker than, the branch behind it (P2, Pocock), retaining a condition seen in the simpler antlers of hog deer or most rusas. Rucervine antlers have almost certainly evolved from such simple patterns by

FIGURE 3. Skull and antlers of sambar, *Rusa unicolor unicolor* (*a*); antlers of Javan rusa, *R. timorensis* (*b*); skull and antlers of Schomburgk's deer, *Rucervus schomburgki* (*c*).

lengthening of the beam, forward-bending of the brow tine, and elaboration of the anterior tine (A2, Pocock), in the thamin *R. eldi* (Fig. 4b), or of the posterior one (P2, Pocock) as well in the other species, the barasingha *R. duvauceli* (Fig. 4a), and Schomburgk's deer *R. schomburgki* (Fig. 3c).

FIGURE 4. Skull and antlers of barasingha, *Rucervus duvauceli branderi* (*a*) and of thamin, *R. eldi thamin* (*b*), with front outer tine of terminal fork branched, and back inner tine almost out of view.

Cervus includes two living species. The sika, *C. nippon*, confined to the Far East, has antlers that have progressed a stage beyond those of *Rusa timorensis*, with antler branch P2 (Pocock) forked so as to produce a total of four times (Heude 1884). The second species, *C. elaphus*, only enters Southeast Asia along the fringes of the main Asiatic mountain chains, where generalized subspecies such as the shou, *C. e. affinis* (Fig. 5b), grade into the more specialized wapiti-like races. Primitively the species has the same number of tines as the sika but

FIGURE 5. Skull and antlers of white-lipped deer, *Przewalskium albirostris* (*a*) and of shou, *Cervus elaphus affinis* (*b*).

with the addition of a large tine (bez) just above the brow tine. Brow and bez extending over the face much increase the proximal armature of the antlers (Bubenik, A. 1982b). Only the more advanced races achieve or rival the distal branch complexity shown by the antlers of the next species.

The white-lipped deer, *Przewalskium albirostris*, is confined to higher elevations of eastern Tibet. The antlers are smooth and flattened, with five tines formed by successive forking of the posterior antler branch, the second and third anterior tines (A2, A3, Pocock) being longest (Fig. 5a). These features are shared by the antlers of the extinct *Eucladoceros* (Adrian Lister, personal communication); whether this is due to common inheritance or to convergence has yet to be determined.

Lastly there is Père David's deer, *Elaphurus davidianus*, an ecological vicar of *Rucervus*, though with hooves more specialized for moving over yielding surfaces and a peculiar dentition: the incisors are the most advanced of any pecoran, convergent upon those of equids.

The Fossil Record

Prior to the late Pliocene, *Cervus* and the related plesiometacarpal genera are not known in Europe or in Asia south of the Tethys (Heintz & Brunet 1982). Probably no more than the one species ever reached America, and then relatively late (Kurten & Anderson 1980). It seems, then, that this group of deer began to diversify north of the Tethys in Eastern to Central Asia. At least eleven genera are known as fossils from Pakistan eastward, but in the absence of an up-to-date revision of the diverse material, it is not yet possible to evaluate how fossils can contribute to an understanding of the evolution of the Cervinae. Nevertheless, some regionalism arising during Plio-Pleistocene diversification is evident: *Arvernoceros*, *Eucladoceros* and *Elaphurus* are known from China; *Megaceros*, *Bohlinella* (= *Nipponicervus*), *Cervus*, *Elaphuroides*, and material referred to *Axis* from both China and Japan; *Rusa* from China and Java; *Axis* and *Hyelaphus* from Java; *Rucervus*—allegedly also in Java—and material referred to *Axis*, from the Siwaliks. *Arvernoceros*, *Eucladoceros*, *Megaceros*, *Cervus*, and *Rusa* are also known from the Pleistocene or late Pliocene of Europe (Lister 1987). The inclusion of *Rusa* is based on the Lower Pleistocene *Cervus philisi* (and apparently related taxa), which had antlers identical in form to those of *R. timorensis* (Groves & Grubb 1987), but conclusive studies to confirm that this resemblance is not superficial have yet to be made. *Axis* sensu stricto is unknown from Europe, in spite of suggestions to the contrary. Conversely, *Dama* is not known from Eastern Asia, the Chinese fossil *Dama sericus* evidently being a species of *Arvernoceros* (Groves & Grubb 1987).

Problems of Homology

In discussing the evolution of antler form in Asiatic deer—however briefly—the issue of homology between antler parts has inevitably arisen. By using Pocock's (1933) terminology for antler parts, neither endorsement nor rejection of his views on homology is intended. Where there is agreement with his hypotheses in this paper, it must be remembered that it applies to the antlers of the Cervinae alone.

The Asiatic deer present a wide diversity in antler form (Figs. 1–5), yet there is reason to believe that each pattern has evolved ultimately from the three-tine condition. Teilhard de Chardin and Trassaert (1937) showed that "inadaptive" and "adaptive" antler types can be differentiated among deer with three-tine antlers. They believed that only the second form had led to further antler evolution, though now it appears that each has independently presaged particular evolutionary styles of increasing antler complexity.

Antlers of *Hyelaphus* (Fig. 1b) and some *Rusa* taxa (Fig. 2b) represent the "inadaptive" type: the back inner tine (P2, Pocock) of the terminal fork is smaller than the front outer tine (A2, Pocock), which is a continuation of the beam. This is the most primitive antler form known in the Cervinae, and it is in relation to this stage that the question of homology should first be raised.

If antlers are found to be very similar in form, then it is possible to regard them not only as homologous as whole organs, but also as homologous part-for-part. For example, brow tines, beam, and front-outer or back-inner tines of the terminal fork must be homologous between different individuals of a three-tine deer species; between different subspecies of the same; and perhaps even between different species, such as *Hyelaphus porcinus* (Fig. 1b) and *Rusa unicolor* (Fig. 2b), when the proportions and dispositions of tines are very similar.

Antlers of *Axis axis* (Fig. 1a) are rather different in proportions, but the positions of tines are sufficiently similar to determine their homology with those of the other taxa just mentioned. Furthermore, *Axis* is generally regarded as a close ally of *Hyelaphus* (Pocock 1943b). Evidently the primitive three-point pattern has become modified in *Axis*.

The only extant deer with "adaptive" three-point antlers are *Rusa timorensis* (Fig. 4b) and some sambar, *R. u. unicolor* (Pocock 1943a). Here it is the back inner tine (P2, Pocock) which is longer and more robust than the other terminal tine, so as to take over the termination of the beam. This large tine is taken to be the homologue of the relatively short, turned-in, back-inner tine in *R. unicolor equina* (Fig. 2b) or *Hyelaphus* (Fig. 1b). And the front-outer tine of the terminal fork in *R. timorensis* is the equivalent of the tine having the same position in *R. u. equina*, though here it is the one which appears to terminate the beam. In *equina* the end of the antler may be said to fork into beam anteriorly and tine posteriorly, with the reverse pertaining in *timorensis*. But intermediate conditions affirm the homology outlined above: In the sambar *R. u. unicolor* the terminal fork may be Y-shaped, so it is not so obvious which tine may be said to continue the beam (Fig. 3a), or it may exhibit intermediate stages between the *equina* and *timorensis* extremes (Pocock 1943a). In this case, homology is established not so much by the close matching of parts as by the existence of intermediate states.

Evidently, the "adaptive" type of three-point antler could have evolved from the "inadaptive" pattern. But it must also be established whether one can infer homologies between antler parts with *different* branch complexity. If the differences are small, this may be possible. For example, we may compare a three-tined antler with a very similar one in which one of the terminal tines forks yet again so that the whole antler is four-tined. Such a comparison may be made between young and old antlers of the same individual, or between different individuals of the same species, or even between antlers of different species. What is a single tine in one instance is a bifurcated antler branch in another. But because one condition can precede the other in ontogeny, and because of the otherwise close matching of the organs and their parts in terms of shape, proportions, and surface sculpture, one is inclined to accept homology between–in the present instance–one tine and a bifurcated antler branch.

Several examples may be given to support these conclusions: *Rucervus schomburgki* usually had normal brow tines, but in fine specimens the brow tines fork (Lekagul & McNeely 1977). Clearly the structure is homologous between individuals whether it is forked or not. Another example come from the closely related *R. duvauceli*. By comparing stags of different ages, it is possible to assemble a complete series from a three-tine stage much like that of *Rusa unicolor*, through four-, five- (Fig. 4a) and up to six-tine stages (Cubitt & Mountfort 1985). It seems reasonable to infer that with successive stages, antler branches become successively more complex, yet the two major branches above the brow tine can always be said to be homologous between individuals. A final source of evidence comes from the sika, *Cervus nippon*. Antlers from young stags can be three-tined (Heude 1884, 1894), closely resembling in their porportions the mature antlers of *Rusa timorensis*. The distal parts of the three-tined or the mature four-tined antler are evidently homologous because the part of the antler which only develops as far as a single tine in one rack may in a later season develop further into a forked structure.

Following these observations, we may infer that from simple antlers more complex ones have evolved. From the "inadaptive" condition, hypertrophy and proliferation of the basal anterior tine (A1, Pocock) has led perhaps to the antlers of *Elaphurus* (but see below); and of the second anterior tine (A2, Pocock) to those of *Rucervus* species (Figs. 3b, 4). From the "adaptive" three-point type, antlers of the sika and white-lipped deer have evolved by a strictly sympodial pattern of branching (Fig. 5a), with the addition of the adventitious bez tine in *Cervus elaphus* (Fig. 5b).

Père David's deer presents a problem, since its antlers (Fig. 2a) are so different from those of other species. Its ability to hybridize with *Cervus elaphus* (Asher et al. 1988) with which it shares the same number of chromosomes (Hsu & Benirschke 1971) suggests that phylogenetically it is close to the other members of the Cervinae. *Elaphurus* also resembles *Rusa*, *Rucervus*, and *Cervus* in the mane and the urine-spraying behavior of the rutting stag (Wemmer et al. 1983). If these indications of its affinity are correctly interpreted, the conclusion is unavoidable that *Elaphurus* has evolved from a three-tine ancestor, a view which accords with Pocock's (1933) opinion that the first bifurcation in the ontogeny of the *Elaphurus* antler is equivalent to the differentiation of brow tine and the rest of the velvet antler in *Cervus*. It is possible that the antlers of *Elaphurus* evolved from a *Rusa*-like condition in which the brow tine was large and erect (Fig. 2b), not having acquired the specialized forward-flexed form (Fig. 5a). The brow tine in the course of evolution enlarged at the expense of the rest of the antler and took on the role of the beam as the most massive element of the antler (compare Fig. 2a and b).

However, Père David's deer is extremely specialized as a "grass and roughage feeder" (Axmacher & Hofmann 1988); its ability to hybridize with red deer may merely reflect the common retention over a very long period of a primitive karyotype, and other resemblances may represent convergences. This opens up the possibility that *Elaphurus* and the other Cervinae diverged at an early date, before the evolution of the three-point antler, when antlers were at a simple dichotomously forked stage, analogous to that of *Dicrocerus*. If the complex antler of *Elaphurus* originated from such a simple state, this might explain the distinctive pattern of axial channels in each main branch (see Chapter 1 by A. Bubenik).

Yet a third hypothesis must be referred to. Extinct species of the genus *Elaphuroides* have been regarded as ancestral to *Elaphurus* (Otsuka 1972). These deer had relatively short, erect, forked brow tines, but a long curved beam ending in a simple fork. If *Elaphurus* evolved from such an ancestor, it would have had to have shortened the beam at the same time as massively enlarging the anterior antler branch (brow tine). Such an evolutionary reversal seems unlikely; unless unique shared specializations of the skeleton are discovered in the two genera so as to confirm their immediate relationship, it seems more plausible for the moment to query any close connection between them.

Evolutionary Trends

Every extant cervine clade with progressive antler form has probably evolved it independently from the primitive "inadaptive" condition (Groves & Grubb 1987). It is likely that increases in body size have also occurred independently in the different lineages. But the two trends have not been in phase. For instance, the very large sambar retains less specialized antlers than chital, sika, or the rucervines, all smaller species. Larger species could be expected to have disproportionately larger antlers: across species, antler size increases allometrically with body size (Huxley 1932). The relationship does not always pertain when particular cases are considered. The Javan rusa *Rusa timorensis* has an absolutely shorter skull than the Malay rusa *R. unicolor equina* yet has absolutely longer antlers (van Bemmel 1949). The chital is absolutely smaller still, as measured by skull length (Pocock 1943b) but has relatively even longer antlers (Dollman & Burlace 1935; Kurt 1978). In these selected examples relative antler size is not positively correlated to body size—indeed the relationship is an inverse one. It would have been preferable to cite measurements of body mass and antler mass, of which linear dimensions may be poor predictors, but these data are not available. Anyway, it is apparent that different species of deer of similar body dimensions could nevertheless differ in relative antler size, due apparently to the intensity of sexual selection determined by their social systems (Clutton-Brock et al. 1980). It is significant that the hinds of Javan rusa and particularly chital forage in more open habitats and form larger social groups (Kurt 1978) than those of the Malay rusa, providing more mating opportunities for stags. Furthermore, the two species with proportionally larger antlers are also those with a derivative antler form, suggesting that change in form can be associated with change in relative size and hence may be of adaptive significance.

Here may lie an explanation for the elaboration of antlers occurring by the proliferation of different antler branches in different lineages. If a species were to increase in body size but did not alter its breeding strategy, its antlers would increase in relative size but change little in form—for example, smaller and larger *Rusa* taxa (Figs. 2b, 3a). But if a species increased both body size and breeding group size, increase in antler size should accelerate and the antlers should enlarge even more disproportionately. Should this acceleration commence on the attainment of different body sizes in different clades, then the state of antler complexity already acquired may determine the pattern of further elaboration and dictate which antler branch should proliferate the most.

This hypothesis could be tested if the fossil record were to become better known, but a fuller understanding of combat style in living deer would also be necessary before other details of antler form could be interpreted.

Conclusion

The great diversity of antler form exhibited by the Asiatic Cervinae can be interpreted as an adaptive radiation stemming from the "inadaptive" three-point antler type and relating to increases both in body size and in breeding-group size in various clades. But this hypothesis depends on assumptions concerning the homology of antler parts which, especially in the case of *Elaphurus*, require further investigation.

Acknowledgments. I am grateful to A. B. Bubenik for comments and much valuable information.

5
Cranial Appendages of Giraffoidea

CHARLES S. CHURCHER

Introduction

Giraffes and okapi belong to the superfamily Giraffoidea. For the purpose of this essay the Giraffoidea includes ruminant artiodactyls from Eurasia and Africa, dating from Miocene to Recent times. They are characterized by skin-covered indeciduous bony horns or ossicones, at least one pair being borne over the frontoparietal suture, with other paired or unpaired frontonasal ossicones or unpaired nuchal or lateral subsidiary bony outgrowths (Fig. 1). The ossicones condense separately within dermal tissue external to the skull roof and lengthen by intercalary growth at the base and superficial to the skull roof. Both males and females may bear ossicones, or females may lack them, or both sexes may be without ossicones. An additional character is the bifid and sometimes trifid crowns of the lower incisive canines.

Ossicones

General

Ossicones are cranial ornaments composed of bone, dermis, and epidermis, which may be penetrated by extensions of the continuous sinuses within the nasals, frontals, and parietals and which are borne on the parietals, frontals, or nasals of members of the Giraffoidea. These extensions of the cranial vault are covered with skin and hair even in adulthood; they are permanent and usually unbranched, although more than a single pair may be normally present in a species. Ossicones are borne by both sexes in *Giraffa*, only by males in *Okapia*, and appear to have been present only in males or to have been absent in both sexes of some extinct taxa.

Ossicones differ from the horns of Bovidae in the lack of a keratinized outer or epidermal sheath that grows constantly from within and elongates the horn. Ossicones differ from the antlers of Cervidae in their permanence and usually in their

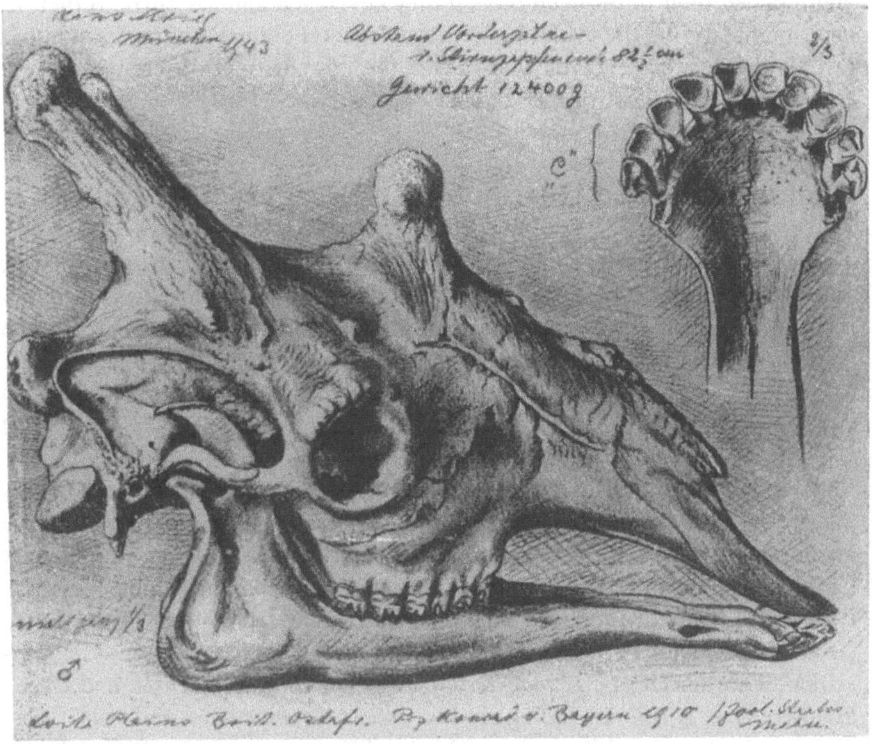

FIGURE 1. Skull and jaw of old male *Giraffa camelopardalis* from the Loita Plains, Kenya, collected in 1910. Note twin parietal ossicones, single nasal ossicone, and nuchal protuberance. Secondary bone has been laid down over the dorsum of the cranium and nasals, resulting in a rough surface with mammilate beading supraorbitally and on nasal, as well as club-ended ossicones. The incisive row shows lower canines with double crowns that resemble two incisive crowns. (From Krieg 1944.)

lack of branching racks. Ossicones also differ from the horns of modern Antilocapridae whose bony core is simple (but the sheath is branched and is shed annually) and whose core is made of spongy cancellous bone rather than a compact cortex and hollow medulla.

The skulls of many living and extinct Artiodactyla have borne bony outgrowths on nasal, frontal, parietal, and supraoccipital bones. In the nonbovid and noncervid groups, many of these outgrowths have smooth or roughened surfaces without strong longitudinal ridges, suggestive of a permanent epidermal covering as in the giraffoid ossicone. These outgrowths included simple, double, and branching conformations; some developed burrs, occasionally more than one, and retained their horns even though they developed annular burrs. It therefore appears that during the evolution of the artiodactyl cranial ornaments, a number of outgrowths were evolved that bridge the now distinct differences between Giraffoidea, Bovoidea, or Cervoidea.

Conformation and Distribution in Recent *Giraffa* and *Okapia*

The main ossicones are usually paired, conical, or subconical structures located over the frontoparietal suture and posterodorsal to the orbits. Less developed median ossicones exist on the nasals of giraffes and still less regular protuberances have been described for other positions on giraffe skulls (Fig. 2A).

Giraffa

Male and female giraffes all possess paired parietal ossicones, with those of males being larger and stronger than those of females. Median nasal ossicones are present in northern giraffes (i.e., Nubian *G. c. camelopardalis*, reticulated *G. c. reticulata*, Sudan *G. c. antiquorum*, Baringo *G. c. rothschildi* and Western *G. c. peralta*). They are reduced or absent in giraffes from south of the equator (Dorst & Dandelot 1970), from East and South Africa (i.e., Masai *G. c. tippelskirchi*, Thornicroft's *G. c. thornicrofti*, Angolan *G. c. angolensis*, and Cape or Southern *G. c. giraffa*, including *capensis, australis, maculata*, and *wardi*; Ansell 1971). Paired occipital horns located behind the ears occur in *G. c. rothschildi* (Fig. 2A), but are really rugose swellings on the occipital crest, and give this race the name of the five-horned Giraffe (Spinage 1968b). Orbital horns are small ossicles that may or may not be fused to the supraorbital ridge. Lydekker (1904) described an "azygous" horn between the orbit and the occipital or nuchal horns, but these may be bilaterally distributed (Spinage 1968b) and are free nodules, without any boss or fusion to the skull. Their origin may be a dermal condensation or an exostotic growth. Whether all azygous growths are homologous or should be termed horns is therefore questionable. Such peripheral horns are often present in older males and uncommon in females.

Skulls of males may have rough surfaces and uneven nuchal, zygomatic or orbital margins. Additional bone is added to the dorsal surface of the skull and ossicones of males (Fig. 2A), where there is no muscle attachment. This additional bone may effectively join the parietal ossicones at the bases by a broad bridge (Dagg & Foster 1976); it may add between 13.0 and 4.5 kg (Dagg & Foster 1976) or as much as 9 kg (Spinage 1968a) to the head, and converts the top of the skull into a battering ram or club. Krieg (1944) considered that the secondary cranial bone resulted from mechanical injury during sparring but, as such bone is occasionally found in females (Krumbiegel 1965) and as females generally do not spar, the explanation for the additional cranial bone may be partly due to sexual genetic differences and behavior!

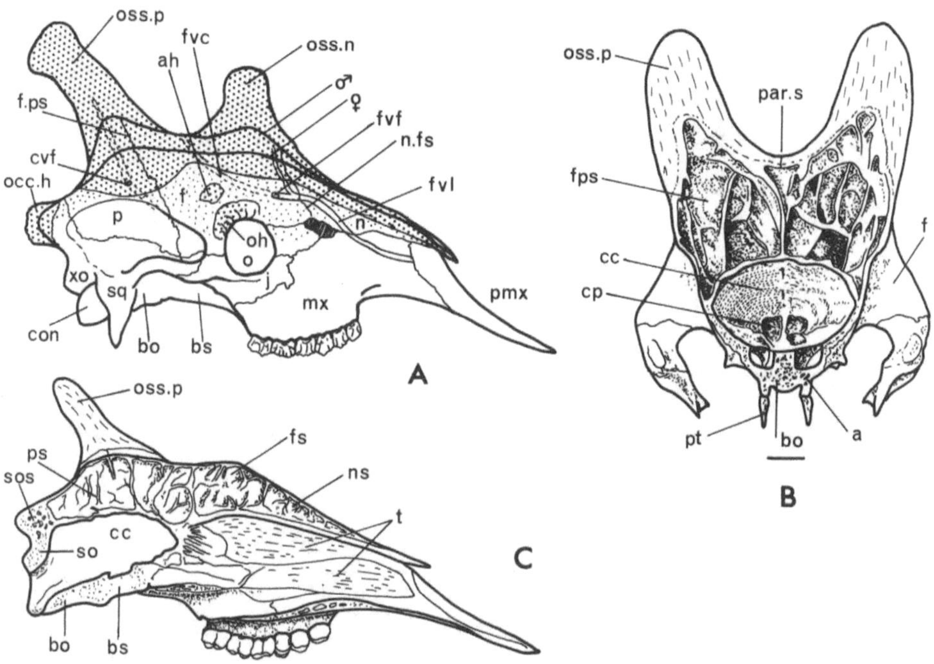

FIGURE 2. Ossicones, secondary ossification, and sinuses of *Giraffa*: *A* Lateral aspect of skull showing courses of veins, areas of dense ossification (heavily stippled) and lesser secondary ossification (lightly stippled), and outlines of skull roofs of male (♂) and female (♀) adults. (After Spinage 1968b.) *B* Anterior view of skull obliquely sectioned posterior to the ossicones. Scale = 30 mm. (After Lankester 1907a.) *C* Sagittal section of young skull showing sinuses within cranial roof. (After Spinage 1968b.) *Abbreviations*: ♂ = outline of male's cranial roof; ♀ = outline of female's cranial roof; *a* = alisphenoid bone; *ah* = azygous horn; *bo* = basioccipital; *bs* = basisphenoid; *cc* = cranial cavity; *con* = occipital condyle; *cp* = cribriform plate; *cvf* = foramen for cornual vein; *f* = frontal bone; *fps* = frontoparietal sinus; *f.ps* = frontoparietal suture; *fs* = frontal sinus; *fvc* = canal for frontal vein; *fvf* = foramen for frontal vein; *fvl* = anterior limit of roofed canal for fròntal vein; *j* = jugal; *mx* = maxilla; *n* = nasal; *n.fs* = nasofrontal suture; *ns* = nasal sinus; *o* = orbit; *occ.h* = occipital horn; *oh* = orbital horn; *oss.n* = nasal ossicone; *oss.p* = paired parietal ossicones; *p* = parietal bone; *par.s* = interparietal suture; *pmx* = premaxilla bone; *ps* = parietal sinus; *pt* = pterygoid bone; *so* = supraoccipital bone; *sos* = developing supraoccipital sinus; *sq* = squamosal bone; *t* = turbinate bones; *xo* = exoccipital bone.

Giraffe ossicones are usually inclined at about 48–58° to the Frankfort Plane (Singer & Boné 1960), are between 100 and 250 mm long in males and slightly less in females, and differ as much as 35% in weight between the left and right ossicones of one individual (Dagg & Foster 1976).

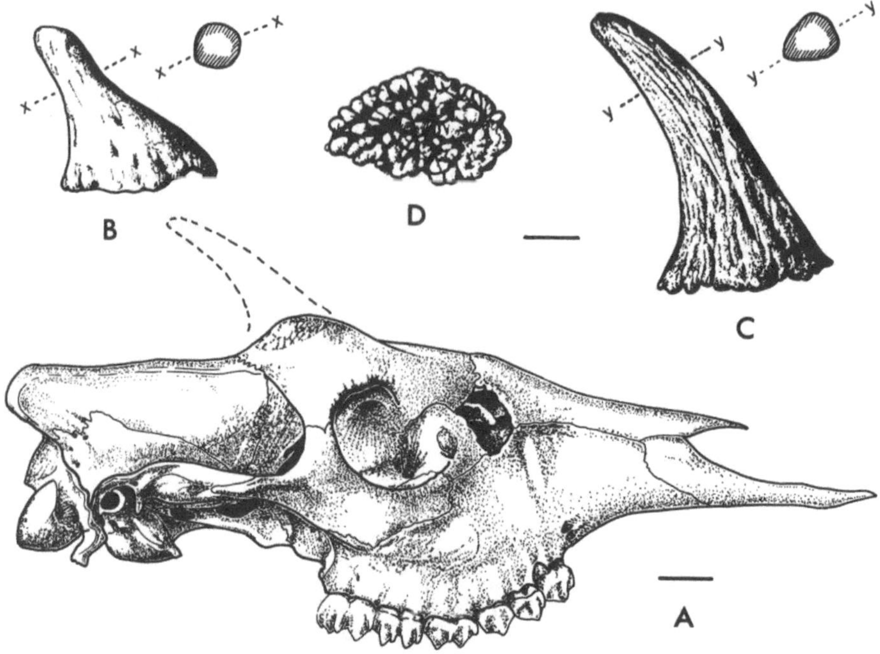

FIGURE 3. *Okapia* skull and detached ossicones. *A* Lateral aspect of skull of young male with position of ossicone indicated by dashed outline, from skull B, British Museum (Natural History), London. Scale = 30 mm. (After Lankester 1910.) *B* Ossicone from juvenile male individual, from skull D, British Museum (Natural History), London. *x---x* indicates cross-section. (After Lankester, 1910.) *C* Ossicone from young male okapi, from skull B, British Museum (Natural History), London. *y---y* indicates cross-section. (After Lankester 1910.) *D* Base of ossicone from young male okapi, from skull B, British Museum (Natural History), London. (After Lankester 1910.) Common scale for *B*, *C*, and *D* = 15 mm.

Okapia

In okapis, only the males carry relatively small ossicones, about 75 mm long, pointed and inclined posterodorsally. The tips of these pointed ossicones are bare bone, polished from use, as the skin covering does not include the tip in adult males (Fig. 3). Some necrosis is often visible in the area where the bare bone and skin meet (Spinage 1968b) (Fig. 4). The ossicones are located above and slightly posterior to the orbit, are fused to the bones, and in adult males overlap the frontoparietal suture posteriorly. Air sinuses in the skull of *Okapia* do not extend over the cranium, and are limited to the median nasal region. There are no recorded instances of additional horns or ossicles in *Okapia*, as occur in *Giraffa*.

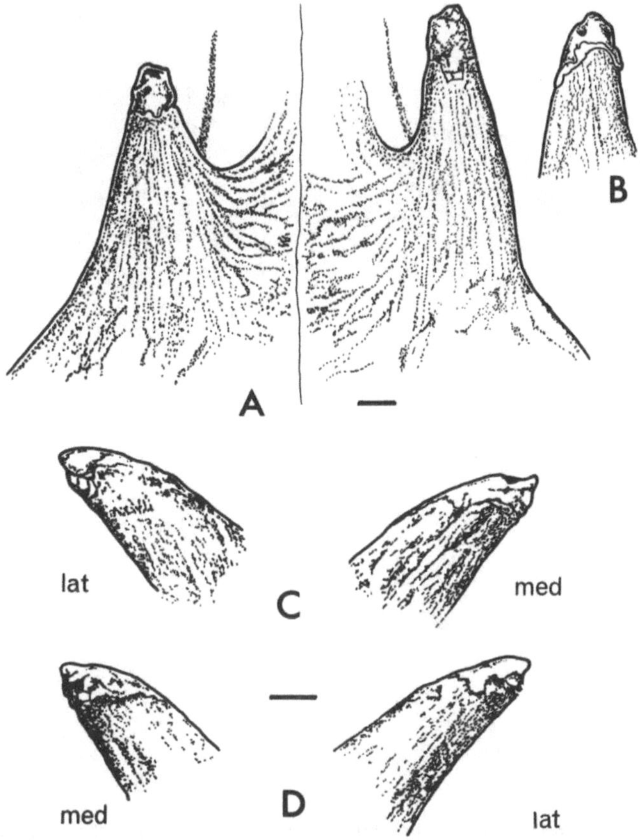

FIGURE 4. Tips of ossicones of *Okapia* showing bare, polished bone separated by necrotic area from skin covered areas. *A* Dorsal aspect of ossicones on mature skull P in the Museum National d'Histoire Naturelle, Paris. (After Lankester 1910.) *B* Lateral aspect of left ossicone on mature skull P in the Museum National d'Histoire Naturelle, Paris. (After Lankester 1910.) *C* Lateral and medial aspects of right ossicone of skull A, Royal Scottish Museum, Edinburgh. (After Lankester 1907.) *D* Medial and lateral aspects of left ossicone of skull *A*, Royal Scottish Museum, Edinburgh. (After Lankester 1907.) Scale bars = 10 mm.

Growth and Development of Ossicones in *Giraffa* and *Okapia*

An ossicone is composed of bone, dermis and epidermis, penetrated by blood vessels and nerves, and by cranial sinuses within. The bone is usually referred to as the os cornu, and comprises the main portion of the ossicone, to which may be added concentric layers of lamellar bone laid down externally on the os in old individuals. The os cornu condenses separately from the frontal and parietal

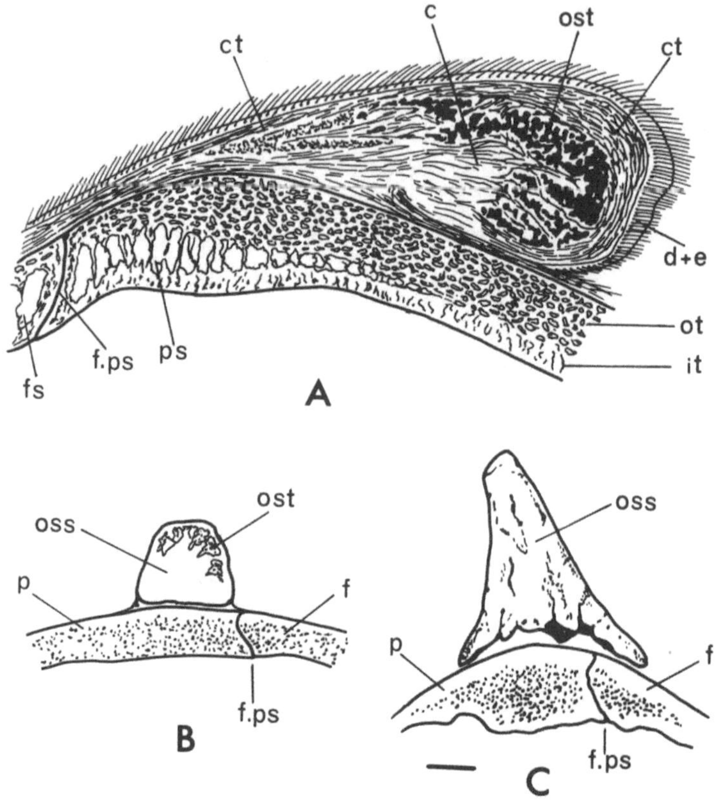

FIGURE 5. Developing ossicones of *Giraffa*. *A* Parasagittal section through an ossicone of a week-old giraffe. (After Spinage 1968b.) *B* Section through tissue of skull roof and ossicone of newborn giraffe. (After Lankester 1907a; from Owen 1840). *Note*: Lydekker, 1904, considered that Owen had the skull roof reversed in his illustration. It is also possible that the section of the ossicone is actually *transverse* or *oblique*, as newly developed ossicones are taller than wide in anteroposterior dimension. (After Lankester 1907.) *C* Section through bony tissue of skull roof and ossicone of a young giraffe. Scale = 10 mm. (After Lankester 1907a.) *Abbreviations*: *c* = cartilage; *ct* = connective tissue; *d+e* = dermis and epidermis with hair follicles; *f* = frontal; *f.ps* = frontoparietal suture; *fs* = frontal sinus primordium; *it* = inner table of cranium; *oss* = ossicone; *ost* = ossified tissue; *ot* = outer table of cranium; *p* = parietal; *ps* = parietal sinus primordium.

bones of the cranial vault, and later fuses to it (Fig. 5). The dermis and epidermis cover the os cornu, extend with growth, and may become cornified and bare of hair at the tip in active males.

Only two living monospecific giraffids exist today, *Giraffa camelopardalis* and *Okapia johnstoni*. Both sexes of *Giraffa* bear ossicones, although those of males are usually more massive than those of females. Only males of *Okapia* bear ossicones, although Lankester (1907a) remarked on the presence of small ossicones in females. The sexual distribution of ossicones in extinct Giraffoidea is mostly speculative, as isolated ossicones, frontlets bearing more or less complete ossicones, and the occasional nearly complete skull do not indicate the sex of the individual. It is likely that cranial ornaments were present in both sexes of primitive giraffids, and may have served in specific recognition or temperature regulation when originally developed. Their use in interspecific male sparring or combat would then have developed later.

Ossicone Development in Giraffa camelopardalis

Lankester (1907a,b) first described the conditions of the ossicones in fetal *Giraffa* and *Okapia*. The fetal giraffe is born with rudimentary ossicones which overlie the parietals (Lankester 1907a; Naaktgeboren 1969). With growth, the base of the ossicone enlarges as a flared cone and, when fusion occurs, becomes integral with the frontal and parietal roof (Fig. 6). The fetal or newborn giraffe's ossicones are oval-elongate in section, with the axis oriented posteromedially-anterolaterally, and covered by long black hair. At birth the ossicones are flexible and lie close to the skull. No ossification is present, and the cartilaginous template is separated from the parietal vault by periosteal connective tissue (Fig. 5A). Lankester (1907a) suggested that "It seems legitimate to consider the lateral ossicones of the Giraffe, and therefore of the Okapi, as originating in a fibrous osteogenetic mass which gives rise to a protrusion of the integument and originates in the connective tissue of the integument rather than in the osteogenetic tissue of the cranial roof, from which it is separated by a dense membraneous periosteum."

Spinage (1968b) confirmed Lankester's (1907a) observations and added that the fetal ossicones are about 25 mm long and that they subsequently grow rapidly. Ossification begins at about 7 days after birth (Spinage 1968b), at several independent centers near the periphery of the distal half of the cartilage (Owen 1849) and at the apex of the ossicone. At the end of 2 years, the ossicone is completely ossified and has elongated through extension at the base where a cartilaginous area extends the bone in a manner analogous to the cartilaginous epiphyseal disc of a long bone (Figs. 5 and 6). The ossicones fuse to the parietals at about 4 years in males and 7 or more years in females. The oldest known individual in which the ossicones were not fused to the parietals was some 11 years old.

Growth in ossicone length slows at about 2 years of age and usually ceases after fusion in females and proceeds very slowly in males. Additional laminar bone is laid down throughout the life of males and occasionally in females. An increase of 25 mm in length and diameter may occur from this peripheral deposition. Parietal ossicones in old males may reach 180 mm in length and 220 mm in circumference.

Ossicones have sinuses within the proximal half of the os cornu that connect with the frontoparietal sinuses (Spinage 1968b; Dagg & Foster 1976) (Fig. 2B). The frontal sinus begins small with a single limited chamber below the median nasal-frontal tumescence. This becomes subdivided by septa and struts (Owen 1868) and becomes a frontal-parietal sinus by extending from the center of the

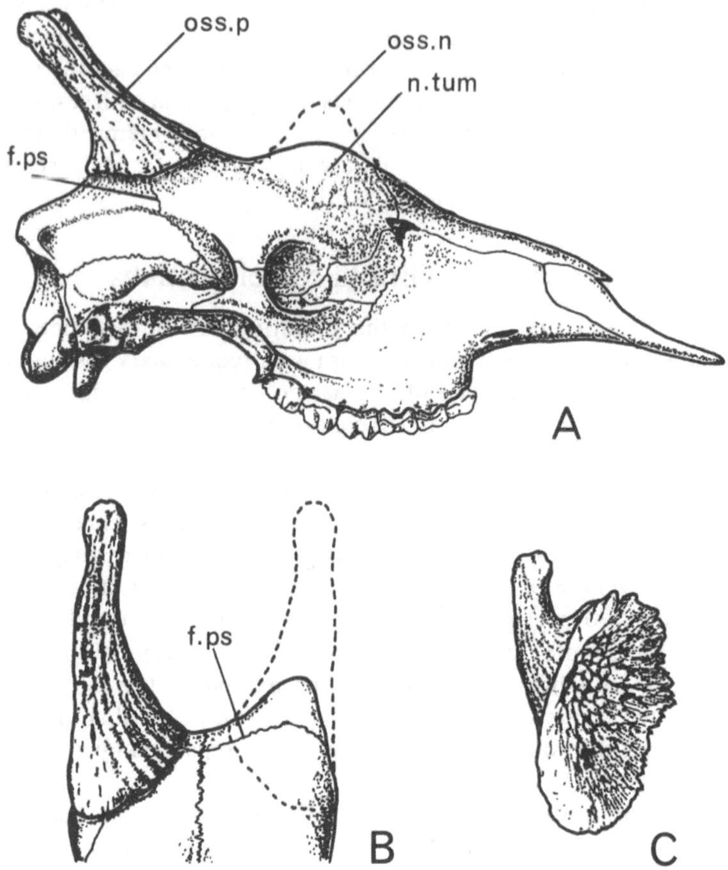

FIGURE 6. *Giraffa* skull and detached ossicone. *A* Lateral aspect of skull of young *Giraffa*, about two-thirds grown. (After Lankester 1907a.) *B* Frontal aspect of ossicones of young *Giraffa*. (After Krieg 1944.) *C* Detached ossicone in cranial or ventral aspect. (After Krieg 1944.) *Abbreviations*: *f.ps* = frontoparietal suture; *n.tum* = tumescence on frontal and frontonasal suture for nasal ossicone; *oss.n* = outline of nasal ossicone; *oss.p* = parietal ossicone.

facial region posteriorly and laterally to the rear of the cranial vault and beneath the ossicones. The purpose of these sinuses is probably a combination of lightening the skull and providing possible cushioning during cranial sparring. During the growth of the ossicones and the development of the sinuses, the centers of the ossicones move medially or posteromedially from above the orbits, alter attitude to incline posterodorsally and slightly laterally, and usually develop somewhat asymmetrically.

Giraffe ossicones are always skin-covered and, in mature males that engage in head and neck sparring in which the ossicones are used as weapons, the apex becomes cornified after completion of basic ossification. The terminal hair is worn away and the keratinized epidermis is anchored to the ossicone by long connective tissue filaments that extend into the bone. The ossicones in both sexes are solid structures until about 4 years when, after fusion with the parietals, the parietal sinuses extend into the base of the ossicone. Ossicones in males are massive, circular in section, terminally knobbed (Fig. 1), and with smaller sinuses than those in females, which are lighter, laterally flattened in section, and bluntly tapered. Superficial blood vessels of the ossicones become buried by the bony laminae laid down in males. The bone is very compact and ivory-like, but softer in young males and females.

The median nasal or third ossicone found typically in *G. c. rothschildi* is often irregular in shape and is variable between individuals. It develops in the same way as the paired ossicones, from a small conical primordium which fuses to the nasals of males but seldom does so in females until very old age (Spinage 1968b) (Fig. 2A).

The other small bone protuberances (occipital, orbital, and "azygous horns") derive from bony rugosities on the skull or from small dermal ossicles.

Ossicone Development in Okapia johnstoni

Lankester (1902) described *Okapia johnstoni* as a new genus and species of Giraffidae. He noted tumescence in the roofing bones of the skull and that the early ossicone is formed as a conical cap that fits over the tumescent bulge on the frontals of the young animal (Lankester 1902). The juvenile horns lie within the skin and are covered by hair that converges at the center of the ossicone tumescence.

Lankester (1907b) described the first fully adult okapi skull and its ossicones and noted that the male's ossicones start as separate conical ossifications which later fuse to the frontals through a boss-like upgrowth. The ossicones are fused to the skull by the time they are 75 mm long, are directed posterodorsally, and are supraorbital in position (Lankester 1902). In the adult, the tip of the bone of the ossicone is free of skin and is polished bone, and the tip appears separated from the body of the ossicone by a small suture (Lankester 1907b) (Fig. 4). Lankester also speculated that female okapi may occasionally have small rudimentary ossicones about 12 mm high and 16.5 mm in diameter, but this seems to be erroneous.

Spinage (1968b) reported air sinuses in the skull, but not over the cranium, as in the giraffe, and that the ossicones lie above and slightly posterior to the orbit. The manner of growth is as in *Giraffa*, and they appear at about 1 year, then fuse with the frontals, with the posterior margin at or overlapping the frontoparietal suture, before the skin at the tip becomes cornified. This skin is lost at about 3 years when the bare bone is exposed. Spinage does not consider that the bony tips may be shed, as suggested by Lankester (1907b). The tips are polished, and there may be areas of necrosis where the bare bone and skin meet.

Use of Ossicones

Giraffa

Male giraffes use their parietal ossicones and sometimes the nasal ossicone on the skull in the manner of projections on the medieval mace. The head is swung on the end of the long neck, the ossicones are oriented toward the opponent with whom the bull is competing for dominance, and the intention is to buffet or collide with the head or neck of the opponent (Spinage 1968a; Dagg & Foster 1976). The shock of the collision is borne by the skull roof, since most motions involve the use of the back of the head with the posterodorsally oriented ossicones, and are powered by the strong dorsal musculature of the neck. The sinuses within the nasals, frontals, parietals, and supraoccipital absorb shock from blows either received or delivered, and the thin walls within the sinuses provide some cushioning for the brain within. Females or cows usually do not indulge in cranial sparring, and their sinus development is less than that in bulls. Continual bruising of the cranial surface during sparring may enhance the deposition of the additional bone noted as typical of male skulls (Krieg 1944). Presumably the extinct two-horned giraffes such as *Samotherium boissieri* (Fig. 7A) or *G. jumae* used their ossicones in the same manner.

Okapia

Male okapis use the points of their ossicones in the same way that antelopes use horns, to pierce the skin of their opponents or to fence (Walther 1962). Okapi males have less developed sinuses in the cranial roof, and this may reflect the different use of the ossicones. Okapi do not use the neck as a flexible handle on the end of which is swung the ossicone-armed skull, but as a stiff extension to support the head when ossicones are locked with an opponent, or when attempting to rake the opponent's flanks. This mode of use of ossicones in living okapi was probably used by extinct forms such as species of *Palaeotragus*, e.g., *P. primaevus* or *P. tungurensis* from the Miocene of Africa and southern Asia.

Sivatherium *and Related Forms*

Sivatheres such as *Sivatherium giganteum* (Fig. 7E) from the Plio-Pleistocene of the Siwalik Beds, India, or *S. maurusium* from similar-aged deposits in Africa are massively built giraffids in which the neck, forequarters, and skull are adapted to compression stresses. The ossicones are hollow at the base, with sinuses extending up to 300 mm outwards from the frontal sinus, and with cancellous centers distally from the sinus (Singer & Boné 1960).

Sivathere ossicones are well constructed for use in sparring, as are antlers, which they resemble in general structure (Spinage 1968b), and their cervical and anterior thoracic vertebrae are massive in comparison to those of *Giraffa*. These adaptations have been interpreted as reflecting cranial wrestling and pushing, or

FIGURE 7. Heads of selected extinct giraffoids. *A Samotherium boissieri*, Samos, Greece, E. Pliocene. *B Giraffokeryx punjabiensis*, Lower Siwaliks, India, E. Pliocene. *C Prolibytherium magnieri*, Gebel Zelten, Libya, E. Miocene. *D Zarafa zelteni* (= *Canthumeryx sirtensis*, Hamilton 1978b), Gebel Zelten, Libya, E. Miocene. *E Sivatherium giganteum*, Upper Siwaliks, India, Plio-Pleistocene. *F Bramatherium perimense*, Middle Siwaliks, India, M. Pliocene. (*A, C*, and *D* after Churcher 1978; *B, E*, and *F* after Colbert 1935.)

even butting, between males (Churcher 1978). Alternatively, Singer and Boné (1960) suggested that sexual dimorphism exists in the ossicones, with males possessing the larger and more twisted or complex and females lighter and simpler ossicones. The recovery of gracile ossicones may indicate that some females may have carried cranial ornamentation (Harris 1976). It is also possible that variation in ossicone robustness represents stages in maturation which are obscured by either individual or sexual variation (Churcher 1978).

Interpretations of other extinct massive giraffoids' cranial ornaments and build suggest that forms such as *Bramatherium* (Fig. 7F), *Hydaspitherium, Vishnutherium*, etc. may have behaved similarly to sivatheres. In these forms, the broad areas and uneven margins of the ossicones may be adapted for locking them when in dorsal or frontodorsal contact. Locking would prevent lateral slipping and possible damage from twisting of the occipital joint or of a joint further down the neck. In antlered male cervids a similar situation is present; if slipping does occur, it is usually towards a repositioning of the antlers in a normal and balanced position.

Zarafa

The use of the ossicones in *Zarafa zelteni* (Fig. 7D) (= *Canthumeryx sirtensis*, Hamilton 1978a) is less easy to deduce. A pushing or butting mode is likely, as serious damage to either contestant would be minimized. However, lateral swings of the head would use the laterally directed ossicones as piercing fids or spikes. Perhaps both modes were used, the former intraspecifically and the latter interspecifically, possibly against predators.

Giraffokeryx *and* Palaeotragus quadricornis

As today there are few four-horned artiodactyls (e.g., the four-horned antelope or chousingha *Tetracerus quadricornis*, or the Soay sheep *Ovis aries*), it is difficult to find an analogue from which to deduce ossicone function for *Giraffokeryx punjabiensis* (Fig. 7B) or *Palaeotragus quadricornis*. It is logical to assume that the four ossicones were used as centering and locking guides, in which opponents' paired ossicones would interlock and apply pressure at the base of the V's. Duerst (1926) stated that in *T. quadricornis* the anterior pair of horns develop only after the posterior pair have begun to be used in sparring, and thus could assist in providing more accurate locking and reduce the risk of vertebral or condular dislocation in sparring individuals. However, no information exists as to the ontology of the ossicones in *G. punjabiensis* or *P. quadricornis*.

Prolibytherium

The flat, "belle epoque", inclined ossicones of *Prolibytherium magnieri* (Fig. 7C) are presumably also adapted as pressure plates used in butting or head wrestling. However, their laminar form and lack of centering adaptations suggest that muscular wrestling would not have been effective or could not have avoided serious damage to the ossicones. It may be that *Prolibytherium* used its ossicones mainly for display to overawe rivals, rather than strength to overpower them. No lower canine of *Prolibytherium* is known, and thus it is Giraffoidea incertae sedis although its cranial ornaments are best interpreted as giraffoid.

Climacoceras

The genus *Climacoceras* bears antlerlike ornaments and comprises two species, *C. africanus* and *C. gentryi* (Hamilton 1978a). It was first described by MacInnes (1936) from Kiboko (= Maboko), Kavirondo Gulf, Kenya and was based entirely on fragments of antlers or ossicones which bore short, irregularly placed knobs or processes. MacInnes identified *C. africanus* as a deer, but Pilgrim (1941b) compared it to *Giraffokeryx* and mentioned large rugose protuberances at the bases of the ornaments. Simpson (1945) followed Pilgrim and placed *C. africanus* in the Lagomerycidae of the Giraffoidea, that is, with the primitive giraffoids.

Hamilton (1978a) reviewed the Maboko materials comprising teeth and antler or ossicone fragments and stated that "the ossicones do not show grooving of the beam" and "the internal structure of the ossicones shows a continuous bone structure from the frontal bone to the ossicone tip." Thus the ossicones were not shed. The rack and tines conformation of the cranial ornaments of *Climacoceras* suggest that they were used both for display and sparring, as in modern Cervidae, as the morphological resemblances between the complete rack of *C. gentryi* and of various species of *Cervus* are too great to be ignored.

Status of Climacoceras *as a Giraffoid*

Hamilton (1978a) described *C. gentryi* from Fort Ternan, Kenya, on a complete left ossicone, left dentary (type), and other dental, cranial, and postcranial remains. The ossicone of *C. gentryi* has a main beam with a long brow tine, two posterior tines and a forked tip, and superficially resembles that of a cervid. No sectioning of the basal part of a beam and the skull roof is reported in the literature for this species but A. Bubenik (1986, personal communication) informs me that the medulla of the beam is cancellous and continuous with that of the cranial vault, as in Cervidae. The accessory tines vary in length and appear to be true apophyseal outgrowths from the main beam, also as in Cervidae. Unfortunately no branching or palmate ossicones exist on living giraffes, but it is presumed that the development of such variants was similar to that in Cervidae.

The type dentary has "the I_3 and canine . . . glued back into" the dentary "and they are more vertical than is natural" (Hamilton 1978a). The canine in *C. gentryi* is bilobed, with larger mesial and smaller distal accessory lobes. Hamilton's Table 1 gives measurements for dimensions indicated "W" for "C", "anterior lobe", and "posterior lobe" as 5, 4, and 1 mm. "W" indicates the buccolingual width for the cheek teeth, so I interpret "W" for the canine to indicate the buccolingual dimension. On Table 26 Hamilton notes "W" for lower canines of *Palaeotragus primaevus* as 10.0 and 11.0 mm, and for *Climacoceras* as 5.0 mm. Churcher (1970, Table IV) gave mesiodistal lengths of milk canines as 7.1 and 8.9 mm, and of a permanent canine as 11.2 mm for *P. primaevus*, buccolingual widths of 3.2 and 3.5 mm for milk canines and 5.5 and 5.7 mm for permanent, and lengths of the distal accessory cusp as 2.1 and 3.6 mm for milk canines and 5.4 and 4.4 mm for permanent. As Hamilton's measurements cannot be correlated with Churcher's, and as Hamilton used different specimen numbers than Churcher and gave no key, it is impossible to compare these two sets of data.

Because there is no evidence for sinuses at the base of the cranial beams and because I_3 and C_1 in the type of *C. gentryi* have been glued back in at an unusual angle, doubt arises whether *Climacoceras* possessed a bifid lower canine and thus could be confidently identified as a giraffoid. As *Palaeotragus primaevus* is present also at Fort Ternan, and as the dimensions given by Hamilton appear to fall within the observed ranges of *P. primaevus*' milk and permanent canines, it may be that the canine in the type dentary derives from that species. On other aspects of the dentition, it is possible that *Climacoceras* is a giraffoid although the dental characters are less obviously giraffoid than are those of *P. primaevus*. Hamilton (1978b) concluded that both *Climacoceras* and *Canthumeryx sirtensis* (including *Zarafa zelteni*) are giraffoids and that the family Palaeomerycidae is probably polyphyletic.

Conclusions

It is apparent that giraffoids have evolved a distinctly original form of cranial ornament in their ossicones, which are developed in a unique manner for Artiodactyla, have adapted it to a variety of demonstrable or deducible uses, and have produced forms that may superficially resemble bovids such as antelopes or protoceratids (e.g., *Syndyoceras* or *Synthetoceras*), or are without parallel within the mammalian class.

Acknowledgments. I thank my assistant, Maryjka Mychajlowycz, for entering the paper and its modifications on the computer, and Mr. John Glover of the Faculty of Arts and Science, University of Toronto, Photographic Facility, for assistance with the illustrations. Support for this paper was provided by Natural Sciences and Engineering Research Council of Canada Grant A 1760.

6
Evolution and Dispersal of African Bovidae

ALAN W. GENTRY

Introduction

While cervids are mainly inhabitants of Eurasia and the Americas, giraffids and bovids are best known from Africa and southern Asia.

Horns evolved in the Miocene ancestors of these pecorans. Early horns were simply lunging or piercing weapons, used in infraspecific dominance-testing encounters. Later came the more elaborate horns, which can be used for guarding and parrying, for ramming, for interlocking, and for display. Janis (1982) has linked the appearance of horns more specifically with territorial behavior. As species with a body weight of about 18 kg or more moved into more open habitats, horns would evolve to be of use to territorial males patrolling their boundaries.

Bovids are the hollow-horned ruminants. They have a hollow keratinized sheath fitting over a separate bony core, which in some cases has internal sinuses but in most consists of spongy bone. In living bovids, neither sheath nor core is branched or seasonally shed. Horns may occur in both sexes or in males only. Bovids have no upper incisors, and vestigial minute upper canines occur in only a minority of individuals. Upper and lower first premolars are also lacking. The cheek teeth are selenodont and the crescentic cusps join to one another earlier in wear than they do in cervids or giraffids. Metapodials lateral and medial to the cannon bone are absent or more reduced than in cervids. Compared with cervids or giraffids, many bovids show more hypsodont teeth, stronger cursorial characters in their limb bones, and territorial behavior.

The bovid classification used in this chapter is shown in Table 1. Names of extant genera and number of species in each are also given. Boselaphini, Bovini, and Caprinae are entirely or mostly Eurasian; Antilopini span Eurasia and Africa; the remaining tribes live mostly or entirely in Africa. Examples of horns of some of the extant bovids of Africa are shown in Figs. 1–4.

TABLE 1. Bovid classification.

	Extant genera and no. species
Family Bovidae	
Subfamily Bovinae	
Tribe Tragelaphini	Kudus, bushbuck, etc. *Tragelaphus* 7, *Taurotragus* 1.
Tribe Boselaphini	Now confined to India. *Boselaphus* 1, *Tetracerus* 1.
Tribe Bovini	Cattle and buffaloes. *Bos* (incl. *Bison*) 7, *Syncerus* 1, *Bubalus* 4.
Subfamily Cephalophinae	
Tribe Cephalophini	Duikers, mostly small forest antelopes, rare as fossils. *Cephalophus* c.16, *Sylvicapra* 1.
Subfamily Hippotraginae	
Tribe Reduncini	Waterbuck group. *Kobus* 4, *Redunca* 3.
Tribe Hippotragini	Roan, sable antelope and relatives. *Hippotragus* 3, *Oryx* 3, *Addax* 1.
Subfamily Alcelaphinae	
Tribe Alcelaphini	Hartebeest and allies. *Connochaetes* 2, *Alcelaphus* 2, *Damaliscus* 2, *Beatragus* 1, *Aepyceros* 1.
Subfamily Antilopinae	
Tribe Neotragini	Dik dik and other small antelopes. *Dorcatragus* 1, *Raphicerus* 3, *Neotragus* 3, *Madoqua* 5, *Pelea* 1, *Oreotragus* 1, *Ourebia* 1.
Tribe Antilopini	Gazelles, springbok, etc. *Gazella* 14, *Antilope* 1, *Antidorcas* 1, *Litocranius* 1, *Ammodorcas* 1, *Saiga* 1, *Pantholops* 1.
Subfamily Caprinae	
Tribe "Rupicaprini"	Goral and serow group, absent from Africa. *Rupicapra* itself might be better placed in Caprini. *Rupicapra* 1, *Oreamnos* 1, *Capricornis* 2, *Nemorhaedus* 2.
Tribe Ovibovini	Muskox, takin. *Ovibos* 1, *Budorcas* 1.
Tribe Caprini	Goats and sheep. *Capra* 7, *Ovis* 4, *Pseudois* 1, *Hemitragus* 3.

Bovid Beginnings

The Gelocidae are likely to include the ancestors and be the nearest relatives of later Pecora (Webb & Taylor 1980; Sudre 1984), but Janis & Scott (1987) propose that this family is polyphyletic.

Leinders (1984) concluded that among Pecora giraffoids are related to bovids because they share: (1) absence of a cervid-like bony bridge across the anterior distal gully on the metatarsal; (2) fewer differences in the amino acids of their ribonucleases than between cervids and bovids; (3) participation of dermal ossicones in horn ontogeny. He had some doubts about character 3, much justified according to Duerst (1926) and Gadow (1902). He did not use at all the single lachrymal orifice of giraffids and most bovids, evidently because he regarded it as primitive to the double orifice of cervids. Leinders & Heintz (1980) had previously used the bridge in character 1 to validate *Moschus* as a cervoid, but if its presence is advanced, then its absence ought not to be used in a cladistic context to relate bovids to giraffoids.

The detailed analysis of Janis & Scott (1987) came to the different conclusion that bovids and cervoids were linked, mainly on the presence in upper molars of

FIGURE 1. Horns in some extant African antelopes. (Not to scale.) A, *Tragelaphus imberbis* lesser kudu; B, *T. strepsiceros* greater kudu; C, D, *T. scriptus* (two views) bushbuck; E, *T. eurycerus* bongo; F, *Syncerus caffer* buffalo; G, *Cephalophus rufilatus* red-flanked duiker; H, *Sylvicapra grimmia* common duiker. (From Bryden 1899.)

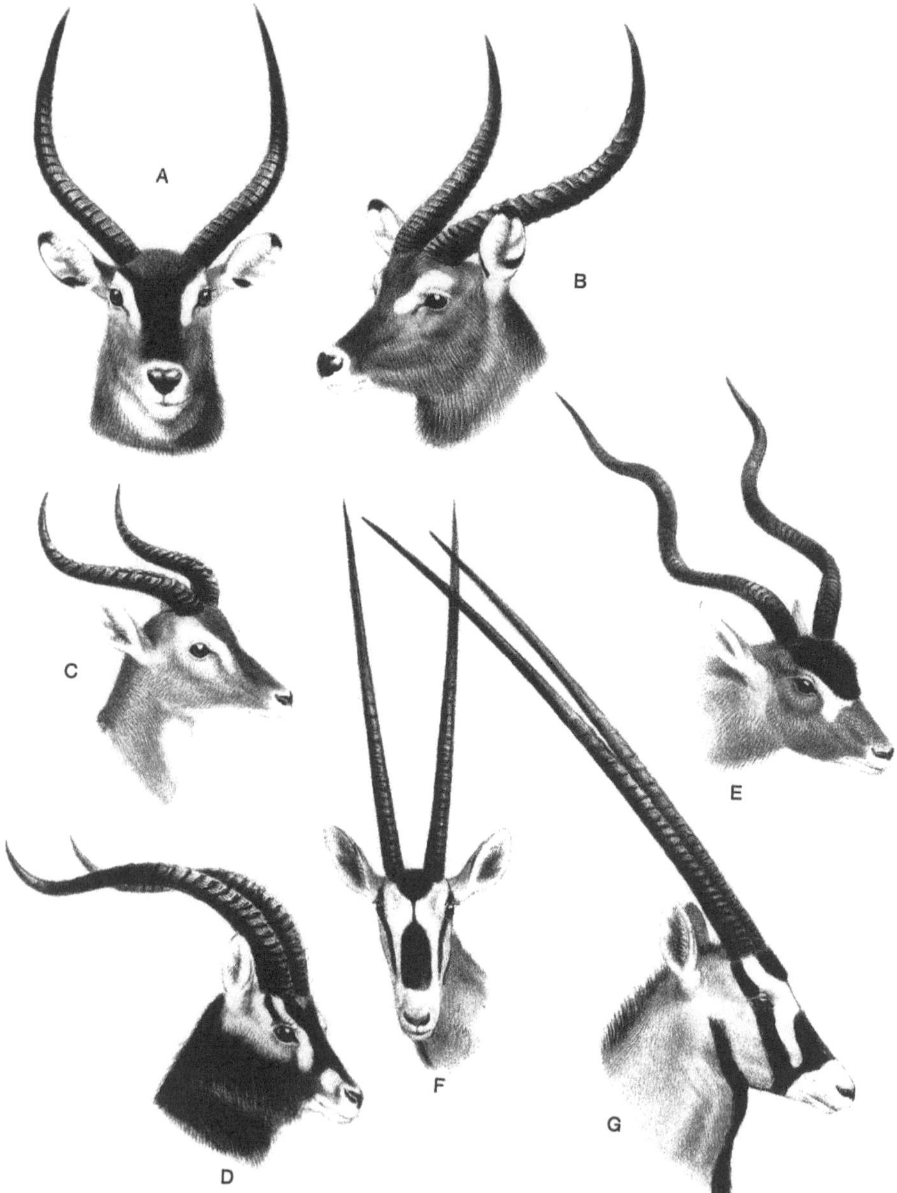

FIGURE 2. Horns in some extant African antelopes. (Not to scale.) A, B, *Kobus ellipsiprymnus* (two views) waterbuck; C, *Kobus kob* kob; D, *Kobus megaceros* Nile lechwe; E, *Addax nasomaculatus* addax; F, G, *Oryx gazella* (two views) oryx. (From Bryden 1899.)

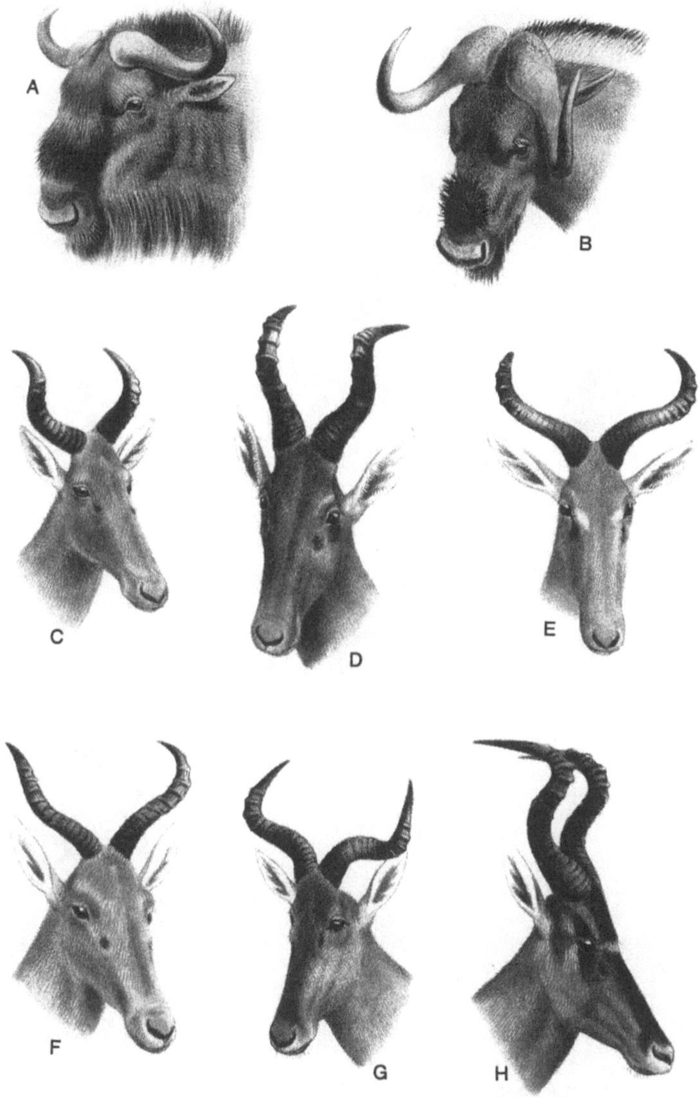

FIGURE 3. Horns in some extant African antelopes. (Not to scale.) A, *Connochaetes taurinus* blue wildebeest; B, *C. gnou* black wildebeest. Remaining figures are some horn varieties within *Alcelaphus buselaphus*, the hartebeest. C, *A.b. buselaphus*; D, *A.b. lelwel*; E, *A.b. tora*; F, *A.b. neumanni*; G, *A.b. cokei*; H, *A.b. caama*. (From Bryden 1899.)

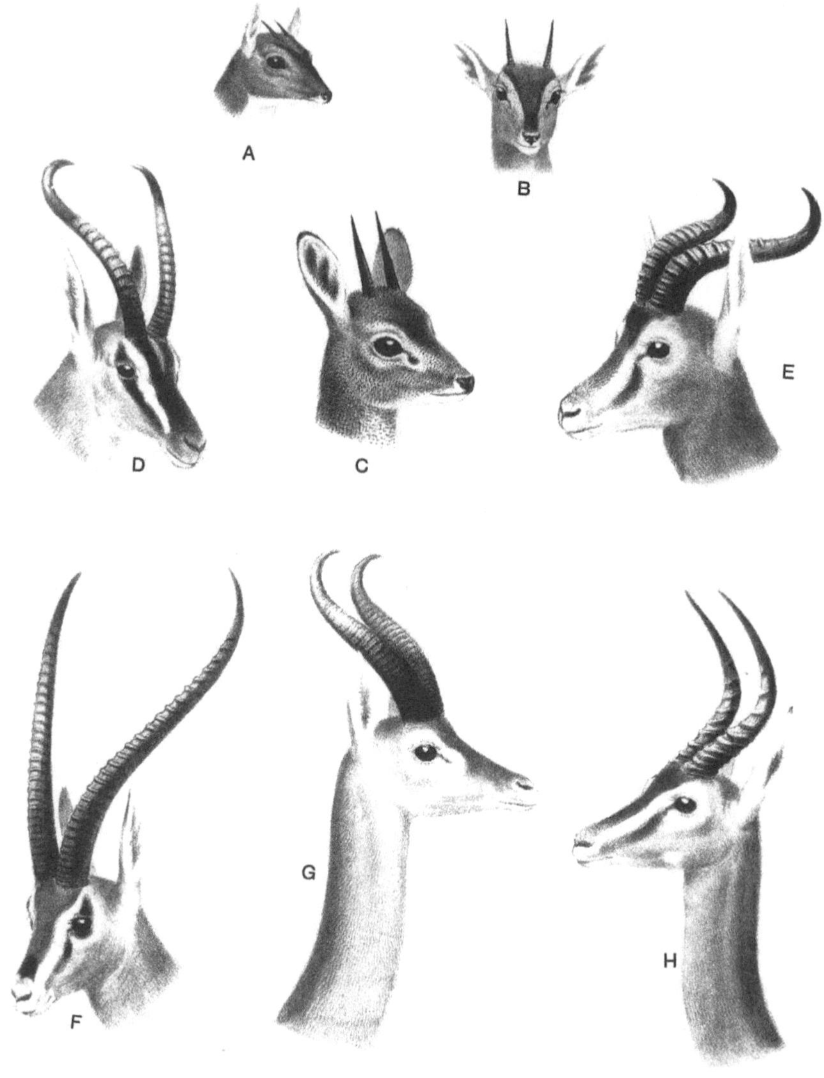

FIGURE 4. Horns in some extant African antelopes. (Not to scale.) A, *Neotragus pygmaeus* royal antelope; B, *N. moschatus* suni; C, *Oreotragus oreotragus* klipspringer; D, *Gazella soemmerringi* Soemmerring's gazelle; E, *G. dama* addra gazelle; F, *G. granti* Grant's gazelle; G, *Litocranius walleri* gerenuk; H, *Ammodorcas clarkei* dibatag. (From Bryden 1899.)

an entostyle derived from the metaconule. This character, however, is not uncommon in Miocene giraffoids in a (?)vestigial state.

If, despite reservations, one accepts a relationship between bovids and giraffids and accepts also that cervoids developed in Eurasia and giraffoids in Africa, then it would be likely that bovids too evolved in Africa.

This expectation is not borne out in the fossil record. The East African Lower Miocene pecoran *Walangania* would be the most obvious candidate for an early bovid relative. It is rather incompletely known except for a juvenile skeleton (Whitworth 1958) and is similar to the contemporaneous or slightly older cervoid *Dremotherium* in Europe. The *Palaeomeryx* fold (a posterolabial diagonal fold on the protoconid of the lower molars) may be less marked than in the European form, and *Walangania* may also be slightly higher crowned. But these are not very certain, and other key characters are unknown or as yet unpublished: the state of the distal metatarsal, the size of its upper canines, or whether it was hornless. If *Dremotherium* is already cervoid on the strength of its metatarsal bridge, then the split of cervoids from bovids and giraffoids must predate the appearance of frontals' appendages. Thus the antler of, e.g., *Procervulus* (see Stehlin 1937) would be a parallel to the horns of the earliest European bovid *Eotragus*. Horns could have evolved up to seven times in ruminants (Janis 1982).

Some early possible bovids are known from Lower/Middle Miocene localities in Africa and Arabia (Hendey 1978; Thomas et al. 1982; Whybrow et al. 1982). They are incomplete and have some puzzling characters; e.g., the bifurcated metaconule on the bovid M^3 from As Sarrar (Thomas et al. 1982). The oldest bovids in the Indian subcontinent occur at 16.1 my (Barry et al. 1985), and they appear to be as old or older than this further north in Asia (Chen 1988).

In Europe all the records of the earliest bovids have been put into *Eotragus*. This form of small-medium size is best known from the classical Middle Miocene locality of Sansan but has never been properly monographed. Females were hornless (Thomas 1977). Earlier horn cores from Bunyol are notably squat and as primitive as can be imagined (Moya Sola 1983). Three African claims have been made for *Eotragus*; at Zelten (Hamilton 1973), Maboko (Thomas 1979), and Fort Ternan (Gentry 1970). The Maboko cranium of *Eotragus*, BM(NH) M15544, has a cranial roof more strongly curved down posteriorly than at Sansan, and this suggests a range of morphology on different continents. Maboko is probably younger than the oldest European localities with *Eotragus* (Fig. 5) but the intercontinental correlation is insecure.

The Sansan *Eotragus* agrees with boselaphines by its horn cores being keeled and by the braincase and facial axes being in line and not angled on one another. However, Thomas (1984) thinks that keels are rare in *Eotragus*, and Moya Sola (1983) excludes the genus from the Boselaphini. A possible complication is that early members of both Caprini and Boselaphini may have been conflated under the name *Eotragus*. The two early Middle Miocene (= late Orleanian?) horn cores shown in Figs. 1 and 5 of Thenius (1952) need not be congeneric, and the latter could be related to the caprine *Caprotragoides*. In general it is evident that early bovid-like pecorans and bovids embraced a diversity of morphologies and therefore of presumed lineages also. When bovids become better preserved in the later Middle Miocene, we can see that the several lineages are each present in both Eurasia and Africa.

African Bovids in the Miocene

Bovids other than *Eotragus* known in Africa during the Miocene are various boselaphines, caprines, and two smaller forms: *Gazella* and the neotragine *Homoiodorcas*. The main characters of Boselaphini are horn cores with keels but no transverse ridges, braincases little angled on the facial axis, strong temporal ridges on the cranial roof, cheek teeth which are brachyodont or at least not very hypsodont, enamel often rugose, lower molars generally without goat folds and premolar rows long. The two extant Indian species are obviously relicts of a formerly more diverse and widespread tribe (Fig. 6).

Boselaphines succeeding *Eotragus* developed an *Antilocapra*-like demarcation on their horn cores. The tips are of decidedly smaller diameter than the bases, and the anterior edges, seen in profile, have a step at the transition. Suggestions that the overlying horn sheaths may have been pronged have not been welcomed by all workers (Janis & Scott 1987). Such horn cores were at first short and have often been referred to *Protragocerus* (see Thenius 1956).

Similar horn cores, referred to as *Miotragocerus*, come predominantly from later deposits (Gentry 1980), although the temporal and morphological overlap make the generic separation a difficult subject. The *Miotragocerus* stock persisted until the latest Miocene or slightly later, and some developed long backwardly curved horn cores and lost the demarcation (e.g., *M. curvicornis* of Samos and *M. browni* of the Siwaliks).

In Africa the most completely known early boselaphine is *Protragocerus labidotus* Gentry (1970) from Fort Ternan. It has much compressed horn cores in males only, with a forward growth toward one another in front of the horn core proper. The horn core tips curve inward, with a strong distal demarcation. Its cranial proportions are low and wide rather than high and narrow. The distal limb bones are longer in the front leg than in either of the two living Boselaphini. It is probably the specialized horn cores which have attracted a lot of taxonomic attention to this species (Solounias 1981; Moya Sola 1983), culminating in the new name *Kipsigicerus* Thomas (1984d). More will have to be said on *Kipsigicerus* because, contrary to Thomas's (1984d) cladogram, it does not share large supraorbital pits with *Helicoportax* and *Selenoportax* (character 10) and it does share both a distal demarcation on its horn cores and an absence of a roughened frontoparietal area with *Protragocerus gluten* (characters 3 and 2). This bosel-

---

►

FIGURE 5. Geochronology for the main localities and geological formations mentioned in this chapter. Divisions of the Pleistocene are not shown, but the Middle Pleistocene begins at 0.7 my and the Upper Pleistocene at 0.13 my. *A* = Algeria, *E* = Ethiopia, *Eg* = Egypt, *F* = France, *G* = Greece, *I* = Italy, *K* = Kenya, *L* = Libya, *M* = Morocco, *S* = Saudi Arabia, *Sp* = Spain, *T* = Tanzania, *Tk* = Turkey, *Tu* = Tunisia, *U* = Uganda. Geological terms: *Fm* = formation, *Mb* = member.

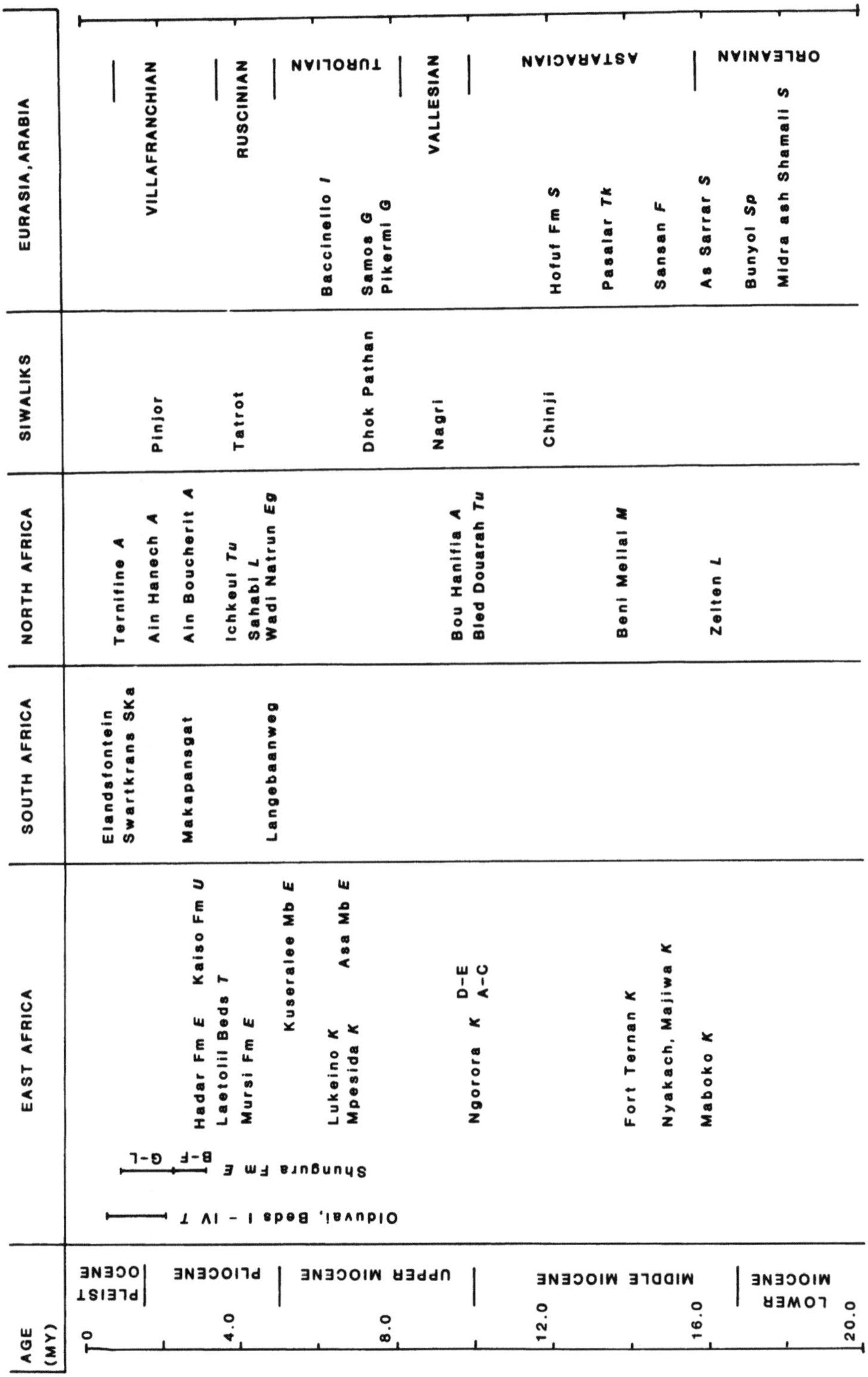
EURASIA, ARABIA
VILLAFRANCHIAN
RUSCINIAN
TUROLIAN
VALLESIAN
ASTARACIAN
ORLEANIAN
Baccinello I
Samos G
Pikermi G
Hofuf Fm S
Pasalar Tk
Sansan F
As Sarrar S
Bunyol Sp
Midra ash Shamali S
SIWALIKS
Pinjor
Tatrot
Dhok Pathan
Nagri
Chinji
NORTH AFRICA
Ternifine A
Ain Hanech A
Ain Boucherit A
Ichkeul Tu
Sahabi L
Wadi Natrun Eg
Bou Hanifia A
Bled Douarah Tu
Beni Mellal M
Zelten L
SOUTH AFRICA
Elandsfontein
Swartkrans SKa
Makapansgat
Langebaanweg
EAST AFRICA
Olduvai, Beds I - IV T
Shungura Fm E
B-F
G-L
Hadar Fm E
Kaiso Fm U
Laetolil Beds T
Mursi Fm E
Kuseralee Mb E
Lukeino K
Mpesida K
Asa Mb E
Ngorora K
D-E
A-C
Fort Ternan K
Nyakach, Majiwa K
Maboko K
AGE (MY)
0
4.0
8.0
12.0
16.0
20.0
PLEISTOCENE
PLIOCENE
UPPER MIOCENE
MIDDLE MIOCENE
LOWER MIOCENE

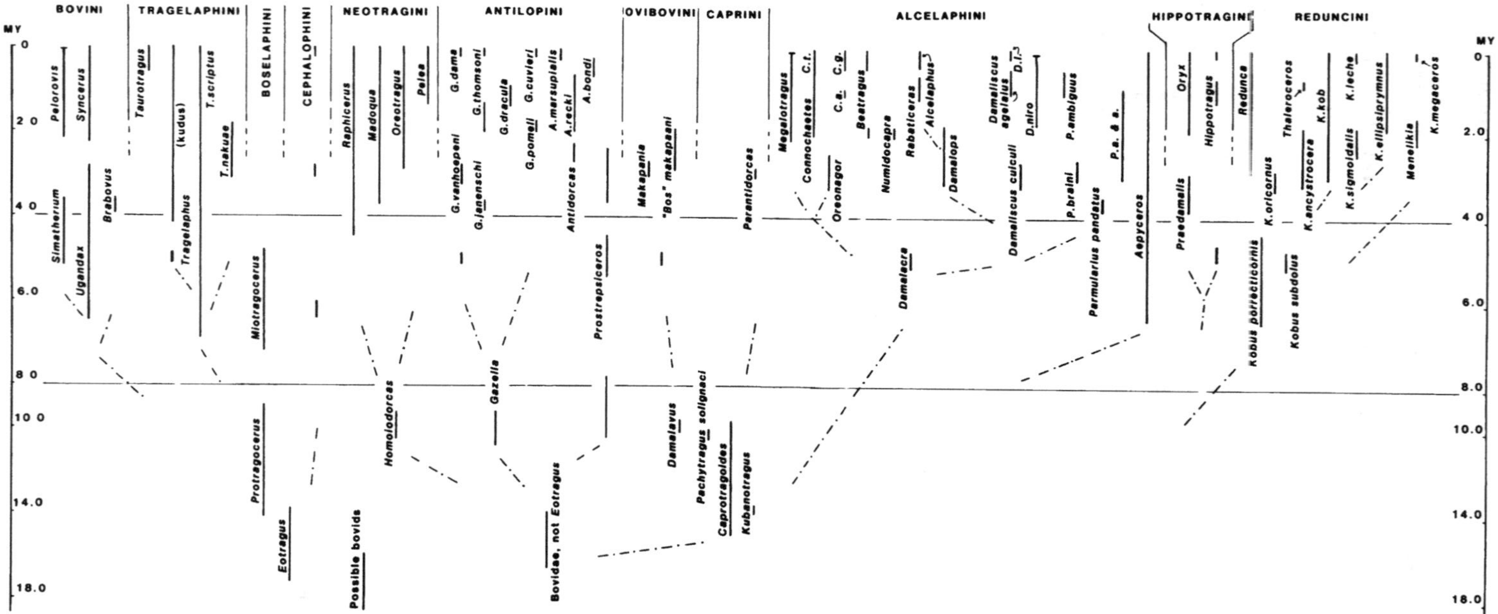

FIGURE 6. Time ranges of African Bovidae. Scale is condensed from 8.0–18.0 my. Tentative lines of relationship or descent are shown. Tribes Bovini to Cephalophini are associated with early boselaphines; Neotragini to Alcelaphini with early nonboselaphine antelopes. The placing of Hippotragini and Reduncini is left open. *C.a.* = *Connochaetes africanus*, *C.g.* = *Connochaetes gnou*, *C.t.* = *Connochaetes taurinus*, *D.l.* = *Damaliscus*, extant species, *P.a. & a.* = *Parmularius altidens* and *angusticornis*. (From Bryden 1899.)

aphine, *P. labidotus*, was still present later in the Middle Miocene along with the poorly known *Sivoreas* (Thomas 1981) and a larger boselaphine.

The latest boselaphines in Africa are *Miotragocerus cyrenaicus* from Sahabi, included in *Tragoportax* by Moya Sola (1983), and *Mesembriportax acrae* from Langebaanweg (Gentry 1980). Unnamed late boselaphine fossils are also known from other localities, e.g., the Asa and Kuseralee members (Kalb et al. 1982). Thomas (1984d) supports the view that horns in the female sex appeared in boselaphines late in the Miocene.

The earliest bovids other than *Eotragus* and Boselaphini are the Caprini *Kubanotragus* (= *Hypsodontus* and many other synonyms; see Whybrow et al. 1982) and *Caprotragoides* (= *Pseudotragus* in part, Thomas 1984a). Both occur in pre-*Hipparion* faunas of Asia, Africa, and Europe, and *Kubanotragus* seems to occur as early in Asia as *Eotragus* in Europe (Chen 1988). *Kubanotragus* has horn cores of rounded cross-section and slight or moderate torsion (clockwise on right side); *Caprotragoides* has horn cores with mediolateral compression and backward curvature. Both are without keels and have upright insertions.

In Africa the most complete and numerous remains of *Kubanotragus* occur at Fort Ternan (Gentry 1970). Females are without horns, although Thomas (1984c) argues for horned females in some Asian populations. *Caprotragoides* was present but rare at Fort Ternan and differed additionally from *Kubanotragus* by the hypsodonty of its cheek teeth, far more pronounced than in Turkish *Caprotragoides* (Köhler 1987). A later *Caprotragoides* species from Ngorora B-D was larger and showed dental advances, such as a more complicated outline of central cavities on its molars and paraconid-metaconid fusion on P_4 (Thomas 1981). An earlier *Caprotragoides* from Nyakach had an approach to a keel on its horn core and thereby suggested a relationship to *Benicerus* of Beni Mellal (Thomas 1984a; 1984c) although *Benicerus* looks as if it had a more inclined cranial roof than did *Caprotragoides*.

Pachytragus solignaci is also present in Africa, in both pre-*Hipparion* and *Hipparion* levels at Bled Douarah. This species has mediolaterally compressed horn cores with an anterior keel, reminiscent of the later and more advanced *P. crassicornis* of Samos. However, its teeth (see Fig. 7) differ from Samos *Pachytragus* by: (1) less marked hypsodonty, (2) enamel more rugose, (3) lingual walls of lower molars more outbowed, and (4) labial lobes of lower molars with better marked ogival symmetry (Thomas 1983). In my opinion, *P. solignaci* should now be removed from *Pachytragus* along with the Ngorora *P.* aff. *solignaci* Thomas (1981). These two could be congeneric with *Protoryx carolinae* of the Pikermi Turolian, which has more primitive teeth than *Pachytragus*, or they could be an independent group. They are certainly not boselaphines, so Caprinae is an acceptable subfamily placing.

An additional complication is the large-sized *Pachytragus ligabue* of the Hofuf Formation, Arabia, related by Thomas (1983) to *P. solignaci* mainly on metaconid-entoconid fusion in the lower premolars and ogival labial lobes and slightly outbowed lingual lobes of the lower molars. The premolar character occurs in P_4s of living Caprinae.

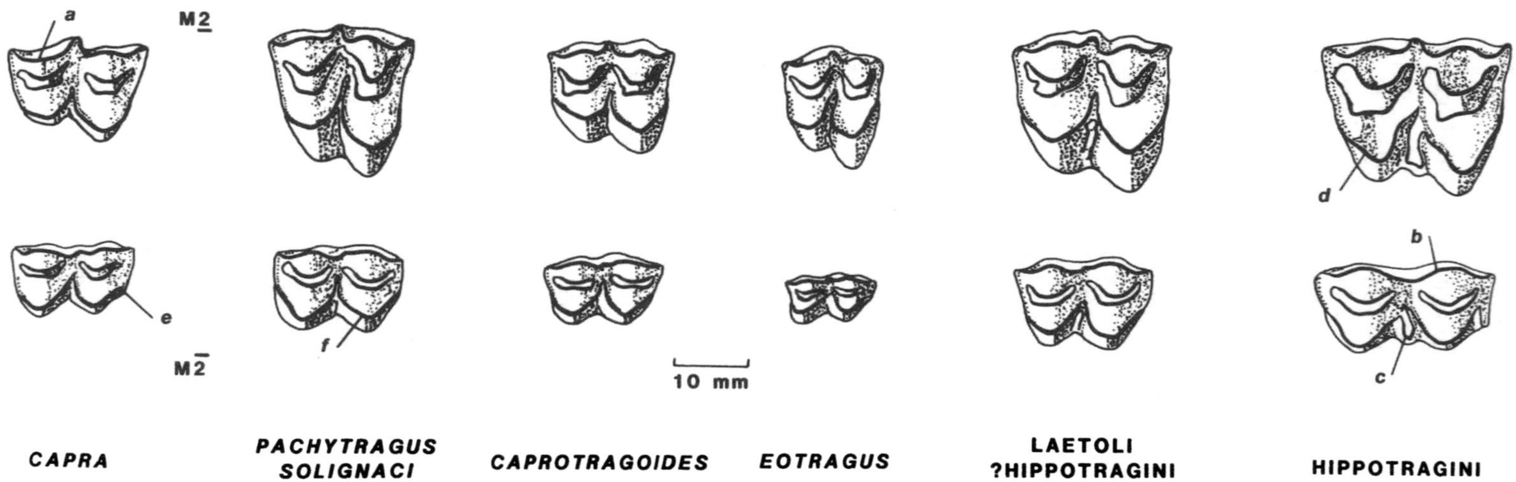

FIGURE 7. Right M^2s (top row) and right M_2s (bottom row) to show contrasting morphologies of aegodont (*Capra*) and boodont (Hippotragini) teeth in extant Bovidae and the intermediate condition of some fossils. Anterior to the right. *a* = flattened or outbowed labial rib between mesostyle and metastyle; *b* = outbowings of lingual walls of lower molars; *c* = presence or absence of basal pillars (entostyles, ectostylids); *d* = constriction of lingual lobes. Also note: *e* = transverse goat fold evolved in both *Capra* and Hippotragini; *f* = ogival symmetry of labial lobes. (From Bryden 1899.)

The African *Pachytragus* differs from African *Caprotragoides* in the inclined braincase roof, less upright horn core insertions in side view, and less hypsodont teeth, and is difficult to envisage as a direct descendant of that *Caprotragoides*. Köhler (1987), however, has advocated a *Caprotragoides-Protoryx* transition in the later Middle Miocene of Turkey. With these many perplexing discoveries, Solounias's (1981) proposal to put *Pachytragus sensu stricto* back into *Protoryx* seems too simple.

Early gazelles have small and little-compressed horn cores without keels and curved backwards; they also have rather brachyodont teeth with long premolar rows and M_3s without an enlarged back lobe.

Small teeth from Maboko not belonging to *Eotragus* have been accepted as *Gazella* (Thomas 1979). Their strong mesostyle (Whitworth 1958) matches undetermined bovid teeth from Beni Mellal (Lavocat 1961). However, they might still belong to the small giraffoid *Nyanzameryx pickfordi* Thomas (1984a), for which Maboko is the type locality. This would leave some fragmentary horn core, dental, and postcranial material from the Nyakach Formation and Majiwa as possible *Gazella* (Thomas 1984a).

The hornless skull of a female gazelle-like bovid from Fort Ternan is distinctive by the close approach of the temporal lines on its skull roof, the back of the cranial roof bent downwards relative to the front part, and the metaconid on P_4 turned forward toward the paraconid. Gentry (1978b) has already suggested a relationship with *Aepyceros* instead, but a chain of later intermediates is needed to demonstrate this.

Some horn cores and other remains from the Ngorora Formation have been named *Homoiodorcas* (Thomas 1981) and placed in the Neotragini instead of the Antilopini. The major axis of the cross-section at the base of the horn cores is oblique rather as if pressure had been applied from the anterolateral side to give a slight inward tilt to the horn core base. Contrary to Thomas's opinion of this as an apomorphy excluding relationship to gazelles, Gentry (1980) reported it in the Langebaanweg gazelle. It also occurs in *Madoqua*.

A small bovid mandible from Zelten is deep posteriorly (Hamilton 1973), as in *Antidorcas*, but this would be a doubtful attribution. Also uncertain is Thomas's (1981) claim for *Antidorcas* at Ngorora. The horn core piece shows thinner frontals than the slightly smaller Reduncini at the same locality. The smallness of P_2, deduced from alveoli, may be mistaken, because gazelle P_2s are often anteroposteriorly longer than their alveoli (Pilgrim 1937).

It is *Gazella praegaudryi* (Arambourg 1959) from Bou Hanifia which seems to be the earliest unequivocal gazelle in Africa. It has little compressed horn cores and no forward turning of the metaconid on its P_4. The back lobe of its M_3 might have become larger in later wear and thus foreshadowed the condition in later gazelles, other Antilopini and Neotragini, and *Aepyceros*.

On the question of horns in the females of Miocene gazelles: Heintz (1969) noted hornlessness in the Turolian of Léberon, France, and Gentry (1970) referred to other instances of hornlessness and to the likely presence of horned females at Samos.

Lehmann & Thomas (1987) have identified some long, noncompressed horn cores with anticlockwise torsion from Sahabi as a *Prostrepsiceros* larger than any previously described (Bouvrain 1982). This is a bold identification, which I accept. More important than the appropriateness of this particular generic name is the finding of a large member of the spirally horned antilopines in Africa. A more normal-sized *Prostrepsiceros* has also been cited for upper Vallesian equivalent deposits at Oued el Atteuch (Thomas et al. 1982).

In a synthesis of bovid biogeographical history, Thomas (1984c) proposed two successive temporal phases. The first, in the later Orleanian and early Astaracian in European terms, contained faunas with Caprini, Antilopini, and *Eotragus*. The second, corresponding to Middle and Late Astaracian, was when Boselaphini radiated.

Of the localities proposed by Thomas for his first phase, Belometscheskaya (Russia) must have been included by mistake since it contains a boselaphine—as he acknowledged later (Thomas 1984c). Among the others, Prebreza (Yugoslavia), Pasalar, and Nyakach contain caprines, but their age relative to boselaphine localities such as La Grive (France) and the Vienna Basin is not known with precision. In Africa it is likely that Nyakach does predate Fort Ternan, but we have yet to be sure that the absence of a *Protragocerus*-like boselaphine there does not arise from ecological or taphonomic causes. My view is that definite *Caprotragoides* and post-*Eotragus* boselaphines appear broadly contemporaneously in the fossil record but not always in company. The remaining localities of Thomas's first phase contain *Eotragus* or some other possible bovid (which may or may not be an antilopine), and these localities probably are earlier than the *Caprotragoides* and boselaphine faunas. *Kubanotragus* is probably already present at this time level in China (Chen 1988) and possibly at Midra ash-Shamali in Arabia as well (Whybrow et al. 1982).

African Bovids in the Pliocene and Pleistocene

Thomas (1980; 1984c) emphasizes the appearance of the modern tribes of African bovids at Lukeino and Mpesida. At Lukeino there are reduncine teeth and horn cores, and definite *Aepyceros* and a fine *Tragelaphus* frontlet leave no doubt about two more African tribes. Bovini also come in for the first time, as in Eurasia.

As noted above for boselaphines, caprines and gazelles, it is probably in the Late Miocene that horns were first acquired in females, and this too may be linked to life in more open habitats.

The diversity of habitats and the gradient between open and closed ones is crucial for African bovids; this is in contrast to Eurasian caprines, which are confined to mountainous or elevated terrain (Schaller 1977). The territoriality of most African bovids is another contrast to Caprinae and perhaps is linked to life in tropical rather than temperate latitudes (Owen-Smith 1977). If it is correct that horns were originally evolved in association with territorial systems as believed by Janis (1982), it is interesting that caprines (and bovines and tragelaphines) continue to bear horns.

Tragelaphini

Tragelaphini have spiraled horns with anticlockwise torsion and two or three longitudinal keels positioned anteriorly, posteriorly, and posteromedially. The braincase is not very angled on the facial axis. The cheek teeth are notable for brachyodonty, absence of basal pillars, and absence of ribs between the styles of the upper molars. Premolar rows are long and P2 and P3 are large. *Taurotragus* and the bongo, *Tragelaphus eurycerus*, have horned females, but other species do not. They are browsing antelopes–broad-spectrum foliage gleaners in bush and woodland (Kingdon 1982).

Horn cores or teeth of small fossil tragelaphines are taken as *Tragelaphus* cf. *scriptus* (bushbuck) or as *T.* cf. *angasi* or *spekei* if they are a little larger. Horn cores of such early tragelaphines are compressed less anteroposteriorly and inserted more uprightly than in living forms. They are present at Mpesida and as a fine frontlet of *T.* cf. *spekei* at Lukeino. A small species is present in the Asa Member (Kalb et al. 1982).

More strongly spiraled horn cores are recognized as kudus. Early possible records are at Langebaanweg and in the Mursi Formation. In Shungura C-G is *T. gaudryi*, which is the size of extant *T. imberbis* (lesser kudu) but with a strong anterior keel. The horn cores are less spiraled than in *T. strepsiceros* (greater kudu), but the accentuation of its anterior keel and increasing mediolateral compression passing up the succession are approaches to *T. strepsiceros*. The appearance of *T. strepsiceros* itself in Shungura G is abrupt, and it is possible that *T. gaudryi* thereupon showed a degree of reversal and evolved into *T. imberbis* by loss of its anterior keel. *T. strepsiceros* is present at Olduvai and reaches a large size in Middle and Upper Bed II.

Unrelated to living tragelaphines is *T. nakuae* of Shungura B-H with anteroposteriorly compressed horn cores inserted at a very low angle. Interestingly, it agrees more with Boselaphini than Tragelaphini in the braincase being set high and not at all inclined to the facial axis, strong temporal ridges, dorsal parts of the orbital rims projecting quite strongly, and supraorbital pits not narrowed or drawn out anteroposteriorly. However, the teeth are fully tragelaphine. A similar species, certainly related and perhaps ancestral, occurs in Hadar DD.

Hadar SH, earlier than DD, has an unnamed *Tragelaphus* (Gentry 1981) which has the peculiarity of lacking keels proximally on its horn cores. It could be a normally archaic tragelaphine except for this specialization, which was later paralleled in *T. imberbis*.

Taurotragus (eland) is as large and heavy as a buffalo but not at all a ponderous animal. Ecologically it is eurytopic (not narrowly adapted; Kingdon 1982). It may be a late evolution from *Tragelaphus*. The large *Tragelaphus algericus* Geraads (1981) from Ternifine (= Tighenif) has horn cores intermediate in appearance between kudu and eland. Geraads was properly cautious about relating this animal to eland because of its relatively late occurrence, especially in relation to the Olduvai *Taurotragus* M 29415 (Gentry & Gentry 1978). However, the latter find was from the surface and perhaps derived from higher deposits. The North African species certainly makes it wise to reconsider earlier occurrences of *Taurotragus*.

Taurotragus arkelli of Olduvai IV (Leakey 1965) has horn cores slightly more uprightly inserted and a braincase less shortened than in living elands. Therefore it is less advanced than the living species.

Bovini

Bovini are large-sized descendants of boselaphines. They have low and wide skulls, horn cores in both sexes emerging transversely from their insertions, internal sinuses in frontals and horn cores, a short braincase and triangular basioccipital, molars with basal pillars and complicated central cavities, upper molars with prominent ribs between the styles, and lower molars without large goat folds. Kindgon describes them as fresh grass bulk grazers. There is one living species in Africa: the buffalo, *Syncerus caffer.*

The African fossil record contains both the *Syncerus* stock and an extinct lineage of long-horned buffaloes, *Pelorovis*. The latter is closer to *Syncerus* than to the Asiatic *Bubalus*, as seen by its shorter face, irregular or absent keels on the horn cores, and a tendency to fusion of paraconid and metaconid on P_4 and other characters. Gentry (1980) holds that the Pliocene *Simatherium* is a probable ancestor of *Pelorovis*. Its horn cores are inserted closer to the orbits, are wider apart, and diverge less toward their tips.

A major find at Olduvai was a herd of *Pelorovis oldowayensis* in Upper Bed II. This species had horns that rise from their insertions (with the head held vertically) and swing round to descend at the tips. It looks as if some individuals would have had difficulties in feeding at ground level or drinking if their horn sheath tips came down to a lower level than the mouth. Earlier examples of *Pelorovis* are less large, and a variety with small, flattened horn cores is known from Shungura D and F. *Pelorovis* is also known from Ain Boucherit and Ain Hanech, as *Bos palaethiopicus*, *B. bubaloides* and *B. praeafricanus* (Arambourg 1979; Geraads 1981).

Bovine teeth at Lukeino have been assigned to *Ugandax*, ancestral to *Syncerus*, and are the earliest record of bovines in Africa (Thomas 1980, 1984c). Kalb et al. (1982) cite a bovine frontlet from the Kuseralee Formation. *Ugandax* is also in the Kaiso (Cooke & Coryndon 1970) and Hadar Formations (Gentry 1981). It has a less shortened braincase than *Syncerus* and horn cores which pass more upward and backward than outward. It is similar to *Proamphibos* from the Tatrot Formation but has less accentuated keels on its horn cores.

Fossil *Syncerus*, as at Shungura and Olduvai, has no large basal bosses like eastern and southern *S. caffer* populations today. The horn cores are internally hollowed only near their bases and are triangular in cross-section, with anterior, upper, and lower surfaces. Skull parts from low in the Shungura Formation have less shortened braincases and stronger temporal ridges than *S. caffer.* In member

C the *Syncerus* was small and short-horned, suggesting that forest or woodland races, comparable to the west African *S. c. nanus* of today, had long been present in the *Syncerus* lineage. The interesting question is the route of genetic descent from two morphs of an ancestral species to two morphs of a descendant species.

An intriguing bovine from North Africa is the Sahabi *Leptobos syrticus* of Petrocchi (1956), a unique early and African occurrence of this otherwise Eurasian late Pliocene genus. Compared with other *Leptobos* the horn core insertions are closer to the orbits, there are keels (anterior and posterolateral) over parts of the horn cores' course and the supraorbital pits are not very wide apart. These are all primitive and make it difficult to retain *L. syrticus* within *Leptobos*. Pilgrim (1937) mentioned the survival of weak keels in a specimen of the Siwaliks *L. falconeri*. Despite its primitive characters, *L. syrticus* does not appear close to *Alephis* (formerly *Parabos*) which precedes *Leptobos* in Europe.

Brabovus Gentry (1987) is an enigmatic skull from the Laetolil Beds. It is taken as a bovine and showed rather a small size, brachyodont cheek teeth and small, equally sized lower incisors and canine. If it is indeed a bovine, it would be significant in lacking such heritage resemblances to boselaphines as keels on the horn cores and temporal ridges on the cranial roof.

Cephalophini

Cephalophini or duikers are solitary frugivores and selective browsers in forests, rather heavily built, and with short spike-like horns. They are rare as fossils but have been important in recent discussions of bovid phylogeny. Compared with *Cephalophus*, *Sylvicapra grimmia* (bush duiker) has more upright horns and longer legs and lives in savannah and woodland.

Thomas (1980) assigned an upper molar from Lukeino to the Cephalophini, and Gentry (1978b) mentioned records from the Late Pleistocene of South Africa and from Makapansgat. Gentry (1978b) doubted some earlier identifications of fossil duikers, and Thomas (1981) expressed similar doubts about a record by Gentry (1978a) from the Ngorora Formation. The *Cephalophus leporinus* from Ain Boucherit (Arambourg 1979) is also suspect; the mandible of Arambourg's Fig. 4 looks like a worn *Parantidorcas latifrons*, common at that locality, and the other specimens are questionable as well.

Reduncini

Reduncines have transverse ridges on the horn cores (which are present only in males), temporal ridges often approaching closely on the cranial roof, a large maxillary tuberosity prominent in ventral view, small cheek teeth in relation to skull size, basal pillars on molars, constricted medial lobes of upper molars and lateral lobes of lowers, upper molars with ribs between styles, lower molars with goat folds, P2s small, lower premolars appearing anteroposteriorly compressed, and P_4 with strongly projecting hypoconid. The three species of *Redunca* are a neat group morphologically but less so ecologically, by reason of the mountain reedbuck, *R. fulvorufula* diverging in habitat choice. *Kobus* embraces a wider range of horn core morphology, but other skull characters tend to be shared with *Redunca* so it is hard to concoct a satisfactory diagnosis of *Kobus*. Living reduncines are grazing antelopes nearly all found in habitats associated with the vicinity of water.

Kobus subdolus Gentry (1980) from Langebaanweg has rather short horn cores with little compression, an approach to a flattened lateral surface, little backward curvature, insertions at a low inclination, and divergence not very great. Similar horn cores come from Sahabi (Lehmann & Thomas 1987) and perhaps Ichkeul, *Redunca khroumirensis* (Arambourg 1979). Somewhat smaller early reduncines have been referred to as *Kobus porrecticornis*, as at Mpesida and Lukeino (Thomas 1980; Gentry 1980).

In general earlier reduncine teeth are less distinctive than later ones, e.g., lower molars with less constricted labial lobes (Arambourg 1979). The teeth from Mpesida and Lukeino nevertheless appear to be truly reduncine, so it is a matter of some difficulty that the Langebaanweg teeth assigned to *K. subdolus* are unlike other reduncines and have primitive resemblances to Tragelaphini.

Gentry (1981) described how the commonest reduncine at Hadar has a P_4 more advanced than later reduncines in that paraconid and metaconid are in contact or fused along the lingual side of the tooth. This form had long, noncompressed horn cores without backward curvature basally, and the lineage may have ended with *Kobus oricornus* in Shungura upper B.

A reduncine of Shungura E-J is *Menelikia lyrocera*, figured by Arambourg (1947). Its most distinctive feature is extensive frontals' sinuses, which in other reduncines are very restricted. Horn core morphology can be seen changing on this lineage; in Members E-G the divergence increases from the base then decreases toward the tip. Later horn cores are squatter and have less pronounced changes in the direction of curvature. A probable earlier species is represented in Member C by some very long horn cores with more gradual curvature. The living Nile lechwe, *Kobus megaceros*, of southern Sudan and perhaps Ethiopia has a conformation of its dorsal face bones reminiscent of *Menelikia*. *K. megaceros* has very long horn cores, so if it were related to *Menelikia* it could not be related to the Shungura E-J lineage.

Also present in Shungura B-J is *Kobus ancystrocera*, with horn cores without backward curvature, compressed basally, much recurved distally, and not very divergent. *Thaleroceros radiciformis* is a unique find from Olduvai IV, large, with massive, nearly parallel horn cores curving upwards and forwards from the base and mounted on a single, united pedicle. Gentry & Gentry (1978) sought to relate it to *K. ancystrocera* via an intermediate morphology at Kanam, Kenya.

Kobus sigmoidalis is abundant in Shungura E-G. Its horn cores are widely divergent, compressed and with basal backward curvature. In member G are horn cores difficult to place in either this species or *K. ellipsiprymnus* the extant waterbuck (Gentry & Gentry 1978), and it looks as if the one evolved into the other. However, something not stated at all by Arambourg (1941, 1947) in his descriptions of *sigmoidalis* is its resemblance to extant *K. leche*. Perhaps some populations of *sigmoidalis* lingered unchanged, adopted a more specialized mode of life, and survive as a restricted relict.

Kobus kob is also well represented in the fossil record but with more variation than seen in the lechwes and waterbuck just considered. In Olduvai II it shows little mediolateral compression, no backward curvature basally, and insertions quite wide apart. In Olduvai III and IV it is rather large and has acquired some compression and backward curvature. In members K and L of the Shungura Formation it is even larger but with shorter horn cores. Thus, if the correlations are well founded, we see geographical variation between the two areas, perhaps clinal or perhaps disjunct. It is also interesting that the evolution in *K. kob* horn cores toward compression and basal curvature is the reverse of what took place in the evolution of *K. ellipsiprymnus*. Perhaps individual recognition and interaction within the two lineages is involved.

Fossil *Redunca* is only common in South Africa. *R. darti* from Makapansgat has horn cores more uprightly inserted than in *R. redunca* or *arundinum* and the basal posteromedially flattened surface lying more medial than posterior. A larger species at Elandsfontein has short horn cores like *redunca* but is closer to *arundinum*.

Hippotragini

Hippotragini are medium to large, rather stocky antelopes. Both sexes have long, not very divergent horn cores without keels or transverse ridges, internallly hollowed frontals and horn core pedicles, hypsodont teeth without much premolar reduction, molars with basal pillars, lower molars with a tendency to have goat folds anteriorly, and P_4s without fusion of paraconid and metaconid. Kingdon describes them as low-density grazers in arid and impoverished zones.

Thomas (1980) refers to a horn core base from Mpesida which has internal sinuses in the pedicle and frontals and could perhaps be of this tribe. More definite is a Sahabi horn core with backward curvature like *Hippotragus* (Lehmann & Thomas 1987). The somewhat later *Praedamalis deturi* from Laetoli and Hadar DD is inserted rather uprightly like the Sahabi horn core and *Hippotragus* but is almost straight like *Oryx* (Dietrich 1950; Gentry 1981). Vrba (1987) founded *Wellsiana* for a distinctive but problematic frontlet from Makapansgat, related by her to *Brabovus* (see above) as a hippotragine.

It is not until the later Pliocene that one can discern definitive *Hippotragus* and *Oryx*. The latter is rarer but known from Olduvai I and Shungura upper G, and also from Ain Hanech and Ternifine in north Africa (Arambourg 1979; Geraads 1981). The Plio-Pleistocene *Hippotragus gigas* Leakey (1965) differs from extant species by less mediolaterally compressed horn cores, shorter premolar rows, a simpler occlusal pattern of its molars, and a short basioccipital with anterior tuberosities localized more as in *Oryx*. It reached a large size in Olduvai II. An extinct South African species, *H. cookei*, occurs at Makapansgat (Vrba 1987). Geraads (1981) recorded *Hippotragus* from Ternifine. The difficulties of distinguishing bovine from hippotragine teeth and the possible presence of primitive hippotragine teeth at Laetoli were discussed by Gentry & Gentry (1978).

Large hypsodont teeth from Sterkfontein 4, South Africa could be Hippotragini at about 3 million years old. They have small but tall basal pillars and rather thin enamel around the outside of the teeth and the central cavities. The upper molars have small styles and broadly rounded medial lobes. They were illustrated by Vrba (1976) and referred by her (1987) to *Hippotragus cookei*. Some Makapansgat teeth as well were assigned to this species.

Alcelaphini

Alcelaphini have horns in both sexes, horn cores often of irregular course or shape, extensive internal sinuses of the frontals, a single large smooth-walled sinus extending into the stalk (= horn pedicle), a short braincase strongly angled on a long face, very hypsodont cheek teeth without basal pillars, evenly rounded lingual lobes of upper molars and labial lobes of lower molars, widely outbowed labial ribs of upper molars, short premolar rows with reduction or loss of P_2, fused paraconid and metaconid on P_4, and markedly cursorial limb bones. According to Kingdon (1982), Estes (1974) and Jarman (1974) *Connochaetes*, *Alcelaphus*, and *Damaliscus* are high-density ecotonal/catenary grazers, rather unselective for plant species but more so for particular parts or growth stages of plants. They live in open, less arid habitats with only poorly defined home areas. They are strongly migratory, which must assist in finding grasses at their optimum growth stages. The territory of a male during the nomadic phase is no longer a piece of ground but a temporary area around a group of females. The precocious young are born in a short breeding season. Alcelaphini are the most abundantly represented tribe among African fossil Bovidae. *Aepyceros* is often not classified in this tribe; it is smaller than other alcelaphines, and females are hornless.

The most completely known early alcelaphines are *Damalacra neanica* and *D. acalla* from Langebaanweg (Gentry 1980). They are a little smaller than living *Alcelaphus buselaphus* or *Damaliscus lunatus*, the skull is rather narrow as in those species and not wide like *Connochaetes*. The completeness of the remains leaves no doubt of their tribal identity and makes the primitive condition of the teeth most interesting. They are less hypsodont than in living alcelaphines, small basal pillars survive on some teeth, upper molars have prominent mesostyles but poor ribs, molar lobes pointed rather than rounded, P_4s less reduced posteriorly and with paraconid not usually fused with metaconid. *D. neanica* is advanced on *D. acalla* by less backward and more outward curvature of its horn cores, insertions further behind the orbits, and a more inclined braincase roof. Gentry (1980) concluded that these and other differences parallel those between *Alcelaphus buselaphus* and *Damaliscus lunatus* today. They constitute a single suite of functionally linked characters which have probably evolved repeatedly in successive pairs of sympatric alcelaphine species.

Damalacra-like alcelaphines were also present in north Africa; a much damaged horn core from Wadi Natrun (Egypt) and the tooth from the same locality were identified by Stromer (1907) as perhaps tragelaphine. In contrast, an Mpesida tooth identified by Gentry (1978a) as possibly alcelaphine is considered by Thomas (1980) to be tragelaphine. Kalb et al. (1982) report an alcelaphine horn core in the Kuseralee Formation.

An alcelaphine at Laetoli has been referred by Gentry & Gentry (1978) and Gentry (1987) to *Parmularius*, an extinct genus related to *Damaliscus*. It is still primitive in its long and little-angled braincase, but it has a low tubercle on its parietal and a localized swelling at the base of its horn cores, as in *Parmularius*. Its teeth are advanced on Langebaanweg and are more recognizably alcelaphine. A large alcelaphine in Hadar SH and DD is also primitive in its not-very-shortened braincase and large preorbital fossa, and teeth are probably still not as hypsodont as in living alcelaphines. It could belong to *Damalops*, an extinct genus notable for occurring in the Pinjor Formation of India and in southern Russia. *Damaliscus cuiculi* Arambourg (1979) occurs at Ain Boucherit. Horn core divergence was wide as in extant *Damaliscus l. lunatus*, but a close relationship is unlikely.

Parmularius becomes the common morphologically specialized alcelaphine in the later Pliocene and Lower Pleistocene, although the speciation seems to be complicated. The type species is *P. altidens* of Olduvai I and lower II, also present in Shungura upper G and as *"Redunca" eulmensis* at Ain Boucherit (Arambourg 1979; Geraads 1981). It has shorter premolar rows than in living *Alcelaphus* and *Damaliscus*, often involving reduction or loss of P_2.

P. angusticornis succeeds *altidens* at Olduvai and is separated from the sympatric *Damaliscus niro* by straight horn cores and a much shortened braincase. *D. niro* has backwardly curved horn cores with the unique character of well-spaced, strong transverse ridges on their front surfaces (Leakey 1965). Some horn core variants in Olduvai Middle and Upper II, taken as *D. niro* by Gentry & Gentry (1978), have an abrupt change of course between base and tip instead of a steady backward curvature. In this they are more like some *Parmularius*, e.g. *P. braini* Vrba (1977) from Makapansgat 2+3. This leads into the complicated question of how to maintain across the range of fossil forms a workable distinction between *Damaliscus* and *Parmularius*. Another interesting problem in this area is that *Parmularius ambiguus* at Ternifine has long premolar rows (Geraads 1981).

In the Middle and perhaps Lower Pleistocene, *Rabaticeras* is known from Morocco, Olduvai III, Swartkrans, and Elandsfontein; it appears to be ancestral to *Alcelaphus*. Its horn cores are inserted uprightly and curve forward with clockwise torsion—as if a backwardly curved *Damaliscus* horn core has been twisted around (Kingdon 1982). The united pedicle of *A. buselaphus* is not yet in being. Nomenclatorial problems will arise if the descent of *A. lichtensteini* is separate from that of *A. buselaphus* back to *Rabaticeras* level.

Beatragus hunteri, the herola or Tana River hartebeest, is clearly a relic species of very restricted present-day distribution. The extinct *B. antiquus* of Olduvai I and II and Shungura upper G was larger, with a wider skull and more uprightly inserted horn cores.

The first wildebeest in the fossil record is the North African Mid-Pliocene *Oreonagor* (P. Thomas 1884), with horn cores less divergent than in later *Connochaetes*. Arambourg (1979) described an *Oreonagor* from Ain Boucherit which was more advanced in increased divergence and in horn core insertions lying further behind the orbits. In Olduvai I and Lower II the *Connochaetes* still has horn insertions closer behind the orbits and horn core tips turned less inward and upward than in living *C. taurinus*, but from Middle II onward *C. taurinus* is present.

In contrast to the straightforward story of the last paragraph, the holotype skull of *Connochaetes africanus* in Olduvai II is problematical by the seemingly feeble development of its horn cores and the primitively short face; it may be related to the South African extant *C. gnou*. *Connochaetes* from Cornelia and Elandsfontein, South Africa, seem likely to be ancestral to *C. gnou* (Gentry & Gentry 1978).

Megalotragus is an extinct alcelaphine, larger than *Connochaetes* and with longer metapodials. *M. kattwinkeli* is known from Olduvai Middle II–IV and the same species or its presumed ancestor from earlier Olduvai beds and Shungura G-L. Harris (1978) has implied that *M. kattwinkeli*, being identical with the destroyed holotype of *Rhynotragus semiticus*, had strongly arched nasals—a specialized or bizarre morphology unlike any other alcelaphine.

It is difficult to classify the perplexing array of fossil and living alcelaphines. Vrba's (1979) scheme deserves scrutiny, partly because it was the first cladistic attempt for a bovid tribe and partly because such classifications are supposed to be uniquely open to correction (Vrba 1979). Cladograms are necessarily precise; for that reason, the more taxa they cover, the less likely they are to be the single historically correct phylogenetic solution. Vrba's cladograms have the additional hazards of the acknowledged parallelism in alcelaphine evolution and the fact that 14% of the character occurrences in her matrix (Table 3) are unknown.

Vrba's placing of *Parmularius* with *Damaliscus* in one megagenus and of *Connochaetes* with *Megalotragus* in another appears reasonable, but the placing of *Alcelaphus* and *Sigmoceros* (= *A. lichtensteini*) in the *Connochaetes-Megalotragus* megagenus is more contentious. Face lengthening is the synapomorphy (no. 51) linking *Alcelaphus* (excluding *A. lichtensteini*) with *Megalotragus*; in them the length P_3-M_3 has declined to 85% or less of the diastema length from C-P_3. However my measurements of this same ratio failed to indicate a significant difference between two samples of 12 *A. buselaphus* and *Damaliscus lunatus* although they are supposed to be in different megagenera. (Respective means = 127% and 121%, Student's T = 0.54.)

At a higher level, *Connochaetes-Megalotragus* and both *Alcelaphus* species are linked by characters 6, 28, and 40. But both 28 and 40 are also present in *Parmularius* and *Damaliscus* in the other megagenus (Vrba's Table 3, although no. 40 is not shown on her Fig. 2). Thus the full weight of this link depends on character 6 alone—clockwise horn core torsion.

Characters 7, 43, and 55 uniting *A. lichtensteini* with *Connochaetes* are all to do with wide skulls and Vrba (1979) herself concedes that the polarity of wide/narrow skulls is debatable. The wide skull of *A. lichtensteini* is presumably linked with the absence of the united horn stalk seen in other *Alcelaphus*, and I would not choose it as a synapomorphy with *Connochaetes*.

My conclusion would be that *Megalotragus* can be associated with *Connochaetes*, although it is hard to see characters in common other than large size and clockwise horn core torsion. *Damaliscus* can be perceived as persistently primitive in many generic-level characters and sharing a common ancestry with the more specialized but now extinct *Parmularius*. *Beatragus* may also have branched from *Damaliscus* as in Vrba's cladograms. It is likely that *Alcelaphus* and *Rabaticeras* are also linked with this stock, separating either in the Lower Pleistocene or at a more remote time perhaps via *Damalops*. Clockwise torsion in their horn cores is thus seen as a parallel evolution to that in *Connochaetes* and *Megalotragus*. Vrba (1979) affirms that the known hybridization between living *Damaliscus* and *Alcelaphus* need not be incompatible with an ancestry separate for 3 to 4 million years.

Both Vrba (1979) and Thomas (1984b) place *Aepyceros* as a sister group of the Alcelaphini, agreeing with my own unsubstantiated opinion at least to the extent of removing it from the Antilopini. *Aepyceros* would thus be a long-lasting independent lineage. Its limb bones, for example, have a number of distinctive characters unlike either of these tribes, and these are presumably linked with the animal's well-known jumping ability during flight from predators. Thomas (1980) described Lukeino fossils as *Aepyceros* sp., and my opinion that the Fort Ternan gazelle could be an early impala has already been mentioned. Pliocene horn cores such as those from the Mursi and Hadar Formations (Gentry 1981) are considerably less lyrated than in the living species. The better known form from the Shungura Formation differs from the living species by its shorter face, supraorbital pits placed less widely apart and vestiges of a preorbital fossa.

Vrba (1980) has postulated that the high incidence of speciation in alcelaphines at macroevolutionary level may be only a by-product of their narrow specialization (stenotopy) at an infraspecific or microevolutionary level. *Aepyceros* on the other hand is a eurytopic or broadly adapted animal and has shown less speciation. Furthermore, the high rate of alcelaphine speciation could account for that tribe's dominance without resort to competitive exclusion or natural selection. This is Vrba's "effect hypothesis" which could result in nonadaptive trends.

These are interesting ideas; however, their hypothetical content remains high for the present, and Vrba herself (1984) refers to the conjectural association between stenotopy and extensive speciation. More details are needed of how the stenotopy is related to speciation events. Does it precede or coincide with them? Could some stenotopes have survived through loss of earlier eurytopy or, conversely, could *Aepyceros* be the product of a former stenotopic radiation and have built up eurytopy later?

Concerning the stenotopy of alcelaphines, this was postulated by Gentry & Gentry (1978) for extinct species at Olduvai, but is more questionable among living alcelaphines. If *Alcelaphus buselaphus*, for example, tolerates poor-quality grazing (Kingdon 1982), is this any less eurytopic than *Aepyceros* with its option of grazing or browsing? As to lack of habitat restriction in *Aepyceros*, a comparable plasticity may be indicated for alcelaphines in distributional data, which suggests that in west Africa *Damaliscus lunatus* ranges (or ranged) further north toward the desert than *Alcelaphus buselaphus*, while in southern Africa *A. b. caama* lives in more arid country than *D. lunatus*. Stenotopy of alcelaphines could be in the category of a hypothesis about extinct species rather than a phenomenon demonstrable in living ones.

A most interesting alcelaphine *Maremmia* occurs in the Turolian of Baccinello in Tuscany. Its teeth are more advanced than at the later locality of Langebaanweg, which supports Thomas's view (1984b) that *Maremmia* is a dentally precocious offshoot from the same ancestry as true Alcelaphini. North Africa must be the best source for the immigration of this antelope to southern Europe.

Neotragini

Neotragini are small antelopes, possibly a heterogeneous group not very closely interrelated. Males alone have small, simple horn cores which are straight or slightly curved forward. Preorbital fossae are large; there are usually no basal pillars on the molars and no outwardly bowed ribs between the styles on the upper molars. Medial lobes of upper molars and labial lobes of lowers are sharply angled except in *Ourebia*, the smallest grazing antelope, which is not a selective browser like the other neotragines (Kingdon 1982).

The earliest neotragine is the large *Homoiodorcas* from Ngorora pre-*Hipparion* levels, already mentioned. Another large neotragine occurred at Langebaanweg and was referred to *Raphicerus*. It showed variably regular keels and ridges on its short horn cores (Gentry 1980). Later South African *Raphicerus*, as at Elandsfontein, is smaller, and only a few of the horn cores show any indication of keels.

The first small neotragine in Africa seems to be the *Madoqua* at Laetoli. It has shorter and thicker horn cores than today, and the molar teeth show more traces of basal pillars. Interestingly, some neotragine teeth from the Laetolil Beds and a horn core from the Kakesio site at Laetoli (KK/82 181) can best be assigned to *Raphicerus* but are no larger than living *Raphicerus*.

An extinct *Oreotragus* comes from Makapansgat (Wells & Cooke 1956). *Pelea*, possibly a neotragine (Gentry 1978), is known from the Transvaal back to Swartkrans (Vrba 1976).

In North Africa, Lehmann & Thomas (1987) record a neotragine at Sahabi, and Thomas (1984b) notes an M_3 from the later Upper Miocene at Menacer, Algeria (Arambourg 1959) as a possible neotragine.

Antilopini

Antilopini are small- to medium-sized antelopes, which Kingdon (1982) categorizes as gleaners in arid or open habitats. They are not notably specialized in morphology, and some characters may also be present in Neotragini or *Aepyceros*. They have complicated midfrontal and parietofrontal sutures, hypsodont teeth without basal pillars, and M_3 often with an enlarged back lobe (hypoconulid). The braincase is not strongly angled on the facial axis nor greatly shortened, a preorbital fossa is retained, and no paraconid-metaconid fusion occurs on P_4.

Gazella is the central and most widespread genus of the tribe, the name *Antilope* having become confined to the Indian blackbuck. The Langebaanweg gazelle is rather large, and a bit larger than the *Raphicerus* at the same locality. Gentry (1980) related it to *G. vanhoepeni* of Makapansgat and through that to the three largest extant African gazelles – *G. granti*, *soemmerringi*, and *dama*. Both Langebaanweg and Makapansgat gazelles were thought to have horned females (Gentry & Gentry 1978).

Another group of gazelles is represented by *G. janenschi* of Laetoli and the unnamed and not very common gazelle of Olduvai and Elandsfontein. These may be related to *G. rufifrons* and *thomsoni* of subsaharan Africa.

North African gazelles cannot be linked very easily with those from subsaharan Africa and look like a south Palaearctic group. The species at Ain Boucherit has quite a wide range of horn core morphology (Geraads 1981); that at Ain Hanech called *G. pomeli* has thickset, backwardly curved horn cores; and *G. dracula* at Ternifine has straight and almost parallel horn cores much compressed distally and very like *G. borbonica* of the European Lower and Middle Villafranchian.

Despite contrary claims by others and myself, I would not now accept as definite *Antidorcas* any remains prior to those horn cores in the Upper Ndolanya Beds (at Laetoli, younger than the Laetolil Beds) which resemble the *Gazella hennigi* horn cores of Dietrich (1950). However, Kingdon (1982) supposes that antidorcines separated early from gazelles and gave rise to alcelaphines. The chief differences of *Antidorcas* from *Gazella* are sinuses within the frontals and a localized backward bend in the horn core shortly above the base. At Olduvai *Antidorcas recki* is a commoner fossil than *Gazella*. In the South African Middle and Later Pleistocene is the small *A. bondi* with teeth so hypsodont that the ventral mandibular margin of late immature animals is seriously distorted by the molar bases (Vrba 1973).

Antilope is undoubtedly known from Shungura C, a very noteworthy record of this Indian genus in Africa. The two horn cores concerned, probably from different individuals, show less torsion than in the living species and a vestigial posterolateral keel.

Problematical horn cores from Olduvai I and II, Shungura K-L, and the Upper Ndolanya Beds (see above) of Antilopini sp. 1 (Gentry & Gentry 1978) probably are distinct from *Gazella* and could descend from or be close to *Prostrepsiceros vinayaki* of the Siwaliks. I am happy to accept Thomas's (1984d) enlightening suggestion of the generic affiliation of this latter species (Gentry 1970). It is interesting that it must have existed at the same time as *Gazella*, *Antidorcas*, and *Antilope* in Africa.

Gentry (1981) drew attention to an antilopine metacarpal, possibly *Gazella*, from Hadar SH which is neither shorter nor thicker than the extant *G. thomsoni* and therefore in strong contrast to *A. recki* from Olduvai I.

Ovibovini and Caprini

Ovibovini inhabited Africa until the Late Pliocene. The best known is *Makapania* from Makapansgat and probably from Sterkfontein 4 (Vrba 1976) closely related to *Megalovis* of the later Villafranchian of Europe. This seems to have been a widespread group in the Old World and also included the earlier *Hesperoceridas* of Spain and *Kabulicornis* of Afghanistan. Another stock is represented by *Bos makapaani* from Hadar, Olduvai, and a locality in South Africa (Gentry 1981).

Other ovibovine remains are known from Shungura and Langebaanweg. There is no reason to suppose that any of the African ovibovines are related to the specialized *Urmiatherium-Plesiaddax-Tsaidamotherium* group that occurred in Turolian and equivalent faunas in Eurasia.

The large *Numidocapra* from Ain Hanech is considered by Geraads (1981) to be an alcelaphine rather than a caprine (Gentry 1978b). *Damalavus* at Bou Hanifia (Arambourg 1959) is very like *Palaeoryx* of the European Turolian and may be an early ovibovine (but see Bouvrain & Bonis 1984).

Parantidorcas of Ain Boucherit (Arambourg 1979) looks more as if it were a hangover from Turolian *Oioceros* of Europe than related to *Antidorcas*. There is no raising of the frontals between the horn core bases, the horn cores are fairly long, and the spiralization is too marked.

Late Pleistocene Bovids

Late Pleistocene environmental changes can be studied in more detail than for earlier periods. Faunal samples also "are more numerous, often larger, and usually under tighter chronological control than Lower and Middle Pleistocene ones" (Klein 1984). This enables the contrast between rich Pleistocene mammal faunas and reduced Holocene ones to be well seen. One or two of the extinct Late Pleistocene African bovids may be considered briefly.

The bovine *Pelorovis antiquus* was originally described from north Africa (Duvernoy 1851), the first fossil mammal to be made known scientifically from the continent. Later on, finds came from East and South Africa; the South African Museum in Cape Town contains much material from the Middle Pleistocene of Elandsfontein. The head appears to have been held vertically and the long horns are downswept and without the basal bosses of larger races of *Syncerus caffer.*

In North Africa the Palaearctic *Bos primigenius* coexisted with *Pelorovis* (Pomel 1894; Hadjouis 1985), but *Syncerus* is not known north of the Sahara.

Especially prominent in South Africa is the alcelaphine *Megalotragus priscus* with long horns like *Pelorovis oldowayensis*. It was the terminal species of its lineage, descendant or successor to *M. kattwinkeli* in East Africa. Klein (1984) hypothesized late-surviving *Megalotragus* in east Africa, and Pickford and Thomas (1984) described *Rusingoryx* from the Upper Pleistocene/Holocene of Rusinga, Kenya. The weird features of this skull are compatible with Harris's (1978) view of *Rhynotragus* as a synonym of *Megalotragus*. *Rusingoryx* is described as if the braincase is not angled on the face and figured as if the occipital surface sloped forward. In fact the occipital has a normal orientation and it is the frontals anterior to the horn bases which display marked inflation; *Rusingoryx* is close to or congeneric with *Megalotragus*.

One could also mention as a final example of a Late Pleistocene extinct form the Melkbos kudu, apparently allied to *T. strepsiceros* (Hendey 1968).

Many examples occur in the Late Pleistocene of extant forms beyond their present distributional boundaries. The lechwe *Kobus leche*, now an inhabitant of periodically inundated floodplains in Zambia, has been found at Florisbad and elsewhere in South Africa. Another reduncine, *Redunca redunca*, is known on the north side of the Sahara (Gentry & Gentry 1978). The evidence for Late Pleistocene *Hippotragus* in North Africa is more equivocal. Teeth from Tamar Hat, Algeria, may be this genus, but *Oryx* is a more likely identification. The styles and ribs of the upper molars are not sufficiently accentuated, the basal pillars of the upper molars are too small, there is no pinching of medial lobes or upper or lateral lobes of lower molars (Arambourg et al. 1934).

Even in Africa, remote from the glacial/interglacial alternations of higher latitudes, cyclical fluctuations of climate are apparent in the Late Pleistocene and Holocene. Arid periods correspond to cold intervals elsewhere and moist periods to warm intervals; in some instances the correlations are on quite a small scale (Hamilton 1982). It has been supposed that alternations caused habitat fragmentation, and that similar fragmentations occurring earlier in the Pleistocene could have caused high rates of origination for mammal species (Gingerich 1984). Paleoecologists working in Africa have been satisfied with alternations as an explanation for species diversity, e.g., Klein (1980) writing on grassland and bushland (fynbos or macchia) herbivores in southernmost South Africa. Paleoecologists in Eurasia and North America, on the other hand, have tried to explain the abundance of large mammals by looking to Pleistocene environments and plant communities unlike any still in existence and also to great diversity within particular plant communities (Stuart 1982; Hopkins et al. 1982).

Bovid Classifications

Inevitably, different authors have proposed a diversity of schemes for bovid classification (Thomas 1984c). Thomas discusses the concept of Schlosser (in Zittel 1911–translated 1925) that bovids be split into Boodontia and Aegodontia. Originally this was a move to counter the unsatisfactory view that antelopes (Antilopinae in the old broad sense) constituted a third natural or monophyletic group alongside bovines and caprines. Thomas notes that later authors have used Schlosser's system indiscriminately as a classification, a phylogeny, and an ecological guide, and this will be apparent in what follows.

Boodonts (Bovinae, Cephalophinae, Hippotraginae) have not exploited such ecologically extreme environments as aegodonts (Alcelaphinae, Antilopinae, Caprinae). They tend to have lower and wider skulls, braincases little angled on their faces, horn cores more frequently keeled, frontals' sinsuses restricted or absent, brachyodont or less hypsodont teeth with well-developed ribs between the styles on their labial walls, rugose enamel and frequently cement, basal pillars often present, and less reduction of premolar rows. Aegodonts have the converse characters, and the two groups were most clearly manifest in the Late Miocene to Early Pliocene. As usual, many objections can be found to this (my own) generalization; three notable ones being the decidedly nonboodont pattern of tragelaphine teeth, the narrow skulls of *Hippotragus*, and the fact that the braincase of *Raphicerus* is not angled downwards.

Kingdon (1982) thought that the aegodont/boodont division of bovids relied too much on teeth, and that dental characters, divorced from an understanding of their function, are unsuitable for bovid taxonomy. [This opinion can instructively be read in conjunction with Gingerich (1977).] He observed that folivorous antelopes such as Neotragini have sharp-angled teeth for slicing and premolars which retain their ancient cutting function. Frugivores such as Cephalophini have more rounded teeth. Grazers have extra folds and rounded edges to facilitate a crushing action, and the premolar rows shorten because crushing can best be done proximately to the fulcrum of the jaws.

In his own classification of bovids, Kingdon (1982) has two subfamilies: the Bovinae constituted by the same three tribes in this chapter and the Antilopinae for all the rest. The molecular phylogeny of Lowenstein (1986) is very similar. Bovinae originated in Eurasia and Antilopinae in Africa. Antilopinae differ from Bovinae in smaller body size, annulated instead of keeled horns, horns used for glancing instead of stabbing blows, facial or pedal glands, nasal panting instead of sweating, often two pairs of mammae, and possession of territorial systems. As always, there are difficulties in making such a generalization. Among tragelaphines only eland fits the Bovinae requirement of being larger than Antilopinae; *Hippotragus* spoils the ascription to its tribe of a preference for arid habitats; the waterbuck sweats (Kingdon 1982); extinct Eurasian Antilopinae (e.g., *Spirocerus*) have keels, and sheep and goats are not territorial.

Within the Antilopinae, according to Kingdon, the Antilopini and the caprine tribes originated early from a neotragine-like ancestor and became widely distributed in Old World arid or inhospitable environments. Hippotragini originated later from Caprini, and Alcelaphini from Antilopini, while the old neotragine stock gave rise to Cephalophini and Reduncini. Thus the main differences from the boodont-aegodont scheme outlined earlier in this section lie in placing the Cephalophini, Reduncini, and Hippotragini with aegodonts.

From what has just been said, one sees that the correct placing of extant genera in the tribes of Bovidae is more or less clear; what is at issue is the placing of older fossil genera and the relationships between the tribes. In Fig. 6 the position of Hippotragini and Reduncini has been left open to maximize the compatibility with Kingdon and with Lowenstein.

While the absence of transverse ridges on the horn cores of Tragelaphini aligns them with Boselaphini, their teeth are very aegodont-like except in their lower crowns. In fact, Thomas (1984c) comments that their teeth suggest a possible derivation from African *Prostrepsiceros* (Antilopini) and considers (1981) that an Ngorora mandible taken by me as intermediate between Boselaphini and Tragelaphini could instead belong to a large antilopine like the Sahabi *Prostrep-*

siceros. The keels and the anticlockwise direction of the torsion of tragelaphine horn cores is of course compatible with relationship to either Boselaphini or Antilopini. Further finds of fossils may be useful in solving this question. An African origin for the tribe looks probable.

The most ancient Bovini in Africa come from Lukeino customarily dated at 6.0–7.0 my, while in the Indian subcontinent bovine remains (apparently *Proamphibos*) come from levels slightly younger than the type Dhok Pathan Formation (Gentry 1980), perhaps around 6 million years (Barry et al. 1982) and possibly younger than Lukeino. It would be premature to deduce from this an African origin for Bovini. Earlier boselaphines likely to be close to the ancestry of bovines, such as *Selenoportax*, *Pachyportax*, and *Samokeros* (Pilgrim 1937; Solounias 1981) are known from Eurasia. Also there is a shortage of information for much of the African Upper Miocene between Ngorora, at which the upper levels may just extend into the Vallesian in European terms, and a later suite of sites such as Mpesida, Lukeino, Sahabi, and Langebaanweg (Thomas 1984c).

The relationship of Cephalophini remains obscure. They are scarcely known as fossils, and their behavior and ecology have been less studied than other African groups.

Reduncini appear in Africa at Lukeino and Mpesida and in the Siwaliks at the same time as the bovine *Proamphibos*, that is, when boselaphines diminish. Reduncines continue into the Pinjor Formation of the Upper Siwaliks wherein some more complete remains are known. Here they show quite strong temporal ridges on the braincase roof, reminiscent of Boselaphini and possess preorbital fossae, as known in *Menelikia* but not in living reduncines. This would all support or be compatible with a boselaphine ancestry for reduncines, but at Langebaanweg the teeth apparently associated with reduncine horn cores and crania are like tragelaphines and the crania themselves are more like aegodont than boodont bovids (Gentry 1980). Could the Siwaliks reduncines be a parallel evolution to those in Africa? In this case the affinities of early reduncines in north Africa might be with these Siwaliks forms rather than with those in the rest of Africa.

Both Thomas (1983, 1984c) and Kingdon (1982) considered the ancestry of Hippotragini. Thomas thought that they (and reduncines) arose from near *Pachytragus solignaci* and Kingdon that they derive from arid-adapted caprines in northern Africa and southern Eurasia. To me the questionably hippotragine teeth in the Laetolil Beds (Gentry & Gentry 1978) suggest that both caprine or aegodont teeth and hippotragine or boodont teeth are opposite specializations and that the primitive structure for bovid teeth is intermediate (Fig. 7). Hence hippotragines would be unlikely to have started from a group already evolving in the caprine direction. Perhaps *Pachytragus solignaci* is sufficiently primitive as a member of the Caprinae to be a feasible ancestor. It is a form about which it would be good to have more settled ideas (see above), and further fossil finds might throw more light on Thomas's suggestion and on *P. solignaci* itself.

The best suggestion for the relationships of Alcelaphini is that they go back to an aegodont origin, probably shared with Caprini. Gentry (1980) drew attention to an early alcelaphine tooth which had been misidentified as tragelaphine–

reasonably so, since tragelaphine teeth too have aegodont characters. It is interesting that the tragelaphine teeth at Lukeino and Mpesida are the same size as Langebaanweg and Wadi Natrun alcelaphine teeth but are perhaps less hypsodont and have narrower labial lobes on their lower molars.

Caprini and Ovibovini are mainly extraneous to Africa and classification problems involve the Middle and Upper Miocene forms. Gentry (1970) placed the definitely nonboselaphine *Kubanotragus* and *Caprotragoides* in the Caprini, but resemblances of these genera to modern sheep and goats are few, apart from the aegodont teeth. The cladistic status of such resemblances as do exist, i.e., whether they are synapomorphies or not, is also in doubt.

The concept of the tribe Antilopini is centered around *Gazella*, but the tribe takes its name from its single living spiral-horned member *Antilope*, the blackbuck of India. *Antilope* brings into the tribe a number of extinct spiral-horned antelopes, mostly Eurasian and some of them of quite large size, e.g., *Prostrepsiceros* and *Spirocerus*. Even if the spiral-horned ones were separated from the rest at tribal level, it would still be difficult to characterize the remaining genera. The African *Litocranius* has strikingly low crowned teeth, more so than in the Upper Miocene *Gazella capricornis*, and it is difficult not to believe that this feature is secondary.

The difficulty in separating Neotragini and Antilopini in the Late Miocene and Early Pliocene suggests a relationship between them. A hazard here is that the Neotragini could well not be a valid group. *Ourebia* has morphological and behavioral features recalling Reduncini, while *Raphicerus* could be in or close to Boselaphini by the absence of annulations on its horn sheaths, wide skull, and rugose surface at the front of its braincase roof.

Concerning the idea of neotragines as the nearest living models to the ancestral bovid, one can take note that *Tetracerus*, the four-horned antelope, is probably even better in this role (Estes 1974). It is a larger animal which fits better with Janis's (1982) ideas of an antelope needing a body weight of about 18 kg to evolve horns. Like *Boselaphus* it lives in India but it is hard to find apparently advanced characters to justify placing it in the Boselaphini; it has no strong temporal ridges, for example. Its unique four-horned state is bizarre and could suggest a relationship to *Miotragocerus*, which may have had forked sheaths in life. Cladistically the four horns would be an autopomorphy, but they could have a broader evolutionary significance as a legacy from a period when pecorans were still experimenting (as it were) with different sorts of horns and antlers. The most similar antelope to *Tetracerus* in Africa is *Sylvicapra*, which shows such features as size, spike-like horns inserted behind the orbits at about the same inclination, temporal lines approaching closely on the cranial roof posteriorly, and a large preorbital fossa. However *Sylvicapra*, in conformity with Kingdon's opinions on the Cephalophini, does have characters which are advanced or nonboselaphine—e.g., annulations on its horn sheaths, horn cores inserted closely together, a more inclined cranial roof, infraorbital foramen higher, and top of mastoid at a low level and others.

Palaeoecology

Ecological factors have been to the fore in recent discussions of bovid origins and diversification. The bovid acme, it is felt, is grazing in an open habitat. But the meaning of open is often unclear in terms of distribution of trees, presence of understory bushes, or presence or absence of grasses. Furthermore, browsing and grazing can each embrace a lot of alternative feeding strategies. Browsers may eat shrub or tree leaves and twigs at various heights or dicotyledons at ground level, while grazers may eat fresh grass shoots or coarse stems in bulk. Among extant Pecora, the brachyodonty of *Litocranius* is an indication of its feeding regime (nipping delicate shoots among the thorns of *Acacia* bushes) but not of a mesophytic or moist habitat, since it lives in dry country or even semi-desert. On the other hand, the hypsodonty of the north American *Antilocapra* indicates nonforest habitats but not grazing. In this animal, browse is important all year long and crucial in winter, although grasses are important in spring.

Grubb (1978) sees primitive characters in forest mammals in Africa and advanced characters in savannah and arid-country inhabitants. Estes (1974) applied this idea to bovids in more detail and listed primitive characters for *Cephalophus* and contrasting advanced ones for *Oryx*. However, both Janis (1982) and Kingdon (1982) claim that duikers are only secondarily small forest dwellers.

Various advanced characters of duikers have been mentioned: high diploid numbers of chromosomes, large brains, long gestation periods, fully pecoran digestive physiology, and style of infraspecific combat. These must make duikers unlike early bovids, but their connection with the secondariness of forest-dwelling is tenuous. Moreover, the gestation period of tragulids is not shorter than those quoted for three duikers in Kingdon (1979, 1982; cf. Geist 1966a). As to body size, there are large as well as small duikers, and it is not definite that the latter are primitive for the group.

The ecological history of duikers is only one approach to the question of the habitat of early bovids. If bovids did differentiate from early Pecora already living in partially open habitats, then modern forest duikers would be bound to be secondary, whatever the state of our understanding of their attributes. Direct consideration of Oligocene and Miocene palaeoecology is still a weak subject. The well-known assumption of the opening of forests in the Miocene is portrayed by Bernor (1983) as laurophyllous evergreen forests giving way to summer drought sclerophyllous vegetation in open woodland across southern Eurasia. This was in response to a climatic trend toward cooler winters and decreased summer rainfall. In East Africa it is supposed that tectonic activity associated with rifting led to "rain shadow" effects and the appearance of more open woodlands. The timing of this is obviously not known, nor knowable, with any precision. It has been unusual to have direct knowledge of both fossil plants and animals from the same localities, and it has been necessary to interpret habitat and vegetation structure from studies of the animals themselves. Andrews et al. (1979) examined quantitatively the ecological diversity of African modern and fossil faunas. From their studies and those of Evans et al. (1981), it seems that east African Lower Miocene faunas derived from forests and Middle and later Miocene ones from woodland-bushland. Hence, in East Africa the coming to prominence of bovids, as at Fort Ternan, does coincide with the appearance of more open habitats.

Hypsodont teeth and short premolar rows were already apparent in bovids as Fort Ternan and kindred sites, but a range of the modern tribes of grazing antelopes was not present until later in the Miocene, as at Lukeino and Mpesida. So in the Middle Miocene there may have been ecological room for only a small number of grazing bovids, or the habitats may have been less extensively open, or some other tough foodstuff may have been utilized by bovids with hypsodont teeth and short premolar rows.

The change at Lukeino and Mpesida appears to coincide in time with the decline of the Clarendonian chronofauna in North America (Webb 1983b). The changes on both continents may be linked with the end of transition to open habitats during the Miocene, but it is interesting that the North American one is manifested as a decline in ungulate diversity as steppe replaced savannah, whereas the African one involves increased diversity, at least for bovids. Knowledge of the African change is very sketchy, and it may of course have taken place earlier than the deposition of the Lukeino and Mpesida beds and have been a consequence of the arrival of *Hipparion* about 3 million years previously. This horse, taken to be a roughage grazer like modern *Equus*, could have had a pronounced impact on vegetation in more open habitats and allowed the exploitation of these habitats by other herbivores, as takes place in the grazing succession in East Africa today. [This is a novel role for a member of the Perissodactyla, once seen as an order which failed in competition with the artiodactyls (Janis 1982).] *Hipparion* was already present in the North American Clarendonian savannahs, and this reinforces the conclusion that the change in Africa at Lukeino and Mpesida should not be taken to indicate the appearance of steppes.

The important aspect of the African change, the recognition of which we owe to Thomas, is that a greater number of bovid tribes are present and that some of them are modern ones for the first time. This is of primary ecological importance. The nomenclatorial changes are more artificial, since the old Middle Miocene Caprini and Boselaphini are as likely to include ancestors of the later tribes as of the Caprini and Boselaphini still extant today. The names of the tribes are less significant than the fact that there was ecological room for more of them by the time of Lukeino and Mpesida.

Conclusion

Bovidae are the hollow-horned ruminants. They have a hollow keratinized sheath fitting over a separate bony core, which in some cases has internal sinuses but in most consists of spongy bone. Neither sheath nor core is branched or seasonally shed.

In Africa the family appears possibly in the Lower Miocene and definitely in the Middle Miocene. As in Europe, these early finds have been put in the genus *Eotragus*. By the later Middle Miocene it is apparent that Africa was inhabited by various boselaphines, Caprinae, and the smaller *Gazella* and *Homoiodorcas*. Where female skulls are known at this period, they are hornless. Nonbovid ruminants continued to be more abundant than in later times.

During the Late Miocene there appeared the first representatives of several modern groups of African antelopes, e.g., Reduncini, Tragelaphini, and *Aepyceros*. This second stage of bovid evolution may be connected with environments becoming more open and/or with the arrival of hipparionine horses.

The fossil record of African bovids shows frequent instances of precocious specializations, for example, the P_4 of the reduncine at Hadar or the horn cores of the alcelaphine *Damalacra neanica* at Langebaanweg.

The present-day fauna of bovids is less diversified than in the Pleistocene.

Tragelaphini, Cephalophini, Alcelaphini, and perhaps Hippotragini could have originated in Africa. The phylogenetic relationships of all tribes remain uncertain; this is especially true of Hippotragini and Reduncini.

In the last 10 years, quite a lot more has become known about fossil bovids in Africa. The increase in factual knowledge (taxa A, B, and C in provenances X, Y, and Z) has not been paralleled by an increased understanding of their evolution; rather, has there come to exist more awareness of alternatives. More new ideas about relationships and events have been advanced than could have come from any one researcher alone; the difficulty is to decide which ones to follow. Cladistics has played a small part in exposing the inadequacy of our previous understanding of the fossils, but less in eliminating it. Convincing sets of synapomorphies indicating one preferred solution for any taxonomic problem may not be forthcoming, and new fossil finds may be a better route to finding links between already known forms. On the other hand, new discoveries may be found to belong to extinct and sometimes mysterious lineages and not to enlighten us about living forms. Moreover, in the early stages of lineage evolution, only minimal change need have taken place from ancestors which may possess attributes intermediate in relation to those of their diverging descendants.

II
Physiology, Genetics, and Behavior

7
The Pronghorn (*Antilocapra americana*)

BART W. O'GARA

Introduction

Some 35 million pronghorns (*Antilocapra Americana*) inhabited North America before the arrival of white men, but by 1924, this population decreased to less than 20,000 animals. From 1924 to 1964, the population increased tenfold; recent increases occurred concurrently with annual harvests of twice as many pronghorns as existed in 1924. Numerically, pronghorns are the second most important big-game animal in North America. Effective law enforcement, habitat improvements, and wildlife management techniques aided population increases to approximately 1 million pronghorns today.

General Characteristics

The average body measurements (in millimeters) of adult female and adult male pronghorns, respectively, of the type subspecies *A. a. americana* collected in Alberta were total length 1406, 1415; height at shoulder 860, 875; length of tail 97, 105; and length of ear 142, 143. Weights varied seasonally: Adult does collected throughout the year averaged 50 kg and ranged from 47 to 56 kg, whereas males in the same collection averaged 57 kg and ranged from 47 to 70 kg (Mitchell 1971). In Oregon, *A. a. oregona* does and bucks killed during autumn averaged 42 and 52 kg (Einarsen 1948), and in Texas, adult *A. a. mexicana* does and bucks averaged 40 and 43 kg during autumn (Buechner 1950a). Weights apparently are not available for the other two subspecies *A. a. sonoriensis* and *A. a. peninsularis*. Data are scanty, but measurements indicate that the weights of these two subspecies should be quite similar to those of *A. a. mexicana*.

The pronghorn's body is rather robust, its legs and feet are long and slender, and its feet lack lateral digits. The head and eyes are large, and the black eyes are protected from mechanical damage by tubular bony orbits and from sun by heavy black eyelashes.

Body colors are contrasting white and rusty brown to tan, and black and dark brown markings occur about the head and neck. The eyelashes and mucous membranes of the nose and mouth are black, and brownish black patches start just below the ears of males and extend downward from 75 to 100 mm. Erectile hairs of the mane, 70 to 100 mm in length, are russet and tipped with varying amounts of black. White hairs of the twin rump patches, about 75 mm in length, almost gleam when erected, fanning beyond the normal contours of the body.

Pronghorn pelage is coarse. The medullary cells are large and air-filled and provide excellent insulation (Murie 1870). Pronghorns molt in the spring. The new hair is sleek and bright by late summer but becomes pithy and dull by mid-winter. Females have two scent glands, one in each rump patch, which function, with the rump patches, in intraspecific communication. Males have two rump glands plus two subauricular glands and one median gland; the latter three are important in sexual behavior (Moy 1970, 1971). Both sexes possess interdigital glands on the forefeet and hind feet that produce sebum, which apparently conditions the hoofs (O'Gara & Moy 1972).

Movements, Groups, and Habitat Use

The timing and length of seasonal movements vary with altitude, latitude, and range conditions. Generally, animals from large wintering herds in areas with substantial snowfall disperse in spring. Deep snow sometimes forces pronghorns to move as far as 160 km from summering areas (Martinka 1966). As the snow melts, the animals move back to summer ranges. In areas with little or no snowfall, seasonal ranges may be dictated by the succulence of the vegetation. Natural barriers, such as steep timbered mountains and large bodies of water, impede the movements of pronghorns and such barriers determine, to some extent, their distribution. Fences, interstate highways, and railways further complicate movements and reduce the carrying capacity in areas where pronghorns must move long distances to procure the year-round necessities of life.

The pronghorn is a plastic species with considerable variation in color, horn shape, and behavior from area to area. Pronghorn herds function as discrete populations. A herd is all the resident pronghorn summer groups on a particular land area that have routine social interactions and often spend the winter as a single group. Bands are subdivisions of herds. Doe bands are composed of female associates that have the nearly exclusive use of a traditional summer range. Summer bands also include bachelor buck bands that associate during the interval between spring dispersal and the onset of rut. Whereas herds are often separated by environmental barriers, bands within herds are isolated primarily by behavioral mechanisms. Variations, especially in configurations of horns, are often seen between herds, probably because genetic exchange between adjacent herds is minimal, a few dominant bucks do most of the breeding, and doe bands are essentially sister groups (Pyrah 1987).

Once they are on summer ranges, does generally collect into bands of a dozen or less, and young bucks form slightly larger bachelor bands. Formation of summering bands of does and fawns takes place gradually over a period of a few weeks. Lactating and nonlactating does band together soon after the fawns are born and utilize the summer ranges of several territorial bucks as well as areas around and between such territories. Thus, bands often appear to have one buck member.

Bachelor bands are usually composed of yearlings and two-year-old males, plus older males that have not acquired territories. Bachelors have larger summer ranges than other pronghorns. Gilbert (1973) observed a bachelor band with a stable hierarchy for seven weeks prior to the rut in Yellowstone Park. The band was socially cohesive although intensive, prolonged displays and sparring were common. The band occupied poorer habitat adjacent to three territories held by large bucks.

After the rut, territorial bucks leave their territories and all the males cast their horn sheaths. The size of wintering herds is determined by geography, range conditions, and the severity of winters. When the snow is deep, pronghorns usually must reach browse or perish. Thus, large herds gather on sagebrush ranges during some winters. The pronghorns in Alberta paw away snow to reach food even when ample forage is available above the snow, probably to take advantage of the better quality of forage protected by snow cover. An established hierarchy at feeding craters reduces the expenditure of energy. Pronghorns select microhabitats with lower wind velocities, less snow, and softer and less dense snow than the average for the area. The animals conserve energy by reduced daily travel and single-file travel in snow and by lying down during postdawn periods when the snow is hard and the temperature is low. They reduce heat loss with clumped bedding patterns during high winds and by lying with their heads curled back alongside their bodies (Bruns 1969).

Yoakum (1972) plotted the 1964 estimated pronghorn populations on a map that depicted the major vegetative communities of North America: 62% were on grasslands (41% short, 21% mixed), 37% on grassland–brushland, and 1% on deserts (hot and cold). Pronghorns are dainty feeders and utilize a wide variety of plants. Many studies of food habits have shown that northern pronghorns depend heavily on browse, particularly sagebrush (*Artemesia* spp.), in winter.

Reproduction

Known gestation periods for captive pronghorns ranged from 245 to 255 days (average 250) at the Wyoming Game and Fish Department's Sybille Wildlife Research Station (Hepworth, personal communication, 1981). The pronghorn's gestation period thus ranks among the longest in the ruminants of North America, exceeded only by that of bison and of elk.

Southern pronghorns have a much longer rut, which begins earlier than that of animals on the northern prairies. Northern pronghorns and those in Oregon generally breed during mid- to late September. The rut continues into October in Colorado and most southwestern states. Breeding activity as early as July and as late as October has been reported by Lehman and Davis (1942) and Buechner (1950a) in Texas. Breeding behavior was observed during late September in Chihuahua (Trevino Fernandez 1978), and Sonoran pronghorns apparently breed during early July along the Mexican border in Arizona (Phelps 1981). Local weather and altitudinal differences influence the time of breeding, as was demonstrated by Yates and Fry (personal communication, 1980) in New Mexico.

Pronghorns are polygamous; females usually become sexually mature at approximately 16 months of age but occasionally conceive at approximately 5 months of age (Wright & Dow 1962; Mitchell 1967; O'Gara 1968). Does on good range usually release four to seven ova that develop for nearly a month before implantation. The walls of each blastocyst elongate and form a tube about 125 mm in length and 0.5 mm in diameter, the thread stage. At that time, the uterus is active and the fragile thread-stage blastocysts are kneaded together and often tangle. One fourth to one third of the blastocysts die of malnutrition during the thread stage because their membranes are so reduced by knotting that absorption is reduced to the point that nutrition is inadequate. Quadruplets often survive the thread stage, and as many as seven embryos have been reported (Mitchell 1965). If only two embryos survive, they both locate proximally in the uterine horns near the body of the uterus. Any additional embryos that survive locate distally near the oviducts. As implantation begins, the fetal membranes of proximal embryos grow extremely fast. When a proximal embryo's necrotic tip reaches the membranes of a distal embryo, the tip usually displaces the distal embryo and it perishes. Pronghorns are the only animals known in which the number of embryos is normally reduced to two during pregnancy by the methods described above (O'Gara 1969a).

The doe begins to lick each fawn immediately after delivery. Neonates right themselves and orient on their mothers immediately after delivery. Both mother and young frequently get up and lie down, which probably establishes mother–infant bonds through touch, smell, and sight. Does eat the afterbirth. At first, a doe licks the entire fawn, but licking of the anogenital region soon takes precedence over generalized licking and occurs at intervals for two to three weeks. During such grooming, fawns assume a distinctive rump-up posture and eliminate; the mother ingests the urine and feces directly, which reduces the odors on the fawn and thus the possibility of a fawn being found in its bed by mammalian predators. Watched by their mothers, young pronghorns select their own bedding sites from the first day. At intervals varying from one to six hours during the next three weeks or so, does return to the area where their young are bedded. Six behavioral features characterize the mother–fawn reunion period: recognition, nursing, anogenital grooming, play, the relocation move, and the move to lie secluded (Autenrieth & Fichter 1975). Young fawns at ease lie with their heads up and appear alert. When they see movement, they flatten out, with their heads on the ground, in the position often seen in photographs.

Females frequently leave their fawns unattended, which leads to the misconception that nearby does are "baby-sitting." Kitchen (1974) observed that does associated with fawn groups only take an interest in their own young in an alert–alarm, flight situation; he frequently saw baby-sitting does leave fawn groups completely unattended for more than two hours.

Fawns develop balance and speed by playing during summer, with exaggerated and excessive movements. Sparring and bunting begins when the fawns are about three weeks old, and the amount of play may reflect the condition of the animals or their environment. Fawns play until early autumn, but play is seldom seen on winter ranges (Autenrieth & Fichter 1975).

Classification

Relationship to Other Ruminants

Taxonomists have disagreed about the classification of the pronghorn in relation to other ruminants for years. Many taxonomists felt that the pronghorn is a possible link between bovids and cervids. Bovids and cervids have more characteristics in common than they have differences, and it has been variously argued that pronghorns belong within the Bovoidia (Vaughan 1978), Cervoidia (Leinders & Heintz 1980), or a subfamily (Antilocaprinae) in the Bovidae (O'Gara & Matson 1975). The lack of fossils, both of early bovids and the ancestors of pronghorns, makes classification difficult. In effect, taxonomists are looking at the twigs of a tree and trying to reconstruct the trunk and its branches.

When pronghorns were collected during the Lewis and Clark Expedition, the similarity to Old World gazelles and antelope was noted, and the pronghorn was given the common name *antelope*, which reflects those similarities. However, the presence of forked "hair" horns was confusing, and when it was finally demonstrated that the horns were also annually deciduous, pronghorns were placed in a family separate from all other ruminants—the Antilocapridae.

Forked horn sheaths are, indeed, unique among present-day artiodactyls. They probably represent a modification of the four-horned condition seen in some fossil antilocaprins. Only two four-horned pecorans exist today—the four-horned antelope (*Tetracerus quadricornus*) and the Jacoba sheep (*ovis aeries*). The latter is a race of domestic sheep that evolved in Scotland during comparatively recent years. Although forked horn sheaths are unique, they hardly seem an important characteristic for taxonomy, especially in view of the spirals, knobs, annulations, and other modifications seen on bovid horns. These modifications apparently serve as aids in fighting and for visual communication with conspecifics—both male and female. The forked horn sheaths of the pronghorn certainly serve similar functions.

Histological studies have shown that the horn-forming process in pronghorns involves the formation of keratin similar to that of bovid horns (Morrison 1961; O'Gara & Matson 1975). Pronghorn horns grow faster than those of other bovids, and considerable hair is trapped in the growing horns. Roots of hairs are embedded in the skin under the basal portions of horns, and hair shafts course obliquely toward the tips of the horns. Hairs do not occur in the hooks and prongs. The base of a horn from a young Rocky Mountain goat (*Oreamnos americanus*) that I examined histologically had about the same amount of hair incorporated into the hardhorn as is incorporated in pronghorn horns. So much for the myth of hair horns.

The shedding of horns by pronghorn bucks is controlled by the annual testicular cycle (O'Gara et al. 1971). As pointed out by Leinders and Heintz (1980), this parallels the growth and casting of antlers in cervids and probably has the same function in relation to the rut. The horns of female or castrated pronghorns are not shed on a regular basis (Pocock 1905; O'Gara 1969b)—further evidence of control by male hormones. Although bovid horn sheaths generally have annual starts and stops in growth, such growth is probably related to annual stress periods brought on by severe winters. Growth rings are more evident in mountain and northern species than in those of lowlands or tropical areas and occur in females as well as males. According to Bubenik A. (1982c), some merycondonts have completely deciduous horns (the horn cores were shed). Merycodonts are generally regarded as ancestral to antilocaprins; however, I have not seen good data either supporting or refuting such ancestry.

The horn cores of bovids arise embryologically as dermal ossifications (os cornu) under the skin that covers the frontal bones (Bubenik A. 1982c). The antlers of cervids arise as outgrowths of the frontal bones (Goss 1983). The horn cores of pronghorns apparently develop in a manner similar to that of cervid antlers (Bubenik A. 1982c), the apophysis of Bubenik A. (1982d). This seems to be a fundamental character, which possibly links pronghorns to cervids.

After considering O'Gara and Matson's (1975) statements that the horn-forming process in *Antilocapra* involves the formation of keratin similar to that of nails and claws, Leinders and Heintz (1980) concluded that "combining these statements with the fact that horn formation is common in mammals, one might wonder if horn formation is a significant character for classification." I agree; compared to the diversity of antlers and horns, only slight changes have evolved in the dental and skeletal characteristics of artiodactyls over the last 20 million years. Characteristics other than horns, and especially the number of lacrimal orifices, the presence or absence of metatarsal grooves, and biochemical findings, are now receiving more attention than horns in attempts to relate pronghorns to other ruminants.

Subspecies

A vast majority of present-day pronghorns belong to the subspecies *A. a. americana*. Goldman (1945) noted that "three somewhat isolated, finger-like southern extensions carry the general range of the pronghorn antelope as a species into Mexico. These peripheral extensions represent geographic races differing from the typical form and from one another only in comparatively slight details of size, color, and structure." The same can be said for the western race *A. a. oregona*. Bailey (1932), in naming that subspecies, stated that "the animals show only slight and gradual variation over their entire range, and no sharp lines of difference between described forms can be found." Whether all five subspecies

FIGURE 1. An adult pronghorn buck in Montana with slightly narrow but quite typical horns. (Photograph by Bart O'Gara.)

are valid is a moot point, which is further complicated by indiscriminate transplants of *A. a. americana* into the ranges of other subspecies.

Horn Morphology

Males

Horns with anterior prongs project from ridges just above the eyes of adult pronghorn bucks. Horn shape, inclination, and curvature vary greatly among individuals. However, horns of a representative buck protrude upward, outward, and slightly forward to form a V; they then curve into hooks, which often bear whitish tips, that usually curve inward or backward (Fig. 1). Prongs are normally single, compressed triangles, but double points, or several small auxiliary prongs,

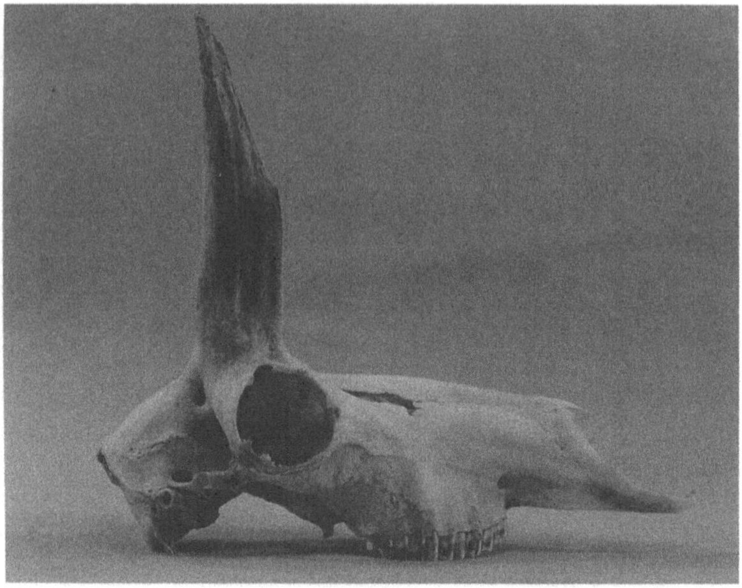

FIGURE 2. The dagger-shaped horn core rising above and slightly behind the orbit on the skull of an adult pronghorn buck. (Photograph by Bart O'Gara.)

are not unusual. Below the prongs, the sheaths are compressed laterally and contain some hair; above the prongs, the hooks are cylindrical in cross section. The hooks normally incline inward or backward and some appear lyrate; occasionally, the tips are directed forward or nearly straight up. Pronghorn horns are usually described as black, but close inspection often reveals streaks of brown, and the bases of some horns are yellowish brown.

Like sheaths on knife blades, the visible horn encases a bony core separated by a living layer of skin. The broad dagger-shaped horn cores, which originate from the frontal bones, stand erect above the orbits (Fig. 2). On mature bucks, the tips are usually 20 to 23 cm apart, and the cores, thickest in back and narrow-edged in front, are about 13 cm long. I have measured skulls of bucks in which the tips of the horn cores were as wide as 30 cm and as narrow as 10 cm. In the latter, the tips of the sheaths touched. The broadest part of the horn cores, where the prongs arise, measures from 38 to 51 mm from front to back. The mean length, circumference, and spread at the tips of the horn cores of nine adult bucks in Alberta were 133, 101, and 224 mm, respectively (Mitchell 1980). The cores are somewhat triangular in cross section, and the prongs develop over the sharp front edges. The cores of male fawns and most yearlings are nearly round in cross section; however, when yearlings grow prongs, the cores are triangular.

The most common anomaly of horn cores is one in which they bend forward. The buck that possessed the skull shown in Fig. 3 held a territory on the National Bison Range, Moiese, Montana, for at least five years. After he had held

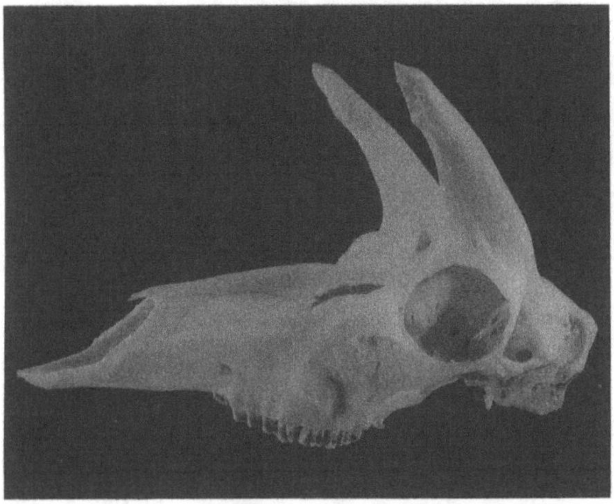

FIGURE 3. The skull of a mature pronghorn buck in Montana with the horn cores tipped forward. The prongs of the horn sheaths were approximately 10 cm above the animal's nose. (Photograph by DeWayne Williams.)

the territory for three years, three does with horns bent forward, apparently his offspring, occupied his territory. Horn cores often tip forward more than those shown in Fig. 3, and the prongs sometimes touch the nose. Such bucks sometimes become quite common in a herd, and they seem to have some advantage over bucks with conventional horns. Hunters selectively pursue such bucks and generally eliminate the strain. This is similar to a situation in Montana in which a strain of nearly white pronghorns was eliminated by hunters.

Lengths of horns vary with individual and area, and maximum horn development generally begins at four or five years of age when the animals reach maximum weight, the teeth are fully grown, and the bucks acquire and hold prime territories. Winter, spring, and summer weather and forage conditions affect the size of the horns. During field and aerial surveys, biologists generally classify bucks with heavy horns, approximately one and one-half or more times longer than their ears, as mature, and bucks with horns nearly the same length as their ears as yearlings. Fawns generally have conical horns about 4 cm in length.

Supernumerary horns are occasionally reported. Bill Hepworth (personal communication, 1979) of the Wyoming Game and Fish Department has seen small horns on the bridge of the nose and as many as four small horns attached to the skin behind the normal horns. Darrell Gretz shot a buck in Oregon that had a small horn on the bridge of its nose (Fig. 4). Another buck with a nose horn, which appeared two or three times as large as that on the buck shot by Gretz, was shot in South Dakota (Anonymous 1983). Jim Gay, President of the Taxidermists, Inc., in Laramie, Wyoming, photographed the skull of a four-horned pronghorn (Fig. 5), and wrote, "We receive several four-horned heads every year, but in almost all cases the second pair [of horns] is small, vestigial, and does not have

FIGURE 4. A supernumerary horn on the nose of a buck shot in Oregon. (Photograph by Darrell Gretz.)
◄

FIGURE 5. The skull of a four-horned pronghorn that was shot in Wyoming. (Photograph by Jim Gay.)
▼

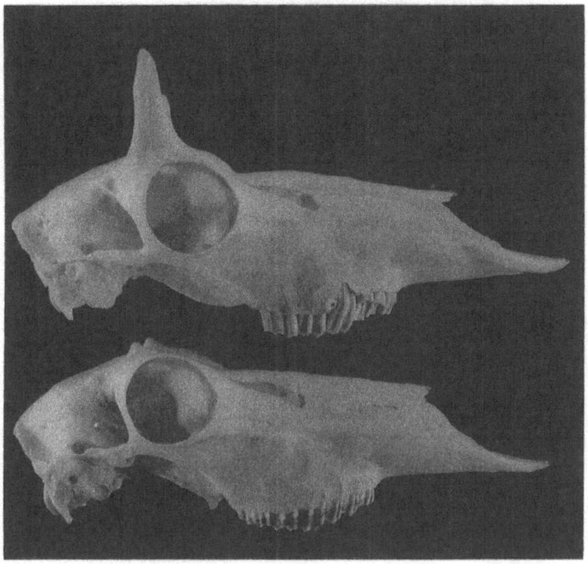

FIGURE 6. The skull of a horned doe (top) with larger than average horn cores and the skull of a hornless doe (bottom). (Photograph by DeWayne Williams.)

a bony core as the regular horns do." Ralph Nichols also picked up a pronghorn skull is the Big Hole Valley of Montana with a small horn core behind each of the normal ones, and Hoover et al. (1959) described a four-horned buck from Colorado with similar horn cores.

Females

Of 95 adult does collected on the National Bison Range, Montana, and in Yellowstone National Park, Wyoming, 29 were hornless. The hornless does had tiny nipples of bone about 3 mm high under whorls of hair. The horned does, however, had bony cores 13 to 77 mm long (Fig. 6). Einarsen (1948) wrote that does in Oregon occasionally have horns comparable to those of mature bucks, and doe horns 30 to 33 cm long are occasionally reported; however, I suspect that many of those animals are pseudohermaphrodites. I have seen two of the latter, each with inguinal testes, with the short penis ending in what appeared to be a vulva, and 30- to 35-cm horns. Although these animals had subauricular and median glands, they would have appeared to be does to anyone not very familiar with pronghorns. Horn sheaths of does in my collection average 40 mm and vary from 8 to 128 mm (O'Gara 1968). Only 5 of the 66 horned does have rudimentary prongs. Representative shapes and sizes of doe horns are shown in Fig. 7. The left and right horns of several does differ in size and shape; some are two to three times longer than their counterparts; in contrast, buck horns normally differ only slightly. The yearling does have comparatively small horns, 8 to 38 mm long. I

FIGURE 7. Representative horns from nine does (approximately life size). Note that three horns have double sheaths and the largest horn is separating from a horn under it. (Sketched by John Eiler.)

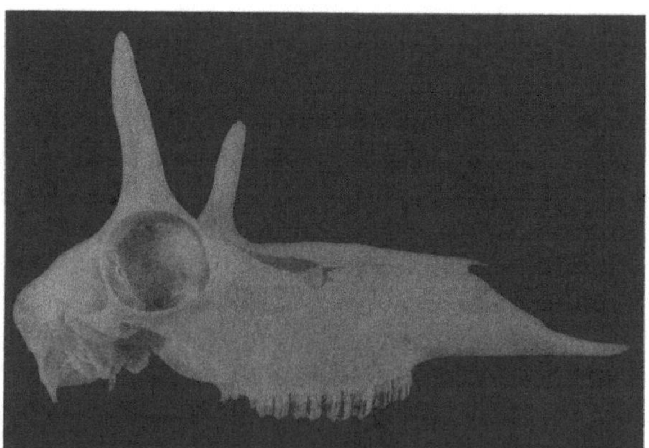

FIGURE 8. The skull and oval horn cores of an adult doe that had 128-mm horn sheaths. (Photograph by Bart O'Gara.)

never saw a doe fawn with horns. The horn cores of almost all does are nearly round to oval (Fig. 8). Four does have double (stacked) horn sheaths and one had a triple sheath. Multiple sheaths are probably retained because they are not large enough or heavy enough to fall as a result of jarring or by their own weight, and also because they usually have no prongs to interfere with stacking.

Horn Growth and Casting

Males

Early investigators thought pronghorn horn sheaths were composed of consolidated hair (Bailey 1920; Seton 1927). However, Morrison (1961) showed that the horn sheath was composed principally of compressed, cornified epithelial cells, and hairs were rare or absent in the upper portions of the sheaths. Hair grows from the skin covering horn cores before it produces hard horn, and hair is incorporated into the horn, mostly near the bases. The roots of such hair are, at least initially, embedded in dermis under the horn sheaths, and the shafts follow an oblique course distally through the sheaths. Although hairs contribute little to the bulk or strength of horn sheaths, sebaceous glands and sweat tubules emptying into the canals of these hairs may prevent horns from becoming dry and brittle, and they certainly must contribute to their odor (O'Gara & Matson 1975).

Hard horn begins to form about the time of the rut, just after maximum testicular development. Horn sheaths are shed about two months later (Fig. 9); thus, the growth and casting of horns are apparently controlled by male hormones in a manner somewhat similar to that of most cervids. Horn formation resembles that of skin. Although the outer layer of any skin feels soft, it consists of a layer of corneum cells, composed of a hard protein substance (keratin) that protects the underlying delicate skin. Claws, nails, hoofs, and horns are specialized extensions of the corneum. Cuboidal or columnar cells in the basal layer of the epidermis, next to the dermis, give rise to cells that flatten as they approach the surface. These flattened cells lose their nuclei and are incorporated into the horn as layers of clear, flat, keratinized cells.

Horn growth ceases before the rut. At that time, the hard horn and the underlying skin are closely associated and somewhat fused to each other. An investigation of 55 pairs of testes from adult and yearling pronghorns in Montana indicated that testicular weight was at an annual low during January and February. Testicular weight increased 2.5-fold by August, but a determination of the exact time that horn growth is initiated is difficult. If one considers the formation of hard horn only, horn growth begins about the time of maximum testicular development, in September (O'Gara et al. 1971). This new growth contributes to the loosening of sheaths, as layers of epithelial cells slough from the old horn sheath.

New hard horn first appears as cones formed distal to the tips of the bony cores. Horn casting occurs approximately two months after the rut, while the testes are declining in weight. Casting may be a mechanical process in which the old horn

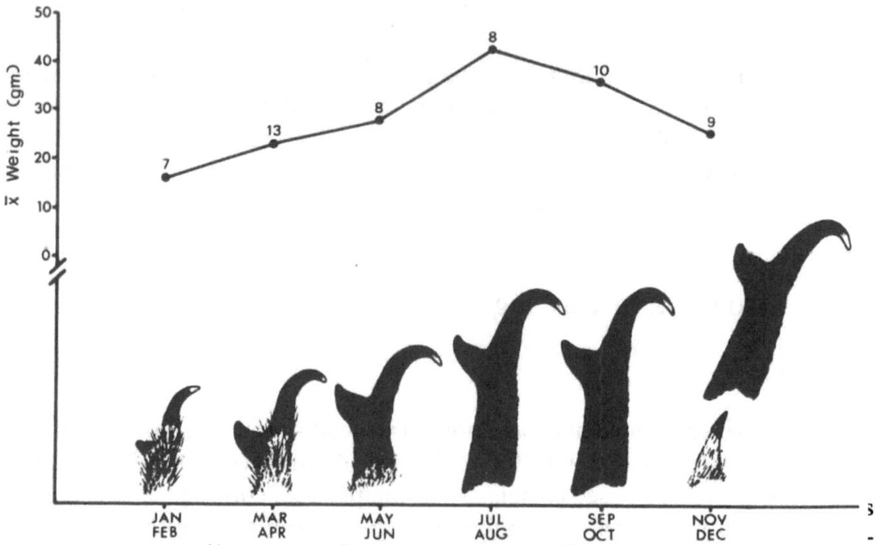

bers above the dots represent the sample sizes. To obtain samples large enough to yield meaningful averages, measurements were grouped by two-month increments. The sketches represent horn development and casting. (Adapted from O'Gara et al. 1971; sketch by Susan Kraft.)

FIGURE 10. A pronghorn doe and buck during late November in Montana. The buck has cast one horn sheath, and will soon drop the other. (Photograph by Bart O'Gara.)

FIGURE 11. A pronghorn buck shot in Montana in mid-November. The left horn sheath came off when the hunter attempted to roll the buck over, using the horn as a handle. Note the smooth, almost hairless skin covering the horn core. About 18 mm of hard horn was present distal to the horn core, including the "ivory" tip. Hair will grow on the basal two-thirds of the skin before it starts to cornify. (Photograph by Bart O'Gara.)

is forced up and off the bony core by the proliferation of tissues beneath it, but long hairs rooted in the dermis and embedded in the sheath hold it in place for some time. The pushing-off process is gradual and almost invariably one sheath falls before the other (Fig. 10). Embedded hairs are usually broken off even though some pull free from the sheaths. Shed sheaths are fringed at the bottom with the ends of the hairs still held fast by the horny material.

Wislocki (1943) compared the periodicity of the testes of the white-tailed deer (*Odocoileus virginianus*) with the seasonal changes of their antlers. He found that the antlers begin their annual renewal at a time when the testes are least active. The antlers become hard and the velvet is shed while the testes are rapidly enlarging, whereas the antlers are lost when the testes begin to decline. The casting of horn sheaths by pronghorns seems to follow a similar pattern (Morrison 1961; O'Gara et al. 1971).

The cones of hard horn are 18 to 25 mm long by the time a sheath is cast and may already display the "ivory" tips seen on many mature horns (Fig. 11). Soft keratin, a forerunner of horn, covers the remainder of the future horn surface. Thus, after old sheaths are cast, new horns consist of bony cores covered by hairy gray skin topped by cones of hard horn, often with ivory tips.

At shedding (Fig. 12, NOV-L), annular swelling of skin can be felt between the tips of the bony cores and the cones of the horns. Shortly after horn casting, the annular swellings, which are warm to the touch and bleed profusely if cut or

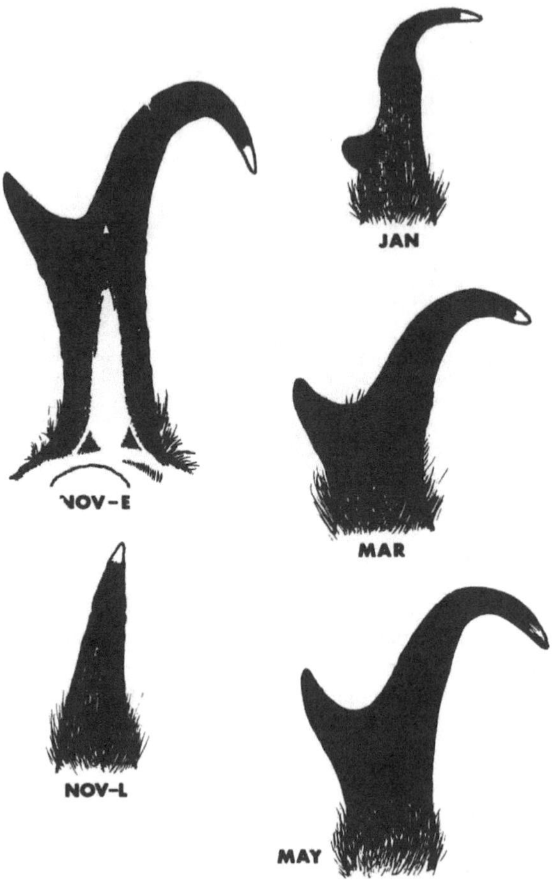

FIGURE 12. Growth sequence of pronghorn horns in Montana. Lateral view of left horns, approximately one-fourth natural size. The relationships of hard horn (black) formation to soft skin (gray) on the bony cores is illustrated. The NOV-E (early) horn is shown in a diagrammatic longitudinal section; the white center represents the bony horn core. Note the wedge of new hard horn growing between the old horn sheath and the horn core. This new material aids in horn casting and is the only hard horn remaining after the old sheath is cast, as seen on NOV-L (late). (Sketch by John Eiler.)

bruised, become more conspicuous than they were at casting. Rapid differential growth from the swellings extrude horn tips upward and usually inward or backward, which forms hooks (Fig. 12, JAN; Figs. 13 and 14). The frontal prongs begin approximately 5 cm above the head and develop rapidly as the formation of hard horn proceeds.

About four months after shedding, the prongs are fully developed and in the position, relative to the hooks, they maintain in mature horns (Fig. 12, MAR). The next phase of development involves a hardening of the soft keratin that covers the bony cores and growth at the bases, like that of most bovids; this

FIGURE 13. Pronghorn bucks during January in Montana. The hooks of the horn sheaths are more than half grown and the prongs are just beginning to grow. (Montana Department of Fish, Wildlife and Parks; photograph by Tom Warren.)

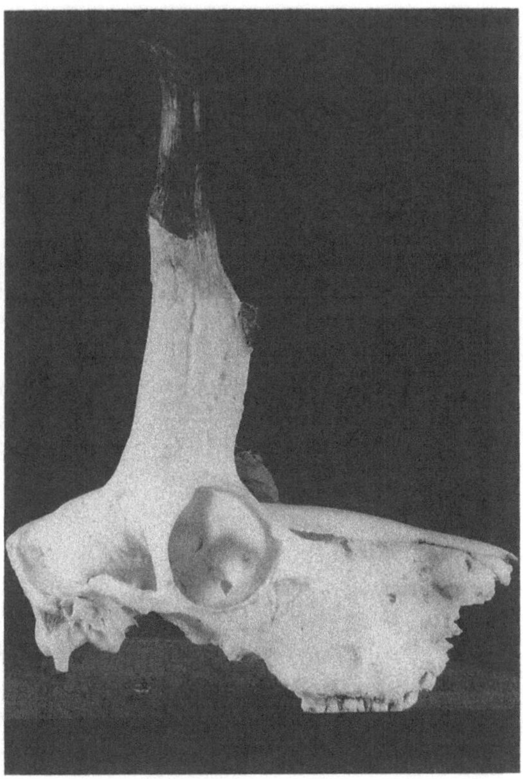

FIGURE 14. The skull of a pronghorn buck that was picked up during spring on the National Bison Range, Montana. The hard horn indicates the buck died during January. (Photograph by Bart O'Gara.)

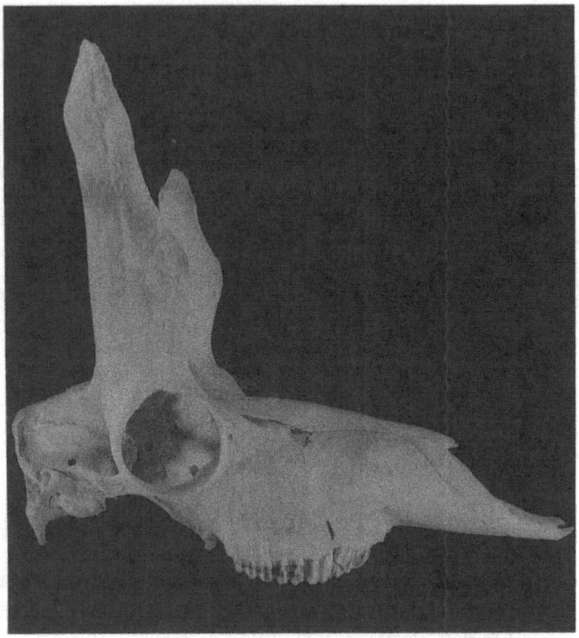

FIGURE 15. The skull of an adult buck collected in January when the hook was being extruded rapidly. Note the swelling and porosity at the tips of the horn cores. Tips of horns collected at other times of year are smooth and slope directly to the tips without the "hip" shown here (see Fig. 7.2). Also note the bulge where the prongs are beginning to grow. (Photograph by DeWayne Williams.)

lengthens the bases and pushes the prongs and hooks upward, away from the head (Fig. 12, MAY).

Growth slows in nearly mature horns, and the formation of hard horn ceases at about 10 months post-shedding. The skin under the horn sheaths remains comparatively inactive until after the rut, when new growth again results in the casting of old horn sheaths.

Horn growth is rapid compared to that in most bovids, possibly because horn formation, unrestricted by outer sheaths, takes place along the entire length of the cores. Nutrients for horn development are probably supplied primarily through the skin, as indicated by the rich blood supply to the annular swellings during the formation of the hooks. However, the horn cores apparently are also involved. During autumn in Montana, when the bucks are in hard horn, the surface of the cores is fairly smooth and dense, and the cores slope directly to sharp tips. From late November through January, when the hooks and prongs are growing rapidly, the bone under the growing areas becomes spongy and porous and enlarges (Fig. 15).

Castration has a marked effect on the horns of pronghorn bucks. Castrated reindeer (*Rangifer tarandus*) are said to continue casting and renewing their antlers (Wislocki 1943). Some pronghorns also continue renewing their horns annually after castration, even though the horns are small and may not be cast. However, the age at which such animals were castrated is unknown. Pocock (1905) described the horns of a castrated buck thus:

> Instead of rising from the forehead as upright, laterally-compressed, bony prominences, the horn-cores bend obliquely forwards in a vertical plane, their axes inclining to the plane of the forehead at an angle of about 45°. They are only about 2 inches long.
>
> The horn-sheaths cover the core to the root, becoming gradually softer proximally, and passing into the hairy integument of the head. They project nearly horizontally forwards in the direction of the nose for a distance of about 3.5 inches, then curve downwards for about 2.5 inches, and then backwards towards the eye for about 3.5 inches, the terminal inch curving lightly inwards and downwards to a point close to the eye. Their total length along the outer or convex curve is thus about 9.5 inches. They also present a spiral twist forming about one-fourth of a complete turn.
>
> . . . .
>
> Although the horn-sheaths have been described above as if each corresponded to a single fully-formed horn-sheath of an adult Prongbuck, closer examination shows that they are in reality composite—that is to say, they consist of a series of horn-sheaths partially severed from each other. The exact number of sheaths involved in the formation of the whole is not easy to determine. There appear, however, to be six.

A castrated buck brought to the National Bison Range from eastern Montana had horns quite similar to those described by Pocock. However, the buck had only three nested sheaths and the horns were hard right down to the skull. He caught his horns in a fence and pulled two sheaths off each horn. I intended to immobilize the animal and see how old it was and whether the number of sheaths corresponded to its age. However, he disappeared, probably killed by coyotes, before I got to it. No one knew how old the buck was when he was castrated.

Another buck on the Bison Range, castrated the day he was born, did not grow hard horn sheaths (Fig. 16). The horn cores leaned forward over the nose and had caps of hard horn at their tips, but most of the cores were covered by skin and hair. Except for tipping forward, the horn sheaths resembled those of a normal buck about two weeks after horn casting.

Females

Variability in horn size prevents casual observers from determining whether a doe has recently shed. Although admitting that critical observations were lacking, a number of biologists have assumed that shedding is similar in does and bucks (Lyon 1908; Skinner 1922; Rush 1944; Einarsen 1948). Horn sheaths of does, apparently ready to be cast, came off during handling in October, November, December, February, and March. A tame doe on the National Bison Range carried double sheaths throughout the spring of 1966 and shed both sets on 15 July 1966 (O'Gara 1969b). The sample size is small, but the above observations indicate that horn casting by does is not as seasonal as that by bucks. Because female reindeer and pronghorns have deciduous antlers or horns, casting and

FIGURE 16. A four-year-old pronghorn buck that was castrated the day he was born. Only 38 mm of the tips were hard horn. The horn cores inclined forward over the nose. This animal was shot in August when intact bucks have very large, actively secreting subauricular glands. Note that the glandular area on this buck is very small, with no sign of secretory material on the dark hair covering the gland. (Photograph by Bart O'Gara.)

renewal by females cannot be controlled by the testicular cycle and must not be under the control of androgens.

Social Significance of Horns

Intraspecific Communication

The pronghorn spends most of its life on the open plains where its conspicuous markings are important for visual communication with other pronghorns. The bold head and neck markings and the horns are important in intraspecific encounters. The prongs of the horns, which extend out in front, serve to engage an opponent's prongs, lessening the danger from the sharp tips as two males test each other's strength. An aggressive buck advances with depressed ears. The blackish horns in combination with the dark facial markings exaggerate the size of the horns (Fig. 17).

Proof that bucks are aware of the horn sizes of rivals is seen in their reactions to decoys. Bow hunters now use pronghorn decoys for "tolling" bucks. These

FIGURE 17. The dominant male (right), his ears back and head slightly lowered, is giving a cheek patch display that exaggerates the size of his horns. The buck on the left is 2.4 years old; the dominant buck is 3.4 years old. (Photograph by Peter T. Bromley.)

decoys are generally silhouettes, that resemble that of a yearling buck, which the archer can hide behind. Mature bucks often approach such a decoy close enough to be killed with an arrow. During the rut, territorial bucks are exceptionally vulnerable because immature bucks circle the harems and try to approach the does. Dominant bucks will repeatedly charge smaller bucks and drive them away. If large horns are placed on a decoy, other bucks will not approach, or they will approach very cautiously.

Courtship and Breeding Systems

A courting buck advertises his secondary sexual characteristics to does by presenting his cheek patch; the horns and face mask form a right angle and the ears are in a neutral position, which accentuates the horns. The mane is partly erected, which increases the apparent size of the neck; the neck is set off by white bands (Kitchen 1974). Thus, the horns of pronghorn bucks are important in interactions with other bucks and with does. Pronghorns have a complicated but flexible breeding system, and much of it is influenced by the bucks' horns.

The early literature and some later literature on the breeding behavior of pronghorns described pronghorns as having a harem-type mating system in which dominant bucks control and defend does without regard to a specific location (McLean 1944; Einarsen 1948; Buechner 1950a,b; Gregg 1955; Folker 1956; Prenzlow et al. 1968; Deblinger & Ellis 1976). Other studies have indicated that pronghorns are territorial and defend a specific area rather than a band of females (Cole 1956;

Cole & Wilkins 1958; Bromley 1969, 1977; Pyrah 1970; Gilbert 1973; Kitchen 1974; Kitchen & Bromley 1974; McNay 1980). Ingold (1969) described a social organization involving both harems and territories in Wyoming.

Kitchen and Griep (1978) reported that another Wyoming population displayed a system in which breeding was restricted to specific areas they called dominions, but the areas were not defended by large males. The females were herded within dominions, but harems were not formed. The does frequently moved between breeding areas during the rut. Large males even moved between dominions with each other, but dominance between bucks shifted as they moved from one area to another. Paris Griep (personal communication, 1978) found that bucks on an overgrazed range did not defend specific land areas. He believed that water availability, poor range, extent of hunting, and perhaps fences played some part in these modifications of behavior. It stands to reason that a buck on a poor range could not defend a territory large enough to support himself and a doe–fawn band throughout the summer.

Deblinger and Ellis (1976) described the behavior of a small group of captive pronghorns in a natural, short-grass prairie in Colorado. Their observations of spatial organization closely paralleled those made about 56 km away by Prenzlow et al. (1968) of pronghorns that displayed a harem breeding system. The animals were tame, and Deblinger and Ellis (unpublished data) conducted an intensive, quantitative study of specific behaviors. These behaviors were consistent with those reported for territorial bucks in South Dakota and Montana by Bromley (1969, 1977) and Kitchen (1974).

These comparisons suggest that differences in pronghorn social organization are not caused by major differences in specific behaviors or displays (these appear to be constant in pronghorn social systems). Rather, social organization is influenced by differences in the density and spatial dispersion of individuals. In other words, the same behavioral repertoire may, under different environmental conditions, result in different social systems. When these results are interpreted in terms of possible genetic and environmental contributions to social organization, they suggest a strong genetic contribution to specific behaviors and displays. They also indicate a major environmental contribution to spatial dispersion, frequency, intensity, and timing of inter-individual interactions.

The variability of the pronghorn's breeding systems obviously has an ecological basis. According to Kitchen and O'Gara (1982), when forage quality, especially succulence, varies between areas and the best resources are clumped, pronghorns tend to be territorial, and the males on the best territories do most of the breeding. As resources become more uniform in distribution and of poorer quality, the system shifts toward dominions. The breeding system shifts to harem formation when unusual resource conditions occur, when population levels are low, or when sex ratios are unusually skewed (ratios of 1 male to 10 females or more). Such flexibility aids in adaptation to specific areas, but the species must also be flexible enough to change systems in a very short time when environmental conditions change rapidly.

Territoriality

Working in parks and refuges where many adult bucks were present and where range conditions were generally good, Bromley and Kitchen (1974) found that territorial bucks did almost all of the breeding. The establishment of territories was related to sexual behavior, but the bucks were territorial for a much longer period (seven to eight months) than was required to breed with females (two to three weeks). All the territories used by does during the rut had open water, and the bucks were most successful on territories where the terrain helped prevent the escape of does (Kitchen 1974). The most clearly defined boundaries were in open grasslands, and often in contested areas, which suggests that the bucks adjusted their use of space to best defend that space against intrusion by other males (Bromley 1977).

As the rut approaches, a territorial buck holds, or attempts to hold, does on his territory. Harassment of does by territorial and bachelor bucks increases as bachelor bucks intensify their attempts to associate with females. Territorial bucks try to keep bachelors away from the females but often leave doe groups on their territories unattended while pursuing single females. The unattended doe groups are then broken up by bachelor bucks that invade the territories (Cole & Wilkins 1958).

A sexually aroused buck announces his intentions to a doe by emitting a high-pitched whine as his body becomes rigid with legs straight, neck vertical, and nose horizontal. If the doe is receptive, she raises her tail, and the buck approaches her rump with short prancing steps. At about 5 m, the buck starts waving his head laterally while flicking his tongue and making a low sucking sound and chewing motion (Fig. 18). The head waving displays the blackish cheek patches and horns to the doe. The display may also waft the odor of the subauricular glands to her. According to Walther (1984), no other species shows such striking and pronounced head waving during courtship.

Does not in estrus move away from courting bucks (Fig. 19). Bucks sometimes chase unreceptive does through a series of loops and curves while uttering deep guttural roars and hooking at them. The bucks occasionally knock does down during such chases. Even does in estrus run from groups of courting bucks, which are usually bachelors. Chases by bachelors are often long and attract other bachelors. When bachelors chase does into territories, the territorial bucks challenge the buck(s) and give the does sanctuary (Bromley & Kitchen 1974). This behavior undoubtedly contributes to the high breeding success of territorial bucks, especially those that are large and have large enough horns to intimidate bachelors.

Bachelors molest females throughout the summer and become very aggressive as the rut approaches. During September in Yellowstone National Park, I saw an estrous doe harassed by seven bachelors. Hair was ripped from at least one third of her shoulders and sides, and blood trickled from several of the bare patches. As a buck mounted her, others tried to dislodge him, and as often as not, the doe received the horn thrusts. When the doe attempted to flee, the bachelors pursued

FIGURE 18. A courting buck displaying his cheek patch and horns to a receptive doe. Does in estrus look directly at bucks and sometimes lick the buck's cheek patches or median glands. (Photograph by Bart O'Gara.)

her, hooking at her with their horns. Twice, I saw the doe knocked down; once she did not get up immediately, and a buck tried to mount her on the ground while others horned her and struck her with their front hoofs.

The pronghorn's territorial system appears to be beneficial to the species in at least three ways. First, it ensures that the largest, most aggressive bucks do most of the breeding (Bromley & Kitchen 1974). Second, the territory provides a haven where does can escape the overzealous courtship of bachelor bucks during parturi-

FIGURE 19. An unreceptive doe moves away from a courting buck. Usually, if she moves around a bush or rock, a territorial buck will leave her alone. Bachelor bucks, and occasionally an excited territorial buck, will pursue does and attempt to gore them. (Photograph by Bart O'Gara.)

tion, lactation, and the rut (Cole & Wilkins 1958). And third, the territories keep bachelors from competing for food with pregnant or lactating does on the best range (Gilbert 1973). These benefits are no doubt real, but the harem breeding system may be equally effective in restricting breeding to the most dominant bucks. Harem breeders should also protect does from harassment by bachelors during the rut. Thus, territoriality is probably most beneficial during spring and summer.

Walther (1984) noted that territorial bucks often separate themselves somewhat from their harems during the rut and stand or rest above them on the slope of a hill

FIGURE 20. Territorial bucks usually face the direction from which encroaching bucks are likely to approach. The dark-colored face and horns warn approaching bucks that they will have to face a territorial buck if they proceed. (Photograph by Craig Jourdonnais.)

in an almost "strategic" position to keep the females as well as the surrounding area under watch. In those positions, their striking neck markings along with their dark-colored face and horns (Fig. 20) no doubt warn approaching bucks at some distance that they will have to face a territorial buck if they proceed.

Bucks on the National Bison Range, Montana, and in Wind Cave National Park, South Dakota, are territorial from late March to October. The territories provide the territorial bucks and their does with sufficient forage. Territorial bucks defend their areas against intrusion by other males, and mark vegetation with subauricular gland secretions and the linked sniff-, paw-, urinate-, defecate- (SPUD) sequence (Fig. 21). Defenses of territories by territorial bucks vary, but most encounters consist of some or all of five phases: (1) stare at intruder, (2) vocally advertise presence, (3) approach intruder, (4) interact (displays) with intruder, and (5) chase intruder (unless he withdraws). Occasionally, fights occur,

Encroaching males that do not leave a territory in response to a snort–wheeze are confronted by territorial bucks. A territorial buck's speed of approach for such confrontations depends on the depth of intrusion into his territory and the age of the intruder, which is indicated by the size of the intruder's horns. Bucks well inside territories, and yearlings, elicit rapid approaches by territorial bucks, which usually lead to running chases as intruders are driven from the area (Fig. 22). However, adult bucks that remain near the edge of a territory are approached cautiously.

Territorial bucks often lope or trot to within 40 to 80 m of an intruder, then lower their heads, depress their ears so the horns show, and walk to within 15 to 25 m. At this point, the walk becomes a slow, stiff, deliberate gait, which ends in a broadside threat that exaggerates the size of the horns, similar to the threat display noted by Prenzlow et al. (1968). If both bucks display, they are usually parallel or nearly so, and parallel displays may lead to parallel walking in a head-low posture, or sometimes to a closer approach. By this time, the intruding buck has usually given ground. During some slow approaches, territorial bucks thrash vegetation with their horns in redirected aggression, mark with their cheek patches, or perform the SPUD. The territorial bucks often grind their teeth during a slow approach and occasionally stop to feed or snort–wheeze. While this is going on, territorial neighbors, and sometimes other intruders, mark, thrash, or feed as a territorial buck approaches. When presenting a horn threat to an opponent, a pronghorn buck does not tuck in his chin as much as do most bovids. The prongs and the entire horns point more forward than do the horns of a bovid.

When a territorial buck and the intruder are within 15 to 25 m of each other, both usually perform the same behavioral acts. Consequently, many acts are used and reused as the two bucks repeatedly feed–thrash–mark–walk–threaten. To intensify an encounter, a territorial buck moves closer to the intruder. In most interactions, the intruder by this time no longer responds to the territorial buck's initiative, and simply watches. This inaction usually leads to a strong horn threat by the territorial buck, which generally results in the intruder being herded away at a walk or chased away at a run. Territorial bucks occasionally utter low guttural roars before or during a chase, especially before they attempt to gore fleeing males. Chases range from a few meters to 5 km, and when the chased intruder is also a territorial buck, dominance is usually reversed as the intruder's territory is entered (Kitchen and Bromley 1974).

The National Bison Range in Montana was the site of two extensive studies of pronghorn territoriality during the 1960s and early 1970s (Bromley 1969; Kitchen 1974). Pronghorns there maintained a territorial mating system from 1965 through 1978. Preliminary observations in 1981 suggested that the mating system had changed, and data from the ruts of 1982 through 1984 revealed a progressive decay in territoriality. Byers and Kitchen (1988) compared data from the "territorial years," 1969–1978, with data from the "decay years," 1982–1984. They found that fewer territories were defended during 1982–1984 than during 1969–1978 because of a smaller proportion of bucks defending territories, not because there were fewer bucks available. Most territorial bucks abandoned their territories early in ruts during 1982–1984 or lost control of the females on them,

A

B

C

FIGURE 21. The SPUD behavioral routine. Not only does a territorial buck execute the SPUD in response to does and intruding bucks, he sometimes even performs it when a human or a predator enters his territory. In that case, he is always perpendicular to the path of the intruders so his size, cheek patch, and horns are evident. (A) A buck sniffing the spot where he will perform the SPUD. (B) and (C) After pawing vigorously with each front foot, urination and defecation on the pawed spot follow in sequence. (Photographs by Jo Meeker.)

following frequent, persistent intrusions by nonterritorial bucks. Many does left territories and mated with nonterritorial bucks. Territorial bucks that maintained control of females did so by shrinking the sizes of their territories.

The mating system change followed a high death rate during the winter of 1978–1979 that removed 75% of the males, including all males older than five years, and all male fawns, from the population. The age structure of the population was skewed toward old animals before the winter of 1978–1979 by coyote (*Canis latrans*) predation on fawns beginning about 1970. During 1982–1984, the number of males present did not differ from the number of males during 1969–1978, but the frequency distribution of male ages was strongly shifted toward younger ages.

The small number of older males during 1982–1984 probably resulted in smaller proportions of males initially defending territories and in less effective territorial defenses. When females clustered on the few territories where defenses were at first successful, they attracted large numbers of nonterritorial

FIGURE 22. The fast action of a pronghorn chase is blurred. The smaller buck is barely evading the pursuing territorial buck. (Photograph by Peter T. Bromley.)

males. The resulting high rates of raids on these territories, coupled with reduced defense radii by territorial males, allowed the females only a 12% reclining time during the rut. This increased energy cost and led to an apparently greater risk of injury on weakly defended territories, which appeared to prompt many females to seek calmer matings elsewhere. Also, if female pronghorns practiced mate selection based upon horn or body size, they might have reduced their efforts to remain on territories during 1982–1984. Males seen during 1969–1978 were larger than males seen during 1982–1984, and their horn size varied more (Byers and Kitchen 1988).

Byers and Kitchen (1988) postulated that at least two conditions influence the tendency of males to be territorial. First, males must be at least three years old before they attempt to defend a territory. Second, the declining proportion of males defending territories during 1982–1984 that coincided with an increased number of males three years old and older suggests that males decide whether or not to attempt territorial defense based upon the frequency of territorial defense in the population.

Horns as Weapons

Occasionally, displays are not sufficient to scare an intruder from a territory, and serious fights result (Fig. 23).

FIGURE 23. Two bucks fighting during the rut. Such fights frequently result in serious injury or death. (Photograph by Peter T. Bromley.)

Seton (1927) watched pronghorn bucks fighting in the Washington Zoo and

> . . . was made to realize how exactly each detail of the apparently harmless horn had a purpose, offensive or defensive, for which it was highly specialized . . . as they thrust and parried, the purpose of the prong was clear. It served the Antelope exactly as the guard on the bowie knife or the sword serves a man; for countless thrusts that would have slipped up the horn and reached the head, were caught with admirable adroitness in their fork. And the inturned harmless-looking points? I had to watch long before I saw how dangerous they might be when skillfully used. After several minutes of fencing, one of the bucks got under his rival's guard; making a sudden lunge, which the other failed to catch in the fork, he brought his inturned left point to bear on the unprotected throat of his opponent, who saved himself from injury by rearing quickly, and throwing himself backward.

Seton (1927) noted further that, in their fights, wild pronghorns are usually struck and sometimes ripped in the throat or neck. The head and neck skin of a male pronghorn is nearly twice as thick as that of the female. Apparently, the thick skin acts as a shield in areas where horn penetration is most probable.

Kitchen and Bromley (1974) observed 15 fights, of which 14 occurred during the rut. All were prefaced by a slow approach, with heads low, ears back, and manes and rump patches compressed. At 0.5 to 1.5 m, the bucks stood facing and staring at each other for as long as six seconds. In nine fights, the bucks clashed suddenly, whereas in five others the bucks slowly engaged their horns. When one yearling presented his cheek patch, his yearling opponent lunged and delivered a serious shoulder injury. Fights between adult bucks consisted of a series of thrusts and counterthrusts that resembled the actions of humans fencing with swords. Because the prongs were not always large enough to engage an opponent's prongs, some fights were horn-to-horn and some head-to-head. Bucks tried to push their adversaries off balance by twisting their necks, and each tried to gain an uphill advantage during battle. Although the average time of the 15 fights was only two minutes, five animals were seriously injured. Kitchen and Bromley (1974) saw 11 other injured bucks, probably victims of fights, and found two large, dead, bachelor bucks with deep puncture wounds in their lungs and hearts.

Thus, the rather inoffensive looking horns of pronghorns are not only important in intraspecific communication, they are also lethal weapons. Heavy winter clothes can be sliced by the horn tips as if by a knife during trapping operations. A partially tranquilized buck once "hooked" me in the arm. The point penetrated about 2 cm and escape was impossible until another person held the animal's head steady. Having handled hundreds of big-game animals in traps, I am fairly casual toward most of them, except pronghorn bucks and Rocky Mountain goats.

Interspecific Aggression

Pronghorns will occasionally present horn threats to, or clash with, other animals. Berger (1985) saw a bighorn (*Ovis canadensis*) ewe and a pronghorn buck threaten each other with their horns at a distance of less than 2 m. Mule deer (*Odocoileus hemionus*) share much of the pronghorns' range, and McCullough (1980) has observed direct aggression between pronghorns and mule deer. From late March through September, the territorial bucks threatened, snort–wheezed at, and chased the deer. During October, the territorial bucks stopped defending territories and became much less aggressive toward each other and toward other species. From October through March, the mule deer were dominant to the pronghorns.

White-tailed deer and pronghorns meet more often than most people realize. The pronghorn's quest for succulent forbs takes it into whitetail habitat in some areas, especially during dry spells. McCullough (1980) believed that the pronghorns on the National Bison Range influenced the distribution of whitetails. The territorial bucks presented horn threats and chased the deer, which generally stayed in or near dense cover during the period when pronghorns were territorial. However, during summer storms, the pronghorns moved into swales and small ravines, or bedded. They were difficult to observe and seemed to be absent from the area. At such times, the whitetails ranged into the open, often long distances from cover. They fed and ran about playfully, only to disappear again when the territorial bucks returned to patrolling their territories. As territorial defense ceased during autumn, the whitetails moved into the open away from streams, and distances between feeding whitetails and pronghorns were often similar to the distances between individual pronghorns in a group.

Pronghorns, especially does defending fawns, often follow or chase predators. Most of those I have seen attacking coyotes struck them with their front feet; however, one horned doe tossed a coyote into the air with her head. Does attacking golden eagles (*Aquila chrysaetos*) on the ground try to butt or horn them, and Roosevelt et al. (1902) reported seeing a yearling buck strike at an attacking eagle with its horns.

Summary

Some 35 million pronghorns once inhabited North America. By 1924, only 20,000 remained. Good management has brought the population back to about one million animals.

Pronghorns are adapted for life on open prairies and once made long movements to procure the year-round necessities of life. Breeding herds are somewhat genetically isolated, which leads to local variations in color, horn morphology, and other characteristics. The timing and length of seasonal movements vary with environmental conditions, but animals in large wintering herds generally disperse in spring to form small summer bands.

Pronghorns are the only animals known in which four to seven embryos are generally reduced to two during pregnancy by intrauterine competition.

The unique horns of pronghorns have determined the classification of pronghorns in relation to other ruminants. Characteristics other than horns are currently receiving more attention in attempts to relate pronghorns to other ruminant taxa.

Horn sheaths rest on bony, somewhat triangular horn cores separated by a living layer of skin. Sheaths are composed of compressed, cornified epithelial cells. The formation of hard horn begins annually just after maximum testicular development, and old sheaths are shed about two months later as the testes decline in weight. Casting of sheaths by males seems to be controlled by androgens in a manner similar to that of antler casting in cervids. Nutrients for horn development are apparently supplied both through the skin and the horn cores. Castration at birth causes the horn cores to lean forward over the nose; also, only the tips of the sheaths harden. The bold head markings and horns of bucks are important in intraspecific encounters with other bucks and with does. Pronghorns have a complicated but flexible social system, and much of it is influenced by the bucks' horns. Populations may have harem or territorial types of breeding systems, the two together, or intermediate types. The breeding system in an area may change from one year to the next. Differences in social organization are not caused by major differences in specific behaviors or displays, which are strongly influenced by genetics. Frequency, intensity, and timing of inter-individual interactions and spatial dispersion are influenced by environmental conditions. Harassment by bachelor bucks drives does to adult bucks' territories and contributes to the breeding success of territorial bucks that are large enough, and have large enough horns, to repel bachelors. Sexually aroused bucks display their blackish cheek patches and horns to does, and bucks occasionally gore unreceptive does. The pronghorn's territorial system appears to ensure that the largest, most aggressive bucks do most of the breeding; provide a haven where does can escape harassment by bachelors during parturition, lactation, and the rut; and keep bachelors from competing for food with pregnant or lactating does on the best range. Defenses of territories involve staring, vocalizations, approaches, displays, and chases. Fights occasionally occur and can lead to severe injuries or death.

Thus, the horns of pronghorns are important in intraspecific communication, and are also lethal weapons.

Acknowledgments. The U.S. Fish and Wildlife Service; Montana Department of Fish, Wildlife and Parks; University of Montana; and Wildlife Management Institute cooperating.

8 Neuroendocrine Regulation of the Antler Cycle

GEORGE A. BUBENIK

Introduction

The antlers are one of the secondary sexual characters of males in all genera of *Cervidae* except *Rangifer* (where antlers are found in both sexes). Their development, seasonal renewal, maturation, and casting are closely related to the activity of the reproductive system. This fact was already well known to Aristotle, who described the alterations of antlerogenesis caused by castration (Goss 1968). However, it took another 2,000 years before progress in chemistry and endocrinology made it possible to investigate the detail roles of various humoral factors in the regulation of the antler cycle.

The basic mechanisms of hormonal regulation of the antler cycle are common to all cervids. However, there may be substantial differences in the seasonal timing of individual phases, the levels of sensitivity of the pedicle or the antler tissues to various hormones, as well as the sensitivity to environmental influences, depending on individuals, age, and species. Except of a long-day breeding roe deer, there appears to be a close relationship of timing and hormonal sensitivity in each group of telemetacarpal and plesiometacarpal deer.

Without exception, all data presented here were obtained from deer living in boreal or temperate regions. The data on antler cycles of tropical or periequatorial deer are at this point too sketchy and often too confusing to be discussed. However, despite the many gaps in our knowledge of antler cycles of tropical deer, the basic mechanisms of antlerogenesis appear to be not different from those found in deer from temperate and boreal regions.

In this chapter, the function of hormones—pineal, pituitary, gonadal, adrenal, thyroidal, and parathyroidal—glands and their relationships to the antler cycle will be discussed. The effects of neuronal stimulations on antler growth are described in Chapters 10, 11, and 12.

The Pineal Gland

The circannual nature of the antler cycle observed in boreal as well as in tropical cervids indicates the involvement of a very precise timing system. It has been established that the seasonality of the antler cycle is determined by the changes

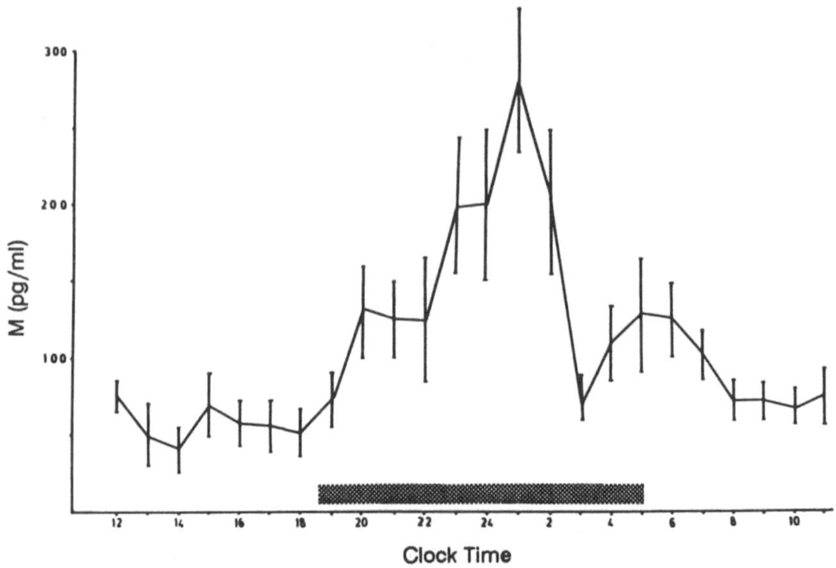

FIGURE 1. Diurnal variations of melatonin in four male white-tailed deer during September. Horizontally, shaded area indicates night time; vertical bars indicate standard error. (From Bubenik, G. & Smith 1987.)

in photoperiodicity (Jaczewski 1954). Experimental alterations of the day length succeeded in inducing up to three full antler cycles in 1 year (Goss 1983) and in causing profound changes in the secretory pattern of gonadotropins, testosterone, and alkaline phosphatase (Bubenik, G. et al. 1987a). The regulation of seasonality of hormonal secretion is dominated by the master humoral transducer—the pineal gland (Bubenik, G. 1986). The removal or denervation of the pineal gland causes desynchronization between natural seasons and the cyclic phenomenons such as the antler growth, food intake, secretion of prolactin, pelage exchange, and sexual behavior (Brown et al. 1978a; Plotka et al. 1981; Schulte et al. 1981a; Snyder et al. 1983; Lincoln 1985).

The perception of seasonal changes in temperate regions is mediated by changes of photoperiodical cues reaching the retina. In the retina, the photic stimulus becomes an electrical message which is further transformed in the pineal gland into a chemical signal, the hormone *melatonin* (Reiter 1980). Melatonin is secreted in a very distinct circadian rhythm; the rapid increase in levels of melatonin coincides with the onset of darkness. The peak values (reaching levels five times higher than during the day) are observed around the middle of the dark period (Fig. 1) (Bubenik, G. & Smith 1987). The timing of melatonin increase seems to be crucial to the regulation of perception of photoperiodicity. Artificial feeding of melatonin during the spring according to a time schedule which simulates the shortening of the day-length, such as happens naturally in

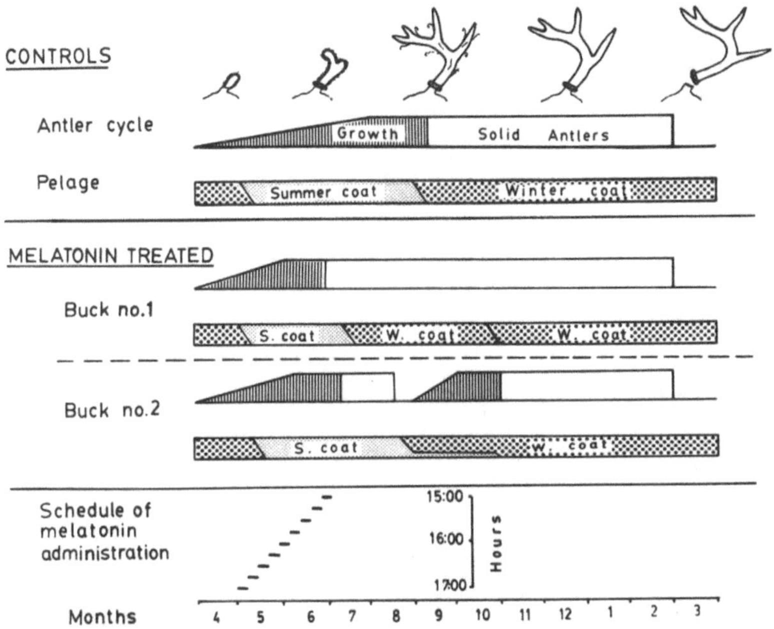

FIGURE 2. Changes of antler cycle and pelage renewal in two adult male white-tailed deer treated orally with melatonin (5 mg/day). Note two winter coats in succession in buck no. 1 and two antler cycles in buck no. 2.

autumn, induces the same events as found at the beginning of the autumn: (a) the mineralization of antlers, (b) change from summer to winter coat, and (c) rutting behavior (Fig. 2) (Bubenik, G. 1983). The early onset of seasonal events was the results of the premature hormonal changes, such as the decline of prolactin and rise of testosterone and follicle-stimulating hormone (FSH) levels, precipitated by melatonin treatment (Bubenik, G. et al. 1986). The experiments with orally administered melatonin have not only confirmed that the pineal gland serves as a transducer for the synchronization of corresponding circannual events with the seasons, but also enabled us to determine which hormones are related to individual phases of the antler cycle. Besides several other hormones involved in the synchronization of antlerogenesis with photoperiodicity, all data point to the overall importance of the pituitary hormone prolactin.

Pituitary Hormones

Prolactin

Prolactin (PRL) exhibits the most pronounced seasonal pattern of secretion. Average peak levels in deer plasma can be as much as 1,000 times higher than

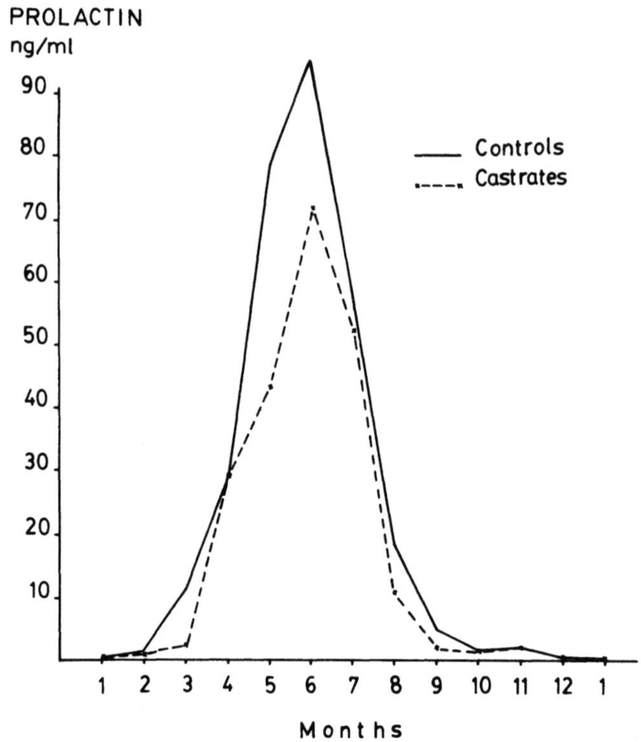

FIGURE 3. Seasonal levels of prolactin (PRL) in 5 intact and 3 castrated adult male white-tailed deer. Note the extraordinary difference between maximal and minimal levels. (From Bubenik, G. et al. 1985a.)

basal values (Fig. 3) (Suttie et al. 1984; Bubenik, G. et al. 1985a; Bubenik, G. & Schams 1986; Barrell et al. 1985) and the annual rhythm of secretion is very closely synchronized with photoperiodicity (Suttie 1980a; Schulte et al. 1981a; Semperé & Boissin 1982; van Mourik & Stelmasiak 1985; Bubenik, G. et al. 1987a). As in most boreal deer, peak PRL secretion coincides with the peak of antler growth. It has been suggested that PRL might be the antler growth stimulus (Mirarchi et al. 1978) proposed by Wislocki (1943). However, blockade of PRL secretion by bromocriptine (resulting in almost 95% reduction of values) had no effect on the extent of antler growth and thus revealed that PRL is not directly involved in this process (Bubenik, G. et al. 1985a). On the other hand, it appears that PRL is involved indirectly in the regulation of antler growth by so-called permissive action. In white-tailed deer, high PRL levels seem to suppress the stimulatory effect of luteinizing hormone (LH) on the secretion of testosterone (T) (Fig. 4), probably by blocking LH receptors on the Leydig cells (Ravault et al. 1982). By keeping T levels low, PRL is preventing the premature onset of antler mineralization and initiation of the rutting behavior during the time

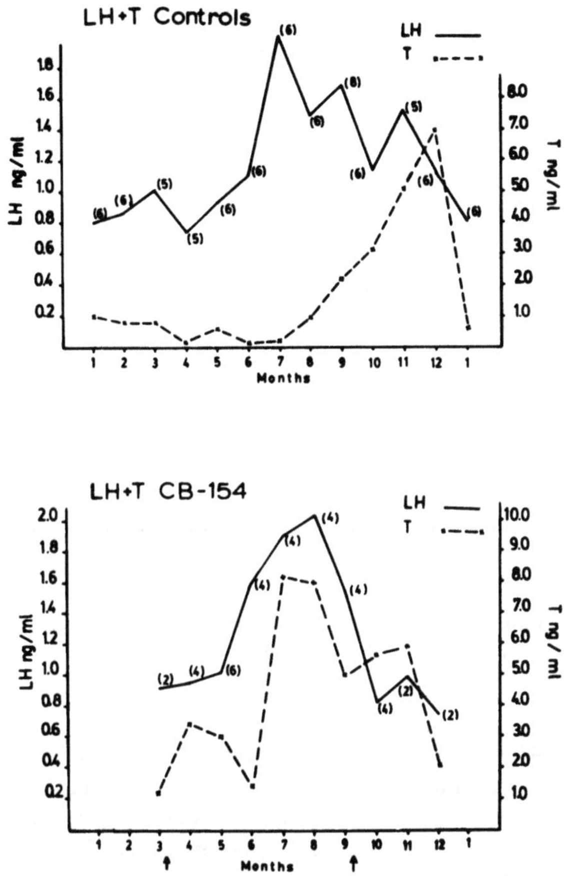

FIGURE 4. Seasonal levels of luteinizing hormone (LH) and testosterone (T) in controls (top) and prolactin (PRL)-suppressed male white-tailed deer (bottom). PRL blockade was achieved by bromocriptine (CB-154). Note the left shift in the position of T peak in bromocriptine-treated deer which resulted in the early rut in July instead of in November. (From Bubenik, G. et al. 1985a.)

unsuitable for successful reproduction (Bubenik, G. et al. 1985a). A similar suppressive effect of high PRL levels on T concentration was reported in the red deer (Barrell et al. 1985; Fennessy & Suttie 1985).

In all mammals investigated so far, the time course of seasonal levels of PRL is almost identical; peak levels coincide with the summer solstice. In white-tailed deer, experimental elongation of the photoperiod initiated at the fall equinox was followed by an elevation of PRL levels, reaching peak approximately 6 months later (Bubenik, G. et al. 1987a). As the major role of PRL is to coordinate seasonal reproductive functions with photoperiodic changes (Bartke 1980), in some species this hormone acts as a progonadotropic, in others as antigonadotropic (Gorbman et al. 1983). The direction of the action depends on the breeding type; in the long-day breeders (such as the roe deer), high PRL levels stimulate reproduction (Schams et al. 1987); in the short-day breeders (such as a white-tailed deer or red deer), they inhibit it (Bubenik, G. et al. 1987a). Therefore any changes of PRL levels will quickly influence the function of the reproductive system. Besides affecting reproduction, manipulation of PRL in

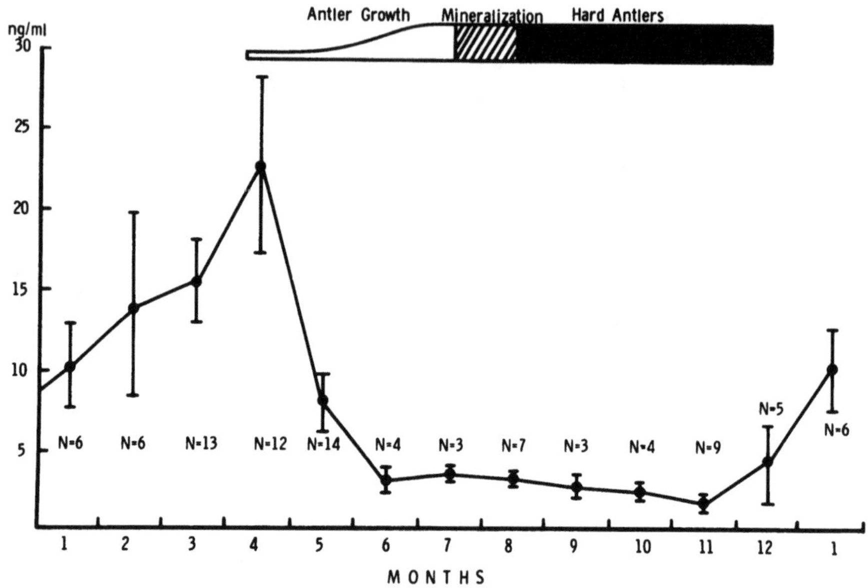

FIGURE 5. Seasonal levels of growth hormone in a group of wild male white-tailed deer. (From Bubenik, G. et al. 1975a.)

white-tailed deer also changed the seasonal timing of thyroidal hormones as well as the timing of pelage exchange (Bubenik, G. et al. 1985a). In view of these facts, it could be hypothesized that, if the pineal hormone melatonin serves as a primary messenger of the photoperiodically mediated seasonality, PRL might act as the secondary messenger (Bubenik, G. et al. 1986).

Growth Hormone

Growth hormone (GH) is another pituitary hormone (chemically and metabolically close to PRL) which has been suspected to be an antler growth stimulus (Wislocki 1943). Maximal levels of GH were detected in deer species just prior to new antler growth (Fig. 5) (Bubenik, G. et al. 1975a; Ryg & Jacobsen 1982; Ryg & Langvatn 1982; Suttie et al. 1989). However, shortly thereafter, the levels began to decline dramatically, and during the period of most rapid antler growth the concentrations of GH are low. In addition, treatment of white-tailed deer by bromocriptine, which is a known depressant of GH levels, did not affect the intensity of antler growth (Bubenik, G. et al. 1985a). GH in red deer is secreted in a pulsatile manner which differs seasonally (Fig. 6). During autumn and early winter, the pulses are frequent and of low amplitude. In spring, the pattern changes to one of high amplitude and high frequency, resulting in high mean levels of circulating GH in plasma. In summer, the amplitude and frequency of

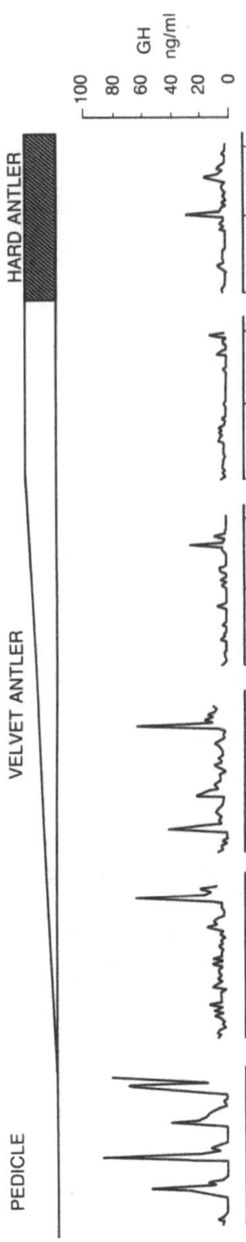

FIGURE 6. Pulsatile pattern of GH secretion in the New Zealand red deer stag. The antler growth season is reversed as compared to Northern Hemisphere. (From Suttie et al. 1989.)

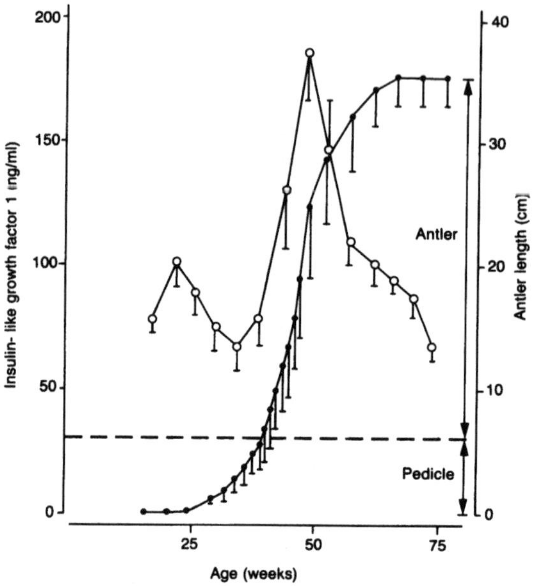

FIGURE 7. Seasonal levels of IGF-1 (open circles) in relation to growth rate of antlers (closed circles). (From Suttie et al. 1985.)

GH declines. The pulse amplitude seems to determine the mean GH levels (Suttie et al 1989).

As it appears that in most instances GH is not directly responsible for promotion of tissue growth, it could be speculated that the high GH levels of early spring, may stimulate production of growth-promoting hormones, called either insulin-like growth factors (IGF) or somatomedins. Plasma levels begin to rise 1 month after peak secretion of GH in the red deer and exhibit peak levels in blood during the most intense antler growth period (Fig. 7) (Suttie et al. 1985a, 1988). Similar findings were also reported for the roe deer (Schams & Barth 1987). Therefore, somatomedins are being implicated in promotion of antler growth. Preliminary evidence indicates that growing antler may be a target organ for one of the somatomedins, IGF-1. Prevention of antler growth in red deer stags resulted in an increase in plasma levels of IGF-1 (Suttie et al. 1988). Authors speculate that the lack of antlers removes the population of IGF-1 receptors. Alternatively, the increase in the circulating levels of somatomedins produced in the liver might have been caused by a negative feedback mechanism involving so far unidentified substance(s) produced in the growing antlers.

Gonadotropins

All current evidence suggests that the gonadotropins—(the pituitary hormones LH and FSH)—are not directly involved in the regulation of the antler cycle. However, as LH and FSH are responsible for the activation of androgenesis and spermatogenesis, their secretion is closely related to the development of the

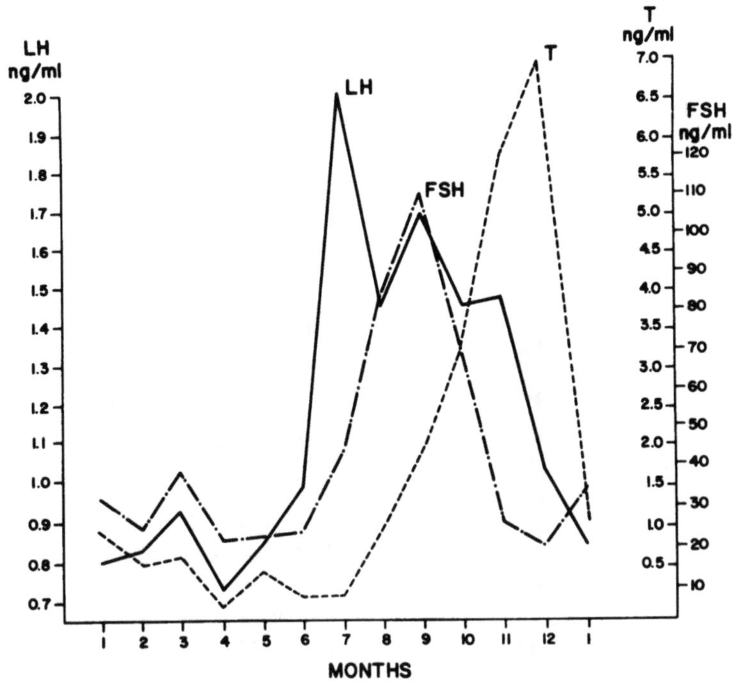

FIGURE 8. Seasonal variations of LH, follicle-stimulating hormone (FSH), and T in intact male white-tailed deer. Note the sequential activation of LH, FSH, and T secretion. (From Bubenik, G. et al. 1982a.)

antler cycle and the activation of reproduction. As in any mammalian species, LH stimulates production and secretion of T by the Leydig cells of the testes. However, the seasonal secretory pattern of LH and T differs substantially in deer species, as the peak LH levels are usually 4–5 months ahead of T (Fig. 8) (Bubenik, G. et al. 1982a; Semperé & Boissin 1982; Suttie et al. 1984). The probable reason for this discrepancy is the changing sensitivity of receptors in the Leydig cells to LH (Sanford et al. 1984). In the spring or early summer, relatively high levels of LH are associated with low levels of T; conversely, in the summer or early fall, even declining levels of LH will rapidly stimulate production and secretion of T (Fig. 9) (Suttie et al. 1984; van Mourik et al. 1986).

Similar change in sensitivity of testes to LH was also observed in deer during the maturation process. In the first year of life, juvenile roe-bucks exhibit relatively high peaks of LH, but T levels are increased only marginally. On the other hand, in the second year, concentrations of LH similar to that observed the previous season will induce a sharp elevation of T (Fig. 10) (Semperé & Boissin 1982).

Preliminary data indicate that the LH:T ratio varies also according to age and social rank. It appears that stimulation of LH and T secretion by the hypothalamic Gonadotropic Releasing Hormone (GnRH) will elicit a distinctly different

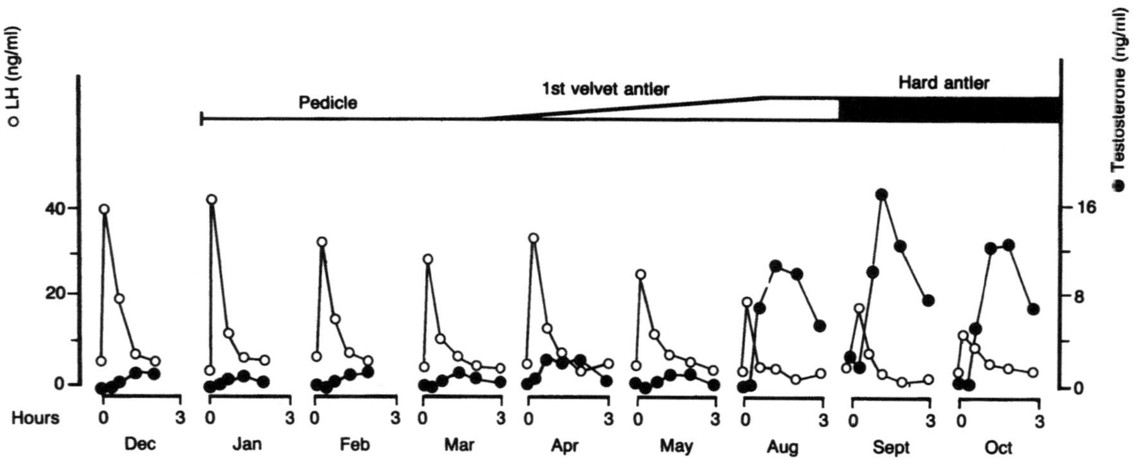

FIGURE 9. Seasonal variation of LHRH-stimulated of LH and T in yearling red deer. All deer were treated once a month with a standard dose of 20 ng/kg or LHRH given i.m. Note the reciprocal relationship between spring and fall pattern of LH and T secretion. (From Suttie et al. 1984.)

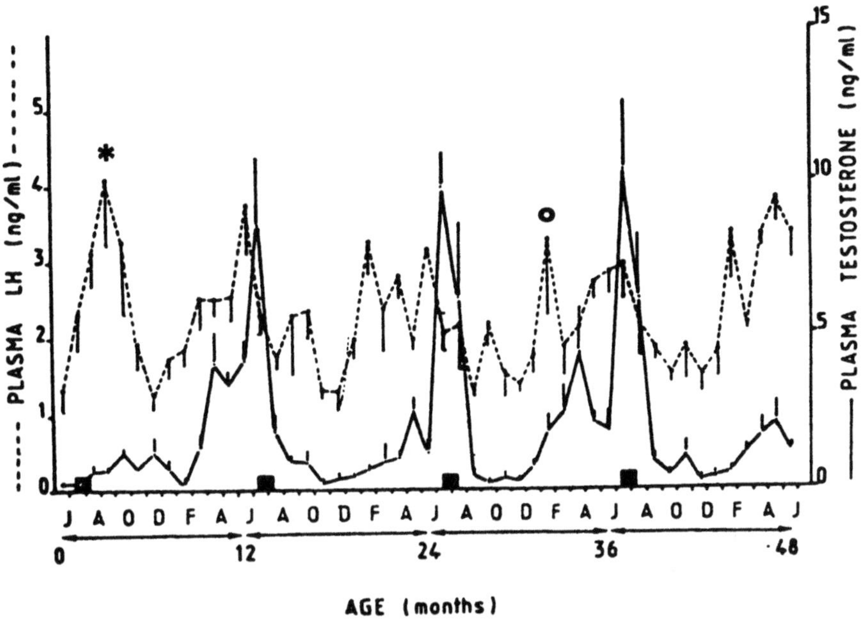

FIGURE 10. Age-dependent seasonal variations of LH and T in plasma of intact male roe-deer. Note the different response of T secretion to LH elevation in the first year (*, gonadal axis immature) as compared to the third year (o, axis mature). Black squares indicate the rutting periods. (From Semperé & Lacroix 1982.)

response in bucks of low rank (e.g., weak antler-producers) as compared to bucks of higher rank (e.g., the prime antler-bearers) (Bubenik, G. et al. 1987b) (Fig. 11). If this is confirmed, a test could be developed which would recognize the antler-producing potential already in juvenile males. The selection of best breeding animals could then start much earlier than is currently possible.

Gonads and Adrenals

Androgens and Estrogens

Sexual steroids of the male deer are produced mostly in two endocrine glands—the testes and the adrenals. Both organs produce a great variety of steroidal hormones; however, the physiological role of these chemicals is known in only a handful of cases.

Testosterone

The effect of T on the antler cycle of the roe-deer was first investigated more than 50 years ago (Blauel 1935). Since then, it was established that in cervids, T is by

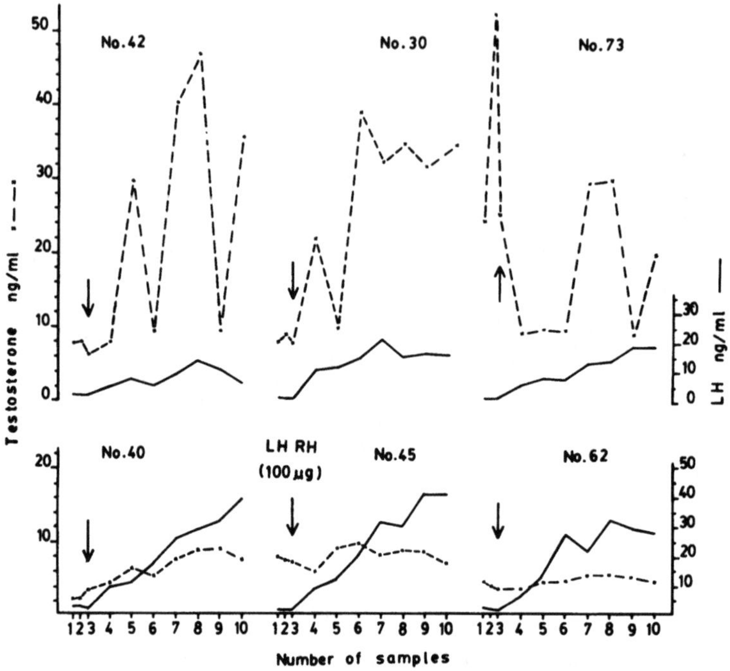

FIGURE 11. LH and T secretion in three "high-quality" (top row) and three "low-quality" (bottom row) white-tailed bucks after a standard dose of LHRH (100 µg) given i.m. in early November. Note the reciprocal relationship of LH to T in "antlerogenetically strong" versus "weak" bucks. (Reprinted with permission from Bubenik, G. et al., 1987b.)

far the most potent androgen influencing the outcome of antler development, mineralization, and casting (Bubenik, G. 1982).

Pedicle Development

The major source of T in mammalian males is the testes. The short-term production of androgens in the fetal testes (around the day 65 in the red deer) coincides with the transitory development of the primordial pedicles (Lincoln 1973). Such pedicles were also detected in early fetal life in both sexes of the roe-deer (Tandler 1913), red deer, fallow deer (Frankenberger 1951; Lincoln 1973), and reindeer (Frankenberger, personal communication). The first postnatal activation of deer gonads occurs during puberty and causes the growth of real pedicles.

Removal of the testes has a profound effect on the development and continuation of the antler cycle. If the orchiectomy is performed before puberty, no antler growth will occur in the castrated male. Similar lack of antler growth was also observed in malnourished young deer males whose T values in plasma did not reach levels sufficient to initiate the development of pedicles (Lincoln 1971b).

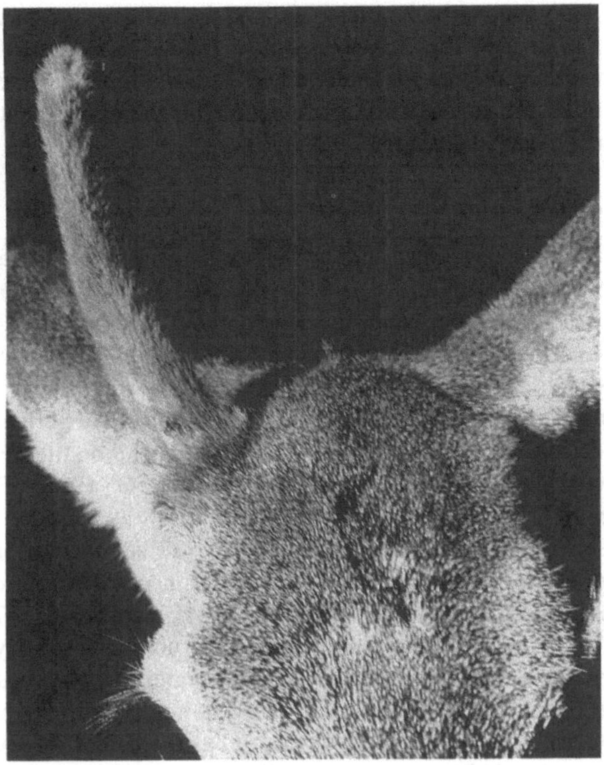

FIGURE 12. Doe "Sara" in which a unilateral antler growth was induced by a surgical trauma to a right pedicle. (From Bubenik, G. et al. 1982b.)

However, the lack of development of pedicles is not a permanent condition, as a subsequent increase of T levels (either by natural means or by a T treatment) will cause the development of pedicles and the beginning of antler growth. Such later stimulation of pedicle development by T can succeed, even in animals which are several years old (Jaczewski & Galka 1967; Lincoln & Fletcher 1976).

A dormant capacity to produce antlers exists in both sexes of cervids. Normally, only females of the genus *Rangifer* (the reindeer and the caribou) grow antlers. However, in females of other deer species, development of pedicles and subsequent growth of antlers has been achieved by T administration (Jaczewski 1976). The lack of antler growth in females might be caused not only by an insufficient amount of circulating androgens, but also by the presence of female sexual hormones, estrogens. Reduction of estrogen production in a white-tailed doe by chemical castration induced a growth of antler pedicles. However, the levels of androgens in a female deer which initiate pedicle growth are probably not sufficient to induce antler growth. A stronger stimulus is needed; this stimulus can be either a high concentration of circulating androgens (presumably T) achieved during the previous rut, or a strong neurogenic stimulus, such as trauma to the pedicle (Fig. 12). Antler growth initiated by these techniques will induce a regular deciduous growth of antlers repeated for many years (Jaczewski 1976; Bubenik, G. et al. 1982b). However, these female proto-antlers will never undergo transformation into real antlers and are usually sequestered with a dry velvet present on the surface. For more details on the experimental growth of antlers in female deer, see Chapter 13.

Effects of Seasonal Variations

Once the pedicles are developed and the first antler growth is initiated, castration cannot prevent subsequent antler growth (Wislocki 1943; Wislocki et al. 1947; Tachezy 1956). On the other hand, the development of antlers in castrated males or androgenized females is never complete. These so-called velvet- or pseudo-antlers of castrates are never fully mineralized; their velvet is never shed and the antlers never cast normally. The characteristic periods of the normal antler cycle (e.g., petrification, velvet shedding, and casting) depend on the seasonal variation of T (Wislocki 1943; Tachezy 1956; Bubenik, G. 1982). Experimental manipulation of T levels will achieve the same result as a natural variation of seasonal levels. Artificial elevation of T (induced by T administration or manipulation of photoperiodicity) will cause out-of-season petrification of the antler bone and the shedding of the velvet (Tachezy 1956; Lincoln et al. 1970; Bubenik, G. et al. 1987a). The velvet shedding is caused by rubbing of the drying or dried velvet on trees, bushes, fences, or other available materials. The cause of the rubbing behavior is not yet known, but it is assumed that it requires a certain elevation of androgen blood levels. Hypogonadic or castrated deer do not rub antlers. On the other hand, rubbing is not necessarily connected to complete mineralization of antlers and the death of the velvet. Still living, heavily bleeding velvet was observed to be rubbed off in normal white-tailed deer, moose, *Alces*, and roe deer, (Bubenik, A. & Bubenik, G., personal observation) as well in white-tailed deer after denervation of antlers (Wislocki & Singer 1946). In addition, a rubbing behavior was also observed in antlerless roe-buck (Hartwig 1972). In view of these data, it can be hypothesized that the levels of androgens which normally induce the death of the velvet—in white-tailed deer, around 1–1.5 ng/mL of T in plasma (Bubenik, G. & Schams 1986)—will also activate the brain centers responsible for the initiation of the rubbing behavior. However, in some instances, either the brain centers are more sensitive to androgens (and will become active prematurely) or, conversely, the threshold of receptors for androgens in the velvet is increased (so the death of the velvet is delayed), as a result, the rubbing of the living velvet will occur. Whether in such case the pain receptors in the velvet (which normally would prevent the deer from damaging the velvet) are deactivated, or the pain threshold is severely increased (by endorphins or other factors) remains to be explored.

Pedicle Versus Antler

One of the unsolved questions in antler development is the fact that T seems to promote growth of the pedicle but retard the growth of antlers. Therefore, pedicle develops during the fall (the time of rising T levels), and the antlers grow during the spring (the time of minimal T concentration). As the antler can be considered an extension of the pedicle (on occasions both will grow in the same fall period in a juvenile male deer) the antlers and the pedicle must differ in the level of sensitivity to androgens.

It is true that androgen levels are generally not too high during the time when pedicles are grown, therefore, it can be argued that T is exercising its growth-promoting effect in low concentration (Thompson et al. 1972). However, during adulthood, peripheral T can reach huge concentrations and still, despite a solid mineralization of the pedicle core, the pedicle will survive unpetrified and will induce another antler growth in the next season.

The reason for this difference might be the genetically fixed boundary of higher sensitivity to T, which determines where the tissue will be petrified and where it will reach only the dormancy stage. This line is usually identical to the baseline of the coronet; however, cases are known where this line was either further below or above the coronet. According to this hypothesis, rising T levels in the fall cause increasing mineralization of the pedicle as well as the antler; however only one of both tissues—the antler—is petrified to the point of death.

Role of T in Antler Growth

Maintaining high T levels during the period of hard antlers will result in prevention of antler casting (Goss 1968; Semperé & Boissin 1982). Conversely, a decrease of T levels by castration or blockade of androgen receptors during the antler growth period will prevent mineralization and cause a continuous antler growth throughout the whole year (Wislocki 1943; Tachezy 1956; Bubenik, G. et al. 1975b; Schams et al. 1987). On the other hand, if castration or receptor blockade occurs during the period of hard antlers, the antlers will be cast within 2–4 weeks (Tachezy 1956; Goss 1961; Bubenik, G. 1982). These data indicate that, once a complete mineralization of antlers and the connecting portion of the pedicle occurs (normally before the rut), reduction of T levels (normally after the rutting season) will cause a separation of the antler from the pedicle.

Because the antler growth occurs during the period of minimal T levels and can be observed also in postpubertally castrated males or androgenized females, it appears that presence of androgens is not required for the process of antler growth. However, because the blockade of androgen receptors by cyproterone acetate (CA) completely stops antler growth in castrated while-tailed bucks (Bubenik, G. 1982) as well as in intact roe-bucks (Schams et al. 1987), it appears that a small amount of circulating androgens (which might be of adrenal origin) is required to stimulate antler growth. Whether such stimulation occurs directly in the antler tissue or is a result of androgen action at the CNS level has not yet been determined. In addition to CA studies, some other data indicate that temporary elevation of androgens might be necessary to maintain vigorous long-term antler growth. In white-tailed deer, the speed of antler growth in castrates diminishes progressively with each additional year following orchiectomy. This decrease in intensity of antler growth can be prevented by T administration to castrates during the fall. Such artificial elevation of androgen levels will cause not only mineralization of antlers (followed by casting) but also tremendous invigoration of antler growth in the following antler growth period (Bubenik, G. 1982). The slow-down of antler growth in castrates might be due to lack of androgen receptors in the CNS or in antler tissues. Cytoplasmic receptors were not detected in the velvet of castrated white-tailed deer; however, in the velvet of the fawns the amount of androgen receptors was close to levels determined in the prostate (Plotka et al. 1983). A small transitory elevation of T found constantly in male white-tailed deer, roe-deer, and fallow deer at the beginning of the antler growth period (Bubenik, G. et al. 1985b; Semperé & Boissin 1982; Rolf & Fischer 1987) might contribute to a rapid proliferation of antler tissue. However, as this transitory increase has not been found in frequently sampled red deer (J.M. Suttie, personal communication), the importance of this phenomenon remains to be elucidated.

Other Androgens

Testosterone is the hormone most intimately associated with the antler cycle; however, the effects attributed to T might be caused by other steroids. T is known as a prohormone (Martin 1978) which is further metabolized to other androgens and estrogens (Fig. 13) (Dorfman & Shipley 1956; Turner & Bagnara 1976); only a few of these steroids have been investigated for their roles in the antler cycle.

5 α-Dihydrotestosterone (DHT). What is known primarily as an androgen influencing male sexual behavior (Beyer et al. 1975) and the function of the integument (Brooks 1975; Bubenik, G. & Bubenik, A. 1985), has been detected in blood of white-tailed deer (McMillin et al. 1974) and fallow deer (Rolf & Fischer 1987). 5 α-DHT, given to castrated white-tailed deer in mid-summer, induced an intense mineralization of the primary bone structure but failed to promote the maturation of bony tissues into Haversian osteons. On the other hand, DHT caused desiccation and darkening of the velvet in a way very similar to that observed during the final mineralization period of normal antlers. However, the treatment with DHT did not result in a complete petrification of antlers and shedding of velvet. Furthermore, next spring the antler bone reverted again to a primitive structure of antlers (Morris & Bubenik, G. 1982). In addition to studies on white-tailed deer, DHT failed to induce polishing of antlers in castrated red deer stags (Jaczewski 1982).

5 β-Androstanediol. It is known to be a less potent androgen than DHT or T (as measured by its effect on androgen target organs such as prostate, levatorani, and/or chicken comb) (Martin 1978). However, its effect on growing antlers of white-tailed bucks was remarkably strong. Given during the early summer (the time of a very intense antler growth) 5 β-androstanediol caused a dramatic slow-down of bone-building activity and formation of mesenchymal and chondrogenous zones and induced replacement of cartilaginous tissue by a partly mineralized trabeculae of woven bone. However, only 4 weeks after treatment, the antler bone

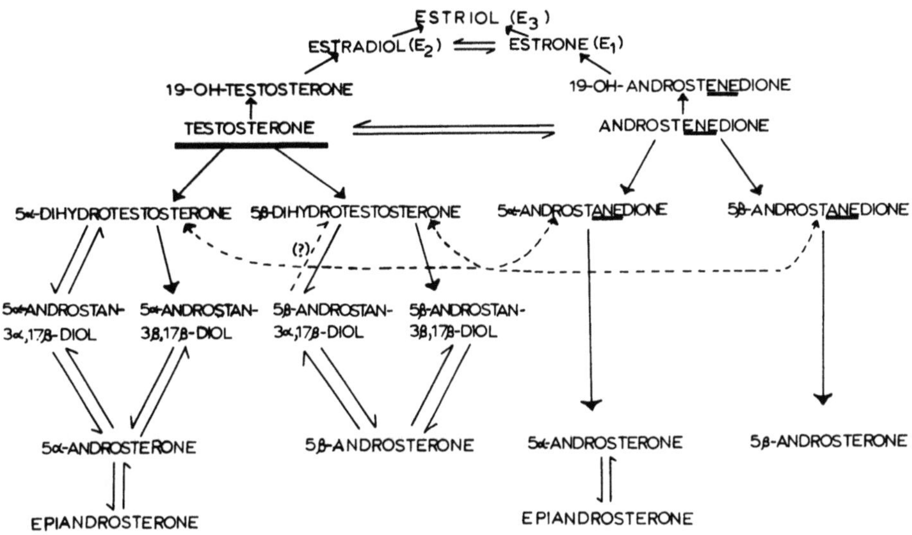

FIGURE 13. Simplified metabolic pathway of testosterone. (From Morris & Bubenik, G. 1983.)

reverted again to velvet-antlers with a rapidly proliferating preosseous tissue in the antler tip and a woven bone abundantly interspersed with mesenchymal and perivascular tissue in the antler core (Morris & Bubenik, G. 1982).

5 α-Androsterone. It is one of the terminal metabolites of T. Despite the fact that in the literature it is labeled as a weak androgen (Martin 1976), 5 α-androsterone given intramuscularly to castrated white-tailed buck caused a transformation of previously rapidly proliferating trabeculae of woven bone into more mature lamellar bone, characterized by little osteoblastic activity in between numerous islands of concentrically built osteons (Morris & Bubenik, G. 1982).

5 β-Androsterone. This other terminal metabolite of androgens had very little effect on the antler bone of castrated white-tailed deer (Morris & Bubenik, G. 1982).

19 OH-T. It is an intermediate metabolite of T into estrogenic compounds. Intramuscular treatment of castrated white-tailed buck by 19 OH-T caused a transformation of a rapidly proliferating woven bone (Fig. 14a) into a fully developed secondary bone characterized by a well-mineralized Haversian system (Fig. 14b) almost indistinguishable from antler mineralized by natural means (Fig. 14c). The velvet of the treated buck became darker and the deer was observed to rub the antler against the fence. However, the effect of 19 OH-T was only temporary. Just 3 weeks after the end of the treatment, the antler again acquired the primitive character of velvet-antler (Fig. 14d). The effect of 19 OH-T was most probably caused by its metabolism into estrogens, as the blockade

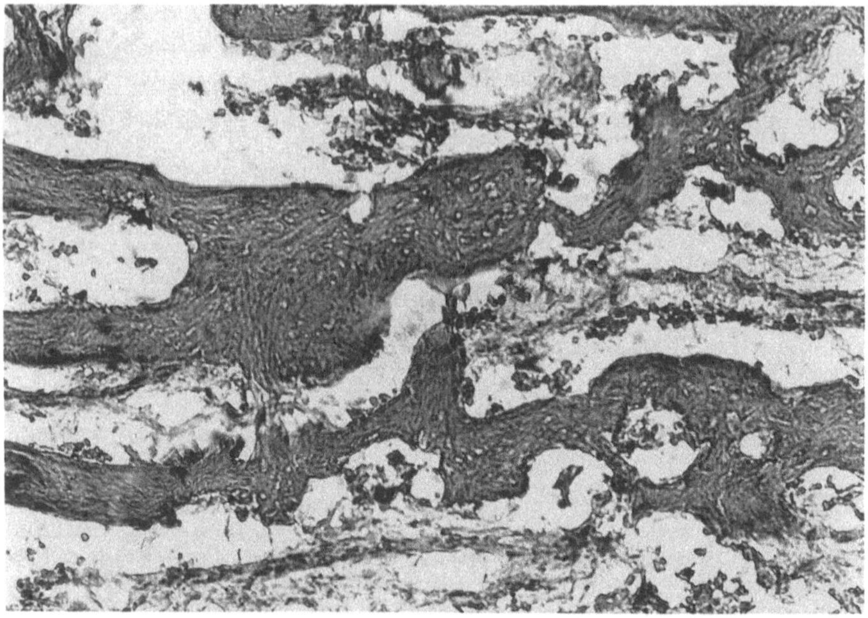

A

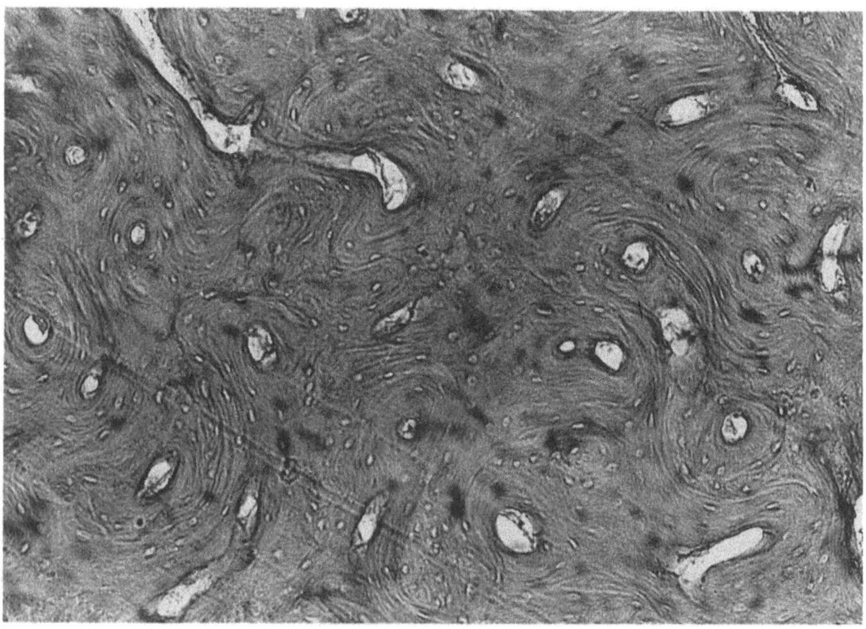

B

FIGURE 14. *A* Rapidly growing "primary bone" of a castrated male white-tailed deer. *B* Antler tissue of the same castrate transformed into mature "secondary bone" after only 3 weeks of i.m. treatment with 19-OH-T. *C* Mature antler bone in the intact buck. *D* Three

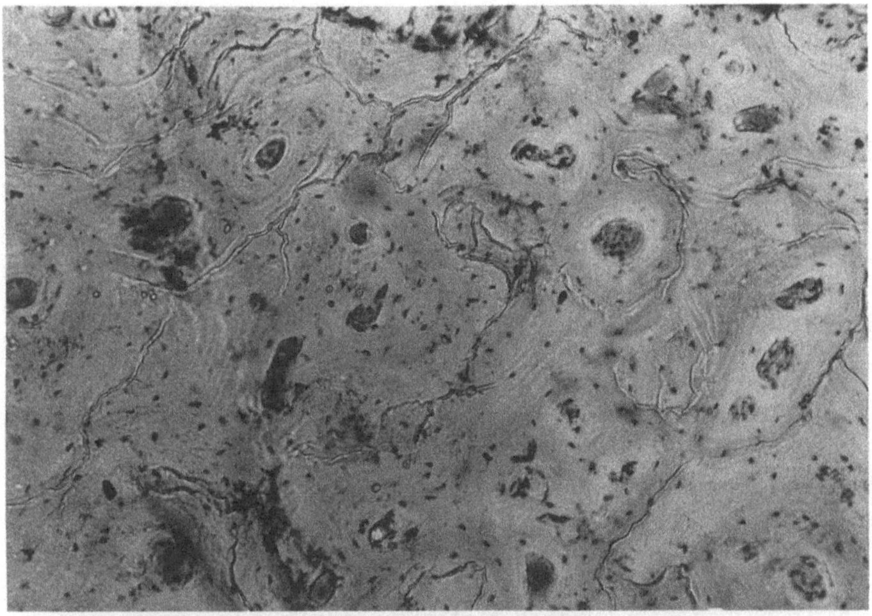

C

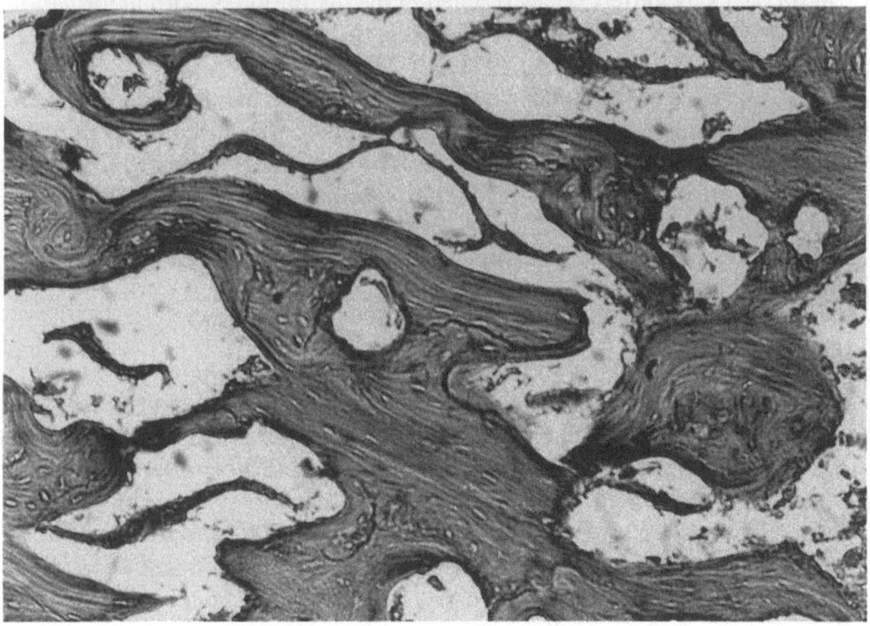

D

weeks after the cessation of 19-OH-T treatment the antler bone reverted again into a primitive spongious bone typical for castrates. (All photographs are from Morris & Bubenik 1982.)

of aromatization of 19 OH-T to estradiol by 1,4,6 androstatriene-3-17-dione prevented all maturation changes caused by 19 OH-T (Morris & Bubenik, G. 1982).

Estradiol (E_2)

This hormone is produced in mammalian males, mostly in the testes. However, a large proportion of male E_2 is produced by conversion of T in the peripheral tissues (Kelch et al. 1972). The role of E_2 in the male deer is not yet entirely elucidated, but in addition to being involved in the feedback mechanisms regulating seasonal activation of reproduction (Bubenik, G. et al. 1979) E_2 appears to participate in the formation of antler bone (Bubenik, G. & Bubenik, A. 1978a; Fennessy & Suttie 1985). Recently, E_2 receptors were detected in the velvet of red deer antlers (Barrell et al. 1987).

E_2 is by far the most potent steroid that will cause quick maturation, mineralization, and velvet shedding of antlers in intact as well as castrated male deer (Blauel 1935; Tachezy 1956; Goss 1968; Morris & Bubenik, G. 1982); however, its mode of action is not yet entirely clear (Bubenik, G. 1982). The antlers of castrates mineralized only by administration of E_2 exhibit a higher proportion of compact bone than comparable antlers of the intact bucks. Conversely, treatment of control bucks by antiestrogens resulted in the disruption of normal formation of Haversian systems and underdevelopment of the compact bone mantle (Fig. 15) (Bubenik, G. & Bubenik, A. 1978a).

However, if E_2 is able to induce a fast mineralization and polishing of antlers and prevent casting, estrogens cannot stimulate other effects of androgens, such as induction of pedicle growth (Waldo & Wislocki 1951; Tachezy 1956; Goss 1970), or initiate all components of the rutting behavior (Fletcher & Short 1974; Fletcher 1978). In antlerogenic events, estrogens have a rather opposite effect: high levels of E_2 prevent the development of the pedicle in male as well as female deer (Goss 1970; Bubenik, G. 1982).

E_2 and DHT

When E_2 was administered alone to castrated white-tailed bucks (10 mg, two times weekly) mineralization and velvet shedding of their antlers was induced within 3 weeks of treatment. This was followed by casting of antlers 2–3 weeks after the end of treatment. However, when the same dose of E_2 (applied to two castrates) was injected together with 100 mg of 5 α-DHT, the antler cortex did mineralize and the velvet was shed, but the cessation of the treatment was not followed by the separation of antlers from the pedicle. When 3 months later the hard antlers were removed surgically, an unusual mineralization process was discovered. The antler cortex, formed by a compact bone collar, was completely mineralized and appeared dead, but the spongious bone core was still living and full of well-filled blood vessels (Bubenik, G., unpublished observation). The lack of casting, which was caused by an incomplete mineralization at the interface between the antler base and the top part of the pedicle, could be ascribed to the

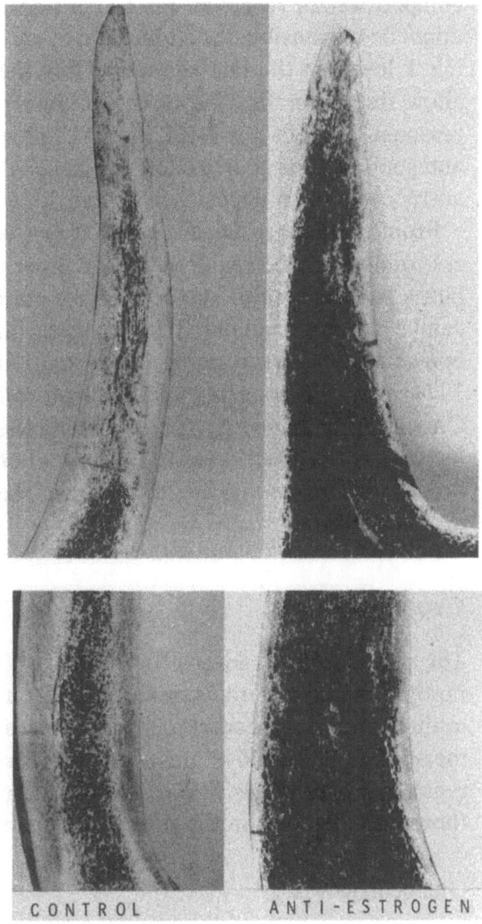

FIGURE 15. Longitudinal sections of the antler tip and the middle portion of the antler beam in control and antiestrogen (MER-25)-treated intact male deer. Note the decrease in proportion of compact bone mantle in the antiestrogen-treated buck. (From Bubenik, G. & Bubenik, A. 1978a.)

effect of DHT. As explained earlier, when DHT was given alone, it induced formation of lamellar bone, but the process of mineralization was never completed. A similar situation (a mineralized and polished antler cortex with the living bone core) was observed when the mineralization process was interrupted by the androgen receptor blocker cyproterone acetate given just after polishing of the velvet. That some parts of antlers were revitalized was confirmed by rapidly rising blood levels of alkaline phosphatase (Bubenik, G. et al. 1987a), an enzyme involved in the formation of antler tissues (Semperé et al. 1986). These observations confirmed that most of the compact bone (which usually comprises more than two thirds of the antler mass) is supplied by blood from the velvet. With the interruption of the blood flow there, the bone cortex will die. However, it appears that a part of the spongious antler core is supplied independently and the blood flow can persist even if the compact bone cortex is petrified and the velvet shed (Brandt 1901). Why DHT prevented the mineralization of the antler core (caused usually by E_2 treatment) is not entirely clear, but some kind of antagonism or competition for receptors can be considered. It could be hypothesized that DHT might be responsible for a blockade of antler petrification during the time of rising T levels in the late summer. Only the decrease of the DHT/T ratio would allow the mineralization of the pedicle resulting in death of the antlers. The seasonal levels of 5 α-DHT in male fallow deer were reported to have an often antagonistic time course (Rolf & Fischer 1987) and therefore could influence the antler cycle in an opposite way than T.

From the data so far available, it can be concluded that both androgens and estrogens might be required for the orderly transformation of living antler bone into a petrified status symbol of deer masculinity. Based on previous results, it can be hypothesized that T (or its metabolites) might promote mineralization of bony tissues by facilitating the formation of bone matrix (Bubenik, G. et al. 1974) as well as speeding up the transformation of primary into secondary bone. On the other hand, E_2 (or its metabolites, such as estriol or estrone) might increase the production of the compact bone by slowing down the bone-building process and promoting the formation of Haversian systems (Bubenik, G. & Bubenik, A. 1978a).

Corticoids

The relationship of adrenal hormones to the antler cycle has been studied only rarely. The major steroids produced in the adrenal cortex of large mammals belong to three groups: glucocorticoids (hormones involved in the stress reaction and the metabolism of carbohydrates), mineralocorticoids (hormones participating in the regulation of minerals in various body tissues and fluids), and adrenal androgens (hormones involved in the process of sexual maturation and sex behavior).

Glucocorticoids

The major glucocorticoid in deer is cortisol, which has an inhibiting influence on growth of large mammals by affecting the growth of cartilaginous tissues (Soyka & Crawford 1965). In a similar way, cortisol administered to roe-bucks substantially slowed down the elongation of antlers (Bubenik, A. et al. 1976). Despite the fact that the seasonal levels of cortisol in calm white-tailed deer began to rise at the end of summer (Fig. 16) (the time of massive antler mineralization) it is not believed that cortisol is directly involved in this phase of the antler cycle (Bubenik, G. & Leatherland 1984). The capacity of the adrenal cortex to secrete cortisol upon adrenicocorticotropic hormone (ACTH) stimulation can exceed several times the normal basal values (van Mourik & Stelmasiak 1985). On the other hand, in a few other bucks there was no significant increase of cortisol even after 40 I.U. of ACTH (Bubenik, G. 1987; Smith & Bubenik, G. 1990; Bubenik, G., Brown & Schams, in preparation). Therefore, it is believed that, depending on individual character of animals, the daily variations of cortisol might exceed the seasonal differences (Bubenik, G. et al. 1977b).

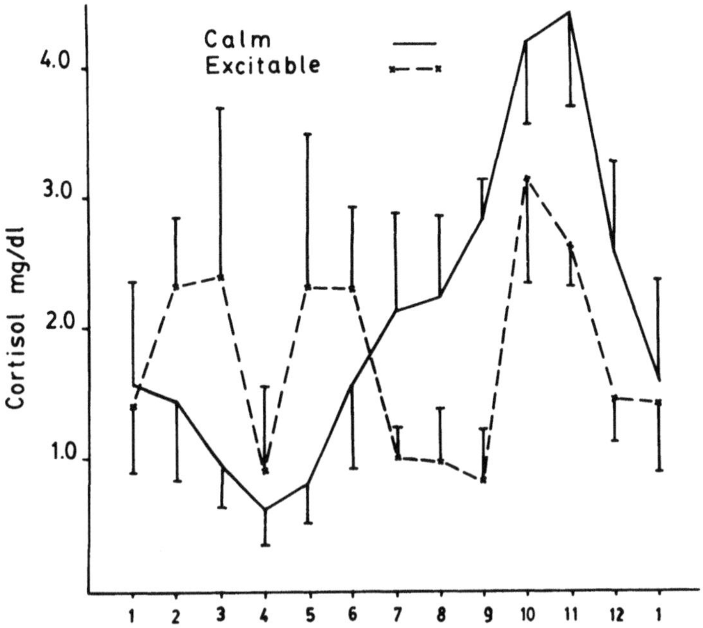

FIGURE 16. Seasonal levels of cortisol in plasma of adult male white-tailed deer. (From Bubenik, G. & Leatherland 1984.)

In addition to cortisol, a smaller amount of another glucocorticoid, corticosterone, is produced in the cortex of the deer adrenal gland. However, the levels of corticosterone are usually only 15%–30% of the amount of cortisol (Seal & Erickson 1969; Bubenik, G. et al. 1975a; van Mourik et al. 1985; Smith & Bubenik, G. 1990) and the relationship to the antler cycle has not been investigated yet.

Mineralocorticoids

It is believed that mineralocorticoids are involved in the regulation of bone mineralization, only indirectly. None of these hormones have been studied so far in deer species.

Adrenal Androgens

Adrenal androgens might play an important role in the induction and control of antler growth in the female deer, as well in castrated males. An antler growth was observed in white-tailed does carrying an adrenal tumor (Doutt & Donaldson 1959). In addition, the development of the pedicles was induced in white-tailed doe by chemical blockade of ovaries by antiestrogens. It is assumed that in this case the blockade of estrogen production enabled adrenal androgens to assume the pedicle-

FIGURE 17. A massive, tumor-like antler growth (the so-called peruke) of a castrated male roe-deer. (From Bubenik, A. 1966.)

stimulating function (Bubenik, G. et al. 1982b). Furthermore, it has been hypothesized that high production of adrenal androgens might be responsible for the growth of antlers in the female reindeer and caribou (Bubenik, G. 1982).

The possible differences in levels of production of adrenal androgens or the different sensitivity of growing antlers to their effect might be responsible for the remarkable differences in proliferation of antler tissues in castrated roe-buck and castrated white-tailed buck. In roe-deer, orchiectomy results in uncontrolled, tumor-like growth (Fig. 17), which will cause the death of the animal, usually within 12–15 months after gonadectomy. On the other hand, castration or hypogonadism in white-tailed or male bucks results in a massive growth of exostoses (known as pearls) on the beams and points (Baber 1987); otherwise, the lack of testicular androgens does not affect the basic antler pattern (Fig. 18), nor is it consistent with other vital functions.

Adrenal Testosterone

The average seasonal levels of testosterone in castrated white-tailed deer varies from nondetectable (less than 0.05 ng/mL) found in the spring and summer, to as much as 0.42 ng/mL found during the autumn. The value of 0.4 ng/mL well exceeds 0.2 ng/mL, the usual basal level of T found in plasma of intact deer during the late winter and the spring (Bubenik, G. et al. 1987c). In castrated white-tailed deer, the levels of T (which is mostly of adrenal origin) has reached on one occasion as much as 1.72 ng/mL (Bubenik, G. et al. 1987c). This amount exceeded 1.5 ng/mL, the average borderline value, which if steadily maintained, induces the petrification of antlers and shedding of the velvet (Bubenik, G. 1982). Levels of adrenal T correlated positively with the intensity of antler

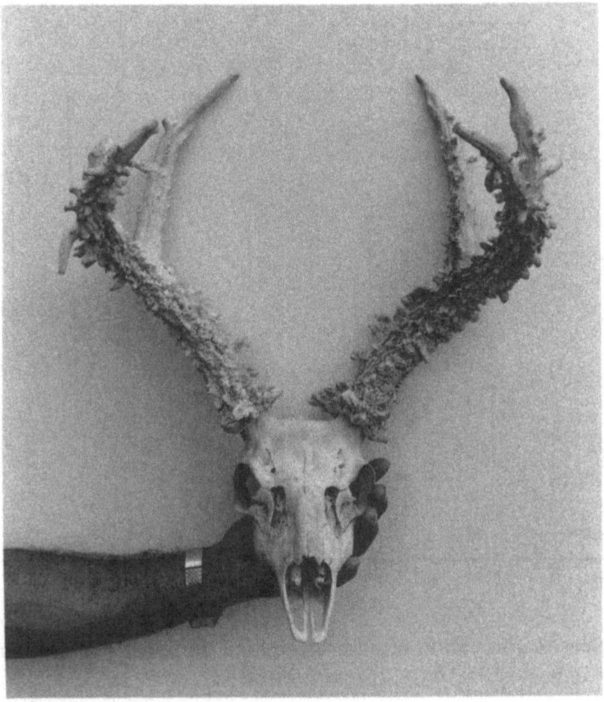

FIGURE 18. The relatively well-controlled antler growth in a castrated mule deer. Note the superficial proliferation of cortical bone on the main antler beam known as pearls. (From Baber 1987.)

growth in white-tailed castrates and their tendency to maintain the proper species-specific shape (Bubenik, G. 1982).

Small amounts of T might be essential for initiation and maintenance of antler growth. In red deer castrates with intact epididymides (which are able to produce a small amount of T) the antler growth and mineralization process were much closer to normal, compared to stags with complete gonadectomy (Lincoln 1975). These findings might be due to the well-known bone-growth-promoting effect of small doses of T found in many mammalian species (Thompson et al. 1972). The dual role of T in regulation of antler growth was first documented in a roe-buck castrated before puberty. Low levels of T initiated growth of the pedicle and supported a vigorous growth of antlers; on the other hand, high levels of T prevented pedicle development (Tachezy 1956).

Androstenedione

In addition to T, another adrenal androgen, androstenedione (A_4), has been investigated in deer. A_4 is produced not only in the adrenal cortex, but similarly to T it is produced also in the mammalian testes—especially during puberty, when the production of this androgen exceed that of T (Rawlings et al. 1972). In

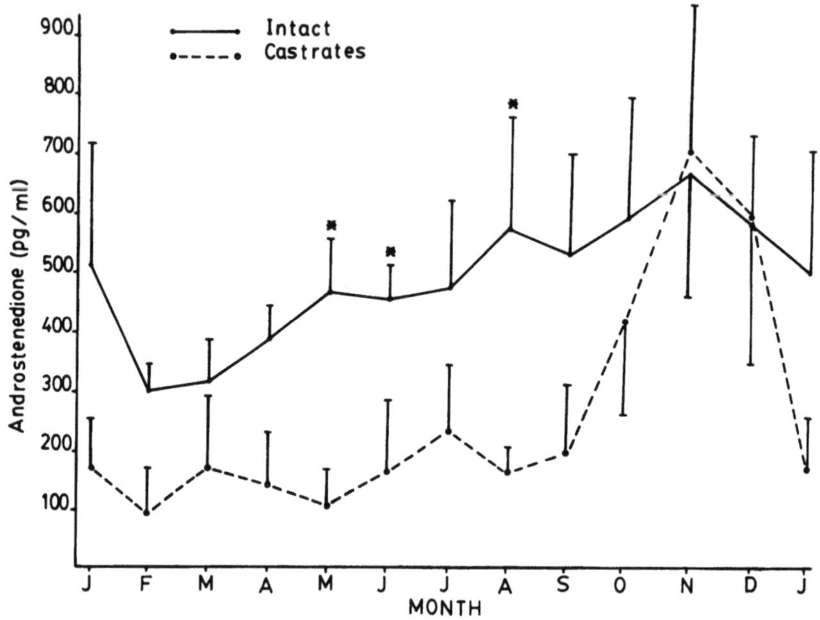

FIGURE 19. Seasonal variation of androstenedione in intact and castrated male white-tailed deer; * indicates $P < 0.05$. (From Bubenik, G. et al. 1987c.)

both organs, A_4 can be either a precursor or a metabolite of T (see Fig. 13). In earlier studies, the variation of seasonal levels of A_4 in the red deer testis (Lincoln 1971a) and white-tailed deer plasma (Mazur 1973) did not differ from the circannual variation of T. However, in more recent studies seasonal levels of A_4 in deer plasma dissociates from levels of T (Brown et al. 1983a,b). In our recent study (Bubenik, G. et al. 1987c), A_4 levels in blood of intact white-tailed deer begin to increase in April (the time of the first noticeable antler growth). After reaching a plateau between June and August, another increase in A_4 followed, until peak levels were reached in November (coinciding with maximal concentration of T). A_4 levels in castrates are elevated only in autumn; that indicated that spring elevation of A_4 detected only in intact bucks is probably a result of increased secretion in the testes. On the other hand, the autumn peak found in intact as well as castrated deer might be due to elevated production in the adrenals (Fig. 19).

As the increase of A_4 levels coincides with the beginning of the antler growth, it can be speculated that this androgen might be the long-sought "antler growth stimulus" of Wislocki (1943), the hormone initiating the proliferation of antler tissues.

A dramatic increase of A_4 levels observed in castrated white-tailed deer during autumn coincides with a similar increase of adrenal T (Bubenik, G. et al. 1987c). These results indicate that production of adrenal androgens in castrates reaches

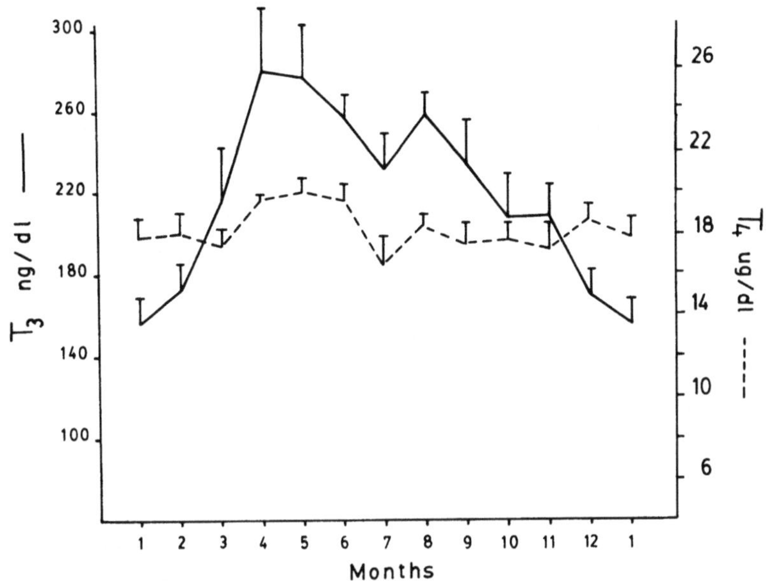

FIGURE 20. Seasonal levels of thyroxine (T_4) and triiodothyronine (T_3) in plasma of intact and castrated male white-tailed deer. (From Bubenik, G. & Leatherland 1984.)

peak levels during the time when intact bucks have their rutting season. In view of this fact, it can be speculated that the circannual variation of adrenal androgens might be responsible for the seasonal changes of velvet-antler growth rate observed in castrated roe deer (Bubenik, A. 1963).

Thyroidal Hormones

Two thyroidal hormones, thyroxine (T_4) and its tissue-active metabolite triiodothyronine (T_3) were investigated in relation to the antler cycle. Because the peak levels of T_3 and in some studies of T_4 were found during the first half of the antler growth period (Fig. 20) (Ringberg et al. 1978; Bahnak et al. 1981, Ryg and Jacobsen 1982; Chao & Brown 1984; Bubenik, G. & Leatherland 1984), thyroidal hormones were suspected to be involved in promotion of antler growth. An early study reported that experimental administration of T_4 to a young roe-buck resulted in an extraordinary increase of antler size 1 year after treatment (Lebedinsky 1939). Similarly, intramuscular (i.m.) injection of T_4 increased antler growth in a fallow-deer (Bruhin 1953). On the other hand, no effect on antler size resulted from either administration of T_4 to a roe-buck by Blauel (1936) or thyroidectomy in a white-tailed and red deer (Hall et al. 1960; Care et al. 1985). In addition, a blockade of antler growth caused by hypophysectomy was not relieved by T_4 treatment (Hall et al. 1966). Conversely, treatment of white-tailed bucks with bromocriptine, which reduced levels of T_4 in blood by 30%, did not affect the extent of antler growth (Bubenik, G. et al. 1985a). Recently, i.m. administration of T_4, which significantly elevated concentrations of thyroidal hormones in blood (oral treatment by T_4 had no effect) resulted in a tremendous improvement of quality of hair coat and increased development of muscles but failed to affect the quality of antlers (Bubenik, G. & Smith 1986).

Because of these contradictory results, the alleged positive effect of thyroidal hormones on the development of antlers is far from certain. Thyroidal hormones are strong anabolics that facilitate metabolic processes and together with GH form a potent combination of bone-growth-supporting factors (Lostroh & Li 1958). However, because increased levels of T_4 and its metabolite triiodothyronine (T_3) in the spring and early summer were found in both sexes (Bahnak et al. 1981; Bubenik, G. & Bubenik, A. 1978b) it is obvious that such elevation is not related to the development of antlers, but rather reflects the availability of nutrient-rich food. It is also well established that levels of T_3 are the best indicator of nutritional status in deer species (Bahnak et al. 1981; Ryg 1984; Chao & Brown 1984).

On the other hand, a recent study indicated that a tissue-active thyroidal hormone T_3 might be utilized in the growing antler bone. The difference between T_3 levels (found higher in the jugular vein as compared to the antler vein) was most pronounced in the early antler growth period and disappeared during the antler mineralization phase (Bubenik, G. et al. 1987d).

In view of these facts, it can be speculated that certain minimal amounts of hormones which improve utilization of energy (such as GH, prolactin, T, or thyroidal hormones) are necessary to maintain normal antler growth. However, it appears than no major elevation of T_3 or T_4 in blood is essential to promote an antler growth; whether an additional dose of anabolic hormones will have demonstrable effect on the intensity of antler growth, remains still open.

Calcitonin and Parathyroid Hormone

Calcium-regulating hormones calcitonin (produced in the ultimobrancial cells of the thyroid gland) and the parathyroid hormone (PTH, produced in the parathyroid gland) are heavily involved in bone metabolism. The hardening of antlers is accompanied by release of calcium from the skeletal bones (Banks et al. 1968a,b; Cowan et al. 1968; Brown et al. 1978b) as well as intense absorption from the food in the digestive system (Eiben et al. 1984). Therefore, it has been postulated that calcitonin and PTH participate in the growth and mineralization phase of the antler cycle (Hall 1978). In an early study, histological investigation of the white-tailed deer parathyroid gland revealed no seasonal variations (Grafflin 1943). On the other hand, calcitonin-producing cells of the roe-buck thyroid gland have been found to be more numerous during the late mineralization period (Pantič & Stošič 1966). Experimental treatments so far failed to affect significantly any phase of the antler cycle. Injections of ox para-

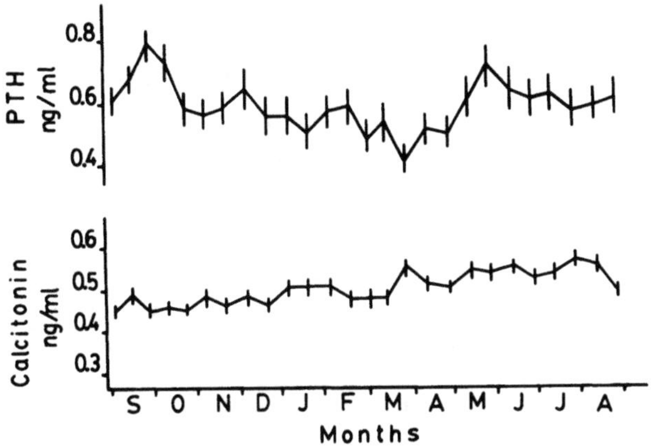

FIGURE 21. Seasonal levels of PTH and calcitonin in plasma of adult intact male white-tailed deer. (From Chao et al. 1984a.)

thyroid extract to white-tailed deer carrying hard antlers did not induce expected casting (Goss 1983). In addition, complete surgical removal of white- tailed deer parathyroid gland (confirmed by chelation experiment) increased blood levels of phosphorus but had no effect on the concentrations of calcium or alkaline phosphatase. Despite the fact that the antlers of parathyroidectomized bucks in this experiment had lower density, the development, ossification, and mineralization were not affected (Brown et al. 1981). On the other hand, in another study, removal of the parathyroid resulted in more pronounced mineralization of antlers as compared to controls (Care et al. 1985). The lack of major disruption of calcium homeostasis by parathyroidectomy could be due to presence of accessory parathyroid tissue, described in white-tailed deer by Mazur (1969), or due to presence of other mechanisms involved in regulation of calcium levels. The relationship between seasonal levels of calcitonin and PTH in plasma and the antler cycle has been investigated only recently. In male white-tailed deer, calcitonin (which causes an increase in calcium content of bones and reduced calcium levels in blood) exhibits high levels at the beginning of the antler growth period (Phillippo et al. 1972; Chao et al. 1984a). PTH (which acts in an opposite way to calcitonin) exhibits a large peak during the late stage of antler mineralization and a smaller one in the first half of the antler growth period (Chao et al. 1984a) (Fig. 21). Calcium-regulating hormones are interacting with vitamin D_3 and the enzyme alkaline phosphatase, levels of which are closely correlated with the individual phases of the antler cycle (Fig. 22) (Eiben et al. 1984; Semperé et al. 1986). Seasonal plasma levels of PTH and Vit D_3 were recently investigated in female reindeer, the only cervid in which females regularly develop antlers. Peak levels of 1,25–$(OH)_2$ D_3 were exhibited in September, the last month of the antler development period. PTH levels were significantly higher during the May to July

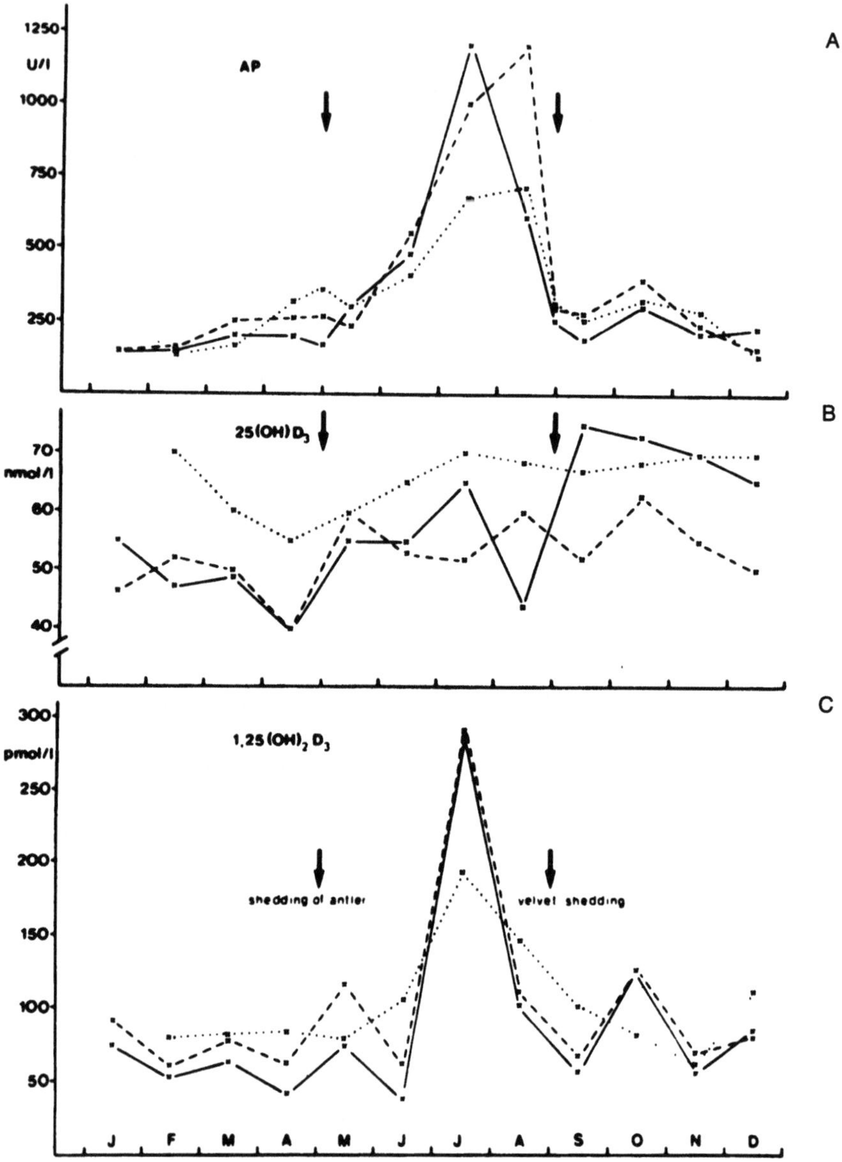

FIGURE 22. Seasonal levels of alkaline phosphatase, 25 $(OH)D_3$ and $1{,}25(OH)_2D_3$, in plasma of adult male fallow deer, *Damma damma*. (From Eiben et al. 1984.)

TABLE 1. Summary of hormonal actions in the antler cycle

Hormone	Action
Melatonin	Synchronizes timing of the antler cycle with seasonal changes. (Primary messenger of photoperiodically-dependent periodicity).
Prolactin	Synchronizes timing of reproductive activity with seasonal changes. Acts synergistically in long-day breeders, antagonistically in short-day breeders (inhibits rise of T).
Growth hormone	Direct effect on antlerogenesis not proven. Stimulates secretion of somatomedins (insulin-like growth factors).
Somatomedins (IGF)	Antler is a possible target organ. Somatomedins may directly stimulate antler growth.
Gn-RH (Gonadotropin releasing hormone)	Stimulates secretion of LH and FSH.
LH	Stimulates secretion of T and T derivatives. The sensitivity of gonads to LH changes with the age and the season.
Testosterone	a) Stimulates antlerogenic periosteum to produce pedicle. b) May stimulate a vigorous, species-specific antler growth through a direct action at the antler tissue levels or also indirectly by acting through a corresponding nerve center. c) Supports progressively intensified ossification and mineralization processes leading to calcification of the pedicle and petrification of antlers. d) Maintains a permanent connection between living pedicle and the dead antler.
Derivatives of T (DHT, E_2, etc.)	May have a role in ossification, velvet growth, and dessication.
A_4	May directly or indirectly support antler growth.
T_4	General metabolic effect through its metabolites (e.g., T_3).
T_3	Actively involved in promotion of antler growth.
Corticoids	Direct action in antlerogenesis uncertain. Indirectly involved through general metabolic effect at the periphery and the CNS. Increase in stress believed to be detrimental to antlerogenesis.
Calcitonin and PTH	Indirectly involved in antler mineralization. Together with vitamin D_3 regulate calcium homeostasis.

period, the time of the most intense antler growth (Baksi & Newbrey 1988). These recent findings seem to support the original assumption of Hall (1978) that calcitonin participates in the antler cycle; however, more detailed studies are necessary to elucidate the mechanisms that regulate the secretion of hormones, enzymes, and vitamins participating in the development and mineralization of bony tissues.

Because the growing antler is the most accessible bony organ in mammals, it serves as a model for studies of bone development, metabolism, and mineralization which in many respects is far superior to skeletal bone. (For details see Chapter 19.)

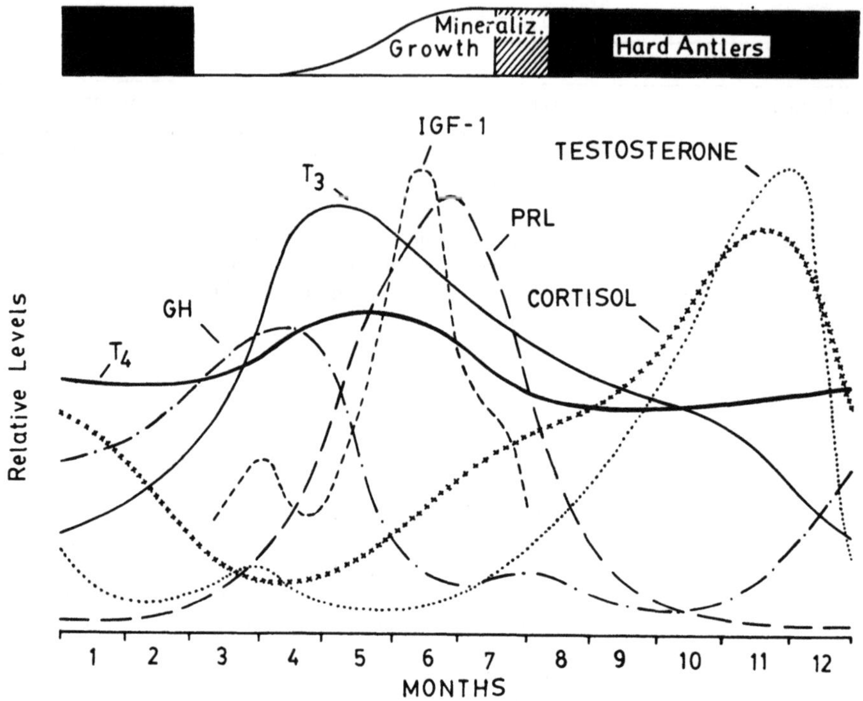

FIGURE 23. Stylized time course of hormonal levels during the antler cycle of a white-tailed deer. Relative levels refer to each hormone only, not to relative levels between hormones.

Summary–Hormone Model of the Antler Cycle

As has been discussed previously, the importance of various hormones in the regulation of the antler cycle varies considerably. At the present time, all hormones known to be involved in the antlerogenesis can be divided into three basic categories:

1. Hormones synchronizing the timing of antler development to the circannual cycles (e.g., melatonin, prolactin, estradiol).
2. Hormones acting on antlerogenesis indirectly through a general metabolic effect or stimulating secretion of other more directly acting hormones (e.g., prolactin, growth hormone, GnRH, LH, glucocorticorticoids, androgens, thyroid hormones, PTH, and calcitonin).
3. Hormones acting directly on the target tissue of the pedicle, growing antler tissues or on the antler nerves (e.g, androgens, estrogens, somatomedins, T_3).

This division is often artificial and arbitrary, as some hormones are suspected to have multiple roles and the effect of others is still putative or controversial. However, the categorization may help us to concentrate on a future exploration of hormonal action as well as on a possible utilization of various hormones in antler production or biomedical research.

The involvement of various hormones in individual phases of the antler cycle is summarized in Table 1. It must be emphasized that the intensity of hormonal action depends not only on circulating levels (for summary see Fig. 23) but also on the amount and sensitivity of hormonal receptors.

9
Of Antlers and Embryos

RICHARD J. GOSS

Introduction

If an embryologist were to investigate the development of deer antlers, he would address such questions as their histogenesis, differentiation, axial determination, morphogenesis, and the inductive mechanisms by which the participating tissues might interact. He would be inclined to explore these problems by the traditional approaches of the experimental embryologist, namely, deletion and transplantation experiments. But he would be working on a scale much larger than that which prevails in embryos. Indeed, it would have to be asked if, in view of the size differences, the same processes that guide embryonic development can also operate in the growth of deer antlers.

Deer antlers are unique in that they are the only mammalian appendages capable of complete regeneration (Brown 1982; Bubenik, A. 1966; Goss 1983; Jaczewski 1981a). Yet the annual replacement of antlers by adults, remarkable though it may be, is preceded by the unusual development of the fawn's first set of antlers—a unique process that deserves more scientific attention than it has thus far received. If the yearly regrowth of adult antlers is to be compared with epimorphic regeneration (the regrowth of amputated appendages), then the outgrowth of a deer's original antlers is comparable to embryonic bud formation, such as occurs when limbs or tails differentiate.

In a sense, antler development is a kind of metamorphosis. It represents a transitional phase between immaturity and adulthood. Similar developmental events occur in holometabolous insects when new legs or wings differentiate in postembryonic stages. Likewise, tadpoles wait until their premetamorphic period to sprout legs. Thus, although de novo antlerogenesis is not without precedent in the animal kingdom, it is a zoological curiosity paralleled among mammals only by the outgrowth of horns in other ungulates.

The Pedicle and the Antler

Deer antlers are actually two structures in one: the antler itself and the pedicle from which it is produced. These two components, while being made up of similar tissues (skin, bone, blood vessels, nerves), are quite different developmentally.

The antler is deciduous, the pedicle permanent. Antler skin, with its very specialized pelage, is known as velvet. The pedicle is covered with fur indistinguishable from that elsewhere on the scalp. Antler bone is more cancellous than the compact bone of the pedicle, at least during the growing phase.

Yet the two structures are intimately affiliated in the sense that the antler cannot grow unless there is a preexisting pedicle from which it can arise. The pedicle develops in the fawn or yearling before the antler itself differentiates. Further, the pedicle does not develop except under the stimulation of testosterone, while the onset of antler generation or regeneration occurs when the levels of male sex hormones are low. Testosterone, therefore, stimulates pedicle formation but inhibits antler growth (Bubenik, G. 1982). Testosterone triggers the eventual ossification of antlers and the shedding of their velvet, which leads to the demise and eventual casting of the antler (Goss 1968). No such changes take place in the pedicle. It is a perennial structure, while the antler is an annual one.

Histogenesis

One of the cardinal criteria of epimorphic regeneration is that there must be something left behind after amputation from which the regenerate can arise. If the entire "regeneration territory" is excised, no regrowth is possible because there are no remaining cells endowed with the developmental potential to replace the lost appendage. In the case of the deer antler, the frontal pedicle represents the regeneration territory. In fact, deletion experiments by Jaczewski (1955) and Bubenik, A. & Pavlansky (1956) have shown that the antler territory may extend some distance into the surrounding scalp and skull. Thus, as in other regenerating systems, the answers to questions about the histogenesis and morphogenesis of the deer antler may be sought in the stump from which it originates.

Attempts to identify the histological source of adult regenerating antlers have involved experiments in deletion, transplantation, and replacement. Since there are only two principal tissues in the pedicle from which the regenerating antler could arise, either the skin or the bone has been removed from pedicles prior to the normal onset of growth in the spring. When the skin was removed by circumcision of the pedicle, it was itself regenerated in time to participate in subsequent antler development (Bubenik, A. & Pavlansky 1959). Except for missing brow tines, the antlers that formed from denuded pedicles were remarkably normal (Goss 1964).

In other cases, the pedicle skin was slit lengthwise and peeled away from the bone. The latter was then sawed off close to the skull and the empty sleeve of skin preserved. Again, the removed tissue was replaced, and regeneration of the antler occurred (Goss 1964). Thus, the capacity for tissue regeneration has frustrated experimental attempts to determine antler histogenesis by the surgical removal of one or another of the potential sources of the antler in the pedicle.

Transplantation experiments have been equally unsuccessful. If discs of pedicle skin are grafted to the outer ear, for example, they may survive and sprout new hairs, but they have never been observed to produce antler tissue (Goss 1972).

Similarly, grafts of velvet antler skin to the hind leg (Billingham et al. 1959), or to the ear or scalp (Goss 1972) have persisted for several years without giving rise to antlers.

Discs of pedicle bone can be grafted subcutaneously to other parts of the body. Although they may remain palpable for prolonged periods, no antlers are induced by them. On the other hand, grafts of transverse slices of whole pedicles, including bone and skin, have yielded a small antler in one case in which the overlying skin had been partly removed in the spring. Other operations of this type, however, have not resulted in antler growth.

It is possible to replace pedicle skin with that from elsewhere on the body to determine if different kinds of skin can take part in antlerogenesis. In view of the technical difficulties in making successful free grafts of skin, the most convenient source is the ear (Goss 1964). The proximity of the outer ear to the pedicle makes it possible to establish what is essentially a flap graft of ear skin to the pedicle. First, the distal centimeter of pedicle skin is excised, taking care not to leave behind any epidermis beneath the basal burr of the as-yet-uncast antler. A full-thickness hole of appropriate diameter is then cut through the ear. After sawing off all but a short stump of the bony antler, the ear is impaled on the pedicle so that it fits snugly around the skinless distal segment. Subsequently, the inner ear epidermis heals to the remaining proximal pedicle epidermis, thus precluding the participation of the latter in antler regeneration. The outer-ear epidermis is unable to complete its continuity because the bony antler is in the way. Once the antler is cast, however, the outer ear skin heals across the exposed end of the pedicle, and normal antler development ensues. This proves that nonpedicle epidermis can differentiate into antler velvet, but it does not shed light on the mesodermal source of the antler. Despite the presence of ear skin at the end of the pedicle, it is probable that pedicle mesoderm (dermis, periosteum, and other types of connective tissues) would be present as a potential source of antler-forming cells.

The Role of the Skin

The foregoing experiments on the adult pedicle have failed to yield the desired information on the histogenesis of the regenerating antler. More definitive results, however, have been obtained by studies on fawns before the pedicle or antler had developed in the first place. In this system it has been possible to investigate the source of the pedicle itself. Although this information does not necessarily reveal the histogenesis of the deer's first set of antlers, it may indeed be relevant to the problem.

Early experiments were carried out by Goss et al. (1964) using weanling fawns of the white-tailed deer, *Odocoileus virginianus*. The object was to determine, by deletion, if the first antlers grown by a deer were derived from the skin or the skull. Excision of the scalp over the presumptive antler site did not prevent the subsequent growth of the antler, which took place after the wound had healed. However, when the low protrusion of the frontal bone was sliced off, with or without removal of the overlying skin, no antlers developed. These findings indicated that the histogenesis of the antler was skeletal, not integumentary.

Studies along these lines were pursued by Hartwig and his colleagues on fawns of the roe deer, *Capreolus capreolus*. In 1967, he showed that the skin overlying the presumptive antler site is incapable of giving rise to an antler if grafted to the hip. On the other hand, removal of the skin that would normally participate in antlerogenesis is followed by wound healing and eventual antler production (Hartwig 1967). These investigations suggested that the integument, while important in antler formation, is not the primary source of the pedicle nor of the antler that it produces.

Interestingly, this was just the opposite of what had been learned about horn growth in cattle and goats by Kômura (1926) and Dove (1935). They showed that removal of the bone did not preclude horn growth, while loss of the skin did. Similarly, transplantation of the skin elsewhere on the head resulted in ectopic horn growth, something that failed to occur following bone grafts.

Hence, horns are essentially epidermal outgrowths that secondarily induce the development of the underlying bone. Antlers are skeletal outgrowths that induce the overlying skin to differentiate into velvet. Each is derived from the tissue that is destined to constitute the hardened component of the final appendage.

The Role of the Periosteum in the Roe Deer

Research on the roe deer has focused on the periosteum. When it was deflected from the prospective site of pedicle formation to a nearby region of the frontal bone, an antler later developed at the latter site, while none formed at the normal location from which the periosteum had been shifted (Hartwig 1968a,b). Even when transplanted to the parietal bone, from which antlers do not ordinarily grow, ectopic antlerogenesis still occurred (Hartwig & Schrudde 1974). Finally, it was discovered that free grafts of frontal periosteum to the foreleg, between the skin and the metacarpal bone, were able to induce antlers at this location, too (Hartwig & Schrudde 1974). Such misplaced antlers do not grow to their normal shapes and dimensions, perhaps because of inadequate vascularization or innervation. However, they are unmistakable antlers as disclosed by the characteristic velvet nature of the pelage, the shedding of the velvet at the end of the growing phase, and their seasonal casting and regrowth in synchrony with the normal schedule of antler replacement.

Subcutaneous transplants of antlerogenic periosteum first differentiate into bony nodules palpable beneath the skin. As they bulge outward, the overlying integument acquires a sparse pelage which eventually takes on the appearance of typical antler velvet. A small outgrowth, often less than a centimeter high, usually develops during the first year. Its true nature is revealed not only by its resemblance to an antler in velvet during the growing phase, but also by its ossification and subsequent loss of velvet yielding a typical dead bony antler on the site to which the periosteum had been grafted.

The implications of these results cannot be overestimated. They demonstrate that the frontal periosteum in the region where the antler is normally destined to grow is the source of the pedicle. They also show that such pedicle bones can induce the overlying skin, whether on the scalp or the leg, to differentiate into antler velvet. The histogenesis of the pedicle, if not the antler which it produces, therefore lies in the frontal periosteum. The overlying skin plays a more passive role in antlerogenesis—a role dependent on the inductive influences of the pedicle bone which develops beneath it.

Periosteal Grafts in the Fallow Deer

The above conclusions raise some provocative questions. How much periosteum is required to induce an antler? Is the size of the resultant antler correlated with the size of the periosteal graft? Does the orientation of the periosteum determine the orientation of the antler it induces? Does the skin on different parts of the body vary in its competence for antlerogenic induction? What is the nature of the inductive interaction between periosteum and skin? In attempts to answer some of these questions, the author has attempted to continue the earlier investigations pioneered by Hartwig.

These experiments have been carried out on the fallow deer, *Dama dama*, using 6–8-month-old fawns whose pedicles were just beginning to develop. The operations involved reflection of the scalp to expose the incipient pedicle on the frontal bone, removal of the periosteal tissue by blunt dissection, and grafting it in the same animal to other locations or in other orientations. Transplants were carried out in December or January and yielded the first sets of antlers the following summer when the deer were beginning their second year of life. Animals were observed for up to several consecutive years to monitor the successive sets of antlers produced as a result of the original operations. Typically, they shed their velvet and were cast and replaced in synchrony with normal antlers.

Graft Size

In order to test the importance of graft size on ectopic antler induction, periosteal discs of various diameters were transplanted subcutaneously to the metacarpal region of the forelegs of a series of deer (Goss & Powel 1985). The transplants, measuring 1.5 down to 0.4 cm in diameter, all differentiated into palpable nodules of bone beneath the skin. In the ten largest grafts, small antlers were induced. Averaging 1 cm in length the first year, these were replaced by larger ones up to 7 cm long in the second growing season.

Smaller periosteal grafts yielded a lesser incidence of antler induction. Periosteal discs 1.05, 0.75, and 0.4 cm in diameter gave rise to overt antler tissue (as diagnosed by the shedding of velvet) in only one out of five cases each. In some grafts, while no antler induction may have occurred the first year, small antlers were sometimes produced in the second or third summers. It has also been shown that wounding an otherwise unresponsive graft by excising some of the overlying skin sometimes promotes antler induction. This is consistent with the essential role that wound epidermis is known to play in other examples of epimorphic regeneration.

For each of the above autografts, there was a presumptive site of antlerogenesis on the frontal bone deprived of some or all of its periosteum. One would expect a reciprocal relationship between transplantation and deletion experiments, and this generally held true. Excision of 1.05 or 1.5 cm discs of periosteum from the frontal bone prevented antlerogenesis thereafter in about 80% of the cases. Removal of discs 0.75 or 0.4 cm in diameter prevented antler production in 20% and 0% of the cases, respectively.

It would appear that the antler territory may be 1–1.5 cm wide, since the removal of that much periosteum abolishes antlerogenesis. Loss of smaller amounts of periosteum, however, may permit the production of pedicle and antler, presumably owing to regeneration of the missing tissues from the margins of the excision area.

Graft Shape

It has also been possible to test the inductive competence of halved discs of antlerogenic periosteum (Goss & Powel 1985). When semicircular pieces of periosteum measuring 1.5 cm in diameter were grafted beneath foreleg skin, the incidence of antler induction varied with the source of the transplant. Lateral halves gave rise to antlers in all three cases, while anterior, posterior, and medial halves were successful in two out of the three grafts of each type.

The frontal regions from which the semicircular grafts of periosteum had been removed produced antlers in all cases, despite excision of the anterior or medial halves. They gave rise to antlers in two out of three cases following loss of the lateral half of the periosteum, and in only one out of three cases when the posterior half was missing. These findings may be taken to indicate that the greatest antlerogenic potential resides in the lateral-posterior quadrant of the presumptive pedicle region. This coincides with the general angle at which the main beam slopes away from the skull.

Minced Grafts

Two experiments were carried out in which the 1.5-cm disc of periosteum was peeled away from the frontal bone and then minced into 2–3 dozen small pieces. These were scrambled, grafted back to the original site, and the skin of the overlying scalp was sutured over them. In one deer only short single antlers 1–1.5 cm

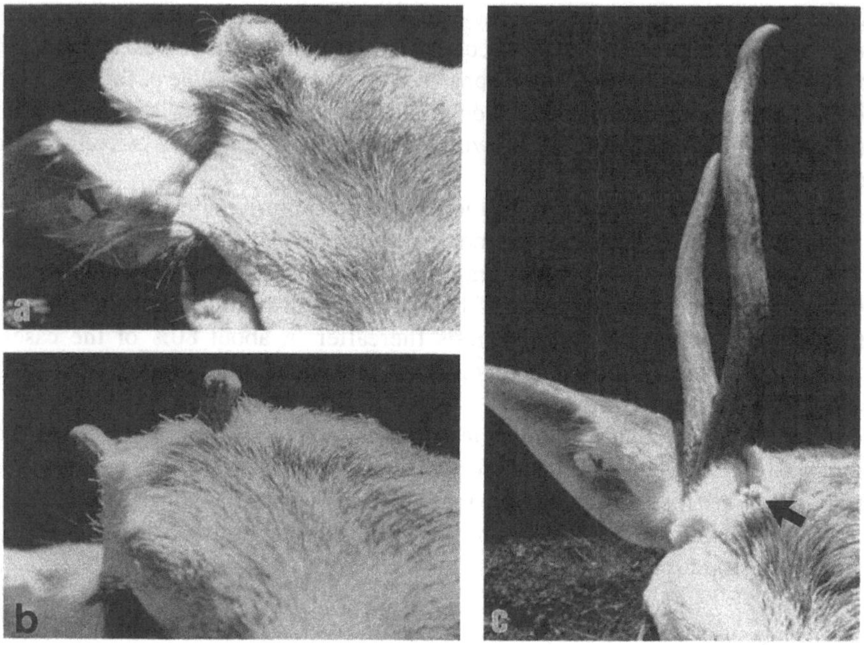

FIGURE 1. Results of mincing and scrambling a 1.5-cm disc of periosteum and grafting it in situ to the right pedicle region of the frontal bone. The operation was performed on January 5, 1982. Two short velvet antlers developed by August 10 (*a*) which later shed their velvet, revealing two separate antlers (*b*). In subsequent years the more lateral antler became long and bifurcate (*c*) while the medial one remained less than 2 cm long (arrow).

high were produced in the following two summers. In the other animal, two separate antlers grew out of the operated region, one medial to the other. During the first summer after operation, both antlers grew to approximately 2 cm in length (Fig. 1a,b). In the next 2 years, the medial one produced only very short antlers up to 1 cm long, while the lateral one regenerated bifurcate antlers approximately 19 cm in length (Fig. 1c).

While abnormalities in antler size and shape resulted from the minced periosteal grafts, there was also a remarkable degree of regulation whereby a rather normal-looking antler was produced in one case. It would appear that the periosteum may not be so fully determined at the time of operation that its separated parts cannot reorganize their morphogenetic capacities enough to consolidate into a whole outgrowth.

Inverted Grafts

In order to learn if the proximodistal axis of the antlerogenic periosteum is determined prior to elongation of the developing pedicle, five grafts of inverted

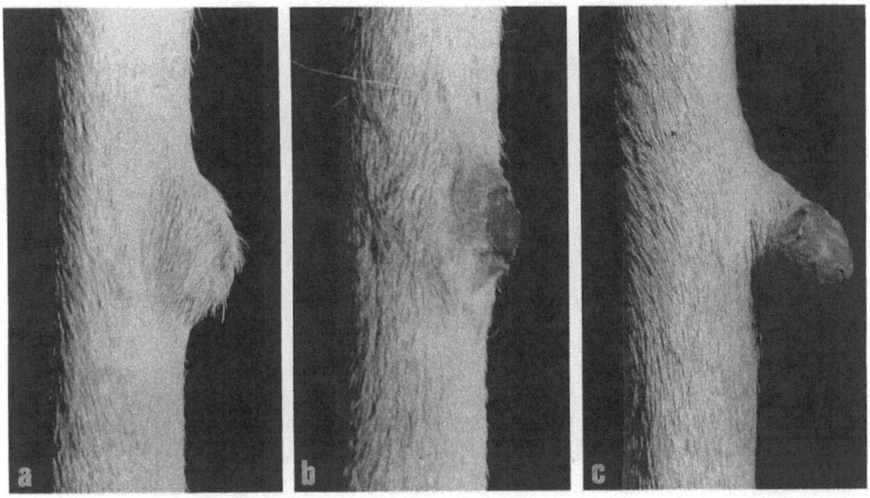

FIGURE 2. Example of an inverted graft of antlerogenic periosteum transplanted beneath the foreleg skin on December 9, 1981. A subcutaneous bone developed (*a*), but failed to induce an antler in 1982 and 1983. On May 2, 1984, the overlying skin was removed (*b*), causing the outgrowth of an antler as photographed on September 19, 1984 (*c*).

(top to bottom) periosteal discs were made beneath the foreleg skin. One such graft gave rise to a small bony antler (0.4 cm high) by the following autumn. This was replaced by a 7-cm outgrowth in the next summer, which later fell out along with its pedicle base and was not replaced in the third year. The four other grafts gave rise to pedicle bones without inducing antlers, although one of them exhibited very sparse hair reminiscent of antler velvet during the first year. This animal died before the future development of the graft could be followed.

One of the above deer that had formed only a subcutaneous pedicle bone about 1 cm high during the first two growing seasons (Fig. 2a) was subjected to wounding. In the spring of the third year the apical skin was removed to promote wound healing over the underlying pedicle bone at a time of year when healing would normally have occurred after casting of the antlers from their pedicles (Fig. 2b). This resulted in the development of an antler 3 cm long (Fig. 2c). Hence, two of the five discs of periosteum grafted upside down to the legs eventually produced antlers, one of which had been stimulated to grow by an integumental lesion.

Inasmuch as leg grafts tend not to develop as well as ones transplanted back onto their normal locations on the frontal bone, six inverted grafts were made to the pedicle regions. In three deer, 1.5 cm discs of antlerogenic periosteum were removed from the presumptive pedicle regions and grafted reciprocally to the contralateral sites. The discs were sutured in place upside down while preserving their anteroposterior axes. The mediolateral axes were reversed with respect to

FIGURE 3. When periosteal discs were grafted upside down in place of the contralateral ones, relatively normal antlers usually developed. In this case, a small accessory antler formed in front of the primary one.

their sites of origin in order to coincide with the axes in the graft sites on the opposite sides of the head.

During the first summer, five of the six grafts gave rise to antlers, the lengths of which varied from 1.5–9.5 cm (average, 5.1 cm). Four of these, which were followed through a second growing season, regenerated branched antlers 7–24 cm in length (average, 18.8 cm) (Fig. 3).

These results, together with those from leg grafts, confirm that periosteal grafts are still capable of inducing antlers even when transplanted upside down beneath the skin. Further, these antlers exhibit normal proximodistal axes. This is in keeping with other experiments in which adult pedicles were grafted to their nearby ears, after which antlers regenerated from both distal and proximal ends of the pedicle segments (Goss 1964). Indeed, it is also consistent with experiments on various regenerating appendages (fins, limbs, tails) of lower vertebrates (Goss 1969c). Whenever there is an amputation stump facing proximally with respect to the adjacent tissues, it will regenerate structures with normal proximodistal polarities. Such regenerates, therefore, are mirror images of the stumps from which they are produced.

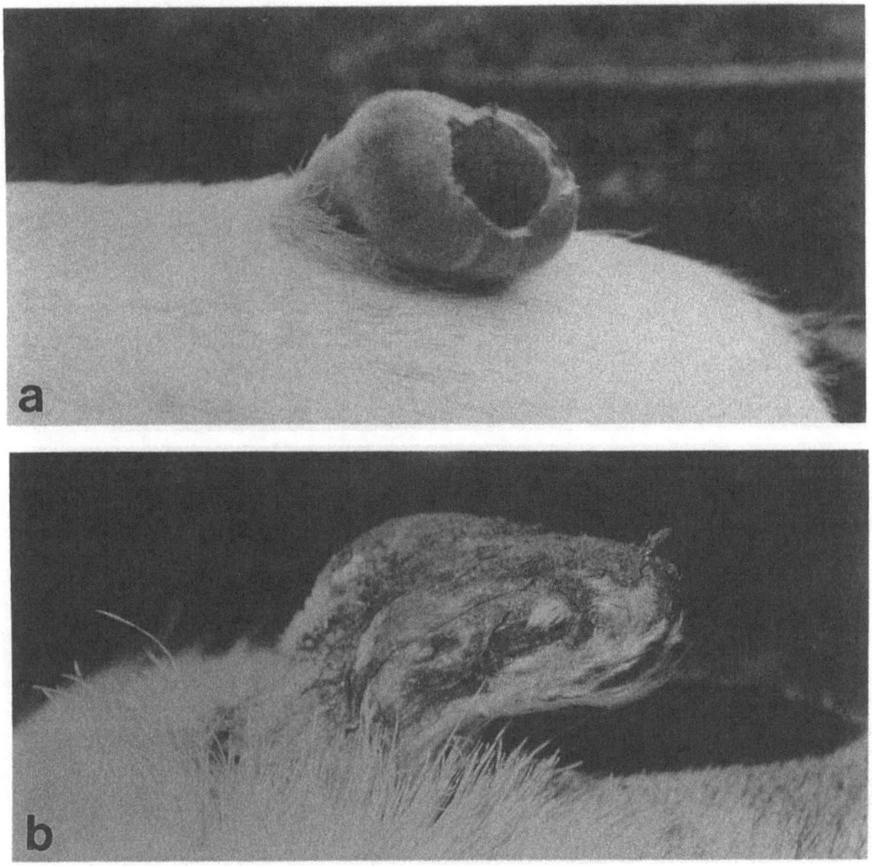

FIGURE 4. Ectopic antler growing from the hip region where a periosteal disc had been grafted 2 years earlier. *a* Antler in velvet on August 5. *b* Same antler after much of the velvet had been shed on October 21, photographed at slightly higher magnification.

Graft Locations

Experimental evidence that antler development may involve skin from such diverse locations as elsewhere on the scalp (Hartwig 1968a,b), the outer ear (Goss 1964), and the foreleg (Hartwig & Schrudde 1974), suggests that any and all kinds of integument might be competent to participate in antler induction. Accordingly, discs of antlerogenic periosteum have been transplanted subcutaneously to a variety of body sites in order to test for antler induction. Most have yielded positive results under favorable circumstances. A few have failed to do so. In some instances, lesions have been inflicted in previously unresponsive skin to enhance antler induction by promoting epidermal wound healing.

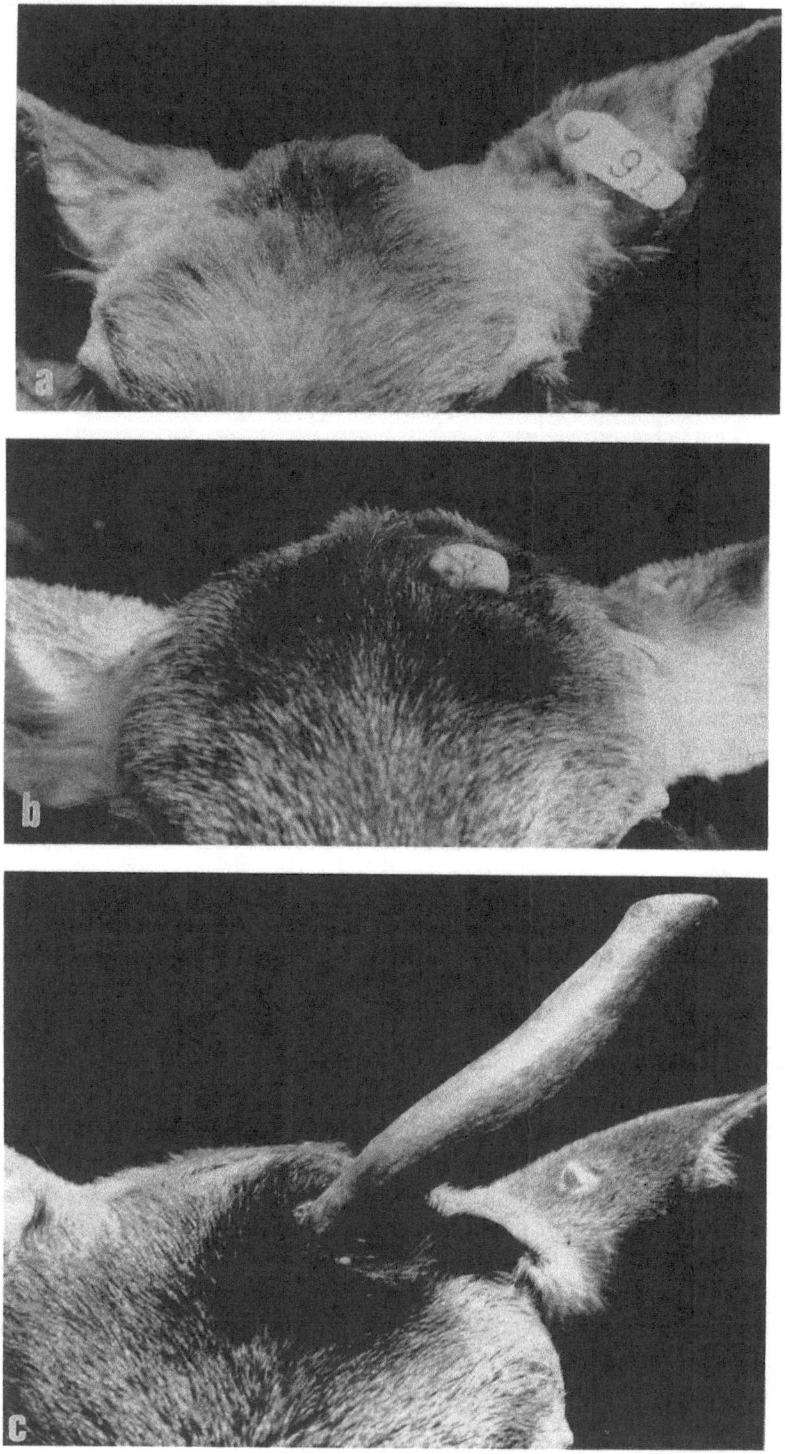
a
b
c

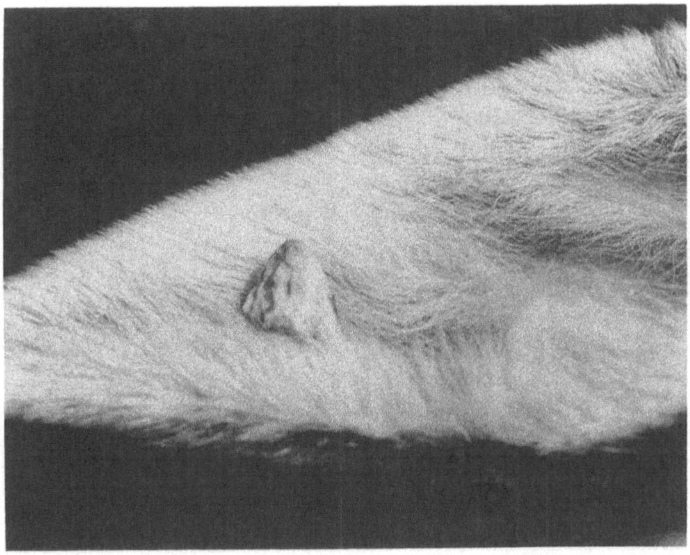

FIGURE 6. A short conical antler developed from a periosteal graft beneath the inner ear skin, as photographed on December 5, 1984. Original transplant operation had been carried out on December 19, 1981.

Varying numbers of grafts have been carried out beneath different skin types (Goss 1987). Although virtually all such transplants to the foreleg have induced antler tissues (n = 10), grafts to other locations yielded lower incidences of positive results. When periosteum was inserted beneath the skin of the hindleg in the region of the tarsal gland, or under the skin of the hip (Fig. 4a,b), antler tissue was induced in one out of two cases in each location. Two of the three grafts to the midline of the forehead gave rise to antlers (Fig. 5a-c). Five grafts to both the outer and inner ear skin resulted in antler development in three and four cases, respectively (Fig. 6). A single graft to the anterior chamber of the eye induced a small antler in the overlying cornea to which it had adhered (Fig. 7a-d). Transplants to other locations failed to induce antlers. None of the five grafts beneath the ventral tail skin gave rise to antler tissue, despite the development of bony nodules in three of the cases. Even when a disc of overlying skin 1 cm in diameter was removed in the spring of the second year from two of the graft sites with larger ossicles, no antler induction ensued.

◄

FIGURE 5. The disc of periosteum grafted to the forehead region on December 31, 1981, gave rise to a velvet-covered outgrowth on May 5, 1982 (*a*), which appeared as a bony button on November 9 (*b*). In 1983, a 14-cm antler was produced (*c*), which was replaced by similar ones in 1984 and 1985.

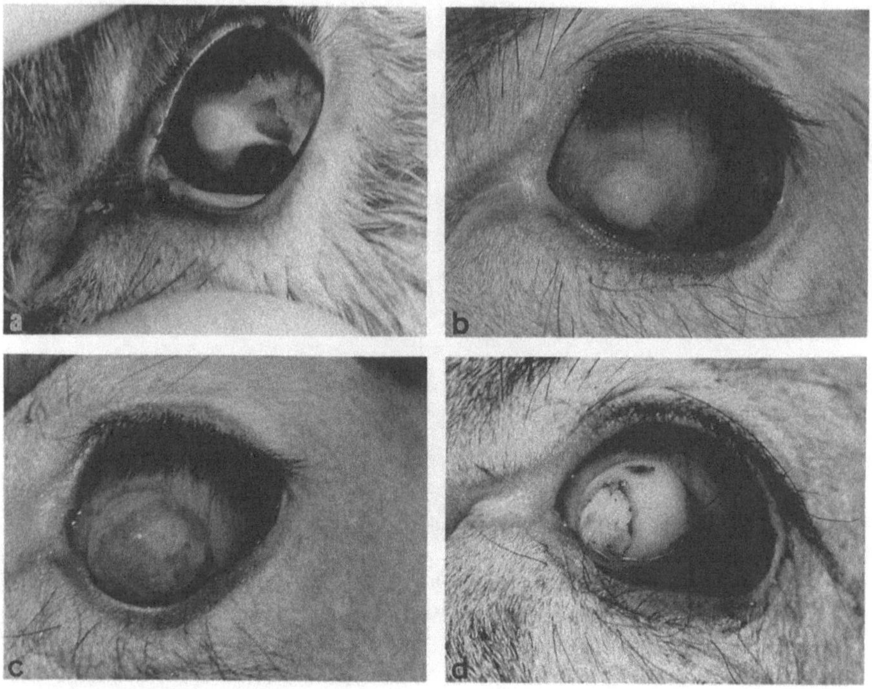

FIGURE 7. Corneal antler induction by intraocular graft of periosteum on December 26, 1981. *a* Transplant adhering to cornea on February 4, 1982. *b* Vascularization of cornea overlying pedicle bone, May 5. *c* Enlargement of outgrowth (without hair follicle induction) on August 1. *d* By September 26, the velvet has been shed, revealing a small bony antler. Graft was lost later in autumn and not replaced in the subsequent 2 years.

Eight grafts were placed beneath the back skin overlying the vertebral column in the mid-dorsal region. Although they all developed into subcutaneous ossicles ranging from 0.5–1.3 cm high, none induced antlerogenesis during the first year of observation. Even when the overlying skin was excised prior to the second growing season, still no antlers were induced.

Another eight grafts were transplanted to the nose. These were inserted through a transverse incision at the junction of the furry region of the snout and the relatively hairless skin of the rhinarium. The periosteal discs were placed into a pocket tunneled out immediately beneath the skin on the front of the nose. In this location, five out of the eight grafts developed into hardened tissue palpable beneath the skin. However, these hardened areas were conspicuously less well developed than the bony nodules usually formed at other subcutaneous sites. None of the grafts gave rise to antler tissue in the overlying nose skin.

A final transplant of periosteum was made to an intramuscular location in the shoulder region. The graft was inserted into an incision in the muscle intentionally to separate it from the skin. It underwent ossification, remaining palpable as a hard, lens-shaped bone throughout the next 2 years. No antler tissue developed in its vicinity.

Inductive Interactions

It is clear from the above results that grafts of antlerogenic periosteum typically differentiate into bony tissue wherever they may be transplanted. Aside from the reduced ossification of nose grafts, which remains to be explained, discs of periosteum form ossicles in the intraocular location, and intramuscularly as well as subcutaneously. These nodules are assumed to represent the pedicle bone which the periosteum normally produces in situ.

Under favorable circumstances, such ectopic pedicle bones may induce antler formation in the overlying integument. All vertebrate appendages, by definition, are enveloped in skin and cannot develop or regenerate in its absence. Antlers are no exception. Unless a layer of skin is in close association with the pedicle bone, antlerogenesis cannot take place. It cannot occur because without the velvet there is no antler. This indispensable association between the mesodermal component (periosteum) and the ectodermal one (epidermis) is similar to comparable relationships in the embryonic or regenerating appendages of other vertebrates. In these examples, the successful elongation and eventual morphogenesis of the structure depends on the operation of inductive interactions between mesoderm and ectoderm. For these inductions to occur, the communicating tissues must be in extremely intimate association. Indeed, in most cases the tissue layers are in actual contact with one another. This facilitates transfer of whatever molecules may be involved in conveying messages from one type of cell to another.

In the regeneration of antlers, such as occurs in adult deer when the old ones are cast and the new ones start to grow, the wound epidermis of the latter establishes a contiguous relationship with underlying mesoderm. Apical thickenings of this epidermis in the early antler bud may reflect important intercellular communications in the crucial stages of incipient antler development. This is significantly different from the histological arrangements that prevail in the generation of a fawn's first set of antlers. Here the thickened periosteum on the presumptive pedicle site of the frontal bone is separated from the epidermis of the scalp by loose and dense layers of connective tissue. The periosteum itself consists primarily of a layer of fibrous tissue about a millimeter thick. If the innermost layer of osteoblasts constitutes the inducing tissue, then the influence must traverse the overlying sheet of dense periosteal connective tissue. Above this is a layer of loose connective tissue which separates the periosteum from the dermis. Control experiments have shown that the transplantation of loose connective tissue from the prospective pedicle region subcutaneously to the leg results in neither bone nor antler development. There is a fibrous ligament connecting the protuberance of the frontal bone with the overlying scalp (Hartwig 1967; Goss & Powel 1985) which transcends the intervening connective tissue. Whether this plays a role other than the mechanical one of holding the inducing and induced tissues together is not known.

The outermost layer of tissue between the periosteum and the epidermis is the dermis itself. On the scalp of a fawn, this is a dense layer of fibrous tissue approximately 1 mm thick. Its epidermis is punctuated with fully differentiated hair follicles of the scalp fur. Despite the differentiated condition of these tissues, together with their combined thickness of up to 2 mm, the inductive influence from the pedicle periosteum somehow promotes the conversion of the typical pelage of the scalp to the specialized velvet of the developing antler.

This is a very gradual process (Goss 1984, 1985). There is no landmark or point in time when the pedicle may be said to have become an antler. Even when the elongating pedicle has attained heights two or three times its width, it may still remain covered with the thick fur characteristic of the scalp. Yet at about this stage of development the hairs begin to become more sparsely distributed on the apical region of the pedicle. Inside, the growing pedicle is capped by cartilaginous trabeculae which grow by the proliferation of precursor cells distally. Presumably it is in this apical region at this period of development that the epidermis is induced to become antler velvet. Unhappily, details of the crucial transformation from pedicle to antler have been described neither histologically nor biochemically.

Conclusion

The production of deer antlers is likened to the development of appendage buds in embryos. Both involve the participation of mesoderm and ectoderm, or the derivatives thereof. Each is mediated by inductive interactions between these two tissues. And in either case, the ectodermal component, while essential for continued outgrowth, does not dictate the type of appendage that is to form. That property resides in the underlying mesoderm.

The mesodermal source of pedicles and antlers lies in the periosteum of the frontal bones, as has been proven by transplantation experiments. This periosteum, grafted beneath the skin of the legs, hip, ear, or scalp, will induce the ectopic epidermis to differentiate into antler velvet. It will do so rightside up or upside down, or if cut in half or minced, but must be or a critical size for induction to succeed. The developmental mechanisms by which this postembryonic induction is accomplished, despite considerable thicknesses of intervening tissues, remains one of the major unsolved problems of antlerology.

Acknowledgments. Research reported in this chapter has been largely supported by a grant from the Whitehall Foundation. The author is also grateful to Peter Brewer and Earl Logan of the Southwick Animal Farm for taking such good care of the deer.

10
Antler Regeneration: Studies with Antler Removal, Axial Tomography, and Angiography

JAMES M. SUTTIE AND PETER F. FENNESSY

Introduction

This paper includes three rather diverse approaches to the study of antler regeneration in red and fallow deer. Each study, however, shares the same goal of attempting to improve our understanding of the process of antler regeneration. All studies took place in the Southern Hemisphere, in New Zealand; this accounts for the reversed seasons.

Serial Velvet Antler Removal

If red deer stags are castrated when the antlers are hard, the antlers are cast some 10–21 days later. The antlers regrow immediately but remain in velvet. If the stag is castrated when the antlers are in velvet, then they remain so and are never cleaned or cast (Fennessy & Suttie 1985). However, if the velvet antler of the castrated stag is surgically removed, further regrowth is possible. (Suttie & Fennessy, personal observations). The aim of the present study was to remove the velvet antlers at intervals from three castrated red deer and then to measure further antler regrowth. A secondary aim was to investigate the effects of pedicle trauma and testosterone priming in restoring the growth potential of the antler.

Materials and Methods

Three young red deer (*Cervus elaphus*) stags, who had grown and cleaned their first spike antlers normally were kept indoors in pens and fed a concentrate diet to appetite. Each stag was surgically castrated under general anesthesia (Fig. 1) in either June (#3), September (#31), or December (#41) 1982.

Nine months after castration, the right velvet antler was removed under Xylazine (Rompun, Bayer Ltd) anesthesia (as was the case for all velvet removal) 2 cm above the antler/pedicle junction (APJ), and the left antler was unmanipulated. This was repeated at 3 monthly intervals (Fig. 1) on three (#41), four (#31), or five (#3) occasions. In June 1984 the left antler was removed 2 cm above the APJ,

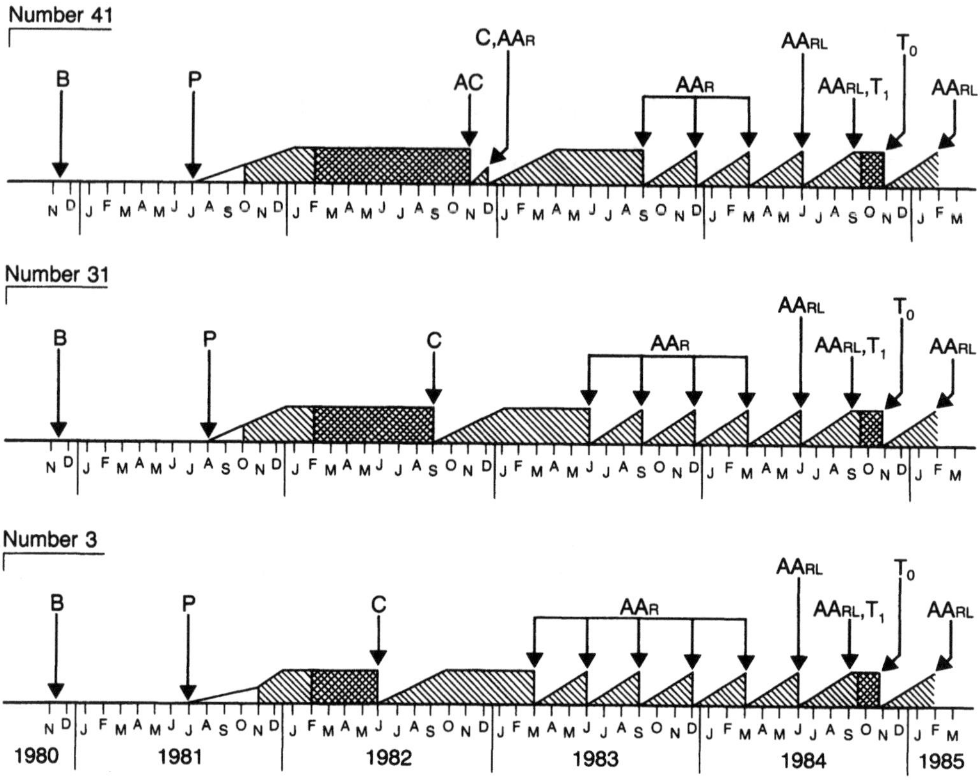

FIGURE 1. The life histories of the three stags used in this study. [box] pedicle growth; [hatched] velvet antler; [cross-hatched] hard antler. Abbreviations: B = birth, P = pedicle growth begins, AC = antler casting, C = castration, AA_R = right antler removed, AA_L = left antler removed, AA_{RL} = both antlers removed, T_1 = testosterone implant in, T_0 = testosterone implant out, followed 3 weeks later by antler casting.

and the right antler (previously serially velveted) was removed 2 cm below the APJ from each stag. In September 1984, both antlers were removed 2 cm above the APJ and a silastic testosterone implant, which previous research had shown maintained plasma testosterone levels of about 10 ng/mL and was sufficient to cause castrated stags to clean their antlers, was inserted subcutaneously. Three weeks after antler-cleaning, this implant was removed and the stags were allowed to cast their antler stubs, which had not regrown, normally. Three months after antler casting, both velvet antlers, which had regrown, were removed 2 cm above the APJ. The timing of the study is shown diagrammatically in Fig. 1.

Each velvet antler was measured with a flexible plastic tape. The portion of velvet antler removed on each occasion was weighed and the number of points was recorded.

Results

Serially Amputated (then Testosterone Treated) Right Antler

Repeated removal of the velvet antlers above the APJ at 3 month intervals led, overall, to a 42% reduction in antler growth rate, a 54% reduction in antler length, and a 72% reduction in weight (Figs. 2–4). Cutting the antler below the APJ significantly increased the degree of regeneration (antler weight and antler length) over the antlers serially amputated above the APJ (Table 1), but not to the size of the first postcastration antler. Similarly, treatment with testosterone, which stimulated cleaning, and, when the testosterone was withdrawn, casting, increased the level of regeneration in terms of both length and weight (Table 2). However, the velvet antler grown after this treatment did not exceed in size that grown immediately after castration. Both techniques were effective in stimulating regeneration. This conclusion is presented tentatively, however, as one stag seriously damaged one of his posttestosterone antlers.

The first antler grown after castration, the antlers grown after cutting below the APJ, and the antler grown after testosterone stimulation were branched; otherwise, simple spikes only were regenerated.

Testosterone Treated Left Antler

The antler grown after castration, which had remained unmanipulated for up to 2 years, was heavier than the antler removed after 9 months (739 g and 372 g, respectively, sed [standard error of the difference] = 141.3). The antler which was regenerated after the first amputation was smaller than the one grown after castration (112 g and 739 g, respectively, sed = 205 g) and had a slower growth rate (0.18 cm/day and 0.38 cm/day, respectively, sed = 0.09) (Figs. 2–4). Testosterone restored growth rate to approximately that of the first antler grown after castration (0.31 cm/day and 0.38 cm/day respectively, sed = 0.07). The testosterone-stimulated antler was significantly longer, heavier, and faster growing than the one immediately preceding it (Table 3). The antler grown after castration and the testosterone-stimulated antler were branched. The antler cut in September 1984 was a single spike.

Discussion

Repeated removal of portions of the velvet antler of castrated red deer stags leads to a reduction in growth rate. Regeneration invariably occurs but is limited in extent. This indicates that either the ability of antler cells to dedifferentiate into actively dividing cells becomes limiting or some stimulating substance is lacking. The fact that removal of the antler by cutting through the pedicle – rather than the antler itself, without any other exogenous administration of stimuli – is capable of increasing antler growth rate tends to favor a hypothesis that the potencies for regeneration of the antler and the pedicle differ.

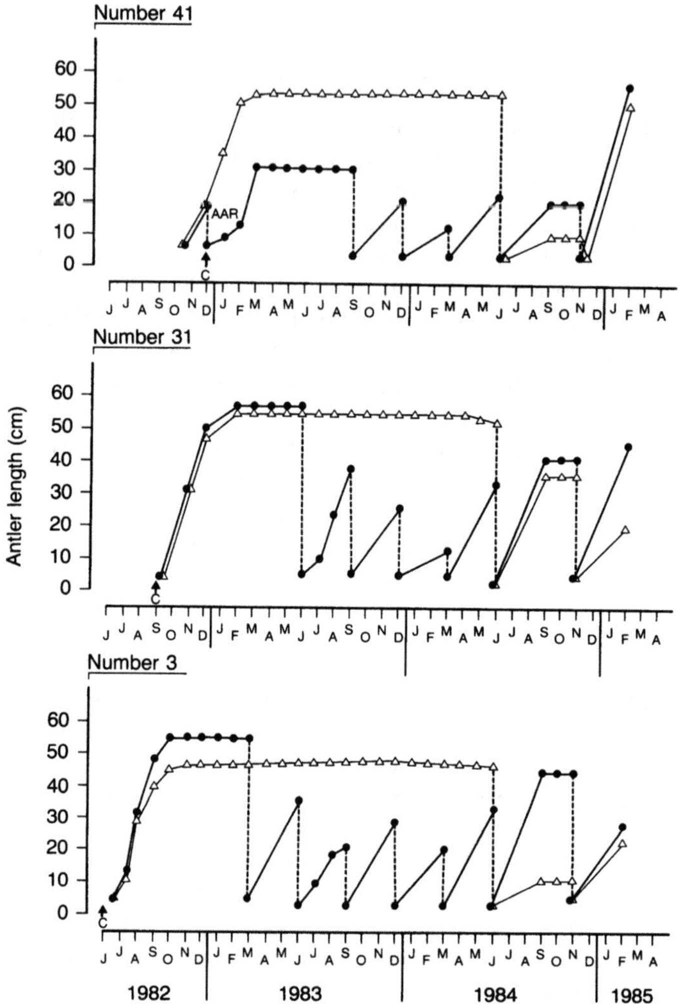

FIGURE 2. Antler length (cm) of postcastration and all subsequent antlers grown by each of the three stags. ●—● serially amputated antler △—△ testosterone-treated-only antler. C indicates the time of castration. The broken lines indicate the abrupt change in antler length following amputation or antler casting.

Long-term castrated deer often show annual increases in antler size, either growing extra points or producing antleromas (Goss 1983). It is not surprising, then, that trauma to the velvet-antler caused regeneration in the present study. Lincoln (1984) carried out a similar experiment to the present study when he amputated the velvet antlers from three castrated red deer stags at irregular 4–6-month intervals for over 2 years. Regeneration was always observed but no diminution in size resulting from repeated antler removal occurred. However, the antler was removed at the top of pedicle (i.e., through the APJ) by Lincoln – not,

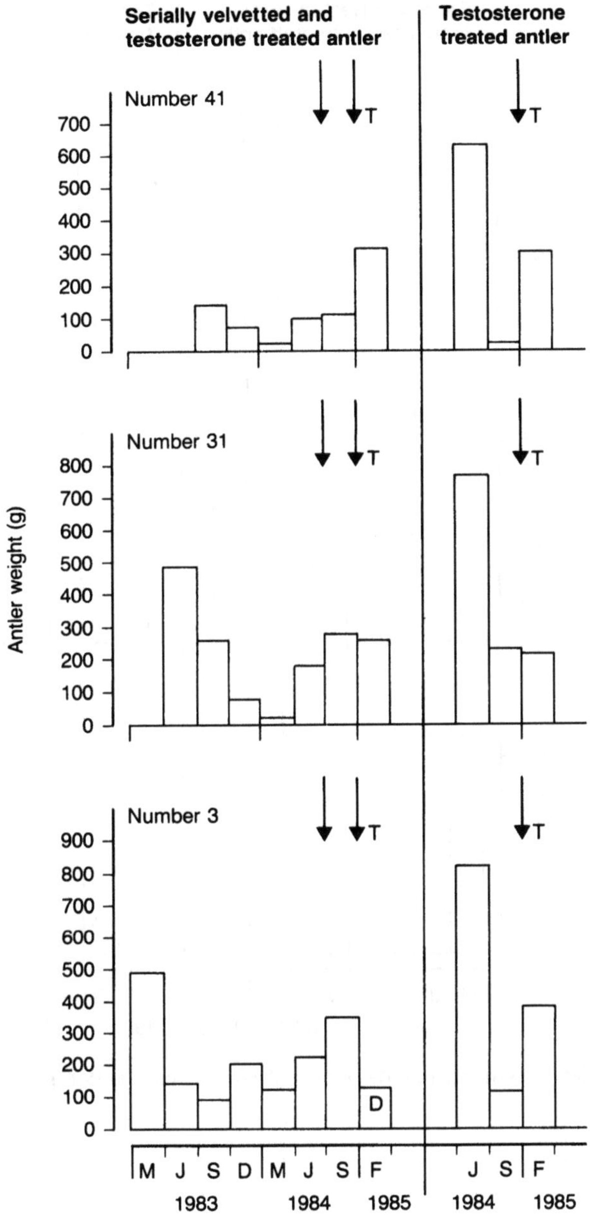

FIGURE 3. Velvet-antler weight (g) after each amputation. Weights refer to antlers removed on the 20th of each month referred to. Arrow indicates timing of antler removal below the APJ. Arrow accompanied by T indicates timing of testosterone treatment. D means antler was damaged and data were not used for the numerical analyses.

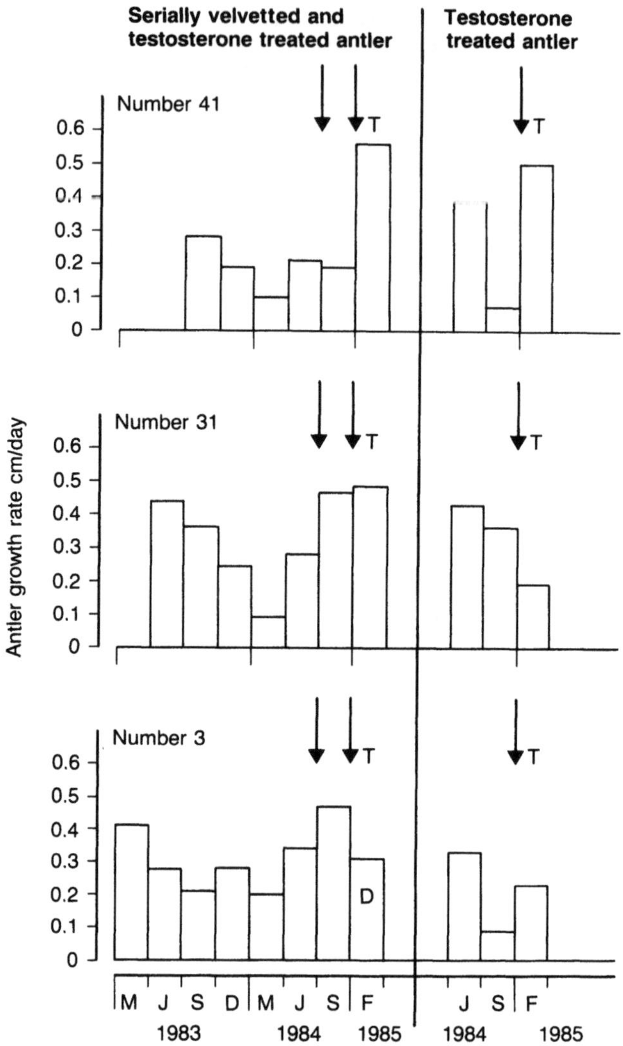

FIGURE 4. Antler growth rate (cm/day) for each stag for each antler. Remainder of legend as for Fig. 3.

as in the present study, 2 cm above or below the APJ. It seems that the regeneration capacity of the antler is restricted, compared with the pedicle, to regrowth of simple spikes or poorly organized antleromas with a low growth rate. In contrast, if the pedicle is stimulated to regenerate, growth rates are faster and branched antlers of the species-specific shape are possible. In Lincoln's 1984 study, although the antlers that regenerated did not get smaller, no branching was observed. It is considered that antler removal along the APJ is somewhat inter-

TABLE 1. Comparison of right antler growth parameters[a]

	Serially velveted antler (above APJ)	sed	Antler after removal (2 cm below APJ)
Antler growth rate (cm/day)	0.22		0.37
sed		0.08	
Antler length (cm)	22.7		36.0
sed		6.8	
Antler weight (g)	108.3		247.7
sed		67.8	

[a] Comparison of mean of first three serially velveted antlers grown after castration with antler grown after removal below the antler/pedicle junction (APJ) (N=3). Sed = standard error of the difference.

mediate between removal above or below this line. It may well be that the antler cells are limited in their ability to dedifferentiate and grow, but the pedicle cells carry this ability.

The antler grown after testosterone stimulation had a growth rate that was either similar to that grown after cutting below the APJ or greater than the previous antler cut above the APJ. One model that was tested by this part of the study was that if pedicle fibroblasts required testosterone priming to stimulate normal antler growth in castrates, then antler growth rates of a testosterone-stimulated antler would be much higher than antlers grown without this stimulation. The antlers did indeed grow faster—but no faster than antlers grown after removal below the APJ. This seems to indicate that testosterone treatment prior to antler development brought about its stimulating effect mainly by causing trauma to the pedicle at the induced antler casting rather than a classical endocrine stimulation during growth. Casting itself would appear to be sufficient stimulus for the pedicle cells to commence regeneration.

The antler that remained intact and unmanipulated on each stag for up to 27 months after castration increased in weight but not in length over this time. The

TABLE 2. Comparison of right antler growth parameters[a]

Serially velveted antler		sed	Antler grown after testosterone stimulation	sed	Antler grown after cut below APJ
Antler growth rate (cm/day)	0.19		0.52		0.33
sed		0.14		0.11	
Antler length (cm)	20		51		36
sed		13.0		10.8	
Antler weight (g)	73.5		286		196
sed		67.8		127.1	

[a] Table compares serially velveted antlers, antlers grown following removal of the previous antlers below the antler/pedicle junction (APJ), and testosterone-stimulated antlers (N=2). This table is necessary, as one stag (#3) damaged his testosterone-stimulated antler and thus could not be included in the comparison.

TABLE 3. Comparison between the left antler grown before and after testosterone stimulation (N=3)

	Before testosterone treatment		After testosterone treatment
Antler growth rate (cm/day)	0.19		0.31
sed		0.09	
Antler length (cm)	19		31
sed		8.9	
Antler weight (g)	112		296
sed		76	

Sed = standard error of difference.

base of the antler became thicker, and progressive bone deposition had probably also taken place.

It is not strictly possible to compare weight or length of the antler grown after castration with subsequent antlers. They were allowed to grow for only 3 months, whereas those grown immediately after castration were allowed to grow freely; in fact, growth took about 4 months. However, growth rates do make for valid comparisons. Neither antler removal below the APJ nor testosterone stimulation consistently increased antler growth rate over that of the first antler grown after castration. Thus, about 0.5 cm/day appears to be maximal for castrated red stags, despite treatment. Similar-aged intact stags would be expected to average about 0.7 cm/day (Suttie & Fennessy, unpublished observations). Why the discrepancy? Presumably, this is not due to lack of steroid priming. Stags treated with medroxyprogesterone acetate, which effectively inhibits luteinizing hormone (LH) release, and thus testosterone, both before and during antler development, grew antlers of similar or only slightly smaller size than controls (Muir et al. 1982; Suttie & Fennessy, unpublished observations). This seems to indicate that lack of testosterone before and during antler growth may not be directly responsible. It is known, however, that in sheep testosterone is necessary for normal pulsatile release of growth hormone (GH) (Davis et al. 1977). As GH results in insulin-like growth factor 1 (IGF1) release and IGF1 has been shown to correlate positively with antler growth (Suttie et al. 1985a), it may be that these castrates have less GH or an abnormal GH release pattern. In support of this hypothesis, castrated red deer stags indeed have lower IGF1 levels than controls (Suttie, unpublished), and testosterone administration stimulates IGF1 secretion in intact red deer stags (Suttie et al. 1985b). A further possibility is that a non-LH-dependent testicular factor is responsible for maintaining antler size at a maximal rate. Although testosterone has been located immunohistologically in the velvet-antler (Bubenik, G. et al. 1974), its role in antler regeneration (as opposed to mineralization) is not clear.

The present study has underlined the importance of the pedicle in controlling the extent of regeneration. Regeneration from the velvet-antler is of limited extent, but near normal regeneration is possible from the pedicle—probably independent of direct stimulation from testosterone.

Computerized Axial Tomography

Computerized axial tomography (CAT) is unlike conventional radiography. In CAT, screens and films are not used to detect x-rays, nor do x-rays penetrate parts of the body outside the section being examined. Instead, an x-ray tube which is aligned precisely with an electronic detector, 180° opposite to it, moves around the subject and measures x-ray transmission through the transverse section many thousands of times. This information is sent to a computer which calculates an array of numbers (called CT numbers) from which it reconstructs a picture. The picture can look superficially like a radiograph and is an image of the structure being examined in a transverse section. Sections of the body can be examined every 2 millimeters to build up anatomical pictures that would not be possible using conventional radiography. Preliminary results of a CAT scan study of fallow deer, (*Dama dama*) antlers are presented in this section.

Methods

Three fallow bucks were slaughtered in either April 1983 (hard antler), November 1983 (velvet-antler, 4 weeks after antler casting) or in January 1984 (velvet-antler, 12 weeks after antler casting). The heads were removed and, within 1 hour, were scanned using a Technicare Deltascan 2020 (Ohio Nuclear Inc.). Scans were made at 120 kV, 50 mA, with 2 mm between sections. The scanner produces images of 512 × 512 pixels, in a square matrix.

Analysis

Although the scanner produces an accurate image of the tissue being examined, it can also print out the actual CT number (an index of density) of any area of the CT scan. This area can be from 1 pixel to 2.6×10^5 pixels, which is from the limit of resolution to the total area of the scan. The area can be set precisely by the operator. For the present analysis the area used was a window containing 53 pixels (0.13 cm^2), because this was the largest area that gave high sensitivity. This window was moved over the scan of the antler or pedicle in rows, and the CT number of each area was recorded. Because the tissue being examined is heterogeneous within each area, estimates of error tell more about the accuracy of positioning of the window than about the variability of the CT numbers of the tissue itself. Nevertheless, typical standard deviations of a range of mean CT numbers are presented in Table 4. This should enable interpretation of CT numbers presented in the following figures.

TABLE 4. Range of CT numbers with typical standard deviations associated with them

CT number[a]	Typical standard deviation
100	15
300	20
500	25
700	50
900	80
1,100	80
1,300	80

[a] Water as a CT number of 0, the densest bone has a CT number of around 4,000, and air has a CT number of −1,000.

Results

A CT scan of the whole head of the buck in hard antler (antlers sawn off above the pedicle) is shown in Fig. 5. The CT numbers at the edge of the antler and pedicle were uniformly high, indicative of compact bone; indeed, the bone was as dense as the braincase. The CT numbers from the center of the antler and pedicle were lower, indicative of cancellous bone. The CT numbers from around the APJ were about 200 units higher, indicating that the bone was slightly denser at the junction between living and dead bone.

Four weeks after antler casting (Fig. 6), the CAT scan reveals that appreciable demineralization of the pedicle must have taken place (as CT numbers in lines E and F were 587-891, having fallen from 960-1371 (lines E-J, Fig. 5)). There was little evidence of compact bone on the outer edge of the pedicle with a cancellous core, as CT numbers at the edge were similar to those of the core. The CT numbers at the tip of the antler (lines A-B), which represents the actual growing point, were low, indicative of a high water content, probably due to the blood capillary network found there and the presence of cartilage. The CT numbers were higher below the tip, where the CT scan becomes markedly darker (line C). This is probably due to calcifying cartilage, which was beginning to differentiate into a more dense outer edge with a cancellous core, as the CT numbers at the edge are higher than those of the core. However, the numbers are too low to be indicative of bone. The CT number of the brain, −51, is indicative of tissue less dense than water (i.e., mainly lipid). The dermis/epidermis of the antler and pedicle appear to be of similar tissue density.

In January (Fig. 7) the pedicle, although still of lower density than the braincase, appears to have begun to redifferentiate into a hard compact outer edge (lines E_b and $_f$, F_b and $_f$, and G_b and $_f$) and a cancellous core (lines $E_{c\text{-}e}$, $F_{c\text{-}e}$, and $G_{c\text{-}e}$). The top of this section of the antler (lines A and B), which was about 40 cm from the growing tip, has high CT numbers, indicative of bone formation. Although CT numbers of the lateral edge (C_g, D_f, and E_f) were higher than the

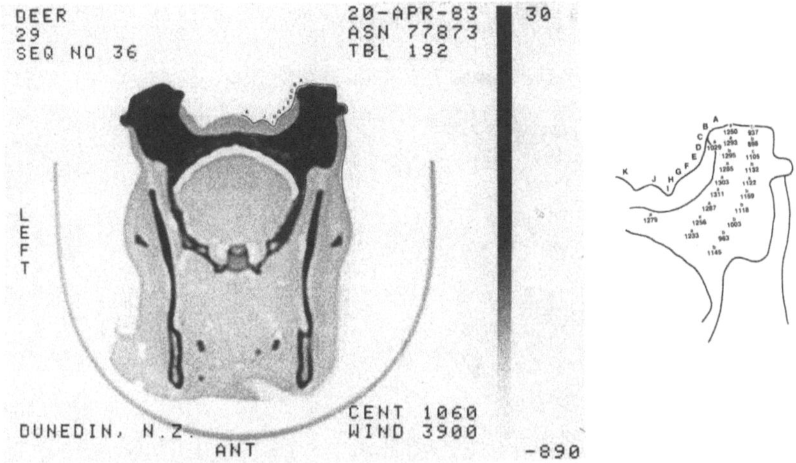

FIGURE 5. CAT scan of the head of a fallow buck (#29) on April 20, 1983, when the buck was in hard antler. The original tomograph has been printed as a negative for illustrative purposes. ASN refers to the serial number of the scan, TBL refers to the position of the head relative to the scanning gantry, and Seq No refers to the position of this scan in the sequence taken on his head. The vertical scale on the right, from 3,010 to −890, refers to the CT numbers that can be visualized in the scan. CT numbers are based on a scale of −1,000 for air, 0 for water, and +4,000 for the densest bone. All tissues have a number which is between +4,000 and −1,000, depending on density. Wind refers to the window width, which at 3,900 is very wide, indicating that almost all tissues should be seen. Cent refers to the center of this scale, 1,060 is a number that gives good detail of hard tissue (a lower number would give better detail of soft tissue). The small numbers on the overlay are the CT values for tissue density at that position. Each line of CT numbers, where appropriate, is labeled with a capital letter and each CT number within each line with a small letter.

core, the medial edge was of similar density to the core (compare $C_{b \text{ and } c}$ and $D_{a \text{ and } b}$ with $C_{d\text{-}f}$ and $D_{c\text{-}e}$).

The development of the trey tine is depicted in Fig. 8. The CT numbers of a representative sample of the scans are also shown. The sequence from 24-42 begins from above the trey tine to below it. Thus, sequence 24 is only of the main beam, 25 also has the very tip of the trey tine, 32 and 33 have the junction of the trey tine with the main beam, and the remainder illustrate the united tine and antler. The main beam above the trey tine, sequence 24, has a hard bony outer edge (e.g., line A) with a cancellous core ($B_{d \text{ and } e}$, $C_{c \text{ and } d}$). Indeed, this varies little throughout sequence 24-42. The tip of the trey tine appearing in sequence 25 was characterized by very low CT numbers, but 2 mm below this (sequence 27), higher numbers, indicative of cartilage, were present in the core (A_b and $B_{b \text{ and } c}$). In

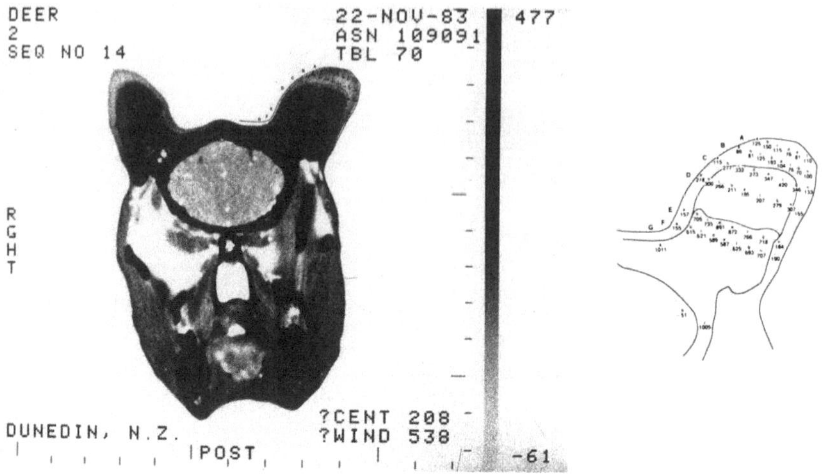

FIGURE 6. CAT scan of a different buck (#2) on November 22, 1983. Legend as for Fig. 5. This window at 538 with a center of 208 is designed to give better soft tissue detail. Note sinuses below the right antler.

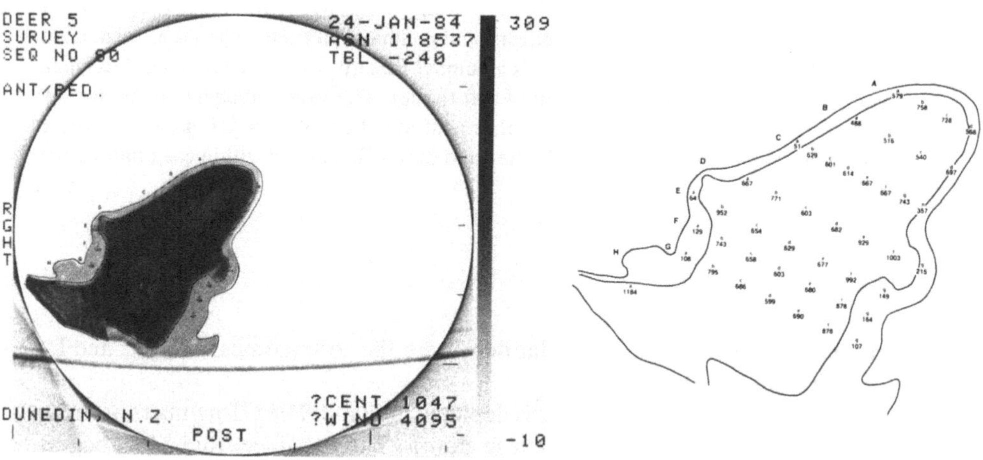

FIGURE 7. CAT scan of the antler base and pedicle from a buck on January 17, 1984. Legend as for Fig. 5. Note the very wide window.

►

FIGURE 8. Sequence of CAT scans from above the trey tine to below it in deer #4 on January 17, 1984. The main beam is in the lower left of each scan. When it appears, in sequence #25, the trey tine is in the upper central part of the scan. Legend as for Fig. 5.

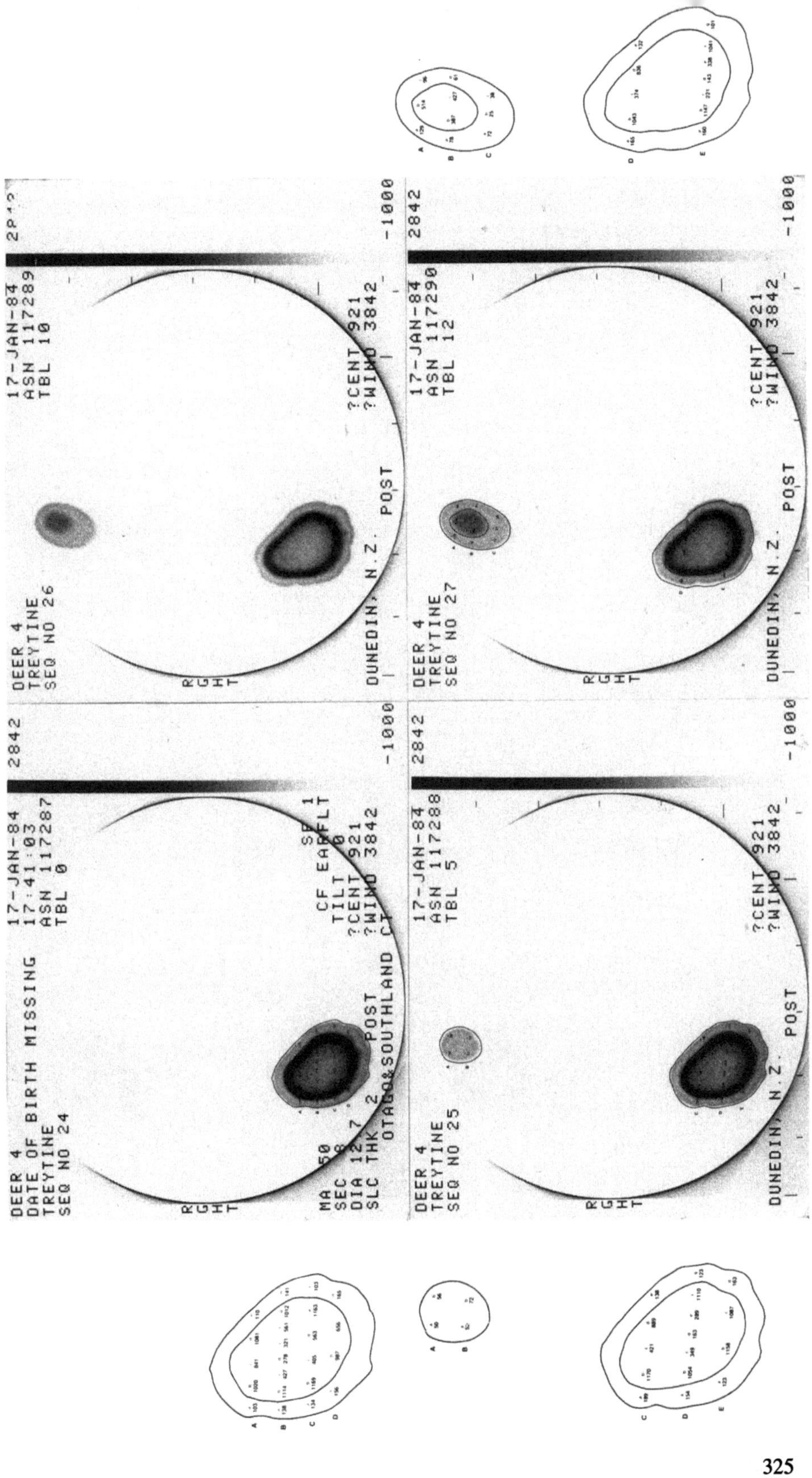
DEER 4
DATE OF BIRTH MISSING
TREYTINE
SEQ NO 24
17-JAN-84
17:41:03
ASN 117287
TBL 0
2842
MA 50
SEC 8 7
DIA 12 7
SLC THK 2
CF EARFLT
SF 1
TILT 0
?CENT 921
?WIND 3842
POST
OTAGO&SOUTHLAND CT
-1000
RGHT
DEER 4
TREYTINE
SEQ NO 26
17-JAN-84
ASN 117289
TBL 10
?CENT 921
?WIND 3842
POST
DUNEDIN, N.Z.
-1000
RGHT
DEER 4
TREYTINE
SEQ NO 25
17-JAN-84
ASN 117288
TBL 5
2842
?CENT 921
?WIND 3842
POST
DUNEDIN, N.Z.
-1000
RGHT
DEER 4
TREYTINE
SEQ NO 27
17-JAN-84
ASN 117290
TBL 12
2842
?CENT 921
?WIND 3842
POST
DUNEDIN, N.Z.
-1000
RGHT

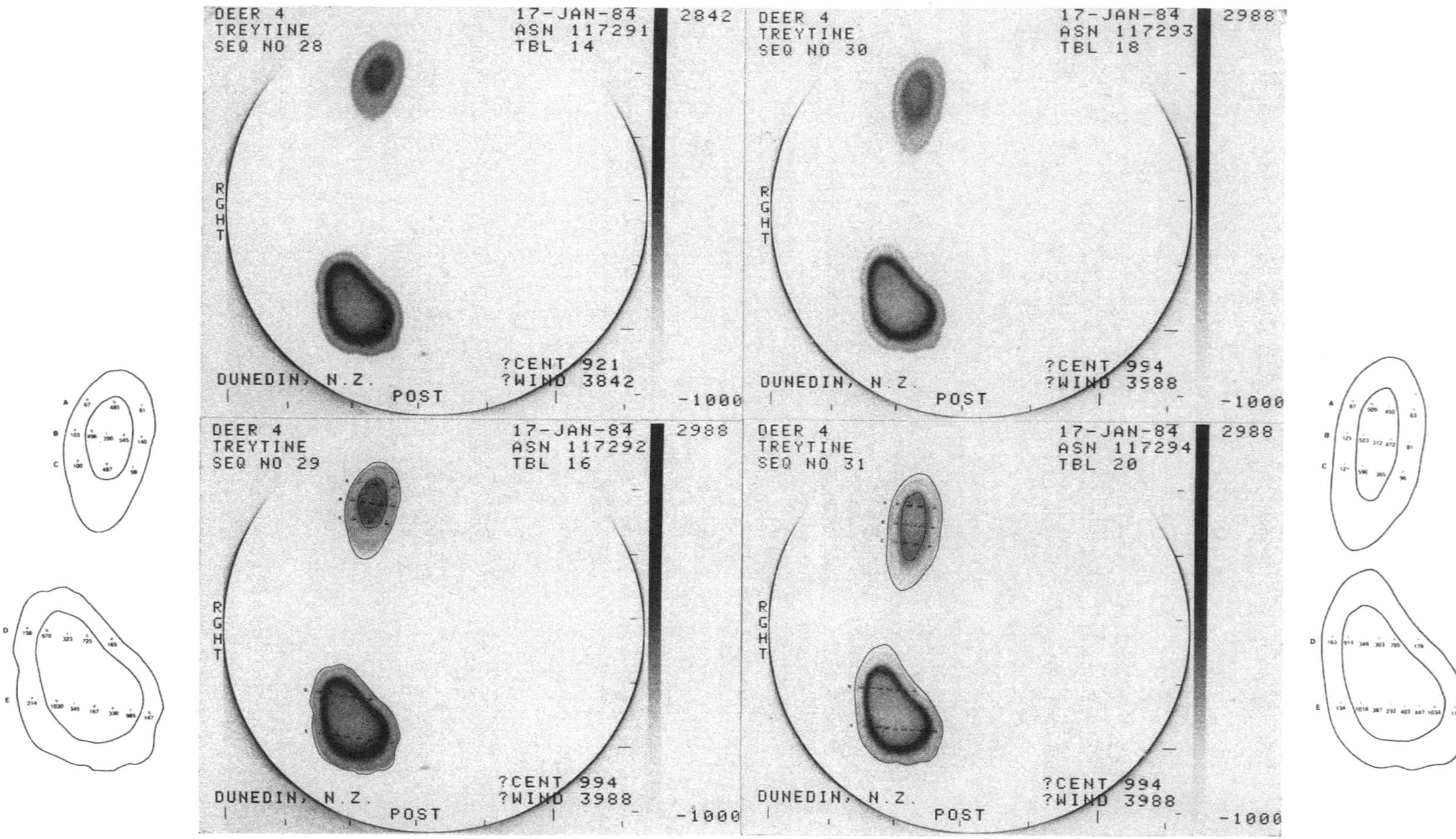

FIGURE 10. (*Continued*)

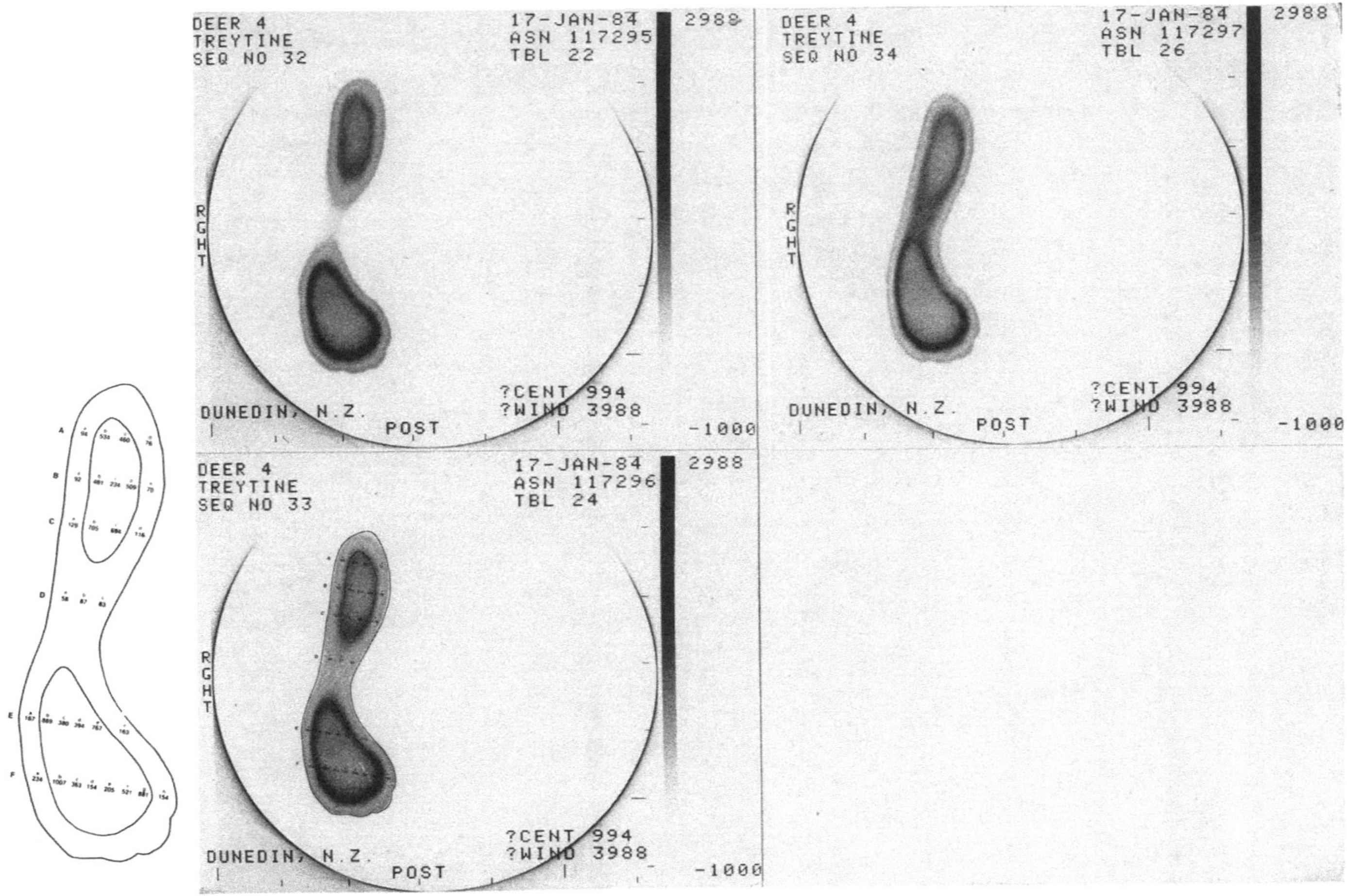

FIGURE 10. *(Continued)*

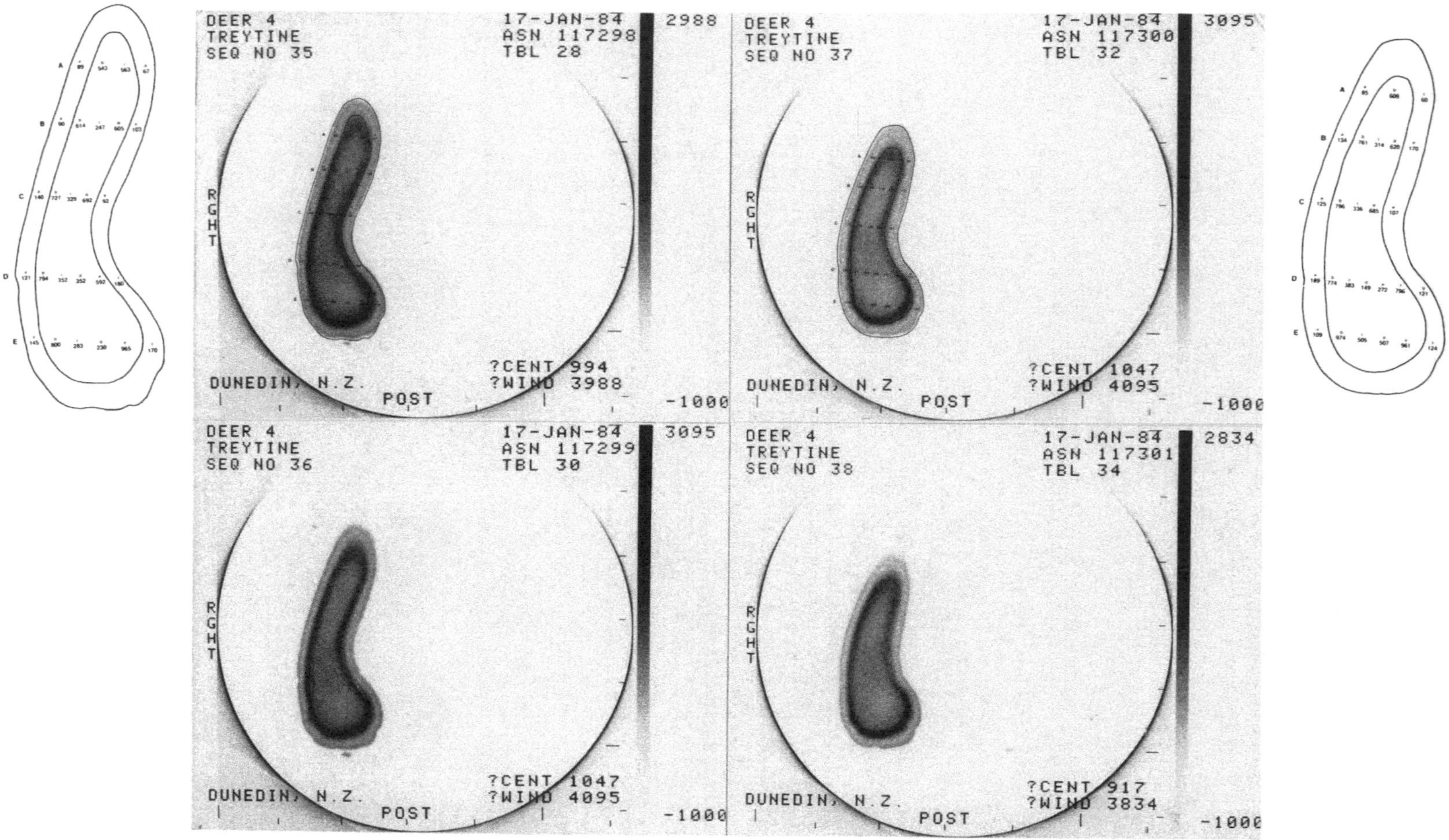

FIGURE 10. *(Continued)*

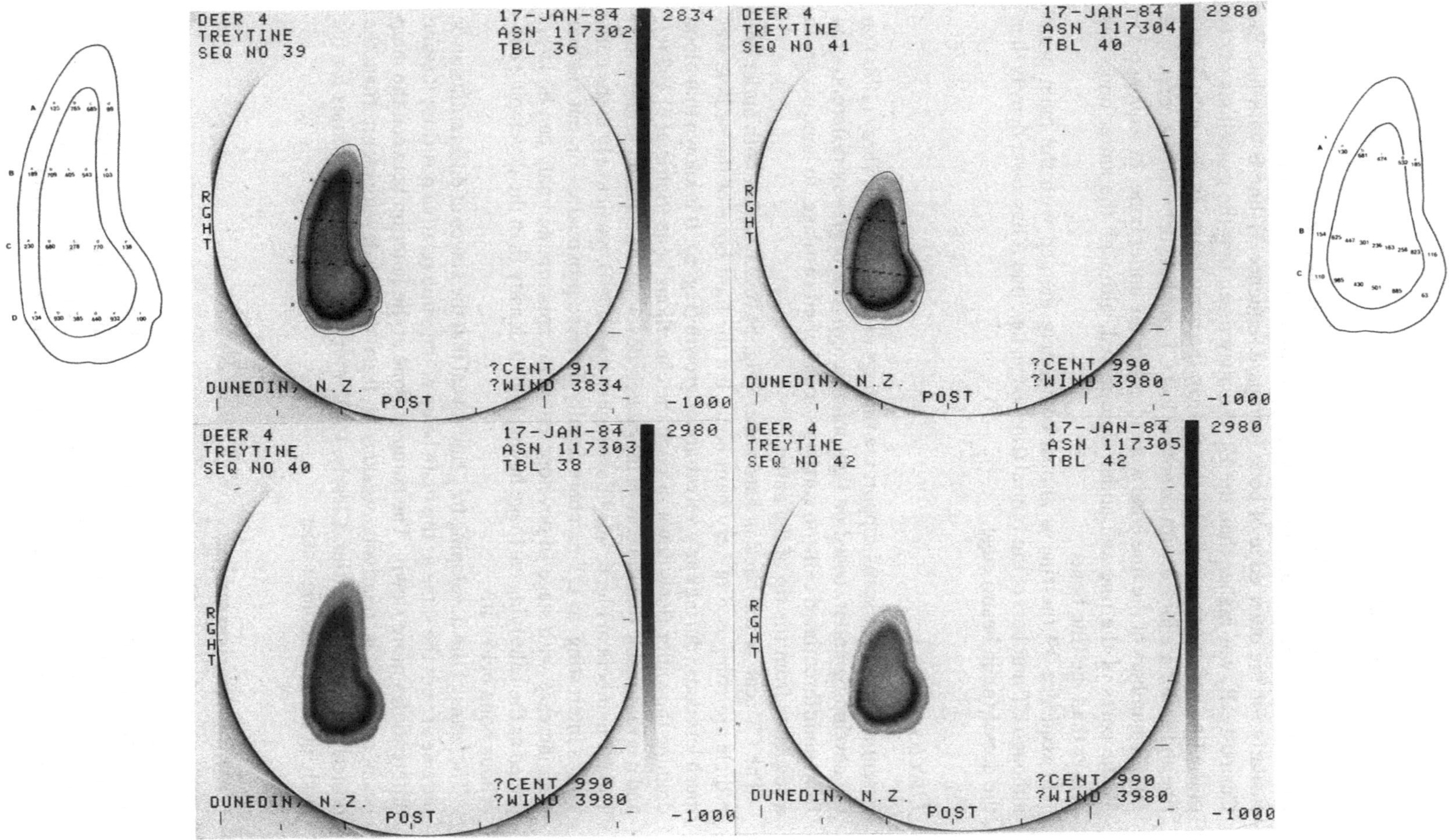

FIGURE 10. (*Continued*)

sequence 29, both main beam and tine have a hard outer edge with a soft core, although CT numbers of the tine were less than the main beam of the antler particularly for the outer edge of bone (compare sequence 31 line B with line E). Curiously, B_c was higher than the E_d; this may mean that the cancellous cavity of the tine was more poorly developed than the main beam. In sequence 33 the tine and antler are joined with an area of low CT numbers between the cores (line D). CT numbers of the tine bone were less than the antler bone. In sequence 35-37 the cores of the tine and antler were joined, although the antler bone was denser than the tine bone.

In sequence 39 the tine is almost completely merged with the main beam although CT numbers of the ventral (tine) edge, i.e., line A has lower density than the dorsal (main beam) edge.

Discussion

Essentially the CAT scans support the findings of Banks and Newbrey (1983) on the developing antler, based on light microscopy. The degree of differentiation from undifferentiated cells, to cartilage, to calcified cartilage, to bone occurs in a sequence from the tip of the antler.

The CT scan of the buck in hard antler (Fig. 5) revealed that both antler and pedicle are composed of very hard bone, but the hard bone of the pedicle was much less dense during early velvet-antler growth (Fig. 6). It is known that blood draining the antler flows down through the cancellous bone (Suttie et al 1985c), and it is reasonable that the core of the pedicle develops the capacity to allow this. The CT numbers in Fig. 6, line C, are higher than line D, particularly in the core. This is interesting, as CT numbers usually increase proximally—or, put another way, increase with stage of development. This apparent anomaly can be interpreted as the calcifying cartilage having higher density than the developing cancellous bone below it.

The scans of the developing trey tine reveal that the stages of differentiation of the tine are not the same as the main beam at the same relative position; the tine differentiates more slowly. The compact bone at the junction between the main beam and tine is relatively weaker, at least during development, than the posterior part of the antler. Clearly, CAT offers important possibilities for the study of the developing antler.

Cranial Angiography

Although the gross anatomy of the blood supply to the growing and regenerating antler has received some study (Waldo et al. 1949; Adams 1979; Suttie et al. 1985c), the relevance of the pattern of blood supply for antler growth has not received any attention. The aim of this study was to use the technique of cranial angiography to trace the pattern of development of the arterial supply during development of the first spike antler and the second branched antler. An attempt is made to relate pattern to function.

Materials and Methods

These are given fully in Suttie et al (1985c). Briefly, 3-month-old red deer stags were fitted with a carotid prosthesis to permit easy, repeated cannulation of the carotid artery (Suttie et al. 1986). At intervals from 3–25 months of age, under general anesthesia, 15–20 mL of 76% radiopaque contrast medium (Urograffin, Schering AG Berlin) was injected manually via the cannula as fast as possible (5–10 seconds). Simultaneously, radiographs were taken of the head and/or antlers. A rapid film changer was used which permitted up to four radiographs per second for 15 seconds to be taken, using XRP1 (Agfa) blue sensitive film.

Results

The antler/pedicle is supplied by the lateral and medial branches of the superficial temporal artery (STA) which arises directly from the carotid artery. Only one branch of the lateral STA supplies the antler (Fig. 9), the remainder supplying only the pedicle or only advancing a short distance up the antler. The medial branch of the STA supplies only the pedicle. The branch of the STA supplying the antler branches itself at the APJ, one branch continuing along the posterior aspect of the antler while the other crosses to the anterior aspect of the antler, before continuing upwards. After the antler is clean of velvet, the STA no longer supplies the antler (Fig. 10). The pedicle continues to be supplied by branches of the artery, however. The branched nature of the pedicle arteries is evident from Fig. 11.

After antler casting, when the growth of the second branched antler begins, it is clear from Fig. 11 that from each branch of the artery in the pedicle, a vessel grows into the new antler. In contrast to the spike antler, the branched antler has several major arteries. The first branch of the STA at the zygomatic arch yields an artery that supplies the posterior aspect of the antler and an artery that supplies the anterior aspect and the brow, bey, and trey tines. The next branch, at the level of the base of the pedicle, supplies the brow and bey tines, and the trey tine and the anterior main beam above the trey tine, respectively. Thus, the arteries supplying these highly evolutionally conserved tines are divided prior to antler growth, and not even in the antler. In contrast (Fig. 12), the arteries supplying the terminal fork branch just before the terminal fork of the antler.

Discussion

Some caution is prudent in attempting to assess the wider implication of the foregoing results, because they are largely based on angiography of the antlers and head of only one stag. However, with this in mind, the antler is supplied by

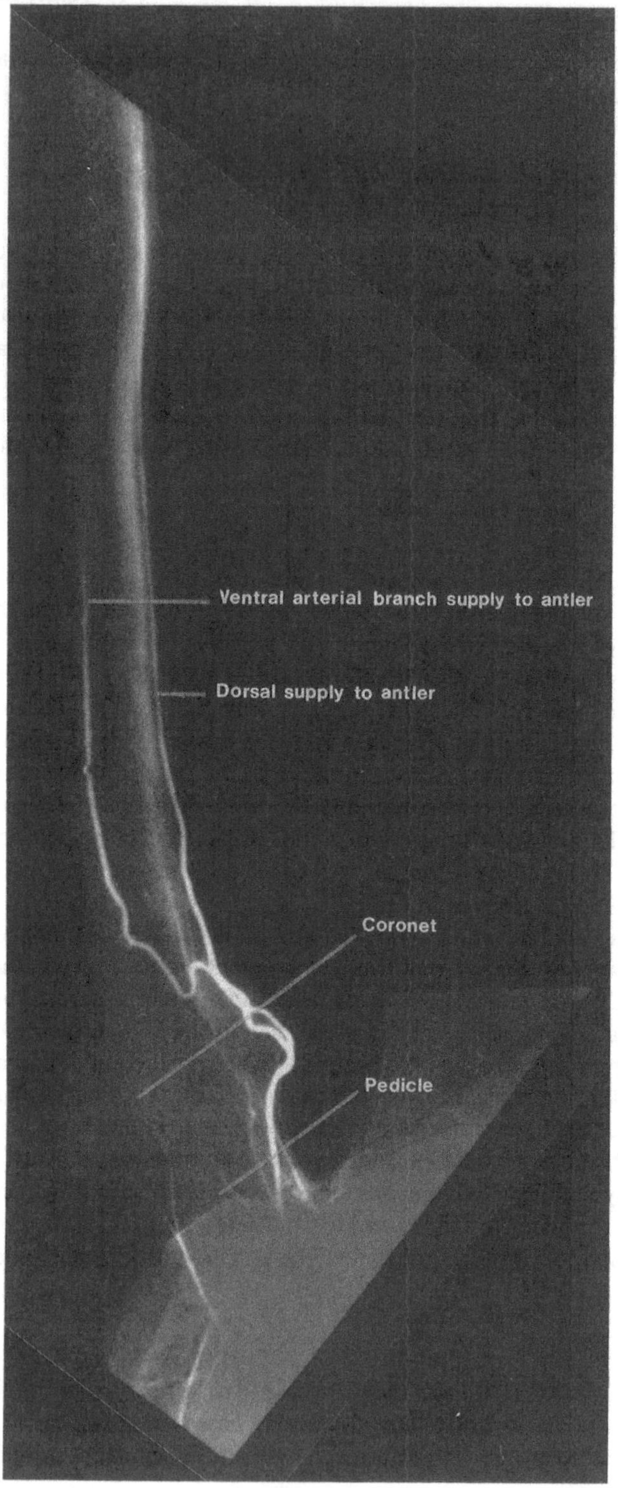
Ventral arterial branch supply to antler
Dorsal supply to antler
Coronet
Pedicle

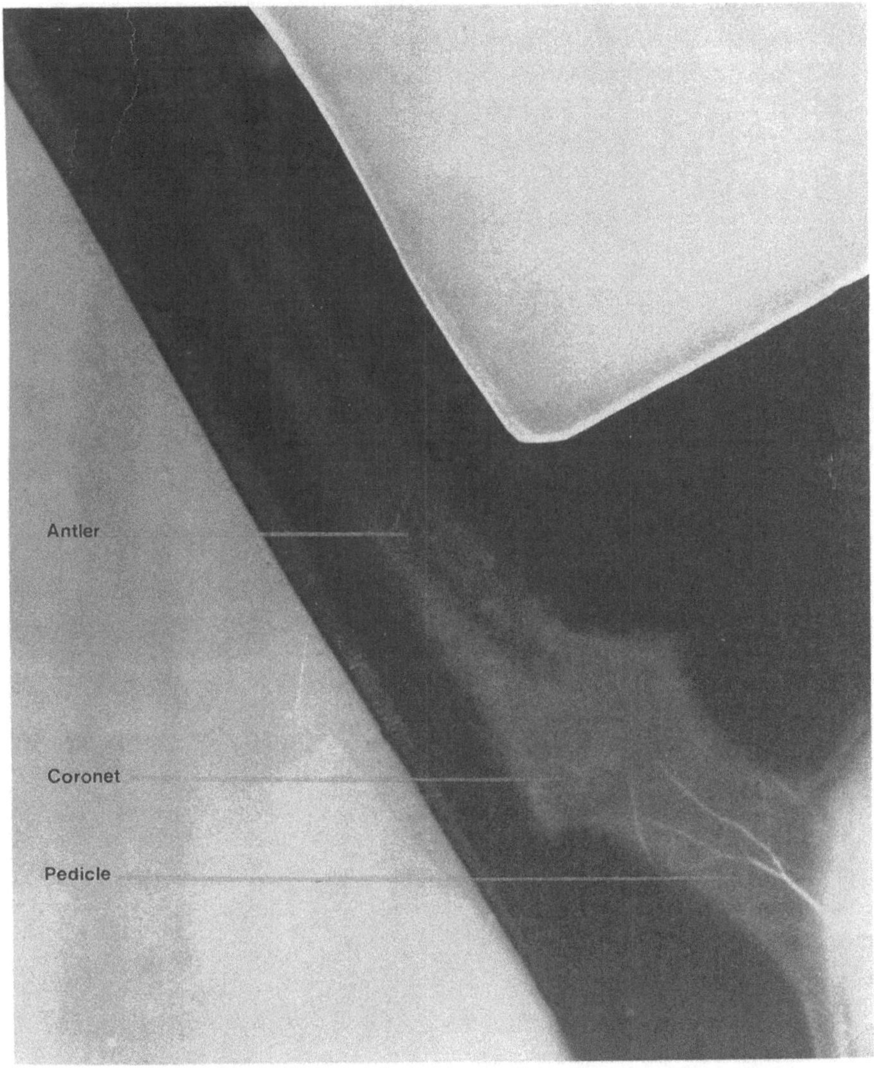

FIGURE 10. Angiograph taken on February 25, 1983, 10 seconds after injection of contrast medium. Antler growth was complete and the antler was clean of velvet; consequently, no arteries supply the antler. Note the branched pattern of arteries in the pedicle.

◄

FIGURE 9. Angiograph taken on November 19, 1982. Midspike antler development 10 secs after injection of contrast medium. One major artery from the pedicle (a branch of the lateral STA, to the right in the figure) branches as it flows round the coronet. One branch supplies the ventral and one the dorsal aspects of the antler. The medial branch of the STA (to the left of the pedicle in the figure) supplies only the pedicle and does not enter the antler.

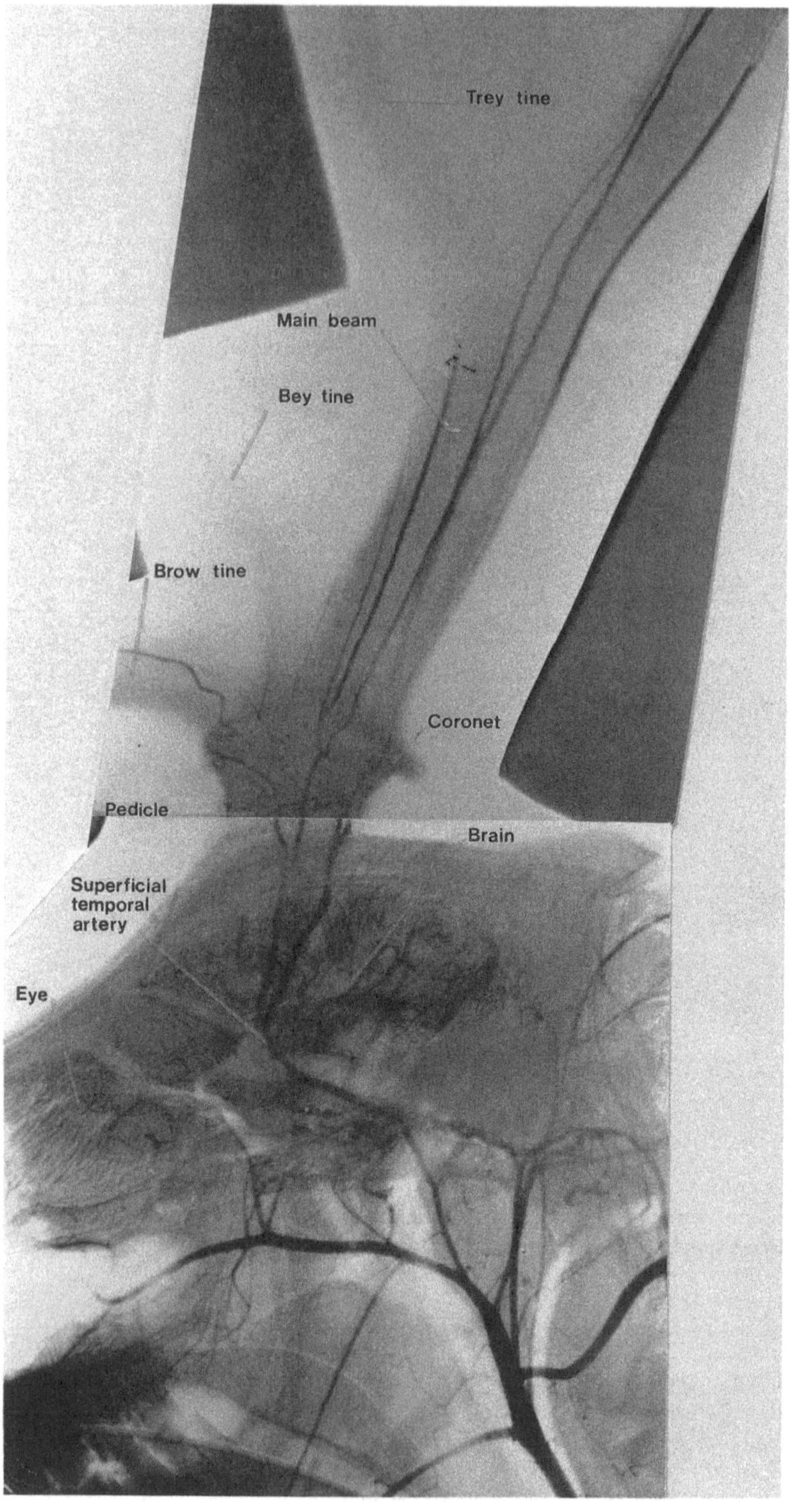
Trey tine
Main beam
Bey tine
Brow tine
Coronet
Pedicle
Brain
Superficial
temporal
artery
Eye

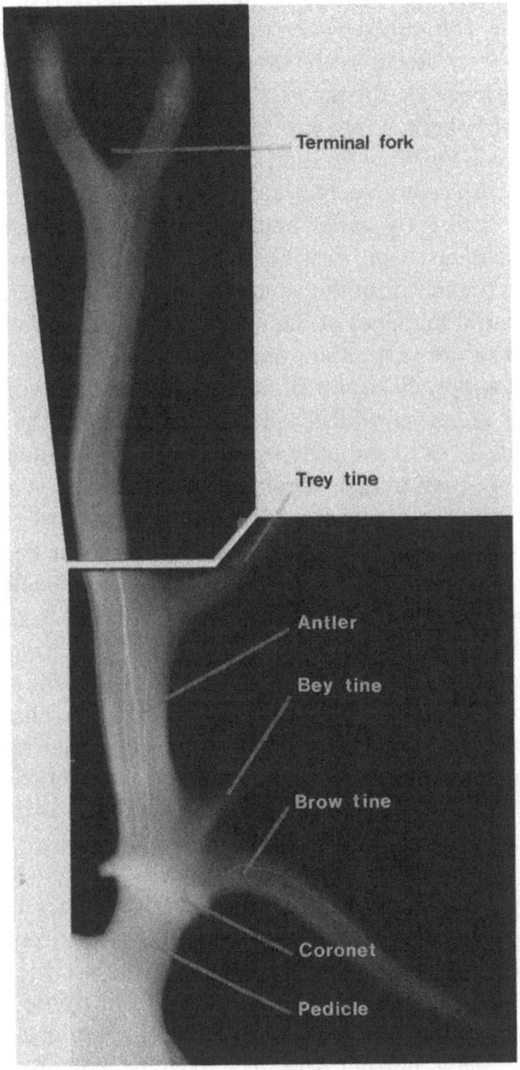

FIGURE 12. Composite angiograph of the antler taken on the same day as Fig. 11 on the same stag. The arteries supplying the terminal fork of the antler branch before the antler branches.

◄

FIGURE 11. Subtractions of angiographs taken in November 1983, when the second branched antler was growing in velvet. The major branches of the artery (the dark lines), to the lower half of the antler, have divided before they reach the antler.

branches of the lateral STA. The spike antler is supplied by one branch, whereas the branched antler is supplied by several. Moreover, the arteries destined to supply the brow, bey, and trey tines are already divided prior to antler growth. Does this have any consequence for theories of antler regeneration?

Antlers invariably branch into two forks during growth. In early growth, the first branch between the brow tine and the main beam is often evident before the previous antler scar is completely healed, say, 2 weeks after antler casting. It is conceivable that such early antler branching requires arteries that are already branched, to invade the tissue, possibly in response to angiogenic factors. As all of the early tines branch from the ventral aspect of the antler, is not surprising that the more ventral branches of the artery supply them. In cases where brow tines fail to grow or are very small, one may ask, has the tine failed to branch, or has the blood supply been insufficient to promote growth? This is rather a chicken and egg situation which can best be answered by asking a further question – namely, do skull and pedicle arteries carry positional information for the antler? Clearly they branch regularly and the number of major branches (four) equals the number of points prior to the terminal fork (where in any case the antler artery branches). Branching is likely to occur because two growth centers diverge; for the antler, the growth centers are likely to be blocks of fibroblasts or generally undifferentiated mesenchymal cells. These mesenchymal cells may secrete angiogenic growth factors, thus drawing arteries which have already branched – however, the fact remains that the "right" number of branched arteries is available for the species-specific branching pattern. It seems the pattern of artery branching could well influence antler branching. If fewer branches are available, then fewer points could be grown. Perhaps if these arteries are artificially stimulated, then more points could be possible. The size of the arteries supplying each tine, by determining blood flow, might partly control tine size by limiting nutrient flow to the developing cells. However, this spatial correlation need not necessarily imply a temporal priority of arterial branching before antler branching. In view of the prevalence of collateral circulation, or the tendency for arteries to enlarge commensurate with the peripheral fields vascularized by them, it is possible that the bud of a tine might be supplied by small arterioles, and as the tine elongates one or more of the arterioles might increase in diameter accordingly. Further work must be done to resolve whether there is indeed a functional relationship between arterial and antler branching; visualization of an already enlarged artery prior to tine formation would provide evidence of such a relationship.

General Discussion

For a general discussion on antler regeneration, a distinction must be made between (1) whether regeneration will occur or not, and (2) the extent of species-specific nature of the regeneration.

Likelihood of Regeneration

Goss (1961), Lincoln (1984), and the present study have shown that antler regeneration is very hard to prevent by merely repeatedly removing portions of the growing antler. That is, regeneration of some sort will tend to occur despite repeated antler or pedicle amputations.

Pattern of Regeneration

The pattern or extent of regeneration after antler amputation or modification is different. Repeated removal of portions of the growing antler in castrated red deer results in slowing of growth rate and unbranched regeneration. Stimulation of the pedicle – by physical trauma or testosterone – induced antler casting in castrates and restored growth rate and shape, but the ultimate size was smaller than would have been expected in an intact stag. This lends support for an argument that (1) the pedicle holds information for the normal species-specific antler pattern, and this can be expressed by trauma, (2) although the velvet and the pedicle both have the capacity to regenerate, this property is more strongly held by the pedicle, and (3) other trophic factors influence antler size. It is not within the scope of this discussion to consider the identity of trophic factors that influence antler size, but they are probably endocrine, paracrine, autocrine, or neurocrine in origin.

The influence of repeated antler removal on the nerve and blood supply to the antler tissue is also relevant. It could just be that the normally "deciduous" antler nerves and arteries have little capacity to regenerate, but those nerves and arteries in the pedicle have a high capacity to regenerate. Reduced regeneration from the amputated antler might then be due to lack of nutrients and/or peripheral or central neural stimulation, rather than indicating a functional disparity in regenerative capacity between antlers and pedicles. However, the healing process typically requires the involvement of both blood vessels and nerves and, as the antler must heal prior to regeneration, it seems that there is ample opportunity for artery/nerve repair or replacement. It is also conceivable that arteries and nerves carry trophic or positional information, and this capacity is lost on differentiation from pedicle to antler arteries or nerves.

A consensus of the above might be that pedicle cells whether bone, nerve or blood are fundamentally different from the same cells in the antler: The cells of the pedicle carry full capacity to regenerate; those of the antler do not. Whether the positional information for the antler is solely vested in the pedicle cells or is centrally located (or both) is, as yet, unknown. It is tempting to reason that, as the regenerative capacity of the antler and the pedicle differ, a local mechanism is responsible for communicating positional information: a central mechanism would have been expected to stimulate antlers and pedicles equally.

The pattern of arterial branching in the pedicle certainly provides many blood vessels to supply the new antler – it is intriguing that even before antler growth the number of major branches of the artery matches the number of forthcoming major early antler branches – despite the fact that the stag in this study had never grown a branched antler before. Whether this effect is due to chance, whether the arteries do carry some positional information, or whether the arteries supply blocks of tissue in the pedicle with positional information must remain speculative. What is clear is that the arteries that supply the major tines are present and do not need to branch in the antler; this is in contrast to the arteries supplying the terminal fork.

In conclusion, antler regeneration is a highly conserved mechanism, although the actual mechanisms governing antler size and shape are more plastic.

11
The Role of the Nervous System in the Growth of Antlers

George A. Bubenik

Introduction

The importance of the endocrine system in the regulation of antler growth, mineralization, velvet shedding, and casting was established a long time ago (for details see Chapter 8). However, there has not yet been an extensive investigation of the role of the phylogenetically younger nervous system, which is more precise in action, more responsive to rapidly changing external and internal conditions, and more coordinated with other body functions.

All individual parts of animal bodies are provided with the nerve supply. The axons of nerve cells, the neurones, grow into their appropriate target organs during embryogenesis or shortly postnatally. According to one hypothesis, each population of neurones is chemically differentiated at early stage of their development. They acquire a distinguishing label, which enables them to recognize a matching label on the surface of their target tissues that they latter innervate. As a development and function of target tissues depends on the nerve supply, in reverse the development and function of the nerve centers depends on chemical information (in the form of trophic factors) provided by the target organ. Without such information, nerve center cannot develop. For example, destruction of one row of whiskers in a newborn mouse results in the disappearance of a corresponding row of neurones in the central nervous system (CNS) (Cowan 1979).

Regenerating antler is an extraordinary organ, as it exhibits morphological and functional characteristics of rapidly developing embryonal tissues in an adult animal. This special property enables us to utilize antler as an excellent model for studies of innervation, two-way communication between target organ and the corresponding nerve centers, as well as for investigation of the role of nerves in tissue regeneration.

The scientific data on the role of nerves in antler growth are still rather modest. Many examples of antler growth alterations ascribed to the effect of nerves are based on observations resulting from various traumas to pedicles and budding antlers. Experimental repetitions of such injuries in immobilized deer were mostly less successful, as the anesthesia blocks nerves and so also the desired effect. Therefore, only partial answers are provided by experiments such

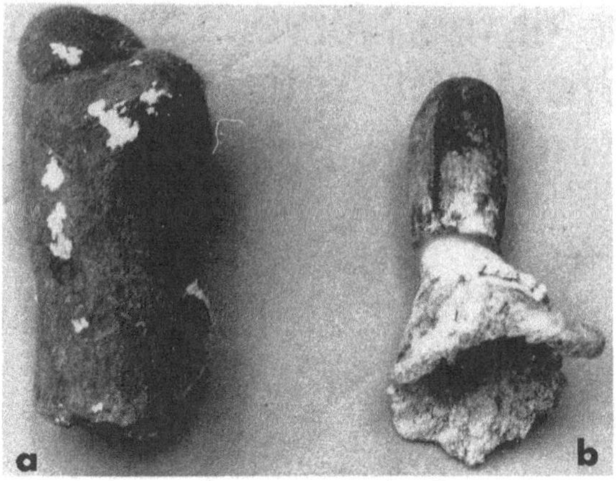

FIGURE 1. *a* In 1980 the right entire pedicle of a white-tailed buck, *Odocoileus virginianus*, Alpha, was fractured. *b* Partly regenerated pedicle on the fractured side was cast next spring (1981), together with a minuscule antler grown there the previous summer.

as the transsections of antler nerves, long-term electrical nerve stimulation, or removal or transplantation of the periosteal antler anlage.

As in many new fields, our data are still sketchy, as we are trying to develop coherent strategies of investigation which would support hypotheses based on only few available pieces of the puzzle. However, as we are investigating a phenomenon which appears to be general to any bone growth and not restricted to antlerogenesis only, understanding of these processes would be much more rewarding than just satisfaction of curiosity of a few interested scientists.

Effect of Injury on Antler Growth

Injuries to antler pedicles or to early developmental stages of growing antlers often result in a temporary reduction of growth rate and alteration of the species- specific shape. The response to injury of pedicles or antlers is more pronounced in developmentally more primitive cervids, such as white-tailed deer, roe-deer, or fallow deer. On the other hand, in more advanced species of deer, such as the red deer, moose, or caribou the response is more limited (for details see Chapter 1).

Although major damage, such as a fracture of the pedicle or a budding antler (Fig. 1a), will usually heal fast, limited antler growth will take place on the injured side during that growing season (Fig. 1b). More limited injury sometimes results in growth of supernumerary points, often in atypical positions (Fig. 2). All these deformed antlers will usually mineralize and cast at the time normal for the species. However, during the next growing season something unusual may happen. The damage suffered in the preceding year is not forgotten. It appears as if

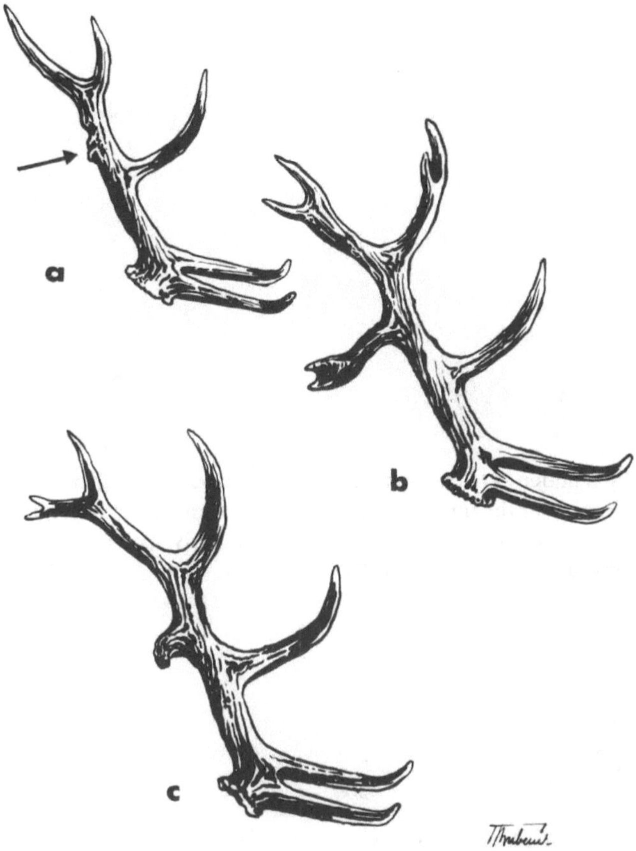

FIGURE 2. Experimental incision on the posterior side of the antler of red deer, *Cervus elaphus*, induced only a slight hypertrophy (arrow) in the first year (*a*), but induced a forked supernumerary point next season (*b*). In the third year (*c*), the response diminished. (From Bubenik, A. 1966.)

the deer may remember his injury and the antler growth in the next season following the accident will reflect the memory of that damage or, more precisely, the memory of the deformation resulting from the healing process.

The massive damage (such as the splitting of budding antlers or a fracture of the pedicle) often result in a remarkable alteration of the antler pattern (Fig. 3), hyperproduction of antler tissues (Fig. 4), or growth of antlers resembling the pattern of another species of cervids (Figs. 5, 6). In most cases the altered shape will persist for years and is extraordinarily similar between antler sets from one year to another. Split points developed as a result of an injury may persist in the same form for 3–4 years. Presence of hypertrophied antlers and royals was

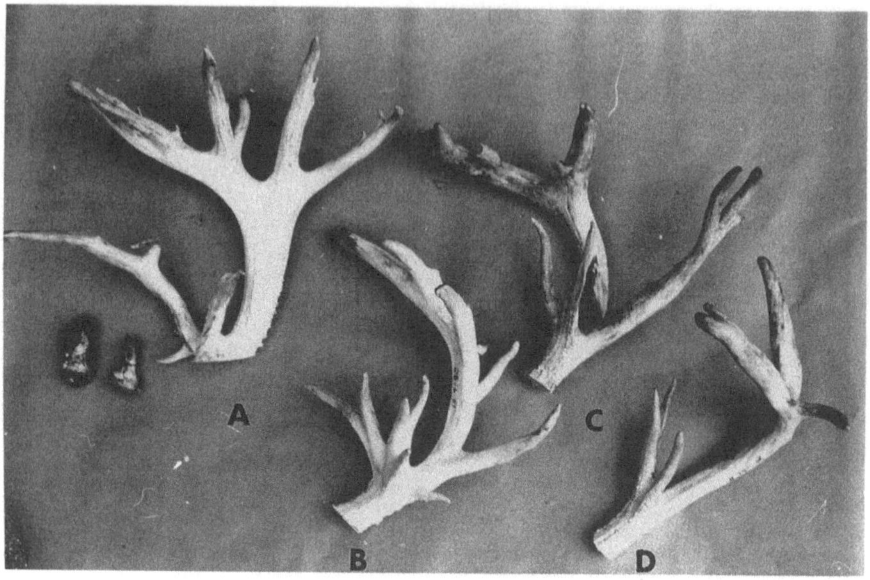

FIGURE 3. Right antler of buck Alpha. *A* Atypical shape 1 year after the pedicle fracture (1981); *B* 2 years after the injury (1982), the main beam started to return to the species-specific shape, except for the first tine, which split into five points. *C,D* Further return to species-specific shape in 1983 and 1984.

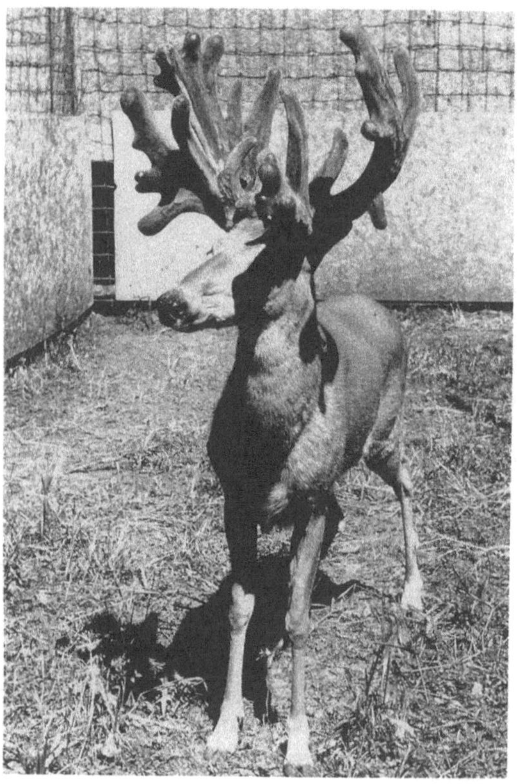

FIGURE 4. The enormous hypertrophy of antlers in the white-tailed buck Billy was the result of an accident in a nonanesthetized animal. The budding right antler was split into three parts. The left antler responded to a smaller degree (see also Fig. 7).

FIGURE 5. Heterospecific antlers of a wapiti deer, *Cervus elaphus canadensis*, from Manitoba, Canada, resembling the palmated antlers of a moose. The exact cause of this aberration in development is unknown.

FIGURE 6. Heterospecific antlers of white-tailed buck resembling antlers of the reindeer, *Rangifer tarandus*. Antler pedicle was fractured on the right side, but the left side responded as well.

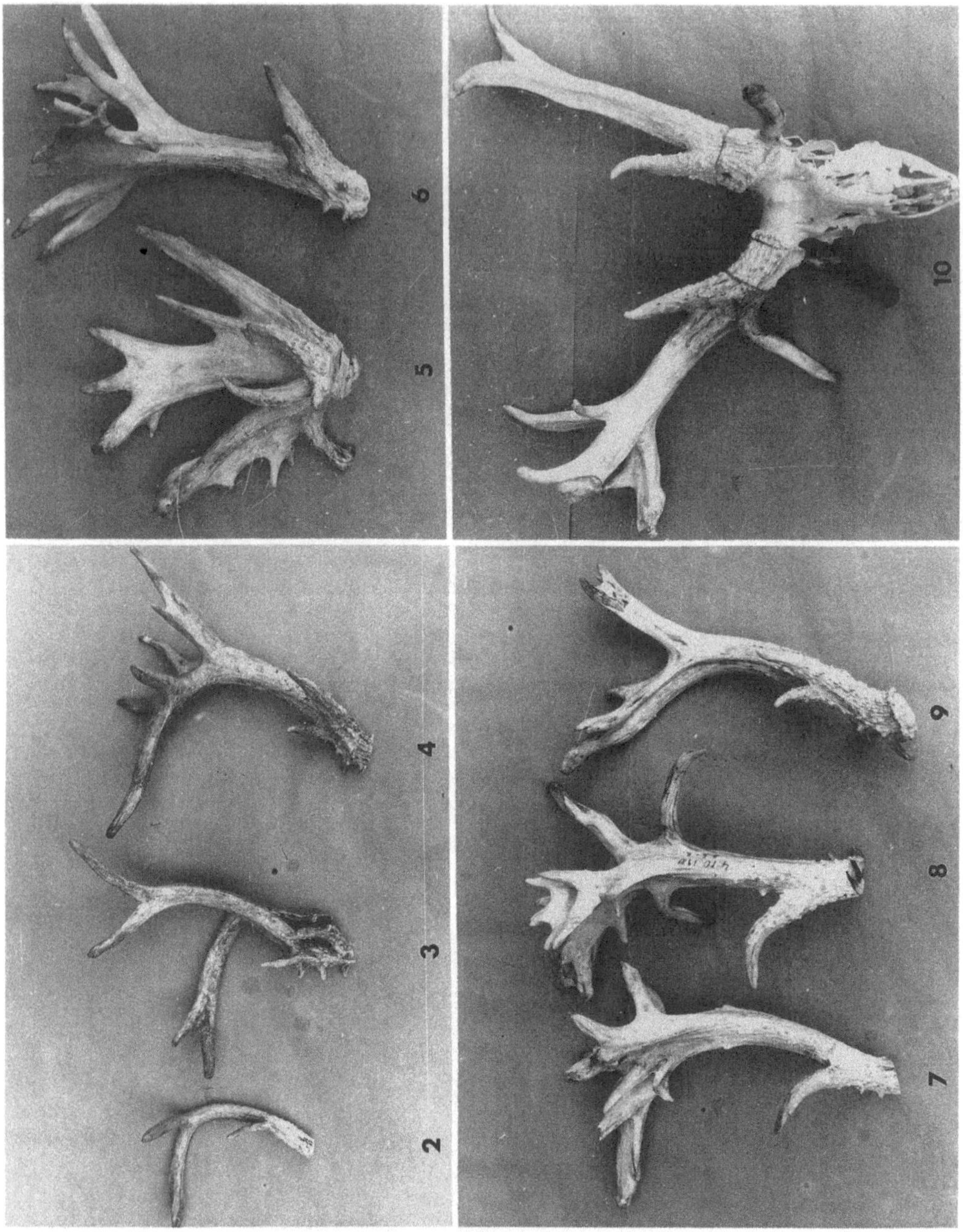

FIGURE 7. Antler architecture of buck Billy, starting from the second cycle (*2*). The first injury (split of the budding antler) occurred in the third year. The second trauma (again, splitting of the budding antler) occurred in the fifth year and caused massive hypertrophy. The bulk of the bone mass that year was located near the antler base. In the following 3 years the bone mass shifted toward the distal regions. The response to the earlier injuries was slowly diminished, but even at the old age of 10 years the buck carried hypertrophic antlers of atypical shape.

recorded even 5 years after the original massive injury occurred in the budding right antler (Fig. 7).

The repetition of the disturbed antler pattern even several years after the original trauma is a remarkable phenomenon, as each year the antler is mineralized, dies, and is later cast. Because the originally misdeveloped body organ (the antler) is discarded each year, the information about the altered shape has to be retained in a nonrenewable area of the body. The information is dormant during the period of hard antlers, until it is reused again during the next antler growth period. Where this information is stored and by what mechanisms it is retained and later reused remain largely unknown. However, detailed, multiannual observations and documentation of accidental injuries in captive deer as well as the results of experimental studies point to the involvement of the nervous system.

The Role of Nerves

The role of the nervous system in the antler's response to injury could be expected. The velvet is heavily innervated (Vacek 1955), and therefore an injury to growing antlers must be painful. A copious nerve supply might be important for antler formation. A complete epimorphic regeneration (replacement of lost appendages, such as the limb in amphibians or the antler in deer) cannot occur or is disturbed in inadequately innervated tissues (Wislocki & Singer 1946; Goss 1984; Suttie & Fennessy 1985). On the other hand, experimentally increased innervation in the limb area resulted in the epimorphic regeneration in amphibian species, where this process would normally not occur (Singer 1964). The first investigation of the role of nerves in the growth of antlers was performed by Wislocki and Singer (1946), who denervated both antlers of white-tailed deer by transecting the antler branches of the trigeminal nerve. This resulted in the growth of severely deformed beams and points. However, as the velvet became desensitized, the denervated deer injured the antlers by hitting fences and other obstacles. Therefore, the authors concluded that the deformation was caused by repeated injuries but that deer can grow antlers without the presence of the nervous supply. As the reinnervation of antlers was not prevented (it is assumed it occurred before the end of the antler growth period) and the effect of the operation on the subsequent antler cycle has not been observed, the results of this experiment was in my view inconclusive.

In a more recent experiment, the role of the nervous system in determining the size and shape of antlers was investigated by Suttie and Fennessy (1985). They reported changes of pattern as well as reduced antler growth on the denervated side, where reinnervation was prevented and success of prevention histologically confirmed. (For details see Chapter 12.) The authors concluded that parasympathetic nerves are not necessary for antler regeneration, cleaning, or casting, but they perform a role in determining antler size and shape.

The results of my experimental studies indicate that in white-tailed deer, *functionally intact* nerves are a precondition for induction of a trophic (growth) response of antlers to injury. Surgical splitting of budding antlers in several totally anesthetized white-tailed deer resulted in a rapid fusion of both halves of beams but failed to produce any changes of the antler shape in first of subsequent antler cycles. On the other hand, the degree of deformation in accidentally traumatized antlers of penned white-tailed bucks which suffered a similar split of budding antlers was directly related to the levels of their consciousness; the most intense trophic response was observed in nonsedated deer (Bubenik, G. et al. 1982b).

The blockade of antler nerves by anesthetics during removal of growing antlers (as performed on hundreds of deer farms around the world to satisfy the market for oriental medicine) might be the reason for lack of trophic response in these deer (see Chapter 12). The timing of antler removal might be another reason why there is no trophic response in deer harvested for growing antlers. The trophic response is progressively diminished with the advancement of development and maturation. The growing antlers are harvested when they reach approximately two thirds of the expected length, and that is usually too late to induce any subsequent trophic response.

Transplantation Studies

Other evidence for the role of the nervous system is provided by experiments where the inductive periosteum of the frontal bone was transplanted. The periosteum from the area which normally gives rise to the pedicle can be grafted to practically any part of the body (e.g., nasal, parietal, and occipital bones of the skull, cannon bone of the leg, skin covering gluteal muscles or anterior chamber of the eye) and will always begin to produce some kind of preosseous or bony tissues (Bubenik, A. & Pavlansky 1965; Jaczewski 1967; Hartwig & Schrudde 1974; Goss 1984; Goss & Powel 1985). (See also Chapter 9.) In one experiment, even interspecific transfer of periosteum on the parietal bone of the nude mouse was successful in inducing the growth of a huge, tumor-like mass of cartilaginous nature (Goss 1983). On rare occasions, a more developed antler tissue (such as a budding antler tip) has been successfully transplanted on the cranial bone of a red deer (Jaczewski 1982).

Despite the fact that some antler tissue which grew from periosteum grafted on different bones underwent most phases of the antler cycle (including velvet shedding and casting), the species-specific shape of antlers was rarely achieved. The further away the transplant was from the original location of the anlage periosteum, the less likely it was to produce a true antler of the species-specific size and shape. The species-specific antler shape was maintained only when the periosteum was transplanted into an area supplied by nerves which normally provide innervation of antlers (Bubenik, A. & Pavlansky 1965; Hartwig & Schrudde 1974; Goss & Powel 1985). The results of transplantation of periosteum and tips of antlers seem to support the idea of the trophic influence of the nervous system on the growth of antlers.

In many respects an antler can be considered a classical target organ (Suttie et al. 1988). Target organ produces trophic factors (such as the nerve growth factor) which stimulate growth and development of nerve cell bodies in the specific area of the CNS. These neurones in turn send axon to innervate the target area (Breedlove 1985, Johnson & Yip 1985).

The stimulatory effect of nerves on growth of skeletal bones is indisputable. Degeneration of nerves in children suffering from poliomyelitis results in a unilateral slowdown of growth rate of the affected limb. A similar detrimental effect of denervation on antler size and form was reported by Suttie and Fennessy (1985). The trophic influence should be weaker at the periphery of the innervation field, therefore, the reestablishment of the connection between the target organ and the corresponding nerve center would be more difficult. In addition, only the anlage (the inductive tissue of the pedicle periosteum) might be predetermined to establish the connection to the nerve center. The differentiated, more mature tissue of the growing antler tip might lack the necessary trophical factors (Johnson & Yip 1985); therefore, successful transplantation of antler tips is rare.

The growth of many classical target organs (such as the secondary sex characteristics) is strongly influenced by sexual steroids (Martin 1978). The antler fits well into that category. The development of antler pedicle in prepubertal males is possible only when levels of blood androgen reach a certain threshold. In addition, all aspects of the annual antler cycle are connected to seasonal variation of sex hormone levels. Because male sex hormones promote differentiation of CNS structures during embryogenesis as well as postnatally (Breedlove 1985), the essential role of androgens in establishment of the deciduous antler growth cycles might be due to their trophic effect in the CNS. (For more details see Chapter 8.) Without the morphogenic effect of androgens, nerve centers responsible for the target area of the pedicle are not able to establish the proper connection.

A crucial experiment for this hypothesis—the denervation of future pedicle areas in prepubertal male deer—has not been performed yet. Therefore, the role of the nervous system in the pedicle development is still undetermined.

Hypothesis of Antler Growth Centers

Each target organ has its own area in the CNS where the neurones innervating the target tissue reside. The so far undetermined area responsible for the neurogenic regulation of antlerogenesis was first postulated more than 30 years ago by Bubenik, A. and Pavlansky (1956, 1959) and named the antler growth center (AGC).

Extensive experimental data summarized in the review of Bubenik, A. and Pavlansky (1965) indicate that there might be two bilateral centers; these could work independently, but they are normally synchronized and regularly interacting. As any other center responsible for regulation of a specific physiological function, it contains as genetically predetermined sequence of commands regulating the species-specific antler growth. However, in addition to that, it also has a flexible component which can respond to any information reaching the CNS via sensory antler nerves. Evidence that the hypothetical AGC on each side of the brain can act independently can be found in numerous reports. Unilateral malformations of antlers, called peruke or wig, (occurring normally only in castrated or hypogondic males) as well as unilateral delay in mineralization or casting has been observed mostly in wild deer shot by hunters (Bubenik 1966).

This unilateral desynchronization of antler mineralization in wild deer could be explained by an inborn lack of androgen receptors in the affected antler. However, a development of unilateral desynchronization of antler cycle was also reported in captive deer after they underwent some surgical procedure or suffered an extensive injury to the head region, pedicles, or budding antlers (Bubenik, A. & Pavlansky 1965; Jaczewski 1982; Bubenik, G., unpublished observation). In these cases the antler cycles on both sides were synchronized before the trauma occurred. Therefore it seems more likely that unilateral disturbance in timing of the antler cycle was the result of traumatic interference affecting AGC. In addition, the neurones of AGC might be affected by various other stressors. A severe psychological stress often resulted in a unilateral delay of antler casting (lasting 2–3 weeks) in deer which previously cast both antlers on the same day (Topinski 1978; Bubenik, G., unpublished observation).

An opposite case, a bilateral effect of unilateral trauma has been documented in captive animals. A severe accidental split of the budding antler on one side resulted in a similar, but less severe, split on the other side (see Fig. 4). Similar splits or occurrence of supernumerary points (see Fig. 6) on the noninjured side was observed in four other white-tailed deer which accidentally injured one of their antlers. The disturbance of antler pattern on the nontraumatized side persisted for several years until both antlers returned to a normal shape.

This phenomenon could be explained by a crossover of innervation. Some medial branches of antler nerves can innervate the opposite contralateral side, and therefore nervous stimulation can reach the opposite AGS. Evidence supporting the transmedial neurogenic stimulation has been obtained in electrophysiological studies. Electrical stimulation of the lateral branch of the antler nerve affected only the ipsilateral antler. However, stimulation of the medial branch of the nerve affected both antlers equally (Bubenik, G. et al. 1982).

Trauma Versus Endocrine System

The trauma is such a powerful stimulus for initiation of antler growth that accidental injuries to the frontal bone area (which normally supports growth of antler tissue in postpubertal males) may result in out-of-season antler growth or antler development in fertile females of cervid species where females are normally antlerless (Blasius 1903; Kierdorf 1985). Similar results also have been achieved by experimental manipulation. A surgical trauma to the periosteum of the pedicle (which had been earlier induced by a postnatal blockade of ovarian

FIGURE 8. Unilateral antlerogenesis in white-tailed doe Sara induced at age 3 by a trauma to the right pedicle which developed as a result of a chemical blockade of ovarian function before puberty. The antler growth was repeated for seven more cycles until the death of the animal.

function) resulted in a unilateral deciduous growth of an antler (Fig. 8) (Bubenik, G. et al. 1982b).

In another experiment an injection of a concentrated solution of calcium chloride, which caused a massive destruction of large area of the frontal bone skin, was such a powerful stimulus, that it induced not only displaced growth of antlers in areas adjacent to the open wound but also induced a bilateral growth of antlers in the intact cows of the wapiti, (*Cervus elaphus canadensis*). The trauma caused by application of $CaCl_2$ resulted in out-of-season growth of antlers during autumn (October to December) (Robbins & Koger 1981). As antlers in wapiti would normally grow only in the spring, it appears that the massive trauma has overridden the hormonally mediated photoperiodic control of the antler cycle. Several more examples of autumn antler growth induced by trauma were documented by Z. Jaczewski (see Chapter 13.).

FIGURE 9. Third antler (arrow) induced by a periosteal traumatization.

Another example of the trauma-induced desynchronization of the antler cycle was observed recently in my captive white-tailed buck. At the age of 3 months, this deer lost the skin on the top of his head by rubbing it repeatedly on wires of the transport cage. The buck developed normal pedicles, but the damaged area was left without skin cover for more than 2 years. This tamed buck became quite aggressive. By rubbing his head against posts he repeatedly reinjured the slowly epithelialized wound area. Finally, during his second antler growth period, the wound became covered by skin. To my surprise, the skin between the antlers was shed during the normal shedding period and a tiny antler appeared between two normal antlers. The miniantler (1.5 cm long, 2 × 2 cm at the base) was cast in mid-March; a new third antler began to grow in its place as soon as the velvet skin grew over the wound. The growth continued quite rapidly, and the third antler reached about 2 cm in length before the two other antlers were cast in the first week of May. At the end of July, lateral antlers reached normal size (around 35 cm), but the median antler was only 10 cm long, exhibiting a slight curvature toward the center (Fig. 9).

The 7-week difference in antler growth activity between tissues localized so close to each other may again point out to a powerful effect of trauma which could

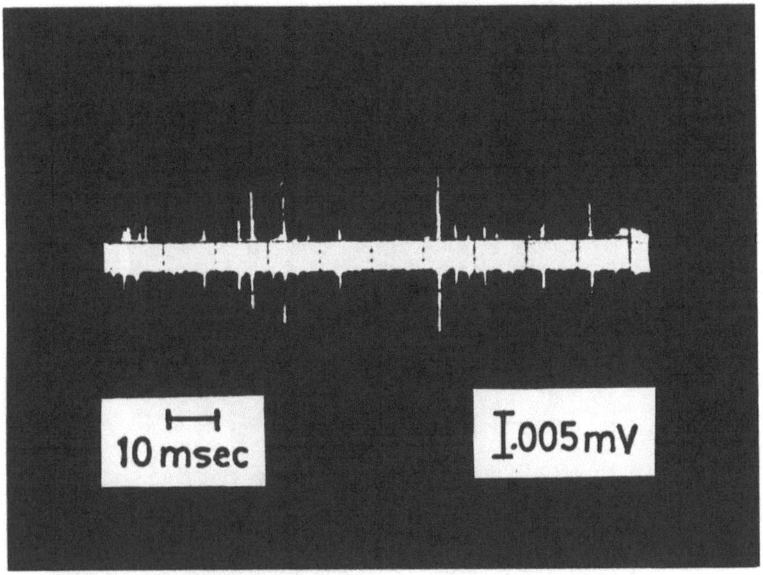

FIGURE 10. Electric recording from the lateral branch of white-tailed deer antler nerve, taken on July 21. (Bubenik, G.A. & E.D. Stevens, unpublished.)

have activated the AGC. In this case perhaps a third AGC has been formed, which induced a premature casting and an early growth of the third antler. The synchronous mineralization of all antler structures indicates that variation in androgen receptor sensitivity is not a probable cause of this rather unusual phenomenon.

The effect of trauma on the induction of antler growth or on the shape of antlers has been also studied experimentally by several authors of this book. For details, see Chapters 9, 12, and 13.

Effect of Electric Stimulation on Antler Growth

Electricity profoundly affects the nervous system of all living organisms (Fig. 10). Several studies utilized various electrical stimulations as means to alter the size and shape of growing deer antlers.

The long-term application of DC current to the growing antlers of a male deer resulted in an abnormal branching pattern and produced antlers that grew in atypical directions (Lake et al. 1982). In addition, electrical stimulation of antler nerves in several white-tailed bucks resulted in severe alterations of antler size and species-specific shape (Bubenik, G. et al. 1982b). Unilateral stimulation of pedicle area by a subcutaneously implanted human cardiac pacemaker did not influence the timing of velvet shedding. However, on the stimulated side, which grew longer than the control side, the spongeous bone of the antler tip was not transformed into an ivory bone, as was the case on the nonstimulated side

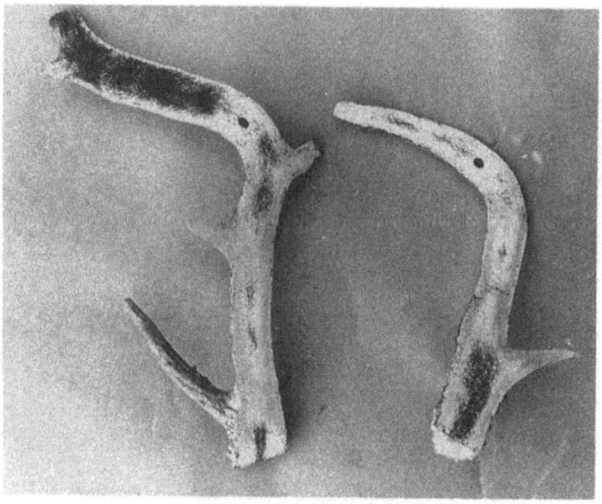

A

B

FIGURE 11. *A* Longitudinal cross-section of the white-tailed antler, which nerves were stimulated with a human cardiac pacemaker (left). Note the massive amount of darkly stained spongious bone in the tip as compared to the ivory bone of the nonstimulated side (right). *B* Atypical antlers in white-tailed buck George in the second year of an antler nerve stimulation (1982). *C* Antlers of George in the next year (1983). Without any new interference, the antlers grew in an almost mirror image of the species-specific pattern (see also Fig. 11d). *D* Top: Cut antlers of George before the beginning of an electric stimulation of the right antler nerve. Bottom: inverted antlers which grow 1 year after the termination of an electrical stimulation.

C

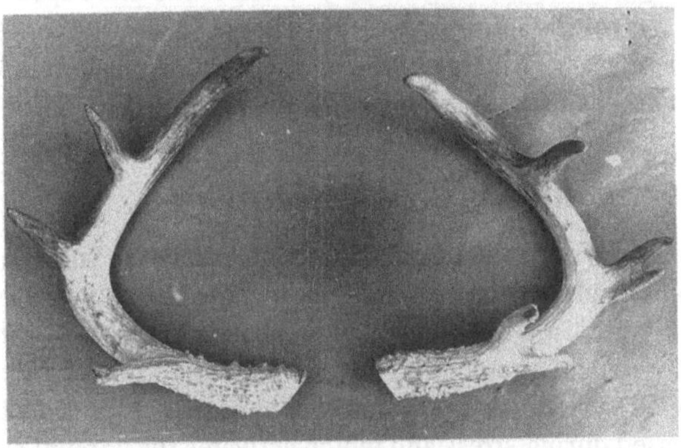

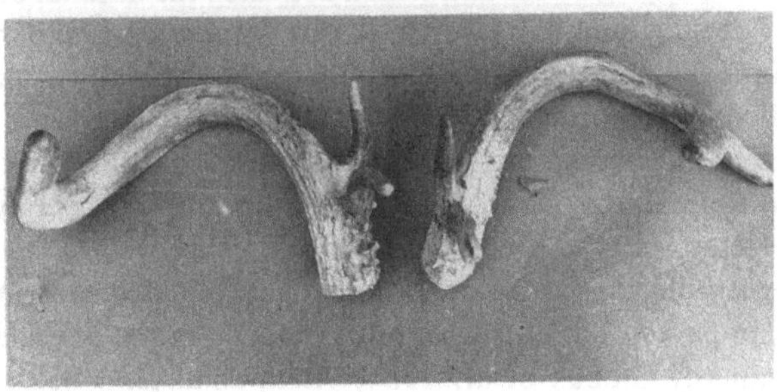

D

(Fig. 11a). In the most interesting experiment, a unilateral stimulation of nerves in the pedicle (performed during two subsequent antler growth periods) resulted in severely malformed antlers (Fig. 11b). However, the surprise came only one year later, when the animal was left to recuperate. Without any stimulation, the antler beams developed a curvature which was opposite to the shape of normal white-tailed deer antlers (Fig. 11b,c,d) (Bubenik, G. & Bubenik, A. 1986). It can be speculated that the deer AGC remembered the disturbed signals coming from the antler area for the 2 consecutive years and so the next season (despite the lack of any trauma or stimulation) the centers guided the growth according to that altered information.

Trophic Memory

The process of delayed response to stimulation resulting from an antler or pedicle trauma was termed trophic memory by A. Bubenik and Pavlansky (1965). The term memory is not ideal (as it involves not a mental process but a somatic one), but nevertheless I have retained this expression because so far no better term has been coined and because there are several similarities between this phenomenon and the memory process: (1) the younger the tissue is developmentally, the greater the response (injuries to pedicle or budding antlers cause the most pronounced alterations of antler shape (Fig. 3,4); (2) the injury must be perceived by functionally intact nerves (hence, no response to injury in completely anesthetized deer); (3) the response is slowly "forgotten," if not reinforced, because with each succeeding year the antler gradually return to the more species-specific pattern (Figs. 2, 7); (4) the more severe the original injury, the longer the memory lasts; and (5) reinjury, which reinforces the memory, causes a massive antler hypertrophy and changes the species-specific pattern (Fig. 7).

Neuroendocrinology of Trophic Memory

The importance of trauma, stress, and intact function of the nervous system in the phenomenon of trophic memory might point out to the neuroendocrine origin of that process. Retention of memory is generally increased in stressful situations. This reinforcement of memory is caused by activation of the hypothalamo-pituitary-adrenal system. A 1–10 amino acid sequence of the ACTH molecule (which is released during stress and which activates the adrenal cortex) has been found to facilitate the memory (Martin & van Wimersma-Greidanus 1979). The noncorticotropic memory-activating sequence is originally a part of a larger molecule called proopiomelanocortin which is produced and stored in the pituitary gland. During stress, several segments of this molecule, such as adrenocorticotropic hormone (ACTH), beta-lipotropin, beta-endorphin, and others are secreted into the bloodstream.

One of these molecules, beta-endorphin, is implicated not only in the suppression of pain perception during injuries but also in the stimulation of regeneration (Vethamany-Globus et al. 1984). Beta-endorphin belongs to a group of brain opiates which might provide the connection between the influence of the endocrine and neuronal systems on the regeneration of tissues. The regeneration potential of vertebrate tissues was found to be highly correlated with the blood levels of beta-endorphin (Vethamany-Globus et al. 1984). The secretion of endorphins is stimulated by pain and acute stress (Young and Akil 1985); therefore, it can be speculated that in deer the nociceptive perception after the injury to antlers is essential to elevate endorphin levels, which in turn will support the antler regeneration. Seasonal levels of beta-endorphin in white-tailed bucks are rising during the antler growth period. In mature bucks, which have bigger antlers than immature ones, the August values were more than twice as high as concentrations in January and February. On the other hand, in young bucks the increase between January and August was only about 50% (Bubenik, G. et al. 1988).

Based on these facts, it can be hypothesized that at least three hormonal substances may be involved in the process of trophic memory: (1) the ACTH "memory sequence"; (2) beta-endorphin; and (3) catecholamines (adrenaline and noradrenaline), which are released during stress and which are also known to increase retention of memory.

Antler–Antler Growth Center (AGC) Pathway

The present data indicate that the growth of species-specific antlers depends on three factors: (1) the presence of the inductive periosteum (this will function in any part of the body, as long as the revascularization is guaranteed); (2) the temporary increase in levels of testosterone (T), which stimulate periosteum to produce the pedicle, the nonrenewable base of the antler; and (3) the presence of specific nerves connecting the pedicle areas with the hypothetical antler growth center, which provide the trophical guidance for the development of species-specific size and pattern (Bubenik, G. 1982).

All three factors must interact simultaneously to secure the development of the pedicle and the growth of the first antlers. The mechanisms by which the first two factors operate in regulation of antlerogenesis were studied to a great extent, and no controversy exists about their presence. However, the role of nerves in antler development is still mostly unelucidated, and the existence of AGC is still largely speculative, based mostly on indirect evidence.

The AGC hypothesis postulated that the connection between AGC and the inductive periosteum is essential for the development of perennial antler growth. Such a connection is possible only when androgen levels are increased beyond a certain threshold. The development of pedicles is permanently blocked if the inductive periosteum is removed before puberty (hence, no sufficient levels of androgens available to stimulate the pedicle development) or the deer is castrated prepubertally (Tachezy 1956; Goss 1964). On the other hand, once the connection is established between periosteum and AGC, the tissue of the pedicle is not essential for the development of antlers. Removal of the pedicle skin, pedicle bone, or even the underlying skull bone (by surgical means or by chemical destruction) did not prevent subsequent antler development. Once the tissues were completely healed, the stimulus for antler development prevailed and new antler growth started somewhere in the area supplied by nerves connecting the original inductive periosteum to AGC. This has resulted in the development of accessory pedicles or growth of supernumerary antlers (Fig. 12) (Bubenik, A. & Pavlansky 1965; Jaczewski 1967; Goss 1964, 1969; Robbins & Koger 1981).

In a case of tissue destruction by chemicals, a permanent blockade of antler growth by administration of $CaCl_2$ was achieved in only two out of eight wapiti calves (Robbins & Koger 1981). As these two animals later exhibited very little

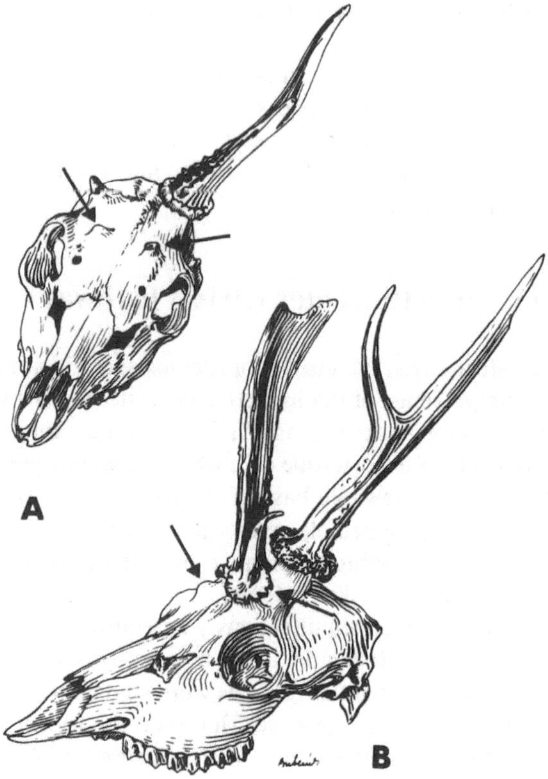

FIGURE 12. Supernumerary pedicles and antlers in the roe buck, *Capreolus capreolus*, formed after surgical removal of the original pedicle (from Bubenik, A. 1966). *A* Year of the operation, *B* 1 year after the surgery. Arrows indicate corresponding structures.

aggression so typical for adult bulls, it can be speculated that these males were hypogonadic and so their level of T in blood during the period of periosteal destruction was too low to initiate connection between the inductive periosteum and the AGC. As there was no connection established, a destruction of periosteal anlage permanently prevented the development of antlers.

AGC Function in Aging Deer

The trophic influence of AGC on the shape of antlers might decline after deer have passed their prime age. The branching pattern is becoming more and more irregular and the direction of beams might turn highly atypical (Fig. 13). It has been speculated that these malformations are due to lack of T, as the average levels of this androgen in blood decline in seniors (Bubenik, G. & Schams 1986). However,

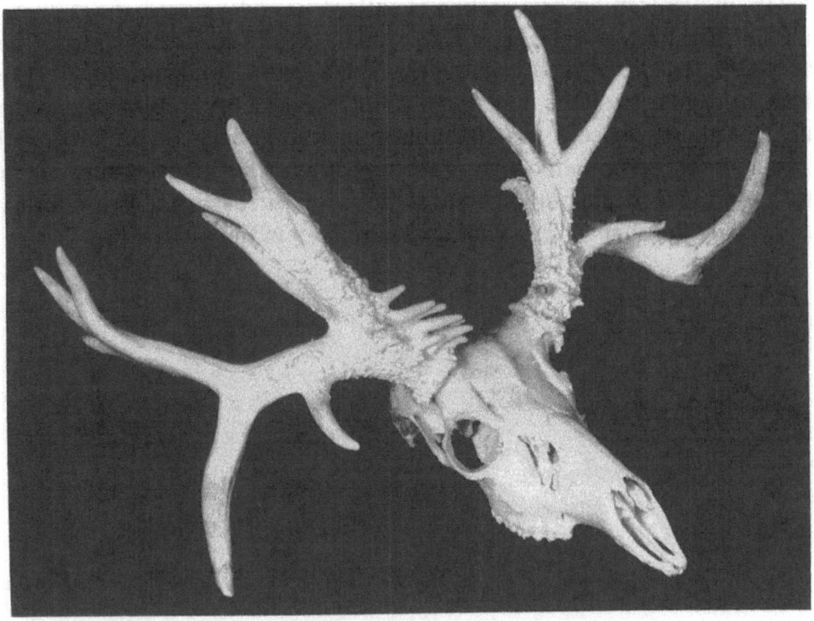

FIGURE 13. Highly irregular shape of the antler architecture in the 10-year-old buck Milan. Note the split of both beams growing laterally, supernumerary points, and hypertrophies of the beam surface above the pedicle, known as pearls.

recently a severe alteration of antler form has been also detected in overaged bucks with plasma T values in the normal range (Bubenik G., unpublished observation). In view of this fact, it appears that the previously mentioned hypothesis (Bubenik, G. 1982) needs to be re-evaluated. The numbers and sensitivity of T receptors can decline with age. It is therefore conceivable that AGC receptors of aging male deer are not sufficiently stimulated during the rut by circulating T. As was explained in Chapter 8, it is hypothesized that, in order to perform its trophical function, AGC requires repetitive seasonal stimulation by T. Therefore, it might be possible that the peripheral levels of T in deer could still be in the almost normal range, but the stimulation of AGC is not sufficient. Without the necessary stimulus, the shape of antlers will not follow the typical species-specific pattern.

Conclusion

The hypothesis of AGC has not been directly verified, but indirect evidence presented here make it plausible. Involvement of the nervous system is indicated by such phenomena as: (1) out-of-season antler growth; (2) an asynchronous growth, mineralization, and casting of antler beams developed after trauma; (3) repeated, trauma-induced specific abnormalities, such as extra tines on beams which are found repeatedly in several cycles after original injury; (4) contralateral responses on the noninjured site; (5) blockade of response to trauma in fully anesthetized deer; and (6) unilateral, trauma-induced antler growth in females. As the nerve cell bodies of the trigeminus (the branches of which innervate antlers) are located in the brain stem, it can be speculated that information determining the shape and size of antlers resides in the same area.

12
Experimental Manipulation of the Neural Control of Antler Growth

JAMES M. SUTTIE

Introduction

The permanent pedicles and growing velvet antlers of deer are richly supplied with nerves. Wislocki & Singer (1946) pointed out that antler nerves were collaterals of nerve fibers from the pedicles, which were destroyed and regrown annually. This means that regenerating nerve fibers grow at up to 1 cm/day in red deer stags during the most active period of antler development.

This review considers direct manipulation of the nerves supplying the pedicle and antler and indirect manipulation by geophysical and traumatic forces.

Innervation of the Antler

The parasympathetic supply is derived predominantly from two terminal branches of the trigeminal (5th cranial) nerve. The supraorbital branch emerges over the dorsal rim of the orbit about 2 cm (in adult deer) from the medial canthus of the eye. The branches of the nerve pass through the orbicularis oculi muscle to supply the frontal region of the head. Major branches run caudomedially to innervate the rostral and medial aspects of the pedicle and antler. The zygomaticotemporal nerve travels caudolaterally through the periorbita to emerge from the caudal margin of the zygomatic process of the frontal bone. The nerve than passes caudodorsally through the retroorbital fat and beneath the frontalis muscle, being positioned ventral to the frontal crest and dorsal to an accompanying artery and vein, before dividing into several branches which disperse on the lateral and caudal aspects of the pedicle (Adams 1979). In some deer nerve fibers from the zygomatic branch of the auriculopalpebral nerve, itself a branch of the facial (7th cranial) nerve, innervate the medial aspect of the pedicle and antler (Adams 1979; Suttie & Fennessy 1985). The sympathetic supply to the pedicle comes from the second cervical nerve (Adams 1979; Rayner & Ewen 1981). The antler normally lacks sympathetic innervation (Wika 1980), the exception being the long-term velvet-antlers grown by castrated male deer (Rayner & Ewen 1981).

Direct Manipulation and Measurement

Antler Nerve Stimulation

Lake et al. (1982) applied a cathodal direct current to antlers of mule deer during the entire antler growth phase. The current ranged from 0.06–3.0 μA, with a voltage of less than 1 volt. Treated antlers were smaller than control antlers, showed abnormal branching patterns, and grew in atypical directions. Further, in three of the four treated deer, premature antler branching occurred; i.e., the antlers did not grow sufficiently to attain the species specific form. The authors concluded that long-term constant low-level DC application caused a reduction in longitudinal growth and altered normal morphology and direction of antler bone growth. In contrast, Bubenik, G. et al. (1982b) considered that Lake et al. (1982) had overstimulated the antlers and that this has been responsible for the reduction in bone growth. Bubenik, G. et al. (1982b) carried out a series of three experiments each on one white-tailed buck; in Experiment 1 they stimulated the lateral, and in Experiment 2 the medial branch of the supraorbital branch of the trigeminal nerve with square wave pulses (0.9 V, 1 msec duration, 175 pulses/min) while in Experiment 3, they used a cardiac pacemaker (5.4 V, 0.7 msec duration, 75 pulses/min) to stimulate the antler periosteum. Electrical stimulation of the lateral branch of the trigeminal nerve stimulated longitudinal antler growth and increased the number of points compared to the control unstimulated antler. The specific gravity values of both antlers from the stimulated buck were reduced compared to unmanipulated control bucks (1.27 for the stimulated, 1.32 for the control, and 1.5 for normal bucks). The stimulated antler was the normal shape for the species. In contrast, the buck in which the medial branch of the nerve was stimulated grew a pair of antlers that were of abnormal shape for the species; that is, stimulation of the nerves to one antler resulted in effects to the contralateral side. Both antlers were about the same size and of similar specific gravity to controls. It may be speculated that stimulation of the medial branch of the nerve to the antler was carried to the contralateral antler either by branches of the nerve crossing the midline or by conduction of the stimulatory current through the skin. Stimulation of the periosteum of the pedicle resulted not only in an increase in antler weight, length, and number of points on the stimulated side, but a gross departure from the species-specific shape and delayed antler mineralization. The stimulated antler grew at right angles to the normal (i.e., outward and downward rather than upward and inward). Remarkably, in subsequent years, despite lack of further stimulation, both antlers (previously stimulated and previously unstimulated) grew abnormally with the antlers curved downward rather than upward, but whether this was an effect of the electrical stimulation is not known.

Taken together, these results show that moderate electrical stimulation of antler nerves results in an increase in antler size and departure from the normal species-specific antler shape. Instances where contralteral antlers have been stimulated indicate that either nervous communication exists between antlers or the stimulation has jumped from one to the other. If long-term poststimulation effects do occur, as is suggested from Experiment 3, it is probable that they reflect large-scale damage to pedicle nerves by the electrical treatment.

Antler Denervation

In the above section it was noted that stimulation of nerves leading to the antler resulted in larger antlers and a disruption of normal antler shape. What, then, are the consequences of depriving the antler of parasympathetic innervation?

Wislocki & Singer (1946) unilaterally exposed and sectioned the supraorbital and zygomaticotemporal branches of the trigeminal nerve leading to the antler in two adult white-tailed deer bucks before antler growth had begun. The control antler regrew normally, but the denervated antler was stunted and deformed because the bucks repeatedly injured the denervated antlers by rubbing them against foliage. The authors concluded that the reduction in antler size and the deformation were consequences of the rubbing due to insensitivity rather than a direct growth-reducing result of antler denervation. The bucks cleaned both antlers at virtually the same time, and there were no significant differences in casting date. This study shows that the trigeminal nerve carries sensory fibers to the antler but does not shed any light on whether nerves actually stimulate antler development.

If the single-spike antlers grown by yearling red deer stags are removed while they are still in velvet, a branched antler similar to that of a 2-year-old stag will often regrow, during the same season, from the cut stump. To test to what extent an intact nerve supply was necessary for this effect, Suttie & Fennessy (1985) used six 9-month-old stag calves, penned indoors and fed to appetite. At intervals from October-December (Southern Hemisphere), according to stage of growth, with the stags under general anesthesia, the antlers were unilaterally denervated using a method similar to Wislocki & Singer (1946) and both velvet-antlers were removed. Success of denervation was demonstrated histologically. Recovery from anesthesia and subsequent healing were uneventful. All stags regenerated their antlers, and on the day after antler-cleaning the stags were castrated to induce premature antler casting. The cast antlers were weighed and measured and the specific gravity determined. After antler casting, exploratory surgery was performed under general anesthesia and pieces of tissue were removed for subsequent histological examination, to determine the extent of any nerve regeneration. At the same time the antlers on the denervated side were removed without local anesthesia; significantly, one of the six stags reacted to amputation despite recumbance under general anesthesia. Subsequent histological examination of pieces of tissue removed at surgery revealed successful denervation in all six stags. It is likely that the antlers of the stag which reacted were innervated by a branch of the facial nerve—a rare individual variation in red deer—and consequently his data were omitted from further analysis along with those from another stag which had severely damaged his denervated antler, presumably due to lack of sensitivity.

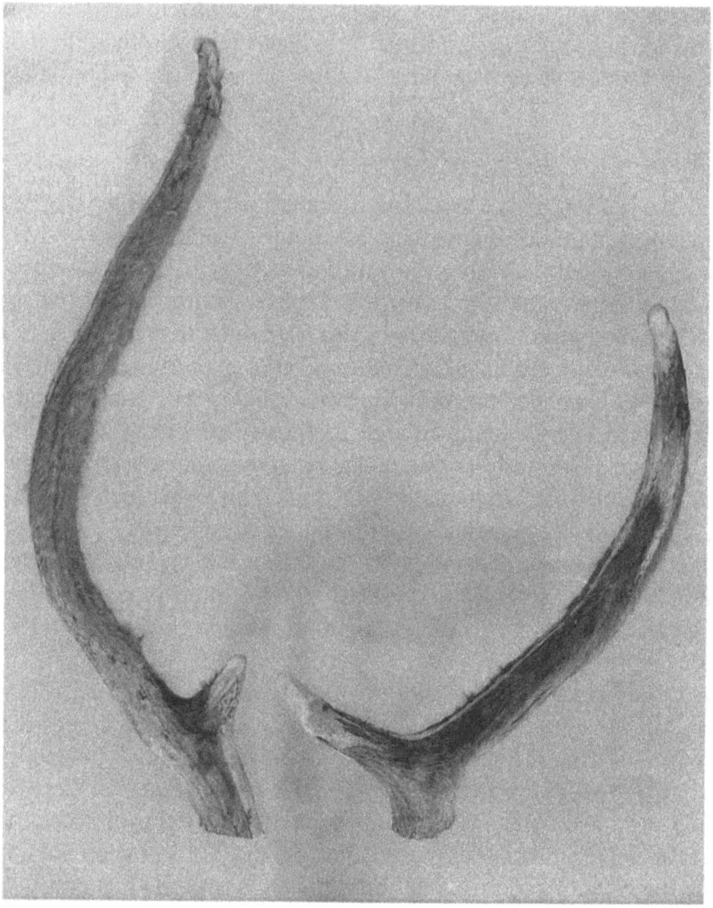

FIGURE 1. Hard antlers grown by a 1-year-old red deer stag after denervation of the left antler and amputation of both previous antlers. Branched regrowth has occurred from both antlers. Although the left antler, to the right of the figure, is shorter than the right control, the brow tine is longer.

No substantial regeneration of nerve tracts took place, though small cutaneous fibers were located on final examination of the stags. The results are summarized in Table 1, and show that the denervated antlers were significantly smaller but of similar specific gravity to the control side, in marked contrast to the predenervation antlers where length and weight were very similar. The shape of the denervated antlers was also altered: the denervated brow tines were relatively longer and the main beams relatively shorter than the controls; that is, the main beam was less dominant in the denervated antlers (Fig. 1). The denervated antlers had fewer tines (1.3 ± 0.22 mean $\pm$ sem) compared to control antlers (1.6 ± 0.35), but this was not significant. Both antler-cleaning and casting were unaffected by

TABLE 1. Primary data and mean ± standard error of the mean for the six pre-experimental amputated segments and four experiment antlers.[a]

		Pre-experimental amputated segments of antler			Experimental antler		
		Control	Subsequently denervated	Difference	Control	Denervated	Difference
Antler length (cm)	$\bar{x}\pm$sem	22.0±3.4	20.9±2.8	1.08±1.08	35.5±5.4	24.3±5.5	11.3±3.4*
Antler weight (g)	$\bar{x}\pm$sem	105.7±19.6	94.6±20.1	11.1±10.6	196.5±34.0	126.5±42.2	70±24.3*
Antler specific gravity	$\bar{x}\pm$sem	1.07±0.01	1.06±0.01	0.003±0.02	1.39±0.04	1.36±0.07	0.035±0.031

[a] Statistical comparisons (two-tailed paired t test) are between the differences in the means.
All antlers were amputated to 5 cm above the head.
*$P<0.05$, otherwise differences are not significant.

TABLE 2. Antler growth rate and electric potential. Retabulated from Lake, Davis and Solomon 1982.

	Early (Apr-May)	Mid (Jun-Jul)	Late (Aug-Sept)
Antler growth rate mm/day	3.4 ± 0.4[a]	4.0 ± 0.3	0.9 ± 0.2
Electric potential mV	4.9 ± 0.7	−8.2 ± 1.4	−0.1 ± 0.6

[a] Mean ± standard error of the mean.

antler denervation. This study showed that although parasympathetic nerves are not necessary for antler regeneration, cleaning, or casting, they nonetheless perform a role in determining antler size and species-specific shape.

Although the antler itself lacks sympathetic (adrenergic) innervation, fibers have been located histologically in the pedicle. The antler can be sympathectomized by removal of the superior cervical ganglion (SCGX). However, Wika & Krog (1980) found that neither unilateral SCGX nor preganglionic cutting of the cervical ramus ganglionarus affected antler growth in reindeer. Lincoln (1984) found that although bilateral SCGX delayed antler growth during the first season after surgery (as the pineal gland had been denervated), there were no obvious abnormalities of antler size or shape.

It thus appears that only the parasympathetic nervous system exercises any control over growth of the antler. Denervation results in shorter, lighter antlers of the same specific gravity, with fewer points and with an altered shape, both in terms of relative size of points and their orientation. This is consistent with the postulate of Goss (1965b), who pointed out that if the eventual function of a structure does not require nerves, then likewise regeneration may be expected to be independent of innervation.

Bioelectricity and Antler Growth

Bioelectric potentials have been measured from a variety of regenerating tissues. In particular, amputation of urodele limbs evokes a large positive peak of voltage of +30 mV. This is followed by a change to −30 mV over the next 5 days. Potentials remain negative until regeneration is complete after about 40 days (Becker 1961). As the antler is an actively regenerating organ, knowledge of the bioelectric potentials associated with growing antlers could be valuable. In this respect, Lake et al. (1978) and Lake et al. (1979) have measured the potential difference between a reference electrode placed on the medial aspect of the antler base and a recording electrode placed on the top in mature male deer (80 recordings on 15 animals) under anesthesia. Clearly, rapid antler growth is associated with significant negative bioelectric potentials (Table 2). In a further study in July, the electric potentials were measured in 14 mule deer between the antler base as a reference and points along the growing antler (Table 3). The antler became significantly more negative with progression toward the tip.

TABLE 3. Potentials found along the antler in July. Retabulated from Lake et al. 1982.

	Position of recording electrode from reference electrode in relation to total antler length			
	1/4 distance	1/2 distance	3/4 distance	tip of antler
Potential mV	+0.6	+0.3	−2.3	−11.1

Taken together, these results show that electronegativity increases as antler growth rate increases, and decreases as antler growth rate declines, while less differentiated antler tissues at the tips are more negative than more differentiated tissues at the base. It is highly likely that these biopotentials play a significant role in initiating regeneration. Such biopotentials in regenerating urodele limbs are skin-driven and occur in the absence of nerves (Borgens et al. 1977). While electric potentials have not been measured in the absence of innervation in antlers, neither have electrical potentials been measured around the time of antler casting, healing, and early regeneration. It is hypothesized that electric potentials will be independent of nerves in the growing antler and that depolarization will take place around antler casting similar to that which occurs in an amputated urodele limb. Another possibility, which leads to an alternative hypothesis, is that the casting of a dead antler may not affect electric potentials, but cleaning of the velvet from the antler at the time of its death during the previous autumn might well cause depolarization. For this hypothesis, the stimulatory effects of depolarization for regeneration would have to be suppressed until antler casting. As casting is under hormonal control, can we invoke an endocrine control of biopotentials?

Leg Amputation

There is anecdotal evidence that the amputation of portions of hindlegs of stags leads to permanent antler asymmetry with the antler contralateral to the amputated leg being smaller and of abnormal shape (Goss 1983). Marburger et al. (1972) amputated the left hindleg of a 4.5-year-old white-tailed buck, who was in hard antler, 7.6 cm below the metatarsal joint. Subsequently, two similar bucks who previously had been subjected to partial tibial neurectomy were amputated in the same way. A single control was retained untreated. All bucks whose legs were amputated grew antlers on the contralateral side which were small and abnormal in shape. The tibial nerve sectioning had in itself no effect on antler shape or size but exacerbated the effect of amputation.

Davis (1982) reported two cases concerning the influence of hindleg amputation on antler growth in a sambar stag and in an Indian muntjak. The sambar stag had his leg amputated at the metatarsus as a result of an accident before he began any antler growth, and although both antlers grew, cleaned, and were cast in synchrony, the right was shorter and lighter than the left and lacked the characteristic shape of the species. The average ratio of right/left over the 5 years of the stag's life was 0.37 for length and 0.31 for weight. For comparison, Suttie & Fennessy (1985) found that the ratio between denervated and control antlers in red deer was 0.69 for length and 0.59 for weight; thus it would appear that leg amputation may have had a more serious effect than specific antler nerve ablation on antler size. In the Indian muntjak which had the right tibia removed 2 months after birth, not only were its antlers shorter but its pedicle was also shorter on the left (contralateral) side.

What then are the lessons (if any) for the study of antlers which can be drawn from the effects of leg amputation on antler size and shape? Some authors (e.g., Fowler 1894) have concluded that antler abnormalities and leg damage are both common in deer and that it is natural for some deer to carry both injuries. Others, such as Morrison-Scott (1960), suggest that a deer licking wounds leaves the contralateral growing antler exposed to danger. Neither of these conclusions explain why perfectly shaped, albeit smaller, antlers are produced by leg amputees; the antlers are not obviously broken or damaged. Although caution is necessary, the following hypothesis might explain how the effects of amputation are mediated. Ascending spinal nerve pathways carrying pain impulses from peripheral tissues synapse with neurons in the grey matter of spinal cord. These axons cross the grey matter to the opposite side of the spinal cord. This axon then travels to the thalamus and synapses with neurons in the brain, which go to the cortex. If these neurons carrying painful stimuli from an amputated hind limb in any way influenced the trigeminal nerve innervating the antler, they would certainly influence the antler contralateral to that in the injured side. The questions remain, however—could communication between a sensory spinal neuron from the hind limb and a cranial nerve (part of the autonomic nervous system) take place at all, and could this communication result in antler size and shape abnormalities?

Indirect Manipulation

Geophysical Forces

Although the left and right antlers grown by deer are normally grossly symmetrical, the reindeer and caribou represent a departure from this norm. The brow tine of this species is a prominent palmate projection which normally only develops on one antler, the other being a simple narrow tine. However, the larger tine does not consistently grow on the left or the right antler, either between individuals or within individuals between years; it does not appear to be a genetically transmitted condition (Goss 1980; Davis 1982). In the Northern Hemisphere, in a survey by Goss (1980), 53% of deer had the left brow tine enlarged, 30% the right side, 15% had both enlarged, and the remainder had neither enlarged. Whereas in reindeer introduced to South Georgia in the Southern Hemisphere, 27% had the left brow tine enlarged, 39% had the right brow tine enlarged, while 9% had both and 26% had neither brow tine enlarged. Thus, the left tine appears dominant in the Northern Hemisphere and the right dominant in the Southern. These observations have been used as evidence that geophysical forces operate on the antler (Davis 1983).

Trauma

There have been few, if any, scientific studies, using an adequate number of animals, concerned with the effects of trauma to the pedicle or antler on subsequent antler development. In addition, it is very difficult to separate the effects of damage to pedicle nerves from damage to pedicle blood vessels, skin, and bone, on antler development. Much of the literature on this subject relies on anecdotal or one-time experiments, which tend to make interpretation of any results difficult.

The Pedicle

Although the pedicle is indispensible for normal antler growth, there is abundant evidence that antlers, albeit of abnormal size and shape, may grow in male or female deer lacking pedicles and hence pedicle nerves (Goss 1961; Robbins & Koger 1981; Bubenik, G. et al. 1982b). Nonetheless, pedicle nerve stimulation and ablation (as has been considered in previous sections) alter antler size, shape, and departure from the species-specific pattern. Thus, two facts seems clear: nerves do not play a role in antler regeneration but they do play a role in determination of size and shape. Although pedicle damage may result in abnormal growth that season and perhaps for several seasons to come, homeostatic mechanisms tend to return the antler size and shape to the species-specific norm as soon as possible.

Antler

In that antler nerves are regenerated annually, it seems unlikely that permanent antler damage can occur due to trauma to a single antler. However, there is a possibility that antler damage can have longer-term effects. Bubenik, A. & Pavlansky (1965) cut the dorsal portion of the antler tip off a red deer stag when the antler was about 40% of its expected final length. Although the injury did not arrest growth, a slight scar remained at the wound site. In the next two antler cycles (but not the third), an additional dorsally pointed tine grew from the site of the former injury but all other tines grew normally in the species-specific way. The authors concluded that this was evidence for a memory of antler injury. They hypothesized that the antler trauma had effected semipermanent changes to a center of antler stimulation that was sited in the central nervous system (CNS). Bubenik, G. et al. (1982b) have extended this hypothesis. They observed the consequences in successive antler cycles of an injury to a growing antler of a white-tailed buck who accidentally split his growing antlers into three parts against a metal grid. The originally split antler became massively hypertrophied, as did the

FIGURE 2. A stag showing regrowth of velvet-antler, about 40 days after velvet-antler removal. The velvet-antler was removed 65 days after antler casting. Three weeks after velvet-antler removal, the stump had healed and regrowth around the lateral edge took place.

uninjured antler. In subsequent antler cycles this gross hypertrophy was maintained. Experimental induction, under deep anesthesia, of injuries similar to those sustained by the buck above failed to reproduce the hypertrophy. The authors claim that the damage to the putative antler growth center in the CNS is proportional to the state of consciousness of the deer, i.e., the pain felt. They state that if the damage to the antler is so severe then the antler growth center (AGC) will remember the effect and perpetuate it. No mention is made of possible anatomical or physiological mechanisms for this effect. The hypothesis—an interesting one—must remain unproven. The case may be that white-tailed deer differ from red deer in their responses to velvet trauma, but any mechanism requires further study.

Each year thousands of stags in New Zealand have their antlers removed approximately 60–70 days after antler casting or when they are about two thirds grown. This is to satisfy an oriental market for the tissue's medicinal properties. Antler removal, which is carried out under the supervision of a veterinarian, is always done on stags whose antlers are rendered insensitive to pain with anesthetics. A tourniquet is applied around each pedicle and a butcher's bone saw is used to cut off the growing antlers some 2 cm above the coronet. Some bleeding may occur when the tourniquets are removed. The antlers frequently regrow (Fig. 2), often only from the lateral edge, where the major arteries and nerves are located. Velveting, as the process is called, does not interfere with cleaning or casting. It might then be thought that such repeated antler ablations would ultimately result in adverse effects in antler size and shape. This does not appear, however, to be the case. The Chinese have carried out repeated velvet-antler removal for centuries, and although until recently no anesthetics were used, no deleterious effects on subsequent antler growth have been noted.

Although no thorough studies have been carried out on the long-term consequences of repeated velvet antler removal. Suttie & Fennessy (1985) found that hard antler weight and shape in 2-year-old deer was not affected by whether velvet-antler was removed when the stags were 1 year of age or not. Each stag had had velvet-antler removed from one side only as a yearling, yet the 2-year-old antler weight was 448 $\pm$ 53 ($\bar{x} \pm$ sem) for the previously amputated side and 472 $\pm$ 90 g for the control side (n=12). This tends to confirm the hypothesis that antler removal leaves antler size and shape unaffected.

Lincoln (1984) removed the growing velvet-antlers from three castrated stags at the line between the pedicle and the antler at about 2–4 monthly intervals for over 2 years. Regeneration always occurred and antlers grew to approximately the same size each time.

The influence of damage to antler nerves on subsequent regeneration is inconclusive, but conservatively it is suggested that the nervous control of antler growth rests solely in the nerves of the pedicle or surrounding area.

Conclusions

The preceding sections have shown by their paucity of hard data that any conclusions must remain conservative and tentative pending further study. Much information is anecdotal; many anomalies are to be found—it would be rather unwise to firmly formulate a hypothesis on this basis. With that in mind, my conclusions are as follows. The annual cycle of regeneration, cleaning, and casting of the antler is independent of innervation. Nerves influence antler size and shape. Stimulation of antler nerves leads to larger antlers while nerve ablation results in smaller antlers. Either way, species-specific shape is altered. Bioelectric potentials exist in the antler; these may be relevant with respect to regeneration, but further study is warranted.

There is no conclusive evidence that trauma to the deciduous portions of the antler which may result in permanent changes in size or shape of antlers are due to changes in nerve function. In contrast, trauma to the pedicle results in long-term changes in antler size and shape. It is suggested that a reason for this could be that the parasympathetic nerve supply has been altered. Antler nerve axons are unique in that they must suffer an annual dieback with a subsequent burst of regrowth from the pedicle. In amphibians, the nerves are known to branch to reinnervate cut limb stumps and thus increase the nerve supply. Were this the case for antlers, it might explain why traumatization of the pedicle stimulates antler growth, by increasing or altering the absolute amount or pattern of the nerve supply to the antler. Whether the parasympathetic nervous system communicates information to the brain about damage to the antler and how this information is stored is unknown. It is not possible to deny the body of evidence collected by Tony and George Bubenik that injury without anesthesia during early antler development has long-term consequences for antler shape. However, strong, ethical experiments are required to fully document and investigate this effect.

The anter must have a genetically determined size and shape. It can be concluded that nerves play a role in controlling size and shape. Whether they operate directly in response to genetic influences or whether they operate via a complex series of growth factors remains to be elucidated. However, the role of local nerves in the pedicle is considered to be of paramount importance.

13
Experimental Induction of Antler Growth

ZBIGNIEW JACZEWSKI

Introduction

When starting my experimental work on the regeneration and transplantation of antlers, I was stimulated by at least three facts. The first was the regeneration of a moose *Alces alces* antler accidentally broken off together with the pedicle (Jaczewski 1954b). The second was the occurrence of antlers or antler-like structures on different bones of the skull in fossil Cervids or Cervoids (Frick 1937). The third was the abnormal occurrence of antlers in females of living Cervidae (except Rangifer) and the occurrence of additional, abnormal antlers in males.

Before discussing the experiments on the induction of antler growth, it seems necessary to present a short recollection of data on the occurrence of antlers in females and of additional antlers in males.

Antlers in Females (Except Rangifer)

There are numerous descriptions of roe deer *Capreolus capreolus* females with antlers (Blasius 1894–95; Rörig 1899; Rau 1931; Wurster et al. 1983; Hartwig 1985).

Also very numerous are descriptions of white-tailed *Odocoileus virginianus* and mule deer *Odocoileus hemionus* females with antlers (Caton 1877; Dixon 1927; Wislocki 1954, 1956; Haugen & Mustard 1960s; Donaldson & Doutt 1965; Crispen & Doutt 1973).

The antlers in females of the subfamily Cervinae occur rather more rarely. They were described in wapiti, *Cervus elaphus canadensis*, and sometimes their length was up to 47 cm (Murie 1951; Buss & Solf 1959). A female sika deer *Cervus nippon* with antlers 10 and 15 cm long was described by Roux and Stott (1948). There are also several descriptions of red deer, *Cervus elaphus*, females with antlers (Prévost 1869; Alston 1879; Rörig 1899; Dulverton 1970; Whitehead 1970). A female cheetal, *Axis axis*, with antlers was described by Dharamjaygarh and Chandra Chur (1950).

Rörig (1899) has described antlered females belonging to different species of deer. However, he indicated that some females, after examination of internal organs, often turned out to be true hermaphrodites or pseudohermaphrodites. He described antlered females (or hermaphrodites) in Rusa deer *Cervus (Rusa) timorensis* and in moose. Prévost (1869) described a female with antlers in a hog deer *Axis porcinus*.

It is almost impossible to conclude what is the percentage of antlered females in any particular species or subspecies of deer. However, Donaldson and Doutt (1965) were able to estimate the median ratio of antlered does to bucks in white-tailed deer in Pennsylvania as 1 per 4,100. However, the authors expressed the opinion that some antlered females killed in Pennsylvania were not reported and therefore the ratio could be considerably higher. According to Ryel (1963) in Michigan the percentage of antlered white-tailed does among adult does varies between 0% and 0.11% in different years. The general opinion expressed already by Rörig (1899) is that most often antlered females occur in the genus *Capreolus*, then in *Odocoileus*, and are much more rare in the genus *Cervus* (Bubenik, A. 1966; Goss 1983). There is an assumption that often the antlers in females occur as a result of hormonal disturbances during multiple pregnancy. In fact, the multiple pregnancy is almost a rule in *Capreolus* and in *Odocoileus* and is extremely rare in *Cervus*.

Already Rörig (1899) had indicated that antlered females after examination of internal organs turned out to be true- or pseudohermaphrodites. Intersexuality in deer was observed many times (Boas 1892; Wislocki 1954). However, the problem is a rather complicated one and was elaborated in some details in domestic mammals only (Hafez & Jainudeen, 1966). The true hermaphrodite (Hermaphroditismus verus) has gonadal tissue of both sexes. There are many kinds of hermaphroditismus verus (see Hafez & Jainudeen 1966). Some true hermaphrodites are fertile and can produce offspring. Therefore the descriptions of roe deer with fawns and antlers or even with regular antler cycles (Boas 1892; Notz 1967; Raesfeld 1977) does not exclude hermaphroditismus. The problem can be solved only by a careful microscopic examination.

An interesting case of a true hermaphrodite in a roe deer was described by Boas (1892). This roe was born in May 1886, and from April 1887 till September had spike-antlers in velvet. These antlers were cast in May 1888 and a new growth was started. During the summer of 1889 this roe was nursing a fawn and in October it was killed. It had antlers 5 and 8 cm long with strong burrs. The post mortem examination revealed normal female genital organs on the right side only. On the left side, a small testicle-like organ was found. According to German authors, the fertile roe deer females with antlers often cast their antlers at calving time (Raesfeld 1977). This fact is similar to the antler cycles in reindeer, *Rangifer tarandus*, females.

In cattle the majority of male pseudohermaphrodites were born as twins (Hafez & Jainudeen, 1966). This fact corresponds with the situation among *Cervidae*. Females with antlers occur most often in *Capreolus* and *Odocoileus*, genera where multiple pregnancy is a rule. The exception is *Alces*, where twins occur very often but females with antlers are rare. The problem of freemartinism among Cervidae was not investigated, but vascular anastomoses were reported in moose (Hamilton et al. 1960; Kurnosov 1962). The most normal polished antlers in females after careful examination with the sex chromatin technique were reclassified as male pseudohermaphrodites (Crispen & Doutt 1970, 1973). Also the tumors of the organs of internal secretion are able to cause the occurrence of polished antlers. Doutt and Donaldson (1959) described a tumor of adrenal cortex probably causing the production of polished antlers in a white-tailed doe. Wislocki (1954, 1956) has described five normal females with antlers in velvet in genus *Odocoileus*. These females were fertile, and some of them were nursing fawns. He also described a hermaphrodite with polished antlers having external female organs but a testis in the abdominal cavity.

The majority of recent authors (Wislocki 1954; Bubenik, A. 1966; Crispen & Doutt 1973; Mierau 1972; Goss 1983) are of the opinion that if antlers occur in normal deer females they are always permanently in velvet. Such females are very often pregnant or lactating and accompanied by their fawns. These velvet-covered antlers can become frozen in winter. Later, the frozen part will be cast during the thaw, and even a regrowth is possible during the next spring. This sequence of events is similar to that occurring in males castrated after puberty and was described long ago (Caton 1877). What could be the cause of antlers in normal females is not known exactly. Some authors (Mierau 1972; Goss 1983) suggested that the ability to produce antlers may be inherited. Other authors suspect a short time imbalance in hormonal regulation (Wislocki 1954; Kierdorf 1985). The possibility of the traumatization as the cause of such antlers was indicated already by Rörig (1899). It is also possible that some subfamilies of Cervidae are more inclined than others to produce antlered females (e.g., Neocervinae).

A very interesting case of a fertile mule deer doe with antlers was described by Mierau (1972). This doe had produced fawns several times and probably had changed her antlers. The length of her antlers was different in succeeding years, and small burrs were present. Postmortem examination revealed completely normal sexual organs. As one of her daughters also had produced antlers, Mierau considered this trait as heritable. A fertile white-tailed deer doe with large antlers in velvet was described by Haugen and Mustard (1960).

Rörig (1899) indicated that small antlers sometimes occur in very old females due to ovarian atrophy associated with senescence. In fact, on the deer farm at Popielno lived a red deer female, born in 1963. This female manifested regularly recurrent estrus till 1984, but repeated mating proved her to be infertile. In the spring of 1985 she produced very small symmetric antlers in velvet, about 0.5 cm long. She died in 1987 (Jaczewski, 1989).

The first indication that antlers in female deer may also occur as a result of accidental traumatization only was given by Blasius (1894–95). He described a tame roe deer doe, which had driven into her scalp accidentally a piece of window glass. This resulted in the growth of an antler about 10 cm in length. The skull of this roe is in a collection of the Natural History Museum in Braunschweig and was described in great detail by Hartwig (1985).

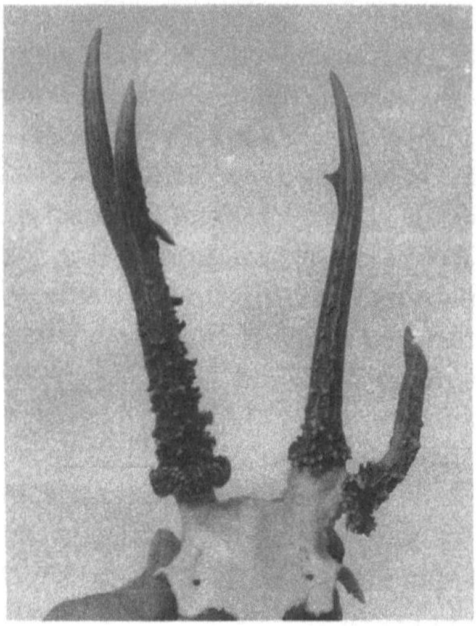

FIGURE 1. Head of a roe deer buck with three antlers. The additional antler probably developed as a result of the injury of the left pedicle. The head is from Oswald-Cerviden- Museum. (Photo by Z. Jaczewski.)

Occurrence of Additional Antlers in Males

The very important work on the occurrence of additional antlers in males of Cervidae was published by Nitsche (1898). Naturally, before Nitsche there were also many descriptions of additional antlers in males of different Cervidae species (Caton 1877; Brandt 1882; Lamprecht 1892; Holding 1896). The skulls with additional antlers are often exhibited in hunting museums (Fig. 1) (Salzle and Schedelmann 1977). Nevertheless, Nitsche was the first to classify these abnormal additional antlers and to give an explanation of their development. According to Nitsche, the additional antlers of Cervidae males belong to the following types: (1) completely independent from the normal pedicle; (2) the pedicle is split into two parts, each developing an antler; (3) the pedicle is normal, but an additional antler develops on its lateral surface; (4) the additional antler corresponds with the branch lacking on the beam, i.e., the main antler and the additional one together form a normal antler. Nitsche was of the opinion that additional antlers in males are caused by the injury to the periosteum of the pedicle or frontal bone. Nitsche had also enumerated the species with hitherto observed additional antlers: roe, red, wapiti, rusa (probably *Cervus timorensis*), white-tailed, reindeer, and cheetal. This list was later extended by barking deer *Muntiacus* sp., (Tilak 1978), wapiti (Bird 1933), and red deer (Raesfeld & Vorreyer 1978). The very interesting cases of additional antler development after the fracture of the pedicle were described by Bosnjak (1898), Jaczewski (1967), and Hartwig (1979).

The majority of more recent authors (e.g., Bubenik, A. 1966; Hartwig 1969; Raesfeld & Vorreyer 1978; Jaczewski 1982; Goss 1961, 1983) are also of the opinion that additional antlers in males develop as a result of traumatization of pedicles or frontal bones. Sometimes the additional outgrowth has a tumor-like structure. An interesting case of an osteofibrosarcoma, including an antler-like structure, was described in a roe buck by Meyer (1979).

Most extra antlers in Cervids originate from frontal bones. However, Dixon (1934) reported mule deer bucks in which the third antler developed on the nasal bones. A similar case in white-tailed deer was described by Wislocki (1952). An interesting case of an antler originating from the zygomatic arch (jugal bone) was described by Nellis (1965).

It should be emphasized that the classification of additional antlers given by Nitsche (1898) is still valid. However, his types Nr. 2 and 4 are probably very similar, both resulting from the splitting of the pedicle. If the splitting is in the lateromedial direction and the anterior part is smaller than the posterior one, then the brow tine may develop separately from the beam (Bubenik, A. & Pavlanský 1959 and Goss 1961).

Regenerative Powers of Antler Tissues

Pavlanský and A. Bubenik (1955) and Jaczewski (1954b, 1955) proved that antlers in moose, red deer, fallow deer (*Dama dama*), and roe deer are able to regenerate after the amputation of the pedicle. The regeneration took place even after the opening of the frontal sinus (Bubenik, A. & Pavlanský 1956; Jaczewski 1961).

These initial investigations were extended further by A. Bubenik and Pavlanský (1956, 1959, 1965), Goss (1961), Jaczewski (1961, 1967) and Pavlanský & A. Bubenik (1960), proving that the pedicle and adjacent regions of the skull have an enormous ability to regenerate antler tissues. Goss (1961) found out that the excision of anterior, posterior, and median halves of the pedicle in sika deer had little or no effect on antler development, but that the removal of the lateral half of the pedicle resulted in almost complete failure of antler renewal. Probably this result is to some degree connected with the fact that the blood vessels supplying the growing antler are situated on the lateral side of the pedicle (Jaczewski et al. 1965). Jaczewski (1967) showed that regeneration after severe pedicle traumatization may be inhibited for several years after an operation; however, the regeneration of an antler can start even 3 or more years after an operation and thereafter will improve every year. Hartwig (1967, 1968a,b) pushed this line of research forward by investigating the significance of different tissues in antler development. He discovered that the periosteum is the most important one (Hartwig 1968a; Hartwig & Schrudde 1974).

It was proved also by Jaczewski (1956a,b, 1961, 1967), A. Bubenik & Pavlanský (1965), and G. Bubenik et al. (1982) that the traumatization of an antler or the amputation of the pedicle sometimes stimulates antler development. In some cases the traumatized antler grows longer and/or heavier than the contralateral one. Traumatization is sometimes used by deer farmers to stimulate the antler growth in spikers (Yerex 1979).

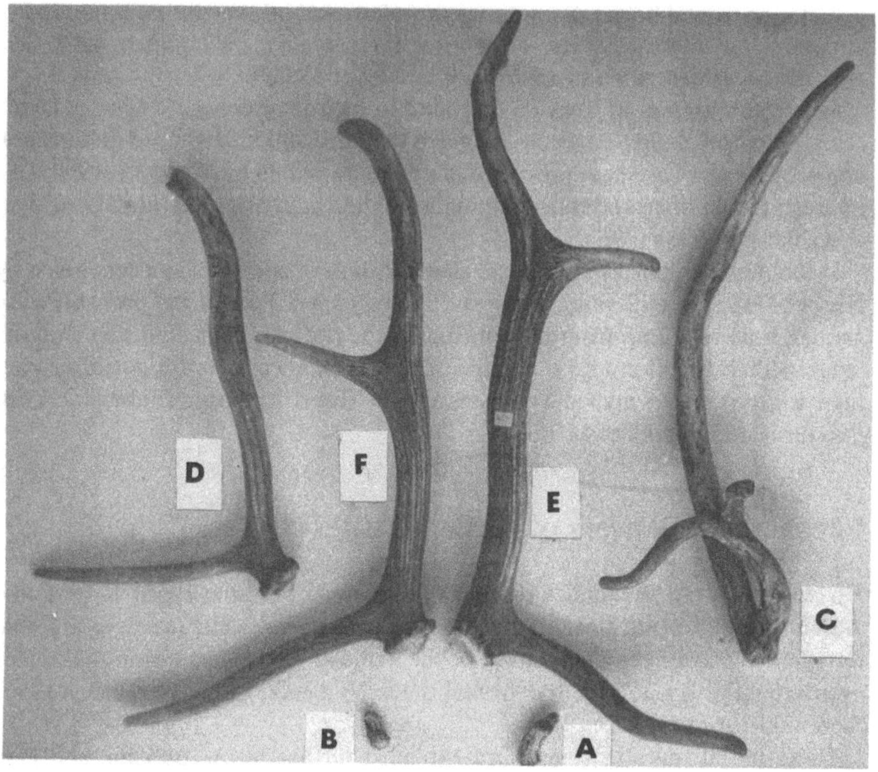

FIGURE 2. Amputation of the right pedicle in a red deer stag stimulated the development of the right antler. *A* Right antler with pedicle amputated on March 21, 1984; *B* left antler cast on May 12, 1984; *C* right antler cast on April 23, 1985; *D* left antler cast on April 21, 1985; *E* right antler cast on April 5, 1986; *F* left antler cast on April 5, 1986. (Photo by Z. Jaczewski.)

Recently at Popielno (Jaczewski, unpublished) the amputation of the pedicle with antler was performed on a red deer stag, born on August 5, 1982. This stag was in poor condition, and the velvet was shed from its small antlers in January 1984. On March 21, 1984, the right pedicle was amputated and a new growth started from the external margin of the wound at the end of April, 1984. The left, control antler was cast on May 12, when the right antler was already about 5 cm long. On May 31, the right antler was nearly 15 cm longer than the left one. The velvet shedding occurred in August on both antlers. The control left antler was cast on April 21, 1985—length 44 cm, weight 282.7 g; the right one on April 23—length 60.5 cm, weight 451.3 g. In the next antler cycle, the antlers were both cast on April 5, 1986: the left one was 56.2 cm long and weighed 549.2 g, and the right one was 65.9 cm long and weighed 707.2 g. Taking into account weight only, it may be concluded that the pedicle amputation caused an increase

FIGURE 3. Skull of a 3.5-year-old red deer stag with three antlers as a result of accidental breakdown of the pedicle when the animal was a little over 1 year old. (Photo by Z. Jaczewski.)

in weight of about 159% in the first cycle and of 128% in the second cycle. This experiment shows clearly that traumatization has a strong stimulatory effect on the antler growth (Fig. 2).

Sometimes an additional antler is produced only by the regenerative powers of antlerogenic tissues. An interesting case was presented by Jaczewski (1967). On September 17, 1964, when the right pedicle with the polished antler of a red deer spike was accidentally broken; it remained hanging on the skin and, not being heavy, became ankylosed to the skull in an abnormal position. In the following year a normal left antler developed, but on the right side, two antlers were produced: one from the broken pedicle's surface and the other in the usual place. The wound to the periosteum probably stimulated the growth. In the next year the stag again produced three separate antlers (Fig. 3).

The Induction of Antler Growth in Males Castrated Before or at Puberty

The effect of castration on the growth of antlers has been known since Aristotle (1910), and many later workers (Caton 1877; Wislocki et al. 1947) confirmed that castration before puberty precludes the growth of antlers. However, our information in this field is based on a few species only (e.g., red deer, wapiti, white-tailed deer, roe deer). It is not known for certain if castration before puberty also precludes the growth of antlers in reindeer, because in that species the antler growth starts very early in life.

In red deer stags castrated before puberty it was impossible to induce the growth of antlers with either prolactin or testosterone in large and prolonged dosage. The implantation of the pituitary gland was equally ineffective. The heavy doses of testosterone caused the growth of pedicles only and some changes in coat and in behavior. It was established that the amputation of such a pedicle induces a strong antler growth response (Jaczewski & Krzywinska 1974; Jaczewski et al. 1976). Pedicle amputation was effective only if bone and periosteum were injured; traumatization of the skin alone was ineffective. The growth induced by severe traumatization was so strong that often the antlers became larger than in normal spikes and branched (Fig. 4).

When castration was performed at the beginning of puberty when pedicles were already visible, the traumatization was effective even without any premedication with testosterone (Jaczewski 1982).

Using this method, Lincoln and Fletcher (1976) were able to induce the antler growth in a hummel (an antlerless red deer stag). All the above-mentioned traumatizations of pedicles were performed in the spring; the antler growth proceeded during spring and summer, imitating the normal antler cycle in red deer stags.

Recently Lincoln (1984) found out that the amputation of antlers in castrated red deer males induces the regeneration of antlers not only in spring and summer but in autumn as well. However, the "autumn antlers" were smaller. Spring and autumn antler growth periods were reported in castrates of other cervids as well; biseasonal growth periods of perukes in castrated roe-buck were reported first by A. Bubenik (1963).

Recently at Popielno we tried to check whether pedicle amputation performed at the end of summer will be as effective in antler growth induction as that performed at the beginning of spring. Two red deer males—Tank, born August 19, 1978, and Pigment, born July 17, 1975—were castrated at puberty. As a result of the amputation of the left pedicle in the spring, they have produced the left antler only (Jaczewski 1982). During the autumn-winter periods, these animals were given testosterone, estradiol, or stilbestrol (S), causing the velvet shedding in autumn and antler casting by the end of winter. A new growth of the left antler only was started each spring and became finished by the middle of summer. On August 17, 1983 (Jaczewski & Bartecki, unpublished), the right pedicle was amputated in both these animals. The amputated pedicle of Pigment weighed 20.6 g and that of the red deer Tank 10.5 g. As a result of that amputation, the right antlers started to grow in the autumn. It was interesting that the left antlers (grown in the spring and summer) were covered with dried velvet and did not grow. On the contrary, the right antlers were covered with "young, fresh" velvet and grew intensively during September and October. These castrates were given 400 mg of stilbestrol on December 3, 1983, and 500 mg of S on January 3, 1984, to cause the velvet shedding. Curiously enough the velvet shedding occurred

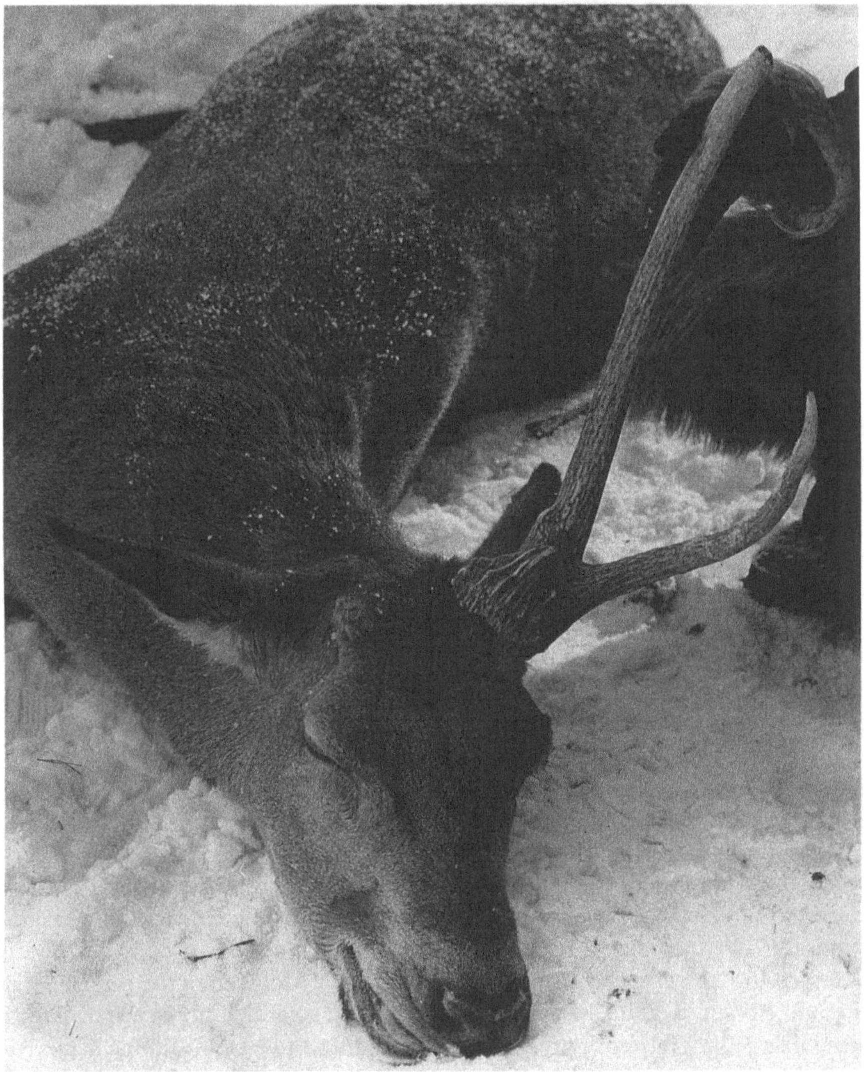

FIGURE 4. A red deer male castrated before puberty. The pedicles were grown after testosterone treatment. The amputation of the left pedicle caused the growth of a big antler. The antler cycles were regulated by hormonal injections each autumn and winter. Six years after the amputation, the left pedicle with antler was very wide but the right pedicle remained unresponsive in spite of hormonal injections. (Photo by L. Bartos.)

nearly a month earlier on the left antler than on the right one. Tank cast his left antler on February 28, 1984 (weight 373.3 g, length 59 cm) and the right one on March 6, 1984 (weight 135 g, length 37.5). Pigment cast his left antler on March 12, 1984 (weight 386 g, length 59 cm), and the right one on March 20, 1984; its length was about 7–8 cm, but it was not found. The next antlers of Pigment were nearly symmetrical: right 58 cm, left 55.2 cm. The next antlers of Tank were nearly symmetrical as well: left 57.5 cm, right 59.8 cm.

These experiments indicate that the amputation of the pedicle is able to induce unilateral antler growth in different seasons of the year. The asynchronic antler cycles after the amputation performed in the autumn indicate that the local factors (regeneration blastema) are probably more important in the regulation of the antler cycle than the general hormonal regulation.

It should be mentioned here that Goss (1972) proved, using the sika deer, that the closing of a pedicle wound with full-thickness skin stopped antler regeneration completely or partially. He extirpated the distal bone part of the pedicle and covered the wound with the pedicle skin. As a result, antler growth was stopped completely or small growth appeared in places where the flap of skin had not healed completely to the opposite side. This result cannot be generalized on other species; A. Bubenik and Pavlanský (1956, 1965) did not prevent regeneration of antlers in roe deer by extirpating the pedicle and closing the wound with the skin.

Antler Grafting Experiments

The first successful experimental grafting of an antler in red deer was performed by Jaczewski (1956a,b). A little later a free graft of an antler on the frontal bones between the eyes was reported in a fallow buck (Jaczewski 1958). In these experiments the whole tip of the chiseled-off antler or pedicle was used as a graft. It consisted of skin, periosteum, and a small piece of bone. During these experiments on the transplantation and regeneration of antlers it very often was observed that the asynchronization of antler cycles occurred in the same animal (Fig. 5). Pavlanský and A. Bubenik (1960) grafted an antler bud in a spike fallow deer, fixing the graft with an orthopedic nail into the base of the pedicle. The grafted bud healed, and its ossification paralleled that of the control side. Curiously enough, this operation induced a second cycle in the same year, because just after velvet shedding, a new antler started to grow from the wounded pedicle. The grafted antler developed during the subsequent antler cycles, but its base fused with the regenerating antler as Nitsche (1898) indicated it would.

Jaczewski (1961, 1967), using the method of grafting the whole pedicle, produced a red deer stag with four antlers: one on frontal bones between the eyes, the second on the parietal bones, and two regenerating in normal places (Fig. 6). The largest grafted antler developed on the frontal bones; its length was 302 mm, weight 266.5 g, and average circumference 90 mm. Curiously enough, this big grafted antler developed during the fourth antler cycle after the operation. Dur-

FIGURE 5. Asynchronization of antler cycles in a fallow deer buck caused by the amputation of the right pedicle. The left normal antler is growing and the right one (grown after the pedicle amputation) is hard and still uncast. (Photo by Z. Jaczewski.)

ing the first three cycles after the operation, this antler was not visible or very small. Its weight in the third was only 1.32 g.

Another very interesting grafting experiment was performed by Goss (1964). He removed the skin from the distal part of a pedicle before the expected time of antler casting. Then the antler was sawed off and a hole was cut in the pinna of the ear to permit it to be impaled on the pedicle. As a result, after antler shedding the newly growing antlers had to use the ear skin for regeneration. These antlers were almost normal except for the absence of brow tine, a common occurrence after pedicle amputation. After velvet shedding, the antler was sawed off above the coronet and the pedicle transected. The liberated ear was then returned to its normal position. However, a segment of the pedicle and a small part of the antler were still grafted to it. After antler shedding, which was hastened by castration, new antler tissue grew in its place. Simultaneously, the proximal end of the

FIGURE 6. A 4-year-old red deer stag with four antlers: two regenerating in normal places and two grafted on parietal and frontal bones. Photo taken in August 1962 just after the bigger frontal graft became broken. (Photo by Z. Jaczewski.)

pedicle graft also regenerated antler tissue, but in the opposite direction. This experiment proved that it was even possible to graft the antler tissue onto the ear.

Hartwig (1968a) and Hartwig and Schrudde (1974) succeeded in grafting antlers on the frontal and parietal bones in roe deer. They grafted the periosteum only (without skin), using the method of a flap graft. However, their most impressive experiment was the free grafting of an antler on the foreleg of a roe buck. In a fawn the periosteum was excised from the place where the pedicle develops and grafted under the skin on the metacarpal bone. In November, when roe deer normally grow antlers, a small antler on the foreleg was produced, and in April and May, according to the typical antler cycles in roe deer, the velvet was shed.

Hartwig (unpublished data) indicated that this small antler was cast and that a new one was produced every year according to the normal antler cycle of a roe buck (Fig. 7). However, in the fourth cycle this antler was not cast, and a small new growth started from the surrounding surface (Fig. 8). The experiments of Hartwig indicate that the periosteum in the area of the pedicle is the most important tissue in antler development. The control experiments of Hartwig showed that the periosteum from other parts of the skull did not produce antlers.

The very important discovery of Hartwig and Schrudde (1974) was repeated perfectly and extended by Goss and Powell (1985) using the fallow deer. They

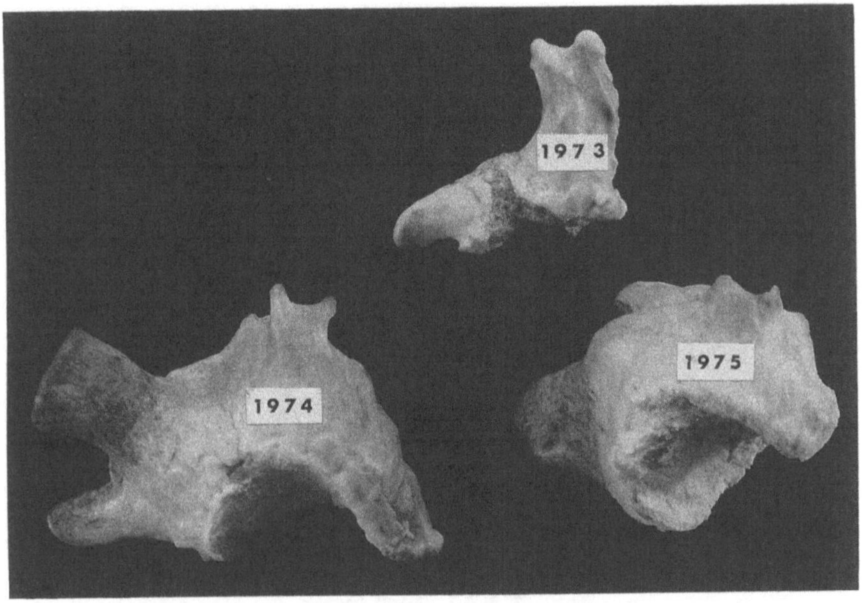

FIGURE 7. Antlers grown on the metacarpal bone of a roe buck—a further course of a famous experiment of Hartwig and Schrudde (1974). These antlers were cast each year at a time typical for a normal roe buck. (Photo courtesy of Dr. H. Hartwig.)

grafted discs of frontal periosteum from presumptive antler sites of 6–8-month-old male fawns beneath the foreleg skin over the metacarpus. The results were different according to the size of the graft. Periosteal transplants 1.5 cm in diameter differentiated always into pedicle bones and induced the growth of small antlers (Fig. 9). These small antlers underwent cycles similar to normal and increased in size up to 7 cm during the next years. The periosteal grafts 1.05 cm, 0.75 cm, and 0.4 cm in diameter did not induct antlers in the first year, but some developed antlers in the second or third year. In general, the smaller the periosteal grafts, the smaller the pedicles and antlers produced from them. Semicircular grafts of periosteum induced antler growth in most cases, especially when derived from lateral and posterior halves of antlerogenic region. Conversely, the donor sites produced antlers in about 20% of the cases following removal of discs of 1.5 cm and 1.05 cm, while 80% and 100% grew antlers after removal of discs of 0.75 and 0.4 cm, respectively. Goss and Powell also observed that the small grafts failing to produce antlers could be stimulated to do so by traumatization. This result confirmed the statement of earlier authors (Pavlanský & Bubenik, A. 1955; Jaczewski 1956a,b; Jaczewski & Krzywinska 1974; Jaczewski et al. 1976; Jaczewski 1981b; Lincoln & Fletcher 1976; Bubenik, G. et al. 1988) that traumatization stimulates antler growth. These experiments proved also that the antlerogenic periosteum in fallow deer is able to convert the normal skin of the leg

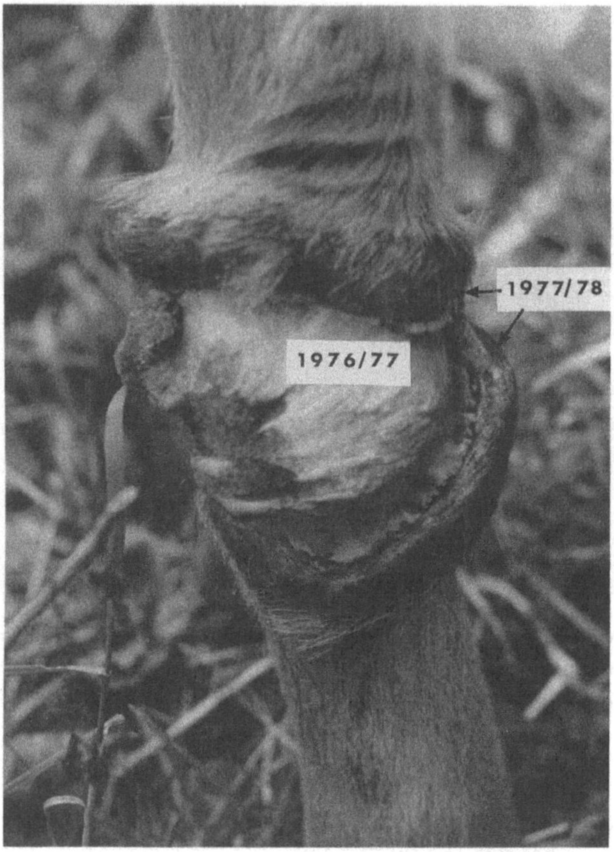

FIGURE 8. Continuation of the experiment of Hartwig and Schrudde (1974). The fourth antler was not cast, and a new growth is starting from the surrounding tissues. (Photo courtesy of Dr. H. Hartwig.)

into the velvet and to produce antlers. The nature of this conversion should be investigated in the future.

Recently Jaczewski and Bartecki (unpublished) used a compound flap graft method to produce a three-antlered red deer stag. They successfully grafted pedicles to the middle line of the skull of 1-year-old spikers. Operations were performed when the pedicles were about 1 cm long. After pedicle removal, in the majority of experiments the wounds remained open without any treatment. As a result, seven three-antlered stags were produced: one antler served as a control,

---

▶

FIGURE 9. The experiments of Dr. Richard Goss on the transplantation of the periosteum from the presumptive antler sites; *A* a graft on the metacarpus; *B* a graft under the skin above the iliac crest. (Photos by G.A. Bubenik and Z. Jaczewski.)

A

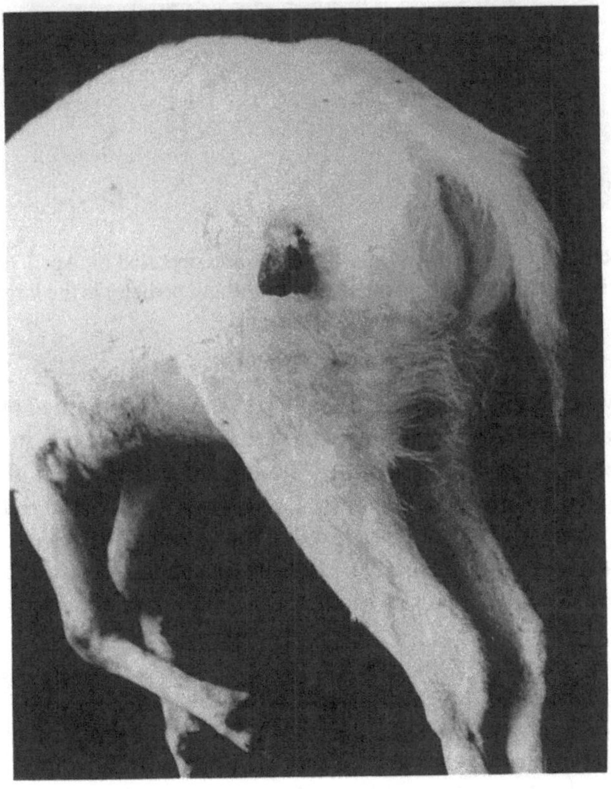

B

A

FIGURE 10. Red deer stag born on June 16, 1987, and operated on April 22, 1988. Photo taken on April 4, 1989. The regenerated antler (without pedicle) is the largest one. *A* Left side view; *B* right side view. (Photo by A. Różański.)

the second was the grafted one, and the third was the regenerated one. As a rule, the regenerating antler grew more quickly than the control one in spite of periosteum removal at this site (Fig. 10). The injured side of the skull always produced two antlers weighing much more (sometimes over 200%) than the control antler. This increase of antler tissue production was very strong even in the third cycle after the operation (Fig. 11). These experiments indicate that the periosteum is not the only important tissue in the induction of antler growth.

Induction of Additional Antler Branches

Among hunters there was a long and intense competition to shoot a male deer with the largest and most impressive antlers. In Europe, North America, Australia, and New Zealand, there are special systems to score the trophy and to esti-

B

FIGURE 10B

mate its value (Jaczewski 1981b; Trense et al. 1981). This competition caused some red deer hunters in Europe to use the rather cruel and dangerous method of shooting stags in velvet with birdshot. They claimed this procedure increased the number of branches. Krzywinski (1974) tried to test this theory by implanting small iron balls in the growing antlers of a red deer stag. He found that very small, additional branches were produced in places where the balls were implanted under the velvet of the beam. Yet, iron implants in the growing tips remained unaffected.

At Popielno, many red deer stags are kept in rather small enclosures. It was observed several times that the injury of the velvet causes additional branches. For example an 8-year-old-stag, Parys, accidentally injured his left brow tine with a wire on April 22, 1983. The velvet was peeled off for a distance of about 12 cm. In that year, this stag developed a four-point brow tine with a crest of bone forming three separate points – 8 cm, 6 cm, and 6 cm in length (Fig. 12a,b). This crest was formed in the injured area. The most important measurements of his antlers

A

B

FIGURE 11. Red deer stag born on May 27, 1986, and operated on April 14, 1987. Again the antler mass on the operated side is much bigger than on the control side. *A* Front view on May 16, 1989; *B* right side view on May 30, 1989. The tip of the bifurcated brow tine is slightly injured. (Photo by Z. Jaczewski.)

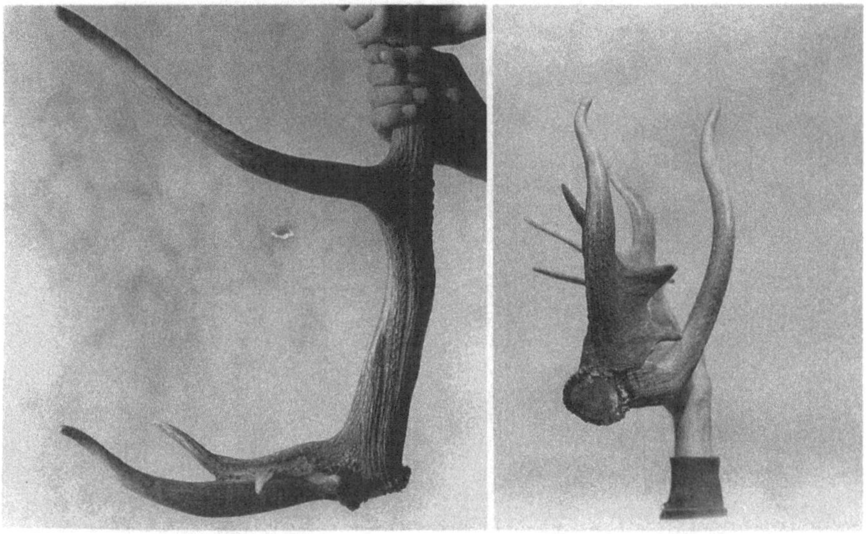

A B

FIGURE 12. A brow tine of a red deer stag after the injury when in velvet. The injury caused the development of three additional branches. In the next antler cycle the brow tine developed normally. *A* Lateral view; *B* bottom view. (Photos by Z. Jaczewski.)

before and after the injury are given in Table 1. As can be seen from the table, this injury did not influence the development of his antlers in succeeding cycles.

The most impressive case of the stimulating effect of injury on antler development was described by G. Bubenik et al. (1982) on a 4-year-old white-tailed buck. The traumatic damage to growing antlers of that buck resulted in the development of enormously hypertrophied antlers having 30 points and weighing almost 4 kg (compared to 8–10 points and 1.3–1.5 kg in control bucks) (Fig. 13). This strong trophic response to trauma is probably regulated by the nervous system.

TABLE 1. Antler measurements of the stag Parys antlers before and after injury of the left brow tine.

	Antlers							
	Left	Right	Left	right	Left	Right	Left	Right
Date of casting	3/15/1983		3/8/1984		3/7/1985		3/6/1986	3/5/1986
Weight (kg)	2.297	2.200	2.379	2.526	2.450	2.600	2.915	3.250
Length of brow (cm)	32.5	34.5	31.5	43.5	39.5	40.5	41.5	44.5
No. points on brow tine	1	1	4	1	1	1	1	1
No. points on whole antler	7	6	9	7	8	7	9	9

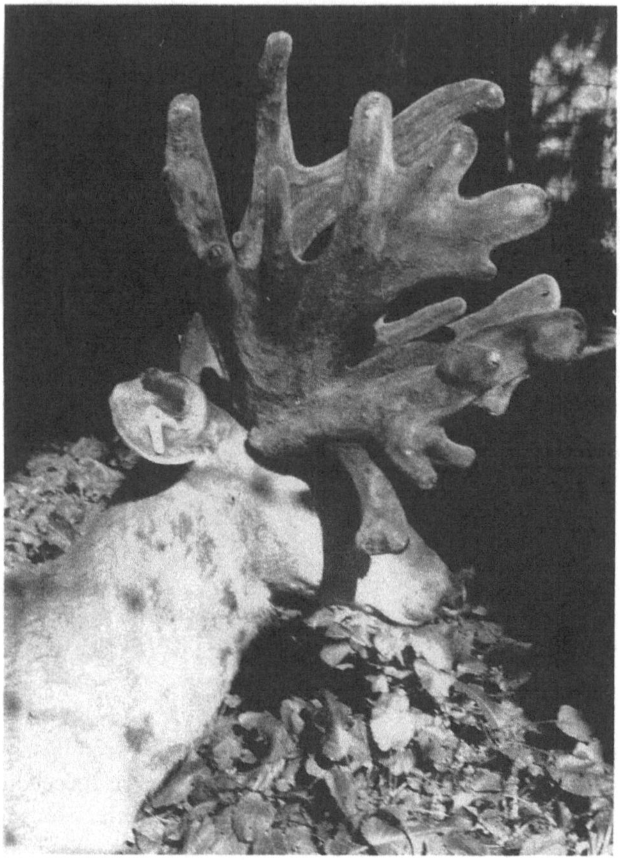

FIGURE 13. The extremely large antlers of a white-tailed buck grown after an accidental injury to the antlers in velvet. An experiment of George Bubenik. (Photo courtesy of G.A. Bubenik.)

In any case G. Bubenik et al. (1982) reported that if traumatization of antlers of white-tailed bucks was performed in deep Rompun anesthesia, the hypertrophy of antlers did not occur. Lake et al. (1982) and G. Bubenik et al. (1982) proved also that electric stimulation of nerves supplying the growing antler strongly influences their development. The response could be different according to the method of stimulation. Sometimes the stimulated antler is even bigger than the control one (Bubenik, G. et al. 1982).

Occasionally, additional branches developed after some pedicle wounds. After a red deer stag suffered accidental pedicle damage, a rather bizarre antler was produced. This nontypical antler developed every year thereafter (Fig. 14).

Deer farmers in New Zealand sometimes cut off the first growth of red deer spikes to induce branching. Using this technique, eight-point antlers were reported to grow in just over 1 year (Yerex, 1979).

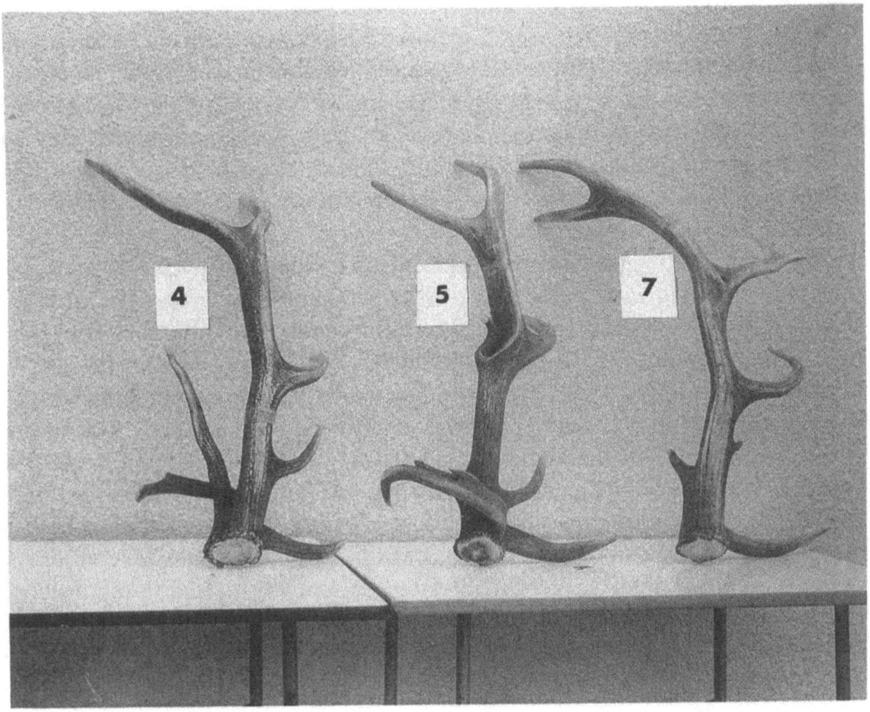

FIGURE 14. The 4th, 5th, and 7th right antlers of a red deer stag. The right pedicle of the stag became injured when the third pair of antlers was sawed off after velvet shedding. This injury caused the abnormal development of the right antler. (Photo by Z. Jaczewski.)

Antler Induction in Females

The first experimental induction of antler growth in white-tailed deer females was performed by Wislocki et al. (1947). One doe produced several fawns; she was ovariectomized and hysterectomized in January 1944 when 6 years old. This doe had small protuberances beneath the skin on her forehead area where antlers are normally located in males. Such protuberances sometimes occur in female deer. In June the doe received 350 mg of testosterone. As a result the protuberances increased slightly in size, but by the end of December they had returned to their original size. In March 1945 this doe received 1,050 mg of testosterone, and by June she produced antlers 5 cm and 7.5 cm in length. By July they were 10 and 12.5 cm long and bifurcated. In July, 700 mg of testosterone were given. The antlers continued to grow, reaching a maximum length of 16.5 and 20 cm. The velvet was shed in September and the antlers were cast in January 1946. They were without coronet and their basal surfaces were perfectly flat. It should be mentioned here that the shape of the basal surface (antler seal) is probably regulated by reproductive status and hormones (Bubenik, A. 1966; Jaczewski 1985; Bubenik, G. & Schams, in press). The other Wislocki's doe did not have any bony elevation on her head. Both ovaries were removed when she was 8 months old. More than 2 years later, in May 1945, the doe was given 1,050 mg of testosterone. By October the pedicles measured about 2 cm, and by November they had receded but could still be palpated on the forehead. Growth was renewed in May without any hormonal treatment; by September the antlers were 2.5 cm and 5 cm in length and still in velvet.

After nearly 30 years, similar experiments with testosterone treatment of females were performed on red deer by Jaczewski and Krzywinska (1975) and on sika deer by Goss (1983). However, the results were quite different. After treatment with large doses of testosterone during autumn and winter the ovariectomized hinds of the genus Cervus produced only pedicles about 2 cm long but without any growth of the antlers. It was determined that the amputation of such pedicles in females at the beginning of May always induced the growth of an antler. If the wound of the pedicle was superficial (skin only), there was no antler growth. The growth was very distinct and sometimes an antler was branched. To avoid frostbite, testosterone, estrogen, or stilbestrol were administered during the next autumn-winter periods. The antlers of females reacted to these steroid hormones with velvet shedding and, after the hormonal level dropped, with antler casting. The next antlers were usually larger. The largest three-point antler was induced in an ovariectomized hind during the second cycle after the amputation; it weighed 705 g and was 63 cm long. During the next cycles this antler became smaller, but an accidental injury to the contralateral pedicle induced a bilateral antler growth. This hind manifested altogether five antler cycles. She died as a result of an overdose of Rompun. Her skull with antlers weighed 2,352 g; the left antler was 51 cm long and the right one 43.5 cm long (Fig. 15) Jaczewski (1976, 1977, 1981a, 1982).

In my further experiments, normal (nonovariectomized) red deer hinds were treated with testosterone during winter; subsequently, their pedicles were amputated at the beginning of May. It was demonstrated that the induction of antler growth is possible even in normal hinds if the injury included bone. With larger doses of testosterone, the hinds became sterile. When pretreatment with testosterone decreased to a dose of 1.5 g, it was possible to induce antler growth even in pregnant and lactating hinds. Antlers in pregnant and lactating hinds developed to about 20 cm in length (Jaczewski 1981a, 1982). It should be mentioned here that female reindeer grow antlers during lactation, but not during pregnancy.

The size of the pedicle is very important in antler growth induction by traumatization: the larger the pedicle, the easier it is to induce antler growth. However, it was possible to induce the growth of antlers in females, not only by pretreatment with testosterone but also with stilbestrol (Jaczewski 1982). Robbins and Koger (1981) showed that it is possible to induce antler growth in wapiti *Cervus elaphus canadensis* females by injection of $CaCl_2$ without any hormonal pretreatment. The necrotic lesions produced by the injection caused a bleeding,

FIGURE 15. Skull of a red deer female with antlers. The velvet shedding was caused by the injection of stilbestrol dipropionate. The left antler developed after amputation of the pedicle and the right one after an accidental injury of the pedicle. (Photo by Z. Jaczewski.)

open wound. The mechanism of antler growth stimulation with $CaCl_2$ is probably similar to surgical amputation of the pedicle.

Also, G. Bubenik et al. (1982) were able to induce the antler growth in a female white-tailed deer without any pretreatment with testosterone. His doe was castrated chemically as a juvenile by antiestrogen treatment and became sterile. Three years later the unilateral antler growth was induced by traumatization only. Traumatization also proved to be successful in inducing antler growth in a hummel (Lincoln & Fletcher 1976).

The majority of authors reported the influence of season in the induction of antler growth. First of all, the artificially induced pedicles increase in size in summer and decrease in winter (Jaczewski 1981a; Jaczewski 1982; Goss 1983). Secondly, the antlers induced by traumatization in the spring became rather larger than those induced by traumatization in autumn (Robbins & Koger 1981; Jaczewski and Bartelcki unpublished).

Artificially induced antlers in red deer females are normally permanently covered with velvet. However, in four cases, a spontaneous velvet shedding occurred without any hormonal treatment. In five cases, an antler in velvet was spontaneously cast off in summer without hormonal treatment. Altogether, 45 antler cycles were observed after artificially induced antler growth in females

(Jaczewski, unpublished). However, I am unable to answer how often the velvet shedding or antler casting occurs spontaneously, because the hormones were administered to hinds rather irregularly. Sometimes their application was started at the end of August and sometimes in November. Nevertheless, I can state that in the majority of cases a spontaneous velvet shedding did not occur.

It must also be mentioned here that in nonovariectomized hinds after several cycles there was a tendency to decreased size in regenerating antlers (Jaczewski unpublished). Mostly the second and third antlers were bigger, but later a tendency to smaller regeneration was observed. I am unable to answer if this phenomenon could be explained by some nervous influences (see Chapter 11, hypothesis of CNS centers and "trophic memory").

Discussion

Experiments with artificial induction of antler growth and the stimulatory effect of traumatization on that growth speak in favor of the theory that antlers in deer developed during evolution as a special adaptation to traumatization. According to fossil evidence (see Chapter 1), the progenitors of the modern deer had only bony protuberances growing from skull and covered by skin. Used as weapons, these structures were vulnerable to repeated injuries. Natural selection among the ancestors of Cervidae probably favored animals capable of healing the wounds on their "antlers" in such a way as to replace lost parts. This hypothesis was advanced by Jaczewski (1955, 1961) and later developed by a brilliant argumentation by Goss (1964, 1965a, 1980a). A little similarity could be observed in a modern okapi, *Okapia johnstoni*, male.

The experiments on the artificial induction of antler growth present several problems which should be more deeply investigated in the future:

The experiments described above proved that traumatization has a stimulatory effect on the development of antlers. This stimulatory effect was observed after the wounding of antlers in velvet (Krzywinski 1974; Bubenik, A. et al. 1982; Jaczewski, unpublished) and also after the wounding of pedicles or periostal grafts (Bubenik, A. & Pavlanský 1965; Jaczewski 1967, 1981a; Jaczewski et al. 1976; Goss and Powel 1985). The experiments of Hartwig and Schrudde (1974) and of Goss and Powel (1985) indicate that the most important is periosteum tissue in the development of antlers. The special features of that particular periosteum should be more thoroughly investigated in the future. This periosteum undoubtedly reacts very specifically to testosterone and probably is able to stimulate other tissues to a very rapid growth.

The second problem is the role of the nervous system in the stimulation of antler growth after wounding. The observation of G. Bubenik et al. (1982) that the stimulatory effect was negligible if wounding was performed in a deep anesthesia indicates the possible role of the central nervous system (CNS). The observations described above—that artificially induced antlers in females became much smaller after several years—speak also in favor of CNS influences.

The asynchronization of antler cycles after wounding of the pedicle was described many times (Jaczewski 1982). As a rule the antler cycle on the traumatized side is more or less retarded in comparison to a normal control side. This retardation could also be explained as a result of nervous influences. In any case, Bubenik G. et al. (1982) proved that asynchronic development of antlers in the same animal could be produced by the electric stimulation of antler periosteum. He proved also that the electric stimulation of the lateral twig of the "antler nerve" produced the asymmetric development of antlers. A strong desynchronization of antler cycles in fallow deer was observed in individuals in permanent stress (Topinski 1975). Naturally, this line of research should be more developed in the future.

The species-specific differences belong to another interesting field. It is a pity that till now nobody tried to repeat the classic experiments of Wislocki et al. (1947) on white-tailed deer females. They reacted to testosterone injections with antler growth unlike the females of the genus *Cervus*.

Conclusion

1. Traumatization has a strong stimulatory effect on antler growth.
2. The most important tissue in antler growth induction is periosteum in the presumptive places of antler growth.
3. Using this periosteum, it is possible to induce the antler growth in other parts of the skull and on other skeletal bones, e.g., metacarpus, as well.
4. The experiments on antler growth induction speak in favor of the "regeneration theory" of antler development in Cervidae (Jaczewski 1961; Goss 1965a).

14
The Annual Antler Cycle of the European Roe Deer (*Capreolus capreolus*) in Relation to the Reproductive Cycle

ANTOINE J. SEMPÉRÉ

Introduction

Deer species living in their natural environments are seasonal breeders. The timing of the rut ensures that births occur in the spring and summer when environmental conditions such as temperature and food availability are optimal for the fawn's survival. Therefore, most temperate deer species breed in the fall (from September to November).

The roe deer, *Capreolus capreolus*, is the exception to this rule. In this species the rut occurs in July-August and implantation of the blastocyst is delayed until early January so that births occur in May and June (Aitken 1981; Short & Mann 1966; Sempéré 1978, 1982). Furthermore, antlers are cast and regrown in winter, whereas in all other boreal deer species antlers grow in spring and summer.

Studies of seasonal testicular activity and antler cycle have been carried out in numerous cervids; e.g., in roe deer, *Capreolus capreolus* (Short & Mann 1966; Barth et al. 1976; Sempéré 1978; Sempéré & Lacroix 1982), red deer, *Cervus elaphus*, (Lincoln et al. 1970) in white-tailed deer, *Odocoileus virginianus*, (Mirarchi et al. 1977; Bubenik, G. 1982). These endocrinological studies revealed that testosterone plays an important part in controlling antler cycle. Investigation of plasma gonadotrophin levels conducted in white-tailed deer (Mirarchi et al. 1978), red deer (Lincoln & Kay 1979), and roe deer (Sempéré & Lacroix 1982; Schams & Barth 1982) have demonstrated a strong seasonal fluctuation of luteinizing hormone. In adults these annual variations seemed to be affected by seasonality of day length (Goss 1969). A first development of embryonic pedicles (Frankenberger 1954) correlates with the rise of testosterone (Lincoln 1973) but does not develop further until postpartum, when in male fawn the pituitary stimulates prepubertal and testicular activity (Sempéré & Lacroix 1982; Foster et al. 1978; Schams et al. 1978). The observation of two antler cycles in roe deer during the first year of life (Tegner 1961; Sempéré 1982) suggests that the first antler cycle in fawns implies early activity of testicular function.

Growth of antlers in cervid species is a repeated event which, in temperate regions, is closely synchronized with photoperiodicity (Bubenik, G. 1986).

Artificial manipulation of day length influenced the onset of antler growth and the timing of mineralization, velvet shedding, and casting (Jaczewski 1954; Goss 1983; Suttie & Simpson 1985; Bubenik, G. et al. 1987a).

The purpose of this review is to define the annual cycle of reproductive activity and to stress the uniqueness of the sexual and antler cycle of the roe deer, as compared to other temperate cervids.

The Sexual Cycle

Prepubertal Roe Bucks

Sexual and Antler Cycles

In fawns, the first antlers appear from September to November when the velvet is shed and hard buttons become visible (Sempéré & Boissin 1982; Sempéré 1982). Hard antler "buttons" in white-tailed deer (Waldo & Wislocki 1951), moose (Peterson 1955), roe deer (Tegner 1961), mule deer (Davis 1962), and reindeer (Meschaks & Nordkvist 1962) were also observed in early fall of their first year of life. In the roe deer these buttons are usually cast in late January. A new growth of second antlers will start in February and become fully developed in May. Depending on the fitness of the deer, they may be spiked or branched.

In roe deer fawns, elevation of plasma testosterone level (1.27 ± 0.18 ng/mL) and a small increase in testis size (2.96 ± 0.28 cm^3) were observed in autumn, at the age of about 4 months (Fig. 1). However, high plasma LH levels (4.09 ± 0.80 ng/mL) were recorded as early as in the first month of age. The pituitary activity decreased in December followed by a regression of testicular activity in January-February.

A parallel increase of both plasma testosterone levels and testis size was observed as early as 2 months after birth. The increase in testis size corresponded to spermatogenetic activity. The diameter of the seminiferous tubules increased significantly from November (109 ± 2 µm) to December (147 ± 6 µm) and some spermatides and spermatozoa were observed. During the decreasing phase of the testis volume in winter, only spermatogonia were present (Sempéré et al. 1983).

These findings support the hypothesis of Levasseur (1976), that high gonadotropic activity occurring after birth initiates the pubertal development of the testis. Similar patterns were observed in lambs (Foster et al. 1978), calves (Lacroix & Pelletier 1979; Schams et al. 1978), and pigs (Colenbrander et al., 1977). These observations lead to the suggestion that unlike in adults, the early pituitary activity of juveniles is the result of an endogenous mechanism upon which external factors such as light would not have a major effect. Data obtained by Goss (1969) confirmed that the growth of first antlers depends on the chronological age and is not affected by simulated equatorial photoperiod. Therefore, in fawns and calves of cervids, the timing of the first antlers is probably endogenous

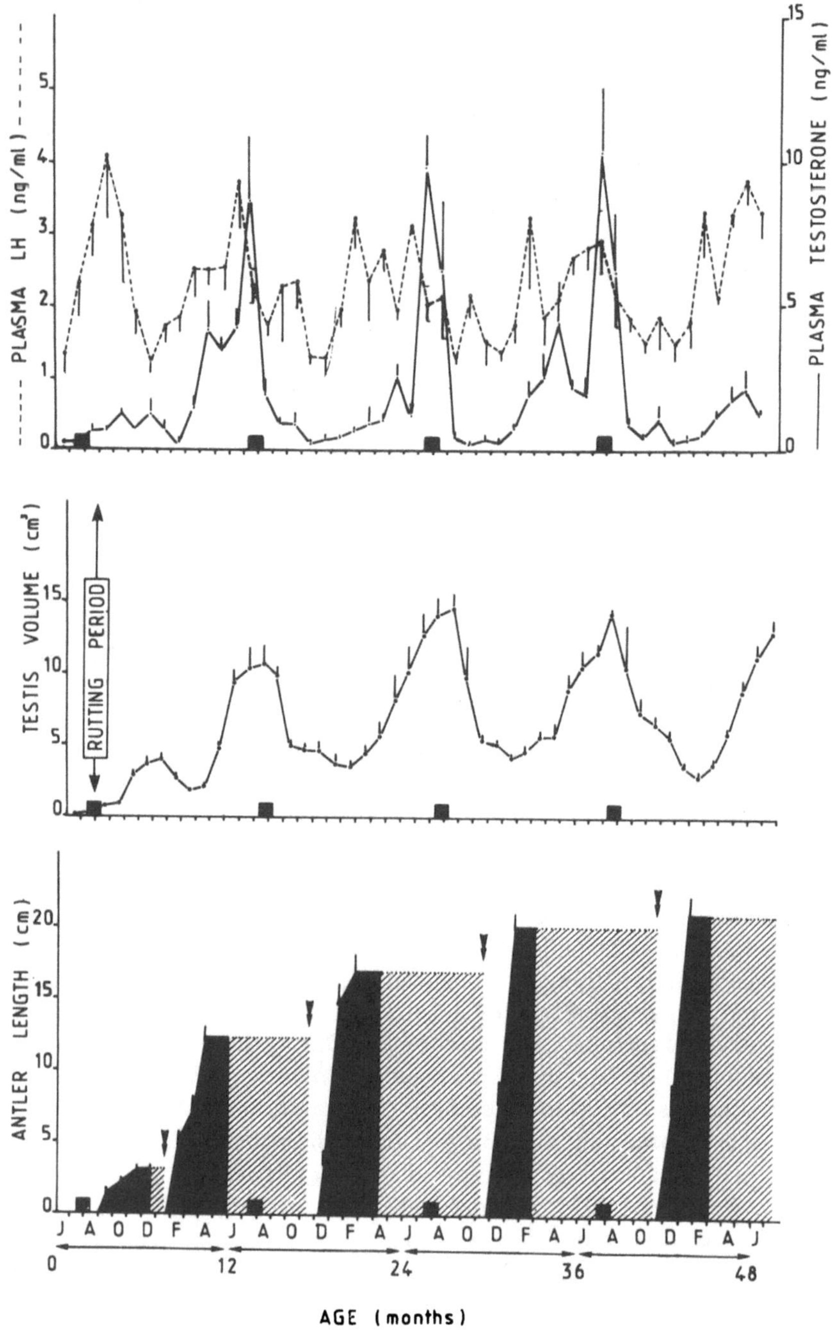

FIGURE 1. Pubertal and seasonal fluctuations of some parameters of sexual activity in roe bucks from birth to 4 years of age (mean ± SEM) (Sempéré & Lacroix 1982). *Black areas*: velvet period. *Shaded areas*: hard antlers. *Black squares*: rutting period.

TABLE 1. Comparative variations of testis volume and testosterone levels in two groups (I, II) of roe deer fawn.

	Group I			Group II[a]		
	n	Testis volume (cm^3)	Testosterone (ng/mL)	n	Testis volume (cm^3)	Testosterone (T ng/mL)
November	15	1.80 ± 0.28	0.30 ± 0.06 (a)	12	3.07 ± 0.29 (c)	0.60 ± 0.01 (g)
December	6	1.55 ± 0.18	0.21 ± 0.02	16	3.40 ± 0.21 (d)	0.39 ± 0.07 (h)
January	15	1.90 ± 0.33	0.15 ± 0.04 (b)	22	2.67 ± 0.19 (e)	0.11 ± 0.03 (i)
February	6	1.68 ± 0.69	0.13 ± 0.08	16	1.80 ± 0.18 (f)	0.10 ± 0.02
March	4	1.66 ± 0.34	0.14 ± 0.05	9	2.45 ± 0.43	0.15 ± 0.06

[a] Animals of Group II have hard buttons only in December.

(a) *vs* (b) $P<0.025$ (d) *vs* (f) $P<0.0001$
(a) *vs* (g) $P<0.005$ (d) *vs* (e) $P<0.005$
(c) *vs* (f) $P<0.001$ (g) *vs* (i) $P<0.001$
(e) *vs* (f) $P<0.001$ (h) *vs* (i) $P<0.001$

(Goss, 1983). Suttie et al. (1984) have shown that young deer stags developed pedicles when they reached a certain body weight.

Similarly, in previously published results, Sempéré and Boissin (1981) have shown that there are two groups of fawns. The animals of group I (GR I) are small and have no hard buttons in winter, whereas a high percentage of fawns (73.5%) constitutes the animals of group II (GR II) which are the heaviest fawns and have hard buttons in December. The testis size of fawns in both groups increased significantly in November. However, the size of the testis was greater in GR II (Table 1: $P<0.001$). Furthermore, the volume of the testis of animals of GR II regressed significantly in January (Table 1). Similar results were also observed by Bejsovec (1955), who found that the weight of roe deer fawns correlates with development of the endocrine system, and thus also with the development of first antlers. By February the testis size was approximately equal in all fawns. In the meantime, plasma testosterone levels were significantly higher in November in GR II than in GR I ($P<0.005$). Androgen levels dropped in all animals in January (Table 1). These data indicated that during the first 8 months of life, the testis activity followed the same pattern in both groups.

Meanwhile, individual variations in testis size and antler cycle indicated that in winter, males of GR II exhibit an earlier androgen peak than males of GR II, whereas the following summer testosterone levels rose in August simultaneously in both groups. In the meantime, the first polished antlers in the GR II were observed in early May and the GR I in mid-June (Fig. 2).

Pedicle development is prevented by castration shortly after birth; however once the pedicle development has been completed, subsequent castration does not prevent antler growth (Tachezy 1956; Bubenik, G. 1982b). These results confirm that the initial testis activation occurring after birth in fawns leads to the induction of growth of the pedicles (Wislocki et al. 1947). The relationship between body weight and the presence of hard buttons in December in a large majority of fawns supported the hypothesis that the growth of body tissues began

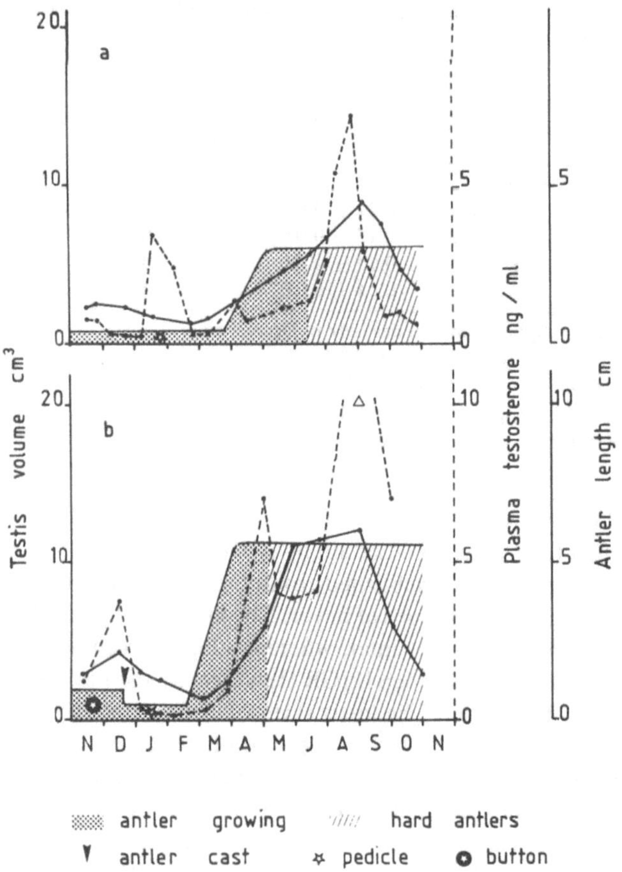

FIGURE 2. Comparative seasonal variations of testis volume, testosterone levels, and antler cycle in two fawns: *a* GR 1, *b* GR II.

in August when these animals (GR II) were sufficiently developed. In the late fall, pedicles and buttons were fully grown in these animals. On the other hand, the beginning of pedicle growth occurred later in the smallest animals (GR I). However, increase in body weight had stopped from November to March in all fawns (GR I and GR II) (Table 2) (Sempéré 1982).

The relationship between body development and the pedicle growth has been reported by Stieve (1950) and Lincoln (1971). In red deer fawns, the age at which pedicle are visible depends on the nutrition and body development (Suttie et al. 1983). However, in this species, normally only one antler cycle was observed during the first year of life even if the fawns were well fed (Suttie et al. 1984). In all deer fawns, the second antler growth period is synchronous with the adult antler growth in the fall, whereas in the heavier roe deer (GR II) such synchronization occurs during their first year of age when they are 8–9 months old. In view

TABLE 2. Comparative variations of body weight and antler length from birth to 4 years in a male roe deer.

	Group I			Group II			
Age (mon)	n	Weight (kg)	Antler length (cm)	n	Weight (kg)	Antler length (cm)	Social rank
0– 8	5	13.7 ± 0.5		17	16.7 ± 0.3	1.0 (button)	Subordinate
8–18	4	17.3 ± 1.2	5.1 ± 0.1	10	21.6 ± 0.6	12.8 ± 0.3	Subordinate
18–36	4	21.3 ± 0.6	18.2 ± 0.3	15	23.0 ± 0.4	19.7 ± 0.5	Subordinate
36–48	3	24.3 ± 0.3	20.1 ± 0.6	7	24.7 ± 0.5	21.0 ± 1.0	Dominant

[a]Body weights and antler lengths are significantly different between the two groups from birth to 36 months.

of these results it can be hypothesized that the animals of GR I were unable to develop their buttons in late fall because of low body weight, which was related to a delay in the development of the endocrine system (Bejšovec 1955). However, testosterone levels in the fall were sufficiently high to promote the pedicle development from which the first set of antlers grew the following spring.

Seasonal Variations of Growth

The body weight of fawns at one year showed very large differences between animals of GR I and GR II (Table 2). Such differences were still observed at the age of sixteen months. However, the animals were fully developed (physically mature) at 4 years and the body weight of old bucks (over 5 years) were approximately equal in the two groups. Furthermore, antler length followed the same pattern, although a very large difference in antler size was observed in both groups at the age of sixteen months (Table 2).

These data indicated that in both groups the males continued to grow until the equilibrium weight was reached at about 4 years of age in spite of the significant body weight differences at the end of the first year. All males (GR I and GR II) exhibited the same aggressivity and reached their dominant rank at the time when the adult body weight had been obtained. These males were able to secure and control their territories.

Sexually Mature Roe Bucks (Older Than 1 Year)

In sexually mature roe deer, large seasonal fluctuations of plasma LH levels were observed. The means of monthly LH concentrations were lowest in December, followed by a sharp and significant increase until February ($P<0.05$). Such higher plasma LH levels were maintained during the spring and summer (Fig. 1). The plasma LH pattern had two phases: January-June and July-December. For 3 consecutive years, the mean plasma LH levels were significantly higher during the first period than in the second (2.42 ± 0.20 *vs* 1.90 ± 0.20 ng/mL, $P<0.05$; 2.64 ± 0.19 *vs* 1.67 ± 0.13 ng/mL, $P<0.001$; and 2.37 ± 0.20 *vs* 2.02 ±

0.10 ng/mL, $P<0.05$). Assuming that LH values over 3 ng/mL represent LH peaks in plasma, frequency of peaks was significantly higher ($P<0.005$) from January-June than from July-December; therefore, the highest values of monthly plasma testosterone levels were recorded in July or August (Fig. 1).

The first peak of androgen levels was observed in early spring. After the rutting period (July-August), testosterone levels decreased quickly; the lowest values were recorded in December.

Similarly, the testis volume showed marked annual variations (Fig. 1). The maximum was reached in summer and the minimum was reached in winter. The seasonal variations of spermatogenesis in roe deer followed the same pattern as the testis weight (Short & Mann, 1966). The exocrine testicular function was totally inactive in December and January and only some spermatogonia were observed in the seminiferous tubules (Stieve 1950; Short & Mann 1966; Sempéré, personal observation). The same seasonal testicular regression was also detected in genus *Odocoileus* (Robinson et al. 1965; West & Nordan 1976). The presence of primary spermatocytes during the reproductively inactive period was also observed in the testis of the reindeer (Meschaks & Nordkvist 1962), fallow deer (Chapman & Chapman 1970), and red deer (Lincoln 1971). In the roe deer histological studies during the late winter testis reactivation have shown primary spermatocytes in February (Short & Mann 1966) and numerous spermatozoa in March (Sempéré, personal observation). Although the maximum exocrine testicular activity was reached in July, spermatozoa were still present in the seminiferous tubules in November (Sempéré 1982).

This study clearly revealed that winter seasonal testicular activity in roe bucks is comparable to the summer testis reactivation observed in other male cervids. LH release in adult roe bucks exhibits pronounced seasonal variations similar to those observed in white-tailed deer (Mirarchi et al. 1978; Bubenik, G. et al. 1982). The increase of mean plasma LH levels, which suggests the onset of activity of pituitary gonadotroph, appears 2 months before testicular reactivation. In most boreal cervids, the increase of pituitary activity is followed by the activation of the sexual function that occurs when day length decreases (Goss & Rosen 1973). In white-tailed deer, the peak of plasma LH was observed in July, whereas the maximum of testosterone levels were observed almost 5 months later in December (Bubenik, G. et al. 1982a) (Fig. 3). This 5-month difference between peak levels of LH and maximum testosterone values in blood is fairly common in other deer species (Bruggemann et al. 1965; Mirarchi et al. 1978; Lincoln & Kay 1979). It appears that maximal levels of testosterone are essential for the rutting behavior (Bubenik, G. et al. 1982). Our present findings indicate that the onset of seasonal pituitary activity occurs in January and the beginning of testicular function follows in early spring. This demonstrates that the roe deer is the only temperate cervid in which sexual activity is initiated by increasing day length and which breeds in early summer.

Therefore, roe deer as a pliocervid (Bubenik, G. 1986) is a typically long-day breeder, whereas white-tailed deer, despite being a pliocervid, is generally consid-

FIGURE 3. Temporal and seasonal relationship among LH, testosterone, antlers, and territorial behavior. *Black areas*: territorial behavior. *Stars*: rutting period—*a* in roe deer (Sempéré 1982), *b* in white-tailed deer (Bubenik, G. et al. 1982a).

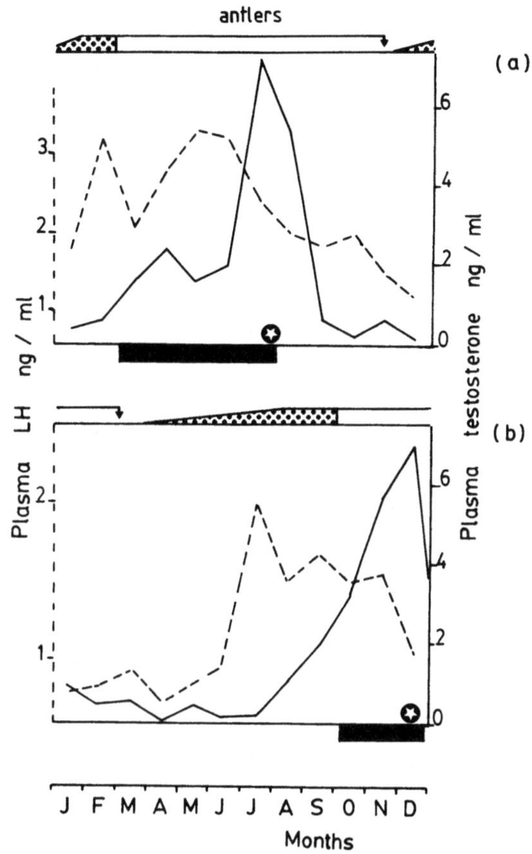

ered a short-day breeder. However, white-tailed deer are photoperiodical in northern latitudes but a continuous breeder close to the equator (Webb & Nellis 1981).

A partial reactivation of spermatogenesis in the spring has also been recorded in several deer species: roe deer (Stieve 1950); red deer (Frankenberger 1954; white-tailed deer (Robinson et al. 1965); reindeer (Meschaks & Nordkvist, 1962). The increase of pituitary-gonadal activity detected in the roe deer in the fall (Figs. 1 and 3) could be considered analogous to the spring reactivation observed in short-day breeding deer (Bubenik et al. 1982a).

It is possible that, among the *Cervidae*, seasonal hypothalamo-gonadic activity may not match exactly the models of short-day or long-day breeders, but can be modified by environmental factors (such as the day length or temperature) or by internal feedback mechanisms.

In roe deer, seasonal variations in the concentration of plasma LH show a biphasic profile. During the period of lengthening days, the peaks of LH which occurred in February and May-June reveal an alternation in stimulation and

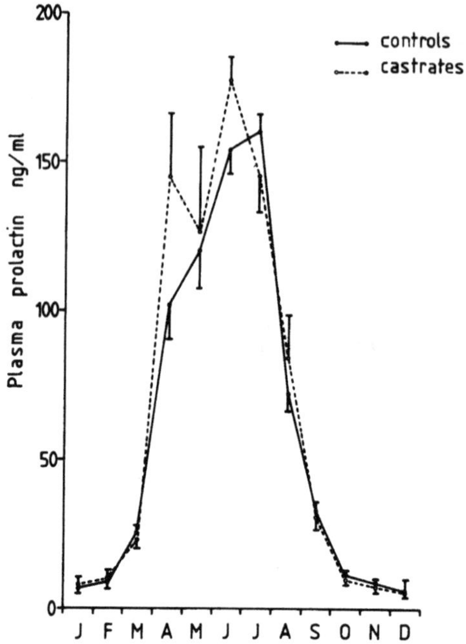

FIGURE 4. Seasonal levels of plasma prolactin in control and castrated roe deer.

inhibition between photoperiod phases. The phase of inhibition of LH coincided with the vernal equinox. Conversely, during the period of shortening days, only one stimulative period was observed, during the autumnal equinox. In other *Cervidae* (which are mostly short-day breeders), an inverse scheme exists, with a stimulative phase occurring during the period of shortening days (Bubenik, G. et al. 1982).

As demonstrated by Goss (1976), the growth of antlers can be initiated by either lengthening or shortening of the photoperiod. G. Bubenik et al. (1982) suggest that by a similar mechanism, the secretion of gonadotrophins might be triggered either by the shift from the declining to increasing day light or vice versa; these gonadotrophin levels might be also modified by feedback mechanisms based mainly on testicular factors (Bubenik, G. 1982).

Prolactin is the hormone strictly controlled by the annual photoperiod. In the adult roe buck, mean PRL concentration showed pronounced seasonal variations (Fig. 4). The lowest values were recorded between October-February; concentrations gradually increased through March and reached their peak between May and July. Similarly to white-tailed deer (Bubenik, G. et al. 1985), the annual PRL cycle was not affected by castration (Fig. 4). Recently, we treated two adult roe bucks with PRL blocker bromocriptine. From March-September two males were injected intramuscularly once a month with 50 mg of Parlodel (long-lasting form of bromocriptine, kindly supplied by Sandoz, Basle, Switzerland).

From March to May, testis size and testosterone levels were affected by the treatment; whereas LH and FSH levels were not significantly influenced by bromocriptine, peak levels of PRL were reduced. However, a short increase of PRL and testicular activity was observed in June (Fig. 5). These results indicate

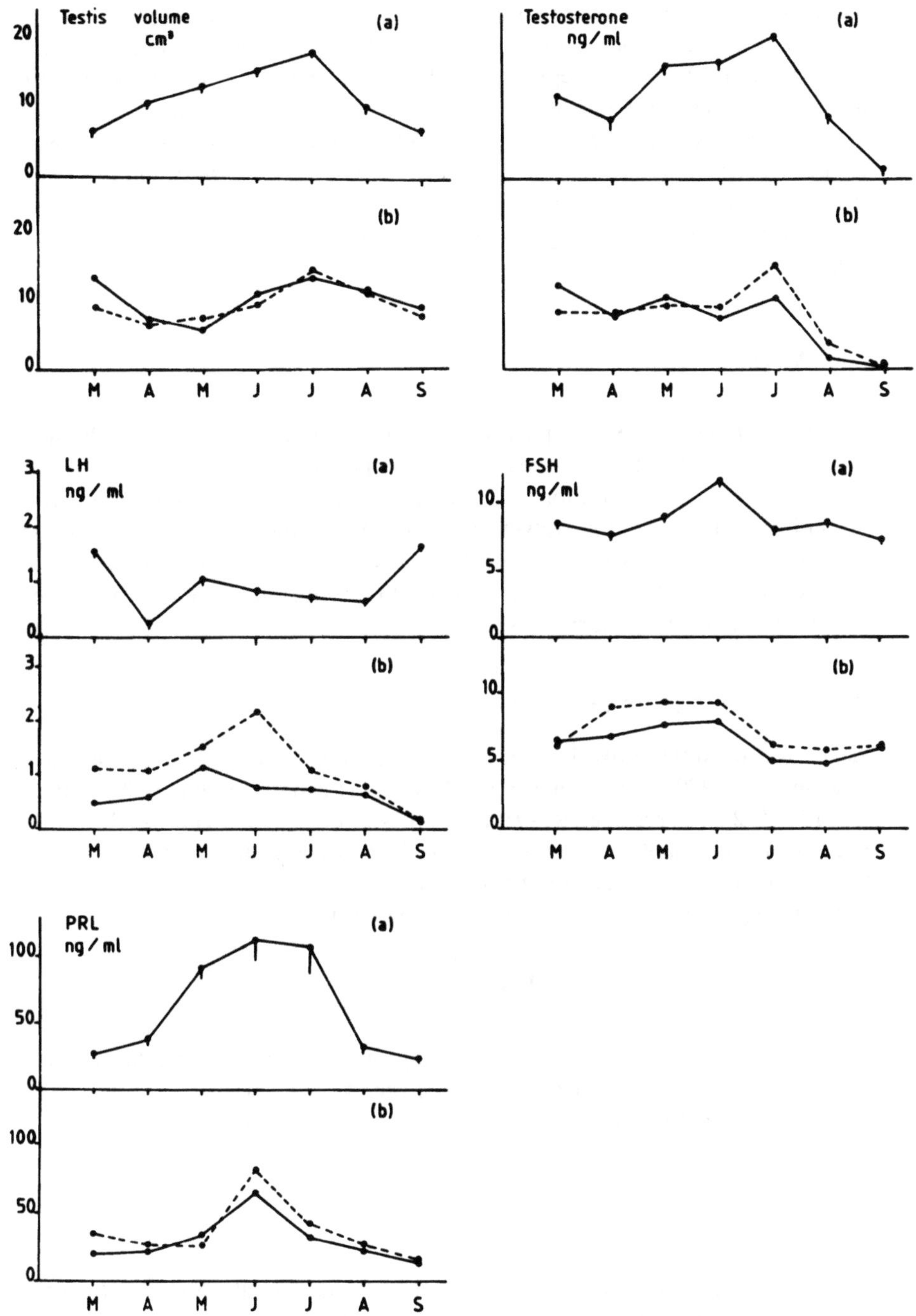

FIGURE 5. Seasonal levels of pituitary hormones (LH, FSH, PRL) and testicular activity (testis volume, testosterone levels) in adult male roe deer: *a* control; *b* treated with Parlodel (bromocriptine).

that in long-day breeders, the lack of PRL in summer can affect testicular activity. Moreover, bromocriptine treatment did not affect gonadotrophic hormones (LH and FSH) but depressed PRL levels. Therefore, it is hypothesized that PRL acts as a proandrogenic hormone in roe deer, which is a long-day breeder (Schams et al. 1986), whereas PRL is an antiandrogenic hormone in short-day breeders (Bubenik, G. et al. 1985; Bubenik, G. 1986). On the other hand, the short-term increase of PRL in June indicates that the monthly interval between bromocriptine injections was too long during the period of maximum secretion.

The Antler Cycle

Growth of antlers in cervid species is a repeated seasonal event closely synchronized in temperate countries with photoperiodicity (Bubenik, G. et al. 1987a). Artificial manipulation of day length influenced onset of antler growth and timing of mineralization, velvet shedding, and casting (Jaczewski 1954; Goss 1983; Suttie & Simpson 1985; Bubenik, G. et al. 1987a). The development and mineralization of antlers in male cervids is closely related to their sexual cycles and variations of sexual hormones.

Duration of Antler Growth

The period of antler growth varies in deer species: red deer, 132–140 days (Lincoln et al. 1970); male reindeer, 98–133 days (Leader-Williams 1979); fallow deer, 122–132 days; roe deer, 81–93 days; and sika deer, 129–141 days (Chapman 1975). The duration of antler growth in roe deer was estimated in tamed animals by a photographic technique (Sempéré 1982).

Antler lengths were estimated by the following formula:

$$LA = \frac{L_o \times l_A}{l_o}$$

LA = Antler length (from the coronet along the beam to the most distal tip)
L_o = Ear length (11 cm, measured before the experiment)
l_A = Antler length (from photo)
l_o = Ear length (from photo)

Antler Growth Phase

Results presented in Figure 6 indicate that the new velvet-antlers started to grow 7–12 days after casting. This phase was followed by a rapid and regular growth phase during 60 days with a growth speed of 3.3 mm/day.

If old antlers are cast in November-December, the new antlers begin to grow almost immediately. In wild roe bucks, the antlers reach their full size in January,

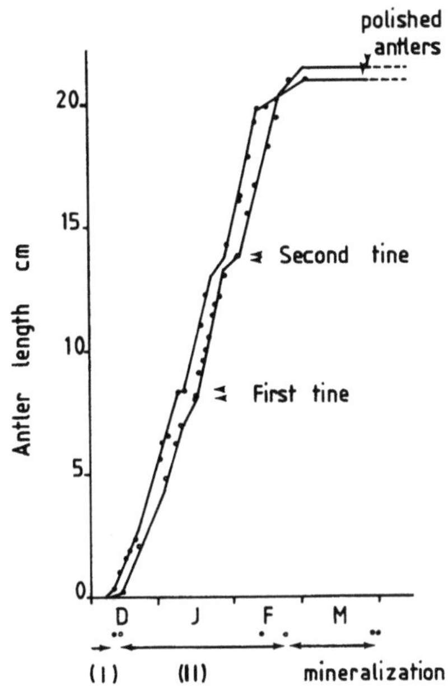

FIGURE 6. Determination of phases of the antler cycle by photograph analysis in two roe bucks: (I) slow growth phase; (II) rapid growth phase.

after about 2 months of intense growth (Sempéré 1982) (Table 3). The intensity of antler growth was linear from November-January ($y = 1.02 \pm 0.31\ x$, $r = 0.9941$). A detailed study of captive animals showed that the antler development period was also around 60 days and the antler mineralization period was 45 days (Table 4). In captive bucks the growth was also linear for the 60 day period of development ($y = 3.50 \pm 0.22\ x$, $r = 0.9975$).

TABLE 3. Mean ± SE of plasma androgen concentrations and antler lengths of wild roe bucks captured in the forest.

Date of capture	No. of bucks	Androgen (ng/mL)	Antler length (cm)
November 15	5	0.40 ± 0.19	1.40 ± 0.87
December 6	3	0.15 ± 0.01	8.83 ± 3.19
December 13	3	0.15 ± 0.01	9.60 ± 0.57
December 20	3	0.17 ± 0.03	11.60 ± 2.24
January 17	4	0.58 ± 0.08*	18.50 ± 0.92*
January 24	6	1.39 ± 0.19*	19.60 ± 0.66
February 7	4	4.61 ± 1.22*	19.90 ± 1.43
February 14	9	2.39 ± 1.28	21.30 ± 0.88

*Value significantly different from that preceding it, $P < 0.05$.
From Sempéré and Boissin 1981.

TABLE 4. Mean ± SE of plasma androgen concentrations and antler length in four captive bucks measured from the top of antler casting in November.

Days	Androgen (ng/mL)	Antler length (cm)
0	0.12 ± 0.025	3.78 ± 0.26
15	0.27 ± 0.10	6.40 ± 0.29*
30	0.40 ± 0.23	10.15 ± 0.67*
45	0.57 ± 0.17	13.60 ± 0.34*
60	0.43 ± 0.16	16.00 ± 0.69*
75	1.99 ± 0.42*	16.60 ± 0.98
90	2.63 ± 1.59	17.20 ± 1.53
105	2.60 ± 1.07	17.20 ± 1.53

*Value significantly different from that preceding it, $P < 0.05$.
From Sempéré and Boissin 1981.

Mineralization Phase

The intense mineralization phase occurred between the end of a rapid antler growth and the period of velvet shedding (Fig. 6). In captive and wild roe bucks, the velvet was polished between the 90th and the 105th day (Tables 3 and 4) (Sempéré & Boissin 1981). During the antler growth period, the androgen concentration was very low but increased between days 60–75 (Table 4).

Factors Affecting the Antler Cycle

Antler Length and Age of Animals

In *Cervidae*, the length of antlers depends on the age of the deer. In roe deer, antler length increased significantly from birth to adulthood (Fig. 7). This difference is more pronounced between yearling and subadult than between subadult and adult (more than 2 years old).

The mean speed depends on the final length of antlers and the age of animals. Mean speed was 3.59 ± 0.23 mm/day for antlers whose length was over 22 cm and 2.63 ± 0.16 mm/day ($P<0.05$) when the final antler length was about 15 cm. Therefore, depending on the speed and the final antler length, the duration of antler growth is comparable whatever the age.

Influence of Hormones on Antler Cycle

The interesting aspect of the developing antlers is the similarity of growth and mineralization processes with the long bones of the skeleton. Therefore, antler development has been compared to a modified form of enchondral ossification in which osteoblasts and chondroblasts are present (Banks & Newbrey 1982).

Numerous studies on developing antlers have been made: bone microstructure, collagen production, organ regeneration, and endocrine and neuronal regulation of bone growth (reviewed in Bubenik, G. & Bubenik, A. 1987).

FIGURE 7. Relationships between the age and the antler cycle (growth speed and length) in wild roe bucks.

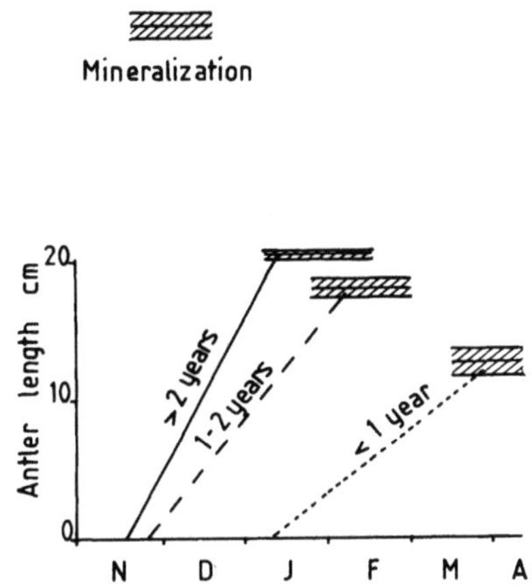

Blood levels of alkaline phosphatase (AP), which is secreted mainly by osteoblasts, are a good indicator of the rate of bone deposition. We recently investigated blood levels of AP and thyroid hormones in white-tailed deer to explore their role in the development of antlers (Sempéré et al. 1986; Bubenik, G. et al. 1987b) by comparing hormone levels in blood from the jugular vein and the lateral antler vein (Fig. 8). The data indicate that AP is produced in large quantities in antler tissue during antler development whereas T_4 exhibited the same levels in antlers and jugular blood. The lower levels of triiodothyronine (T_3) in antler blood compared to jugular vein revealed that T_3 is utilized in the growing antlers, especially during the production of preosseous tissues in May-June. In addition, Eiben et al. (1984) reported peak levels of calcitriol (1.25 $(OH)_2D_3$) in plasma of fallow deer, which were concomitant with peak levels of AP.

Recent studies conducted in roe bucks (Sempéré et al. 1989) have shown that a large increase of plasma AP and 1.25 $(OH)_2D$ from the jugular vein have been observed during the antler growth period in December (Fig. 9). However, blood sampled in the velvet vein exhibited higher levels of AP and calcitriol than in the jugular vein. At the same time, inversed fluctuations of 25 (OH)D and 1.25 $(OH)_2D$ were observed between jugular and antler veins. The decrease of plasma jugular and plasma antler vein 25 (OH)D from November-January would indicate that growing antlers could metabolize 25 (OH)D to 1.25 $(OH)_2D$, a process which generally occurs in the kidneys (Kumar 1984).

Action of Testosterone

Except in *Rangiferins*, antlers are secondary sex characteristics of *cervidae*. Therefore, they are primarily affected by the annual fluctuations in the secretion of sex hormones (Tachezy 1956). Androgens appear to be the hormones which are most directly affecting the antler cycle (Mirarchi et al. 1978; Bubenik, G. et al. 1982). Casting and regrowth of the antlers is dictated by changes in

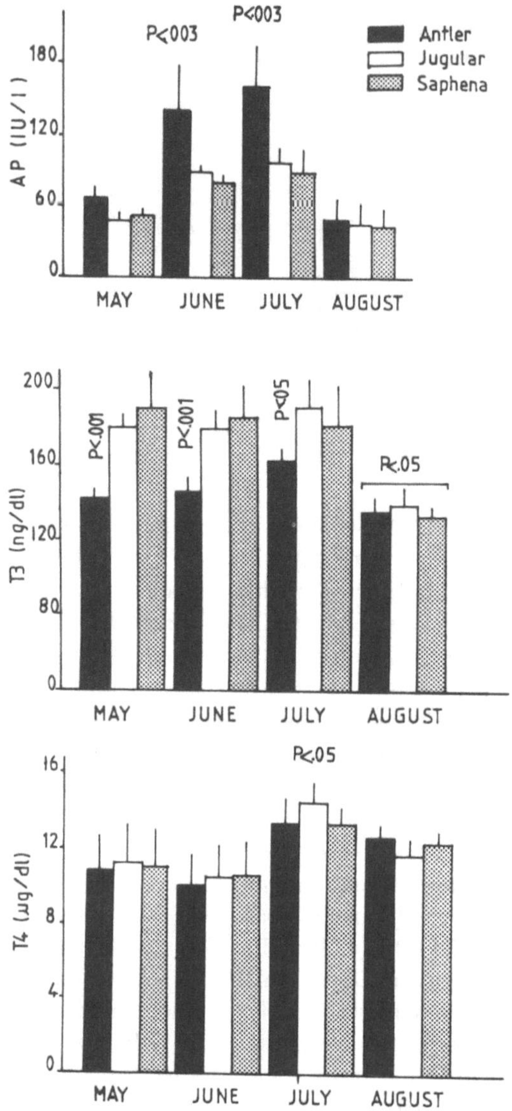

FIGURE 8. Concentration of AP, T_3, and T_4 in white-tailed deer antler (A), jugular (J), and saphenous veins from May-August; $P<0.003$ indicates significant differences between A and J (Bubenik, G. et al. 1987d).

testosterone and occurs when androgen levels are very low. Maturation of the hard antler begins when testosterone levels increase (Wislocki et al. 1947; Lincoln et al. 1970; Bubenik, G. et al. 1975, 1977; Sempéré & Boissin 1981).

The relationship between antlers and gonads has been confirmed by castration in cervids (reviewed in Chapman 1975). Administration of testosterone before old antlers have been dropped may postpone casting for as long as the hormone injections continue (Jaczewski & Galka 1970; Goss 1968; Sempéré & Boissin 1982).

FIGURE 9. Concentrations of 25 (OH)D, 1.25 $(OH)_2D$ and AP in the roe deer antler and jugular veins from November-March.

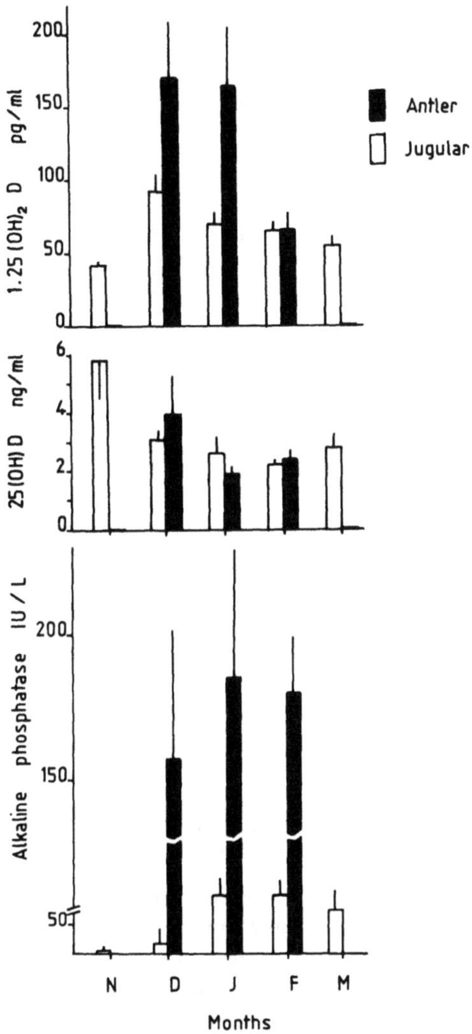

In my recent experiment, injections of 1 ml of hexahydrobenzoate of testosterone (125 mg/ml) twice a month to two groups of three adult roe bucks, before the antler casting, delayed the cast of antlers for the length of the treatment. The first group was treated from the beginning of November to the middle of December. The treatment of the second group started also at the beginning of November but continued until February 1. The antlers were cast and the growth of new antlers began when testosterone injections ceased (Fig. 10). The time of full antler development (until the velvet was shed) was the same in the first group as in controls, but it was shorter in the animals of the second group. This is probably related to a shorter mineralization phase (22.0 ± 1.0 days in the second group *vs* 36.3 ±

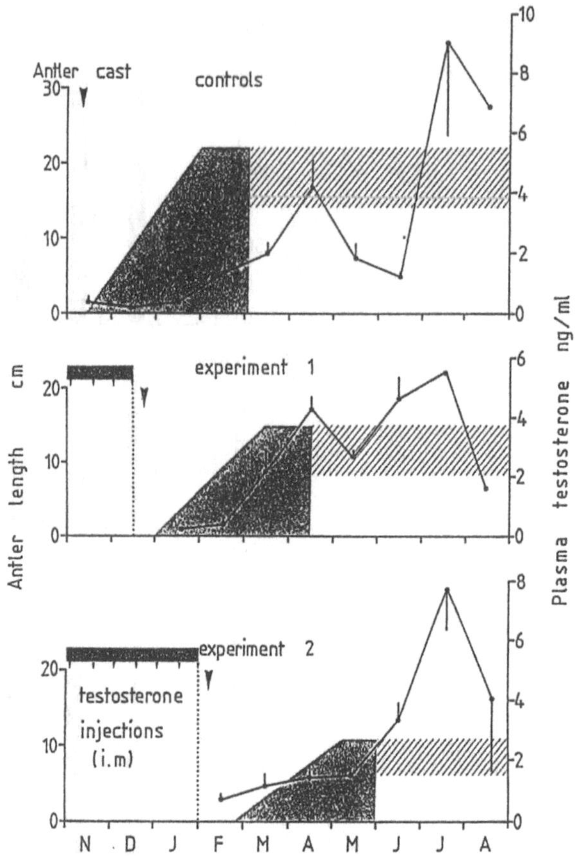

FIGURE 10. The influences of testosterone hexahydrobenzoate injected for 1.5 months (experiment 1) or for 3 months (experiment 2) on the antler cycle in roe deer; the relationship between plasma testosterone concentrations and the antler growth and mineralization in controls and experimental bucks.

2.3 days in controls). The length of antlers was also reduced in both groups of treated animals. The testosterone levels of controls and the first treated group increased significantly during the mineralization phase. In the second group this significant increase in androgen levels occurred already during the growing phase ($P<0.01$).

In the first experiment, the delay of antler casting for 1.5 months did not alter the length of antler development. In the second experiment, the delay of antler casting for 3 months reduced the speed of antler growth and the length of the mineralization phase but did not affect the duration of antler growth period.

Action of Adrenal Androgens

The castration effect is variable in different deer species. In roe deer, castration induced uncontrolled growth of poorly mineralized antlers (peruke); castrated white-tailed deer, fallow deer, or red deer have partly mineralized antlers; castrated reindeer exhibit completely mineralized antlers which keep the velvet (Bubenik, A. 1966).

The role of the pituitary adrenal axis in antlerogenesis has been investigated in normal and castrated roe bucks (Bubenik, A. et al. 1976). The cortisone administration (22 consecutive days with a daily dose of 25 mg of cortisone) induces premature shedding of the velvet. In castrated animals the cortisone treatment delayed the growth of the peruke, which demonstrated that antler growth can be independent of gonadal steroids. The exogenous glucocorticoids also depress ACTH secretion. Therefore the authors suggest a possible role of adrenal androgens in the evolution of the growth and mineralization process of the antlers.

Action of Estrogens

In addition to androgens, female sexual hormone estrogens also exercise a powerful influence on the antler cycle.

It has been demonstrated that administration of estrogens delayed the casting of mature antlers. A subcutaneous implant of 100 mg of estradiol-17B was extremely effective in preventing red deer from casting their antlers for at least 2 years (Chapman 1975). In roe deer, blockage of antler casting was also observed after treatment with 20 mg of estradiol-17β implant (Sempéré, unpublished). Finally, in roe deer implanted 1 week after casting, new antler growth was inhibited at the time of the implant (Schams et al. 1986). These results indicate that estradiol-17β has a direct effect on the growth and casting of antlers.

Influence of Photoperiod

In boreal deer, the annual replacement of antlers is particularly sensitive to seasonal changes in day length. Jaczewski (1954) has shown that in red deer which were abruptly switched from the increasing day length to the decreasing one, the antlers ceased to grow and became mineralized. Goss (1983) demonstrated that the antler cycles can be affected by changing the day length.

Adult roe deer kept in a room with an artificially altered photoperiod (providing two complete cycles per year) exhibited periods of antler growth which correlated with the photoperiodicity (Fig. 11). The antler casting coincided with winter solstice. The first antler cycle was not affected by a new light regime because the velvet antlers were already growing at the start of the experiment, but the second antler cycle was modified. The antlers of the second cycle were smaller than those in the first, had rounded tips and were not completely polished. The rapid increase of testosterone levels during the antler growth

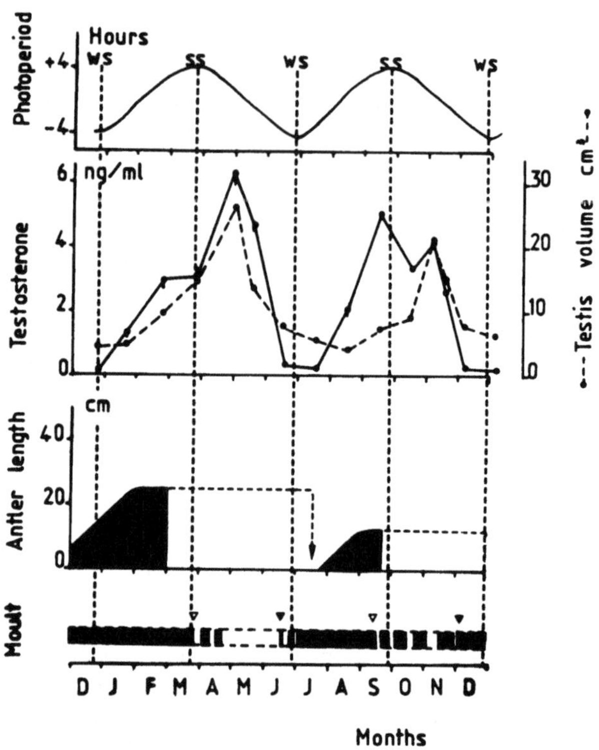

FIGURE 11. The effect of a two-fold acceleration of the annual photoperiod on the sexual cycle, the antler cycle, and the moult of a male roe deer. *Arrows*: casting period. *Open triangles*: summer coat. *Black triangles*: winter coat.

period had probably caused premature ossification before the development of cartilage matrix was completed (Suttie & Simpson 1985).

Conclusions

The rutting season occurs at the time specific for each deer species. The timing of the rut is regulated in a way to ensure that births occur in the spring: the larger species with the longest gestation have the rut earliest (Lincoln 1985). The roe deer which is a small species has a rutting period in July-August. In this deer the 5-month delay in the implantation of the blastocyst ensures that the fawns are born in the May-June period (Bischoff, 1854). This mechanism, which appears to be a secondary adaptation of the female sexual cycle, also may explain the special characteristic of the roe buck sexual cycle: the long-day breeding period. Despite the fact that, in the roe deer, blood levels of LH are normally elevated during the period of increasing day length (January-June), an increase in plasma gonadotrophin hormones was also observed in the fall (September). A comparable pattern of spring and fall gonadotrophin secretion was also described in white-tailed deer (Bubenik, G. 1986). It was suggested by West and Nordan (1976) that the equal ratio of light to dark periods during the fall and the spring equinox provides for the gonadotrophin stimulation. Roe buck may respond briefly to photoperiodic stimulation in the fall, but reproduction did not occur at this time because gonadal activation is incomplete. The interesting aspect of the developing antler is its similarity of growth and mineralization processes with the long bones of the skeleton (Banks & Newbrey 1983). We have shown that alkaline phosphatase, which is a good indicator of the rate of bone deposition, appears to best reflect the intensity of antler growth. Studies conducted in cervids have shown that an increase of 1.25 $(OH)_2D$ has been observed during the antler growth phase (in winter in roe deer, in summer in fallow deer). Moreover, we believe that the velvet of growing antlers can metabolize 25 (OH)D to 1.25 (OH)D, a process which generally occurred only in the liver and kidneys (Kumar 1984).

15
Endocrine Mechanisms and Antler Cycles in Rusa Deer, *Cervus (Rusa) timorensis*

SIMONE VAN MOURIK AND TEODOR STELMASIAK

Introduction

The Acclimatisation Society of Victoria, Australia, introduced at least 17 species or subspecies of deer in the last century (Bentley 1978). Six species became established in Australia: fallow deer, *Dama dama*, chital, *Axis axis*, hog deer, *Axis porcinus*, red deer, *Cervus elaphus*, sambar, *Cervus (Rusa) unicolor*, and rusa deer, *Cervus (Rusa) timorensis*. Information about their history is sparse, apart from a record in Acclimatisation Society's report (1868) of the import of rusa deer ("Batavian deer"). However, it appears certain that more than one variety of rusa deer was released in Gembrook, Victoria (1890), the Royal National park, New South Wales (1907), and Torres Strait Islands (1912). Rusa deer populations exist in the Royal National Park and on the Prince of Wales Islands but were eradicated in Victoria.

An extensive culling program in the Royal National Park in 1970 resulted in rusa deer being available for farming in Victoria. These animals formed the nucleus for a herd that now numbers approximately 1,000 animals. The rusa deer are maintained for breeding, venison production, and collection of velvet-antlers on a farm located in Southern Victoria (38° 42'S 146° 6'E) with improved pasture (ryegrass and clover) subdivided in paddocks of variable size.

Although rusa-like and axis-like deer are known from continental Europe between the Pleiocene and Pleistocene, it is unlikely that they were the ancestors of the deer present in the tropics and Australia nowadays. All these deer with rusa- and axis-like antlers disappeared from Europe before the first glacial period, but there is no explanation at this stage.

Because of geographical barriers, it seems impossible for *R. timorensis* to have reached Java and surrounding islands from Europe. Therefore it is more likely that Java-Rusa deer evolved in the Szechuan (China) region together with other Rusa subspecies. Later on the change in climate (Holocene: postponed glaciation in Eastern-Siberia, North-China) initiated the migration southwest over the numerous land bridges to Borneo and Java (A. Bubenik, personal communication), the more susceptible subspecies first, Java-Rusa later. Dr. Cao Keging (personal communication to A. Bubenik) has confirmed the presence of *R. timorensis* in Taiwan during the late Pleistocene. Research is being conducted to elucidate its presence in China where four different species of rusa, incl. *R. unicolor* have been found since the early Pleistocene.

If rusa deer evolved outside the tropical region and migrated late to the tropics, this might explain the observed seasonality in reproductive processes after transfer back to a temperate zone (Peking Zoological Gardens, personal communication). No data on endocrinology of rusa deer under tropical or temperate zone conditions were available, apart from general descriptions of rusine deer—e.g., Muller & Schlegel (1839–1844), Mohr (1918), van Bemmel (1949, 1973), Fraser Stewart (1981)—and more recently, specific accounts on antler structure (Acharjyo & Bubenik, 1982).

Over a 4-year period, blood samples have been collected from manually restrained or immobilized stags (Van Mourik & Stelmasiak 1984) and analyzed by radioimmunoassay for prolactin, luteinizing hormone (LH), and testosterone (T) (van Mourik & Stelmasiak 1985; van Mourik et al. 1986). These data were evaluated in relation to observed seasonality in reproductive behavior, physiology, and antler growth. Besides, data and yield of velvet antlers were collected to establish an annual pattern of antler growth in stags of different age.

Annual Plasma Prolactin Concentrations

The cervid species investigated previously originated from the temperate zones, where distinct photoperiod changes exist and the animals exhibit well-synchronized annual patterns for reproductive activity and antler growth. These patterns are closely correlated with the annual changes in prolactin, T, and LH concentrations (Fig. 1).

Circulatory plasma prolactin levels have been reported for cervids. White-tailed deer, *Odocoileus virginianus*, have been investigated by Bubenik et al. (1983, 1985), Mirarchi et al. (1978), Schulte et al. (1980a,b, 1981a,b), and roe-deer, *Capreolus capreolus*, by Schams & Barth (1982), Sempéré & Lacroix, (1982) in the Northern Hemisphere and red deer, *Cervus elaphus*, by Kelly et al. (1982) in the Southern Hemisphere. An annual pattern in the secretory activity of prolactin-producing cells is indicated by prolactin concentrations in plasma obtained by radioimmunoassay as well as light microscope cytochemistry (Schulte et al. 1980b) and ultrastructural changes in the prolactin secreting cells in the hypophyses of deer (Schulte et al. 1981b). The zenith of activity occurred during midsummer and the nadir occurred in winter, suggesting the photoperiod as a prominent factor in the regulation of the annual cycle of prolactin secretion as well as antler growth.

In contrast, rusa deer, *Cervus rusa timorensis*, which originated in a tropical region with negligible changes in the photoperiod, do not have such a well-defined antler growth cycle and reproductive season. This lack of synchrony with the photoperiod exists in the native habitat of rusa deer and even persists after their transfer to northern latitudes (Mohr 1918; van Bemmel 1949).

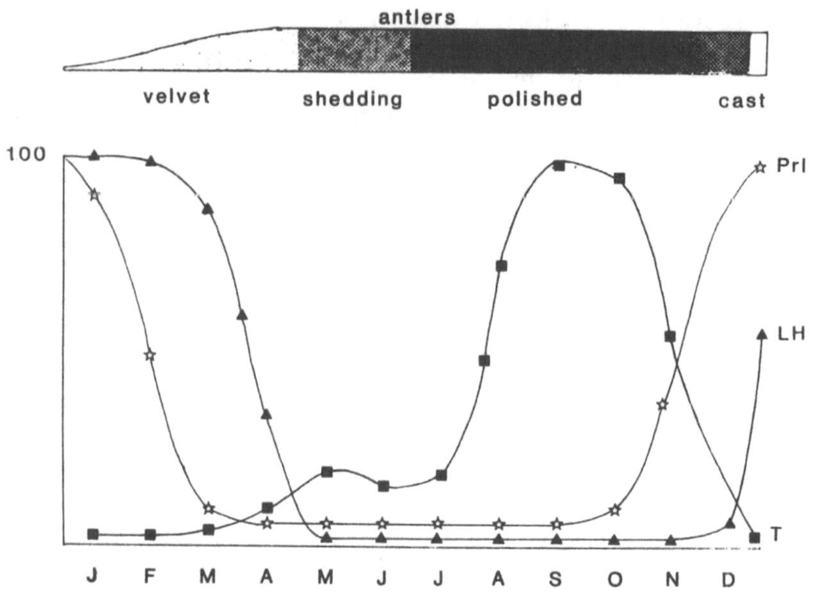

FIGURE 1. Seasonal levels of prolactin (Prl), luteinizing hormone (LH), and testosterone (T) in male rusa deer.

However, in agreement with previous reports on prolactin levels in cervids, rusa stags showed elevated levels of prolactin during the summer months (October-December). Our results are in contrast to the statement by Goss and Rosen (1973) that "tropical species of deer are evidently unresponsive to fluctuations in daylength." It is more likely that responsiveness to a changing photoperiod is an inborn mechanism only to be expressed in changing prolactin levels of deer living outside the equatorial region. This also supports the finding that the increase of prolactin is controlled strictly photoperiodically and can be observed in all mammalian species investigated, irrespective of their seasonality and sex (Karg et al. 1973; Ortavant et al. 1978).

Although the annual prolactin pattern for rusa deer resembles that of other cervids in respect to photoperiod changes, the reproduction and antler-growth patterns are totally different. Breeding commences in July (late winter), calves are born in April (autumn), and antler growth starts in January for the majority of the population (van Mourik 1986).

All temperate zone species of deer produce antlers in the spring and summer months, with the exception of roe-deer, which grow antlers in the winter. The roe-deer, however, is the only ungulate to exhibit delayed implantation so that the calving season occurs during favorable climatic conditions (spring).

Prolactin stimulates limb generation in amphibians and caudal regeneration in reptiles (Turner & Bagnara 1971); these regenerative properties may carry over to some degree in mammals. In northern temperate zone latitudes, prolactin levels in white-tailed deer peak in May-June and coincide with the maximal rate of antler-growth (Bubenik, G. et al. 1983, 1985a; Mirarchi et al. 1978; Schulte et al. 1980a,b). However, the early proposal by Wislocki (1943) that prolactin was a major hormone in relation to antler growth could not be substantiated by later research. Pedicle or antler development could not be initiated by administration of ovine prolactin in red deer stags castrated before puberty (Jaczewski & Krzywinska 1974). Also, a white-tailed buck whose antlerogenesis was disrupted by gonadotrophin blockade did not respond to prolactin (West & Nordan 1976a).

The rusa stags in this study are velveted each year at the appropriate growing stage (6 weeks) of the antler, i.e. before bifurcation of the main beam. The fast-growing stage of the antler coincides with the declining concentrations of prolactin. This is in direct contrast to observations on white-tailed deer (Mirarchi et al. 1978), which showed lowest prolactin concentrations when the antlers were hard and polished.

Blockade of prolactin secretion in white-tailed deer by intramuscular administration of bromocriptine (2-bromo-ergokryptine) reduced spring prolactin levels by more than 90% but did not affect antler growth (Bubenik, G. et al. 1985a). Following this blockade of prolactin no significant changes in the seasonal cycles of LH and follicle-stimulating hormone (FSH) were observed. However, T reached peak values in July (summer) rather than November-December. Bubenik, G. et al. (1985a) have speculated that in normal untreated bucks the high prolactin levels observed in June and July delayed the elevation of plasma T by LH. If the increase of prolactin was prevented, T levels rose simultaneously with rising levels of LH. This interrelationship among LH, T, and prolactin does not seem to exist in rusa deer. Plasma LH increases in January and T begins to rise significantly around July (winter) (van Mourik et al. 1986), while plasma prolactin concentrations increase from September onward.

Annual Plasma Testosterone, LH Concentrations, and Seasonal Variation in LH-RH Responsiveness

T and LH concentrations have been reported for white-tailed deer, *Odocoileus virginianus*, red deer, *Cervus elaphus*, roe deer, *Capreolus capreolus*, and reindeer, *Rangifer tarandus* by Bubenik, G. et al. (1982a, 1983); Mirarchi et al. (1978); Haigh et al. (1984); Lincoln (1971); Suttie et al. (1984); Schams & Barth (1982); Sempéré et al. (1982, 1983); and Whitehead & McEwan (1973).

We were able to establish an annual pattern in testosterone and LH plasma concentrations in manually restrained stags or stags immobilized by Rompun and Ketalar (van Mourik & Stelmasiak 1984) (Fig. 1). LH was only detectable in plasma samples collected between January and April, while T showed a peak in September.

Endocrine and histological studies in white-tailed deer, roe-deer, black-tailed deer, and red deer indicate that the seasonal rhythm does not have just one peak. Bubenik, G. et al. (1979) showed that two peaks of estradiol-17 occur in mature white-tailed deer—one in October-November just before the rut and the other in April-May at the beginning of antler growth. Bubenik, G. et al. (1982a) also

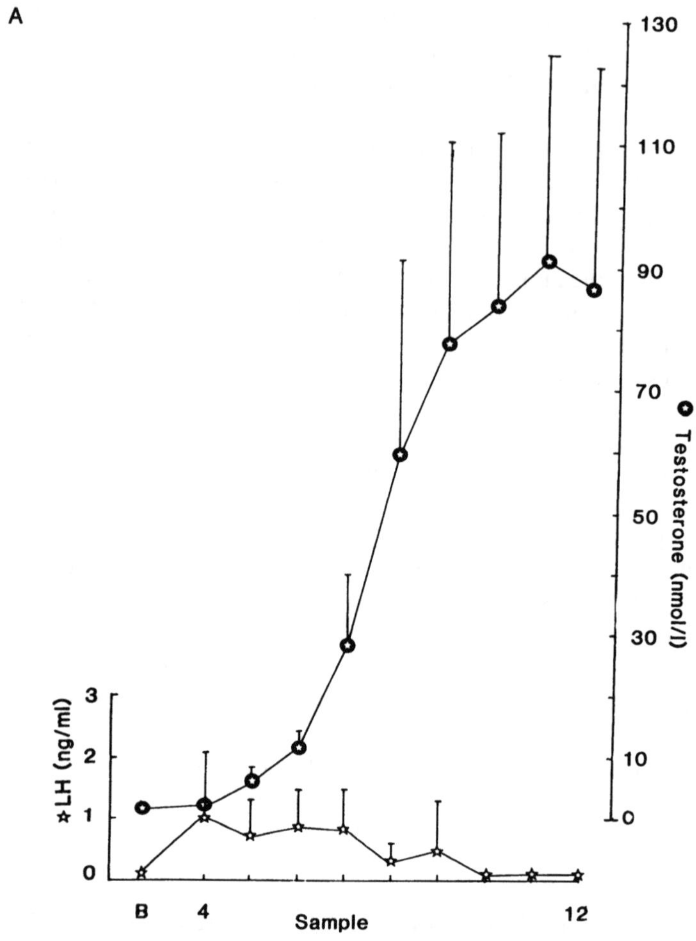

FIGURE 2. *A* Testosterone and LH levels (means and SD) in male rusa deer after intrajugular administration of 500 ng of LH-RH in August. *B* Testosterone and LH levels (means and SD) in male rusa deer after intrajugular administration of 500 ng of LH-RH in April.

demonstrated two peaks in gonadotrophin activity in white tailed deer—first an LH peak in July and FSH peak in September and subsequently synchronous increases in March, although there was no rise in serum testosterone related to the March rise in gonadotrophin levels. In roe-deer an annual biphasic elevation in testosterone concentration has been reported (Gimenez et al. 1975; Schams & Barth, 1982). West & Nordan (1976b) noted that spermatogenic activity in black-tailed deer was at a peak in November and declined sharply through January-March. However, in May spermatogenic activity was observed in nearly all tubules, but it had declined again by July. Frankenberger (1954) observed changes in the proportion of absolute number of testicular interstitial cells in red deer shot throughout the year. Peaks occurred before the rut in late August followed by a decline and another peak in January.

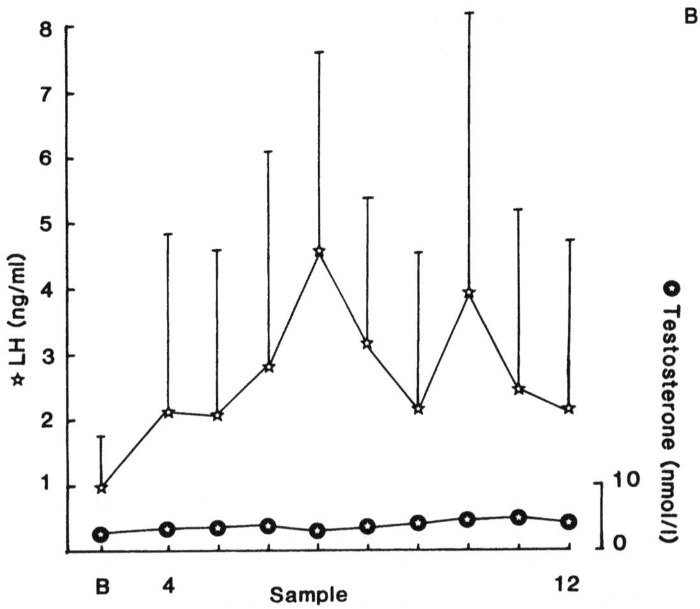

FIGURE 2B.

Despite the occurrence of an elevation of testosterone concentrations in May (autumn), the major rutting period for rusa stags starts at the end of July and extends to October. The May T rise stimulates male sexual behavior, expressed at "flehmen" toward hinds with calves at foot (van Mourik 1985, 1986). This can be considered as autumn rutting, in contrast to springtime rutting behavior, which has been reported in red deer (Fletcher and Short, 1974) and caribou (Whitehead & McEwan, 1973). Occasional births are recorded well out of season, indicating that a rusa stag population is capable of breeding nearly the whole year round. This is in agreement with G. Bubenik et al. (1982a), who stated that maximal concentrations of T are not required for the process of spermatogenesis and shedding of velvet in white-tailed deer.

The absence of a detectable annual pattern is basal concentration of LH and T in pretreatment samples in immobilized stags for LH-RH (Luteinizing Hormone-Releasing Hormone) responsiveness tests indicates an inhibition of hypothalamic secretion by Fentaz-Rompun mixture. It might be assumed that the drugs inhibited LH-RH, LH and subsequently testosterone secretion. The inhibitory effect of Fentaz-Rompun mixture on hypothalamus provides a convenient model for study of seasonal changes in responsiveness of pituitary-testicular axis to LH-RH. The maximal responsiveness, expressed in testosterone concentrations, was recorded in August, coinciding with the rut (Fig. 2a). No alterations in testosterone concentrations were noticed in April after LH-RH administration, but LH concentrations reached highly elevated levels in contrast to other times of the year (Fig. 2b).

TABLE 1. Influence of age on velvet weight and date.

Age	N	Mean weight (g)	SD	95% Confidence interval
2 Years	29	625.52	176.4	558.40–692.63*
3 Years	48	874.58	171.7	824.73–924.44
Mature	44	931.55	203.2	869.76–993.34
All	121	835.60	219.7	796.06–875.14
Age	N	Mean days in 1982 (Jan 1 = 1)	SD	95% Confidence interval
2 Years	29	80.76	14.6	75.20–86.33
3 Years	48	76.48	16.9	71.55–81.41
Mature	44	59.72	22.5	52.89–66.56*
All	121	71.41	20.62	67.70–75.12

*Less than all other groups ($p < 0.05$ Duncan multiple range test).

The unusual nature of the sexual cycle of rusa deer (long-day breeder) as compared with other *Cervidae* (short- day breeders) further illustrates the differences in sensitivity to photoperiodic stimulation (van Mourik & Stelmasiak 1985) and indicates that the endogenous mechanisms regulating the gonadal functions may differ between species.

Velvet-Antler Growth

The annual hormonal pattern described in the previous sections agrees with the observed pattern of antler growth in mature stags. In general, stags cast their antlers from November onwards and growth starts immediately after healing of the pedicle wound. In 6–8 weeks' time, the velvet antlers have reached a stage with a rounded or slightly flattened main beam. For commercial purposes, the velvet-antlers are removed at this point.

Influence of Age on Velveting Date and Yield

Data of velvet weight and collection data were analyzed in relation to the age of rusa stags in the 1982 velveting season (Table 1). Analysis of data reveals, not surprisingly, a higher velvet weight yield for stags 3 years and over in comparison to 2-year-old stags. However, despite the nonsignificant difference in velvet weight for older stags, the mature animals reach the appropriate velveting stage earlier in the calendar year. Since no casting dates were available, no growth per day could be calculated.

Unfortunately, no velvet weight data are available for the 1983 season, apart from some of the stags born in 1980. All animals showed an increase in velvet weight (mean increase 321.6 g) and were velveted 26 days earlier in the year. In 1984, data for different age groups and some individuals were collected and analyzed. The same pattern as in 1982 is evident with earlier velveting and

TABLE 2A. Mean velvet weight (g ± SD) for different age groups.

	Age (years)				
	2	3	4	5	6
1984	503.09 114.2	760.82 139.9	828.07 136.9	1004.20 206.8	999.33 214.6
N	71	61	27	25	15

TABLE 2B. Mean velvet days (from Jan 1 ± SD) for different age groups

	Age (years)				
	2	3	4	5	6
1984	84.4 10.0	59.2 18.0	45.9 11.0	40.6 12.0	30.5 17.0
N	66	57	26	25	15

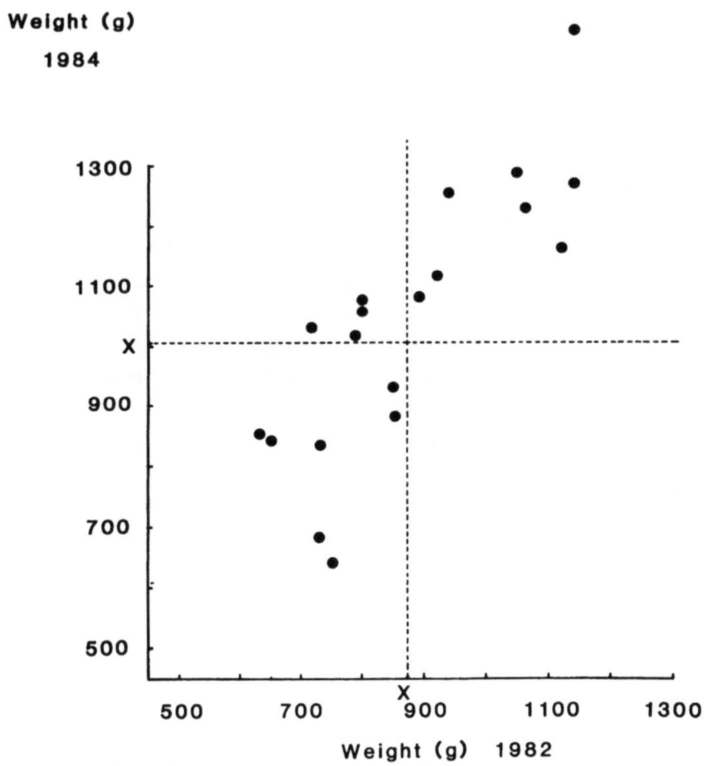

FIGURE 3. Comparison of weight of velvet-antlers at 3 years of age (1982) with the weight of velvet antlers at 5 years of age (1984).

higher yields for older animals (Tables 2A and 2B). Some individual animals (n=44) out of three age groups in 1982 could be followed up to the 1984 season. The velvet weight in 1984 has been plotted against the yield in 1982 (Fig. 3). With the exception of 5 stags, all animals increased their velvet production at least up to 6 years of age.

A comparison of velvet yield of individuals with the average population yield (± SD) in a certain age class indicates high and low producers under identical environmental conditions. However, selection of stags for velvet production at 2 years of age does not seem to be a reliable selection technique, since future velvet production is variable. Selection of 3-year-old stags with equal or higher velvet production than the total population is reliable. Stags with above average production in 1982 produced in 1984 1238.12 g velvet (n=8), against 893.18 g (n=11) for the low producers in 1982.

Although based on few results, the above data indicate that the deer farming industry (i.e., income from velvet) could benefit from monitoring individual performance of stags. At present no breeding records for stags or hinds are available, so no data can be analyzed for offspring from particular stags for an indication of heritability of velvet production.

Conclusions

Reproductive performance, behavior, and hormonal data of rusa deer clearly show a long-day breeding pattern with rut from late-winter to spring and calving in autumn, in contrast to other boreal cervids with a short-day breeding pattern.

Although calving under adverse climatological conditions (autumn) does not seem to be beneficial for the survival of the species, in relation to annual changes in vegetation quantity and quality (i.e., nutrition), this adaptation or "reestablishment" of evolutionary-based long-day breeding pattern of rusa deer after transfer from the tropics seems to be the best solution.

After a period of increased fat-deposition in autumn, both sexes of boreal cervids show a voluntary hypophagia over winter as well as during the rut for males. The hypophagia is accompanied by a shift to fat metabolism, increased recycling of urea nitrogen to maintain positive nitrogen balance, and a change in reticulum volume and total surface due to alterations in villi structure (Bamberg 1985; Hofman 1978; McMillin et al. 1980; Simpson et al. 1984). The underlying mechanism of the hypophagia is unknown.

Pasture growth in Victoria takes place in spring (September); it dries up over summer, resulting in minimal food availability in February-March. Under mild, wet conditions during autumn, a short growth period might occur before the winter quiescence.

Mature animals can utilize the spring and summer pasture for maintenance and fat deposition. This adaptation further confirms the sensitivity to increasing day length, putting this species apart from all other boreal cervid species, which deposit fat during decreasing day length. During the winter, hinds do not show such as dramatic a drop in food intake as stags, and fat reserves can be metabolized for body maintenance and lactation. Energy demands for the lactation decrease over winter as the calf gets older. The stags have an additional drop in food intake during the rut (end of winter), but if they enter the winter in good condition, fat reserves will last into spring.

If a rusa calf survives the first 5–7 days after birth in autumn, it is unlikely to succumb at a later stage despite the cold winter weather. At the time of weaning (spring), food availability and uptake are optimal so that calves enter their first winter in good condition.

Conception during autumn and calving in spring would be disadvantageous for the hind and calf. The hind would have to use fat reserves for her own winter maintenance and growth of the fetus. It is likely that not enough reserves are available for fetal growth at the end of gestation, resulting in smaller calves. Even though the calf would encounter more favorable climatological conditions at birth, no food will be available at weaning in winter; besides, this period coincides with low voluntary food intake.

If the observed reproductive pattern would have been detrimental for the survival of the population, it can be assumed that they would have adopted another pattern. This does not seem to happen. Although some hinds over the 4-year observation period calved late 1 year, this was not characteristic, and the next calf was usually born well within the range of the calving period for the population.

The fact that rusa deer have been able to adapt to or express a long-day breeding pattern makes this species very well suited for farming purposes under Victorian conditions.

Acknowledgments. The authors are grateful to Willow Ware Australia Pty. Ltd. for provision and care of experimental animals. This study was supported by grants from the Reserve Bank of Australia, Rural Credits Fund, Commonwealth Special Research Grant, and The University of Melbourne.

16
Nutrition and Antler Development

Robert D. Brown

Introduction

Of all of the wild mammalian ungulates, the cervids may well be the most intensively investigated. Scientific articles and books abound on the ecology, behavior, and physiology of deer. No doubt the white-tailed deer *Odocoileus virginianus* of North America and the red deer *Cervus elaphus* of Europe and New Zealand have been the most studied. Interest by hunters, biomedical researchers, and those who sell antlers for medicine has led to numerous articles and two earlier books on the phenomenon of antler growth (Brown 1982; Goss 1983). One would suspect that the nutritional requirements of antler growth have been well established.

Unfortunately, such is not the case. Studies of the overall nutritional requirements of deer for maintenance and body growth have been reviewed (Cowan & Clark 1981; Fennessy 1982; Robbins 1983), but detailed investigations into the specific requirements for antler growth have been lacking. Research into the physiology and nutritional requirements of antler growth has been hampered by the lack of qualified investigators interested in the subject, inadequate funding, limited numbers of animals or length of experiments, the difficulties of working with wildlife, and the seasonal nature of the cervid's reproductive cycle and metabolic needs. Hopefully recent findings using deer and deer antlers as models for human bone disease, the growing commercial market for venison and velvet antlers, and the amazing growth of the game ranching industry both in North America and abroad will provide the impetus and funding to take antler studies out of the realm of hobby research and into the spotlight of credible scientific endeavor. It is the purpose of this chapter to review the research which has been done and to suggest areas which need further inquiry.

Mineral Requirements of Antler Growth

Composition of Antlers

It would seem obvious that to estimate the nutritional requirements of antler production, one should first determine the composition of antlers, then extrapolate the nutrients required to produce such structures. This approach is not as simple as it would seem, since the final structure, the mature hardened antler, is far different in composition from the growing antler. Ullrey (1983) reported that the composition of a growing antler of a yearling white-tail in June was 20% dry matter, of which 80% was protein and 20% was ash. Kay et al. (1981) sectioned growing red deer antlers (n=6) in June and found the tips to average only 8.2 ± 0.5% ash with a calcium (CA)/phosphorus (P) ratio of 0.4 ± 0.14, whereas the bases of the antlers averaged 34.5 ± 0.7% ash with a Ca/P ratio of 2.2 ± 0.24. Quite likely the mineral requirements for antlers depend on the stage of growth, since the ossification process is not uniformly distributed over the antler growth period. The collagenous architecture of the velvet-antler ossifies from the tip back by endochondral and intramembranous mechanisms. Muir et al. (1985) reported that 65% of the mineral found in the mature antlers of red deer was deposited in the last 10 weeks of growth. Three kg of mature, hardened antler contained 536 g of calcium, 348 g of which was deposited in the latter period.

Caton (1877) reported that the dry matter of mature, hardened antlers of a variety of deer species averaged about 60% ash and 40% organic matter. Kay et al. (1981) sectioned hardened red deer antlers three weeks after velvet shedding and found them to average 51.7 ± 1.9% ash with a Ca/P ratio of 2.2 ± 0.02. In his review of the literature, Robbins (1983) reported that on the average, mature antlers contain about 11% P and 22% Ca (Rush 1932; Chaddock 1940; Bernhard 1963; McCullough 1969; Hyvarinen et al. 1977). Miller et al. (1985) analyzed 18 sets of antlers from seven white-tailed deer. They found a mean density of 1.63 g/ccm, but with significant differences between deer and portions of the antler analyzed. These were similar to densities reported earlier (Bernhard 1963; French et al. 1955). Interestingly, Miller and his co-workers (1985) found that overall antler density decreased with age. They attributed this to the theory that older deer, which have higher circulating testosterone levels, possibly mineralize their antlers faster and thus less completely than younger deer. Miller et al. (1985) analyzed the chemical composition of their antlers and found the following on a dry matter basis: 103.3 ± 0.5 mg/g P, 190.1 ± 1.3 mg/g Ca, 10.94 ± 0.09 mg/g Mg, 4.98 ± 0.04 mg/g Na, 900 ± 20 μg/g K, 55 ± 2 μg/g Fe, 6.6 ± 0.2 μg/g Mn, 116 ± 3 μg/g Zn, 128 ± 3 μg/g Al, 92 ± 1 μg/g Sr, and 129 ± 2 μg/g Ba. Despite varying concentrations of minerals in different sections of the antlers, a Ca/P ratio of 1.88 ± 0.01 was fairly constant throughout.

Determination of Mineral Requirements

Unfortunately the mere analysis of the composition of the antler at its various stages of growth tells us only the amount of nutrients which must be deposited, not the amount of dietary nutrients which must be consumed by the animals in order to make such deposits. Lack of dietary mineral has been implicated as the cause of broken antlers in Tule elk *Cervus elaphus nannodes* (McCullough 1969) and Roosevelt elk *C. e. Roosevelti* (Bower 1983). The first scientific investigations into the mineral requirements of white-tailed deer were begun in 1952 at the Pennsylvania State University (French et al. 1955, 1956). French and his co-workers (1955) felt that "a preponderance of spike bucks and antlerless males is a direct sign of malnutrition." To test this theory, fawns were captured from around the state, bottle-raised, and weaned on to a variety of experimental diets, some of which included low calcium (0.09%), low phosphorus (0.27%), low calcium and phosphorus, and the latter ration with supplemental vitamins A (3,000 USP) and D (300 IU/kg). The deer were penned in pairs, which unfortunately precluded measurement of individual intake and led to the possibility of feeder dominance.

Although the sample sizes were too small for statistical analyses, the deer on the deficient diets seemed to do less well in body growth and antler development than the controls. The combination of deficient calcium and phosphorus had a worse effect than either of the minerals deficient alone, although this result may have been caused by poor feed consumption. The addition of vitamins A and D did not compensate for the lack of minerals. In the second year of the study, one deer of each pair was switched to the control ration in April. Interestingly, the deer stunted in their first year did not catch up with the controls during the second year. French et al. (1955, 1956) concluded that the mineral requirement for antler growth must be somewhat higher than 0.09% calcium and 0.27% phosphorus.

In the third year of the study, intermediate levels of minerals were added to the rations. Again the sample size was quite small, and the results were hampered by antler injuries. The best antlers produced in these studies came from deer on diets of 0.64% calcium and 0.56% phosphorus (Magruder et al. 1957; McEwen et al. 1957). Later Cowan and Long (undated) of Penn State reported a study in which 14 3-year-old bucks were divided into two groups, one fed only 0.2% phosphorus and one fed a mineral supplement. There was no difference in antler development between the groups. They suggested that while younger deer might have higher requirements, mature white-tails could exist on a diet of no more than 0.2% phosphorus. More recent work on fawns by Ullrey et al. (1973, 1975) has shown 0.45% and .30% phosphorus to be optimum for young deer. Jacobson (1984) suggested phosphorus levels as low as 0.14–0.29% to be adequate for deer in Mississippi. As Robbins (1983) points out, however, one must be cautious of interpreting mineral requirements determined by regression, as the technique assumes that the availability of the mineral is independent of its dietary source.

The seemingly low mineral requirements for antler growth might give one reason to pause, particularly in light of the much touted popular notion that antler growth causes a severe mineral drain on the deer. One answer is that the deer used in the Penn State studies and Ullrey's work were relatively young, small deer with comparatively small antlers. Huxley (1931) reported Scottish red deer stags producing antlers weighing 1.56 kg, which would have contained 350 g Ca, 160 g P, and 5 g Mg. This is less than the total minerals produced by a red deer hind during her full lactation: 370 g Ca, 300 g P, 20 g Mg (Arman et al. 1974). Hyvarinen et al. (1977) also reported that minerals contained in the antlers of their penned or grazing red deer were no more than those produced in the milk of lactating hinds.

A second explanation of the relatively low mineral requirements of antler growth is the possibility that deer deposit mineral in their skeletons before antler growth, transfer those minerals to the antlers during ossification, and then replace the minerals in their skeletons from the diet. In a series of experiments with ^{45}Ca and ^{32}P, Cowan et al. (1968, 1969) showed that male deer deposit the radioisotopes in their skeletons from November to March, then transfer the minerals to the antlers as necessary. They found that the mineral of the non-weight bearing bones (skull, ribs, sternum) to be the most labile. These observations are refuted by Řeřábek and A. Bubenik (1963), who injected ^{32}P into fallow deer *Dama dama* and found deposition in the antlers but not in the skeleton. Nonetheless, the existence of a cyclic physiologic osteoporosis in deer has been confirmed gravimetrically, histologically, and radiologically (Banks & Davis 1966; Banks et al. 1968a, 1968b; Hillman et al. 1973; Brown et al. 1978b; Brown & Pritzker 1985).

Another phenomenon which could explain the deer's flexibility in meeting its mineral requirements is the possibility of seasonally differential absorption rates. Robbins (1983) warned of the dangers of estimating mineral requirements as a percent of the diet. Both diet composition and the amount of intake fluctuates dramatically in deer by season. Wetzel (1968) conducted calcium balance trials in male white-tails from ages 6–27 months. He found that calcium retention was higher from August to October, the period of ossification and velvet shedding, in yearlings but not in 2-year-olds. Calcium and phosphorus retention decreased as the deer grew older. Stephenson and Brown (1984) combined calcium balance trials with calcium kinetic studies on two 2-year-old and two 4-year-old white-tailed bucks. Calcium absorption from the gut increased as the antlers grew. Near velvet shedding, skeletal accretion increased. The younger deer had greater calcium absorption and lower endogenous fecal and urinary calcium losses than the older deer. These authors interpret their kinetic data cautiously, as the technique did not allow the mathematical separation of the growing antlers from the skeleton as a separate calcium pool.

There is currently no explanation for this apparent ability of deer to increase the rate of calcium absorption from the gut during times of antler growth. In fact, it may not happen in the wild with deer on mixed diets to the extent that it does in penned deer on pelleted diets of constant composition. In the wild, changes in diet composition may change mineral availability, and oxalates and phytates or calcium:phosphorus imbalances may hamper absorption. It has been found, however, that the circulating level of 1,25 dihydroxy vitamin D increases in fallow deer during antler mineralization (Eiben et al. 1984). Similar results have been found in white-tailed deer (van Der Eems and Brown, unpublished). If this is the case, there may well be a dietary requirement for vitamin D precursors in male deer. Requirements for antler growth for other vitamins or minerals have not been established.

Protein Requirements of Antler Growth

General Protein Requirements

If the mature antler requires minerals largely drawn from the skeleton and diet, then the growing antler, composed largely of collagen, may require substantial dietary protein. Even the hardened antler is nearly 45% protein, and this has led Robbins (1983:225) to comment: "the incremental protein requirement [for antler growth] is potentially far more significant than the energy requirement." The maintenance protein requirement for white-tailed deer has been established at about six percent (Murphy & Coates 1966; Holter et al. 1979b), whereas requirements for growth are about 16–17% (Verme & Ullrey 1972).

Determination of Protein Requirements

The studies conducted by French and his coworkers (1955, 1956) included experiments designed to determine the dietary protein level for optimum antler growth in white-tails. Four control fawns were fed a diet of 16.8% protein, whereas a pair of male fawns were fed diets of medium protein (9.5%), medium protein plus methionine, low protein (4.6%), and low protein plus sulfur. As yearlings, the deer on the medium- and low-protein diets did less well in body growth than the controls. Addition of methionine and sulfur produced no improvement. Antler growth seemed stunted on the deficient diets, but again sample sizes were small. The deer on the medium and low protein rations began their first antler growth two to four weeks later and shed their velvet about two weeks later than the controls.

In the second year of the study, those deer which were switched back to the control diet in April did not catch up with the control deer in body weight or antler development. The researchers concluded that body growth must take precedence over antler development, and that a ration of 13–16% protein is required for optimum antler development. In the third year, eight bucks which had been maintained on complete or intermediate quality rations gained an average of 10.9 kg of body weight and 322 g of antlers over the previous year. Nine deer shifted from adequate diets to those deficient in protein or minerals lost 4.5 kg of body weight and 16 g of antler. Four deer which had remained on deficient diets fared better, gaining 8.2 kg in body weight and 56 g in antlers. In fact, one 3.5-year-old deer produced forked antlers on only a 7.3% protein diet. This data suggested that the shifting of deer from one diet to another might decrease feed intake. Again it was noted that the deer on the deficient rations began antler growth later, shed velvet later, and cast antlers earlier than the deer on complete or intermediate rations (Magruder et al. 1957; McEwan et al. 1957).

Influence of Genetics

In 1975 the Texas Parks and Wildlife Department began a series of studies into the relationships between nutrition and genetics of antler growth in white-tailed deer

(Harmel 1979). Male fawns were bottle-raised, then weaned onto diets of high (16%) or low (8–10%) protein which were believed adequate and equal in energy and other nutrients. After the first year, half of each group was switched to the other diet, and in subsequent years, diets and groups were switched again. Unfortunately, as in the Penn State work, these divisions caused the resulting groups to be too small for statistical analyses, but definite trends were evident. Although there was little difference between yearlings in antler development, those deer maintained on higher protein diets for 4 years had heavier bodies and better antler development than those kept on lower protein diets. Bucks switched from high to low rations and vice versa produced antlers which fluctuated with the diet, although their antlers sometimes exceeded those of deer maintained on the 16% protein diet. This indicates at least that antler growth may respond to an improved diet later in the life of the deer. A decline in body weights during the last year in three of the groups confounded the data and was not explained (Harmel 1977, 1978, 1979).

In another study Harmel (1979) maintained five deer which had forked antlers as yearlings and six that were spikes on a 16% protein ration until age 5.5 years. The bucks which had "forked horns" as yearlings had consistently better antlers in subsequent years. At age 5.5 this group averaged 10.4 kg more in body weight than the spike group and produced antlers which were 63% heavier, had a 19.8% greater inside spread, 16% greater main beam length, and 12.3% greater basal circumference than their counterparts. The "forked" group averaged 8.8 points per deer at age 5.5 as compared with the "spike" group's 7.0 points. One can infer from this work that perhaps the antler production of the yearling spikes was limited by their genetic potential. We cannot rule out, however, that an 18 or 20% protein ration, or one higher in energy or other nutrients, might have allowed the yearling spikes to catch up.

Finally, in another genetic study at the Kerr Area, deer which were spikes as yearlings were mated to randomly selected does. Their offspring were compared to those of a single buck which had 9 points as a yearling, and later developed into a clearly superior deer. The bucks which were spikes as yearlings produced 19 F-1 male offspring, 26.3% of which were spikes as yearlings. Their doe offspring were backcrossed to their fathers and produced 17 male backcross offspring, of which 58.8% were spikes as yearlings. The "superior" buck produced six male fawns, none of which were spikes as yearlings. All of these deer were maintained on a 16% protein ration.

While this study indicates the possibility of a strong genetic influence of the sire in the antler development of its yearling offspring, it is by no means definitive. The spikes used as sires came from unknown backgrounds; their lack of antler development could have been nutritional or genetic in origin. While 26.3% of the F-1 generation were spikes, 73.7% were not—a figure of perhaps greater importance. Backcrossing seemed to concentrate the phenomenon, and the results serve as evidence of the possible existence of a "spike gene." On the other hand, comparing any of these offspring to those of a clearly superior sire may not be realistic—what might have been the offspring of 4-point or 6-point sires?

While it is not the purpose of this chapter to discuss the genetic influences on antler growth, it is important to realize the potential effect of genetics on nutritional experiments. A superior white-tail sire (15 points at 4.5 years) in our pens produced eight male offspring, half of which were spikes and half of which had from 4–11 points as yearlings. The sire, the does, and all of the offspring were on an identical 18% protein ration. Clearly nutrition of the fawns had little to do with their resulting antler growth as yearlings (Brown 1985). Yet in another study, we caught wild pregnant does and placed them on diets of varying quality May–December. Even though the resulting fawns were weaned onto a high quality ration, their body and antler development as yearlings seemed proportional to the diets of their mothers. Genetic influences can override nutritional experiments, especially if animal numbers are small, or if the experiment is not continued for a number of years.

Timing of Protein Availability

Ullrey (1982) attempted to simulate the seasonal changes in deer diet quality in Michigan by feeding male white-tail fawns a 7% protein diet from October until a simulated "green-up" in March (early) or April (late). The early green-up group also had access to red oak acorns, only 7% in protein but high in energy. At the time of simulated green-up, the groups of five deer each were offered a 16% protein diet. The early green-up group tended to have better body weights, antler weights, beam diameters and main beam lengths. The only statistically different value ($P < 0.05$) was more antler points in the early green-up group. This data suggests that the additional energy provided by the acorns, or earlier availability of protein may be important for optimum antler development.

However, as in the mineral requirements for antler growth, it may be erroneous to view the protein requirement of antlers as merely that which must be consumed, absorbed, metabolized and deposited in the final structure. What may be more important is a level of protein intake which will allow the animal to meet is maintenance requirement and its genetic potential for body growth. It may be that while that the amount of protein that goes into the antler may be relatively small, the amount that maximizes body growth and allows a surplus for antler development may be dramatically greater.

Energy Requirements of Antler Growth

The question of the importance of energy and its relation to protein intake was raised by Robbins (1983). He proposed that the expression of protein requirements as a percent of the diet was incorrect, because the ratios of protein to energy vary widely in free-ranging deer, and "such ratios could impact heavily on body growth and antler development."

Since the mature antler contains only about 1% fat, the energy deposited in the antler is admittedly low, especially when compared to, say, the energy contained in the milk of the lactating doe or hind. Nonetheless, energy must be required to absorb, metabolize and deposit the materials of the antler, not to mention the activity factor costs of carrying the weight of the antlers for several months.

Determination of Energy Requirements

Energy requirements are by nature difficult to determine in wild animals. Unlike the materials which go into the construction of the antler or the deer's body, energy is not a nutrient–it is a property of nutrients, and it is measured as a unit of heat expressed in calories of joules rather than in grams. The maintenance requirement for energy in ruminants is called the fasting metabolic rate (FMR), and includes that energy required for basic body functions, minor voluntary activity, and maintenance of the body temperature in a thermoneutral environment. The FMR can be determined by a series of balance trials, or more accurately, through the use of direct calorimetry–the measurement of the deer's actual heat production. This latter technique is difficult and requires expensive equipment, so that most energy determinations are determined by indirect calorimetry, the measurement of gas exchanges in chambers or with masks. Excellent reviews on energy metabolism of ruminants are found in the classic texts of Brody (1945), Kleiber (1961) and Blaxter (1962).

In addition to the FMR, animals have additional energy requirements for body growth, fattening, fur production, pregnancy, lactation, thermoregulation in both hot and cold environments, and such activities as browsing, walking, climbing, and running. Obviously, the environment plays an important role in determining the energy requirements of a cervid. The effort the animal must expend to find, consume, and digest its food, the nature of the topography for locomotion, and the ambient temperature, wind velocity, and snow depth are mitigating factors. The determination of the energy requirement of cervids has been difficult under this maze of contributing factors. Nonetheless, determinations of the energy costs of maintenance and activity factors have been determined for white-tailed deer (Silver et al. 1959; Ullrey et al. 1970; Thompson, C.B. et al. 1973; Holter et al. 1977, 1979a; Moen 1978, 1985; Mautz & Fair 1980), black-tailed deer, *O. hemionus columbianus* (Mautz et al. 1985), mule deer, *O. h. hemionus* (Baker et al. 1979; Parker et al. 1984); wapiti, *C. canadensis* (Gates & Hudson 1978; Robbins et al. 1979; Parker et al. 1984; Hudson et al. 1985), reindeer, *Rangifer* (White & Yousef 1977), caribou (McEwan 1970), roe deer, *Capreolus capreolus* (Weiner 1977), and red deer (Fennessy et al. 1981). This is by no means an inclusive list; excellent reviews of the energy requirements of cervids are found in Moen (1973), Robbins (1983), and Hudson and White (1985). Unfortunately, there have been no studies to determine the specific requirements for antler production in any cervid species. The reason for this is obvious–while the deer is growing its antlers, it is also gaining weight due to both growth and fattening, foraging for food, and participating in a number of activities which would contribute to its energy requirement.

Antler Growth and Body Weight

Fortunately relationships exist which will help us determine the optimum energy intake for maximum antler development. These are the relationships between body weight and energy requirements, and body weight and antler size. Body weight is clearly the most relevant factor affecting a deer's energy requirement; the larger the animal, the greater the FMR, and the more energy needed for maintenance, locomotion, etc. Kay et al. (1980) stated, "the most compelling requirement determining the quantity and quality of food ingested is the need for energy, which in turn relates to the size of the animal."

It is important to note that the relationship between body size and energy requirements is not a direct one, but rather a logarithmic one. Kleiber's famous theorum that the relationship across species is FMR = 70 body weight $(kg)^{0.75}$ is still considered the standard, although wide variations exist both between and within species (Kleiber 1961). Metabolizable energy requirements for maintenance of white-tailed deer have ranged from as low as 97 $kcal/kg^{0.75}$ for adults in the winter (Silver 1959) to 334 $kcal/kg^{0.75}$ for maintenance and growth of fawns in the autumn (Holter et al. 1979a). If one considers activity, the requirement easily doubles (Robbins 1983). One must also be cautious about considering body weight as the sole criterion for the determination or estimation of energy requirements. At the same weight, one deer may be fat, another one lean, one pregnant, another open, one gaining weight, another losing. The change in body constituents may be an important, though not always a measurable mitigating factor (Robbins 1983).

Huxley (1931) found that the antler weight of European red deer was proportional to the body weight raised to the power of 1.6. Hyvarinen et al. (1977) found similar relationships in red deer stags of the same age within the same herd. In fact, this general relationship exists across species, from the smallest, the muntjac, *Muntiacus reevesi* to the largest, the extinct giant Irish elk, *Megaloceros giganteus*. A. Bubenik (1985) reviewed similar information and found that in small species of deer, antlers represent about 1.5 ± 0.4% of body weight, while in larger species, the ratio was about 3.4 ± 1.4%. He also found that in 32 species of cervids (excluding *Alces alces cameloides*) there was a direct relationship between antler length (r^2 = 0.95) and beam length (r^2 = 0.88) and metabolic body weight. For more information on these relationships, see Chapter 1.

Antlers as a Condition Index

The relationships discussed thus far raise an important point—as body weight increases, metabolic rate and feed intake per unit of body weight decrease but antler size increases. Thus the demands of the antlers upon the nutrients in the feed consumed must increase sharply (Huxley 1931).

Clearly this would indicate that if a deer's body takes precedence over its antler growth, and feed supplies are limited, then the stunting of antler growth will most effect the larger deer. There is little evidence of this, however, most probably for two reasons. First, as mentioned previously by Cowan and Long (undated), one often finds that the nutrient (energy) requirements of large deer are often less than those of younger deer due to the fact that younger deer have additional requirments for growth. Thus younger deer on restricted diets may be stunted in both body size and antler growth. In addition, limited nutrition can effect the timing of the antler cycle, thus effecting the nutrient drain on the animal.

General undernutrition was blamed on the number of red deer with malformed and broken antlers in the Cupola Basin of New Zealand in 1962–63 (Clarke & Batcheler 1972). Undernutrition has also been blamed for the fact that 99% of the wapiti shot in Fiordland, New Zealand were spikes in 1965–1970, and the preponderance of spikes among yearling elk in North Yellowstone Park (Greer 1986) and in Colorado (Boyd 1970). This general relationship between antler size and feed availability has been suggested as a means of accessing the general nutritional condition of the herd. Severinghaus et al. (1951) found that antler beam diameters and point development in deer aged 1.5–3.5 years old were more dependent on range conditions than age, and that differences were most pronounced in yearlings. Anderson and Medin (1969) suggested the use of mean antler weight of yearlings as a criterion of physical condition. Ullrey (1982) reported a relationship between the antler bean diameters for yearling white-tailed bucks and the bone marrow fat concentrations of 1-year-old does in the same areas of Michigan. Ozoga & Verme (1982) found better beam size but not number of points in a supplementally fed herd of white-tailed yearlings. Smith et al. (1982) found that antler measurements were correlated with both kidney fat index (KFI) and testis weight. Thus antler beam diameter of deer of the same age class may be a useful index of the nutritional condition of the herd as suggested by Anderson and Medin (1969).

Seasonal Feeding Cycles of Deer

One cannot begin to discuss the effect of energy restriction on body and antler development in deer without an important caveat. Unlike domestic animals, deer undergo an annual cycle of body weight gain and loss quite separate from growth. In addition, the diets of the deer vary seasonally, both in availability and in quality. Whether or not the deer has adapted its digestive abilities to meet such cycles is a matter of disagreement among wildlife nutritionists. Cowan and Clark (1981) argue that deer have digestive systems similar to those of cattle and sheep, similar energy requirements, and similar digestive efficiencies. In fact, they state that "a ruminant is a ruminant" and dismiss efforts by others to define the digestive systems of deer as somehow different. Our limited experience (Wheaton & Brown 1983b) has shown little difference between the in vivo digestive efficiency of white-tails and sika deer (*C. nippon*) on pelleted rations or of white-tailed bucks and does over several seasons, or between in vitro values (with cow inocula) and in vivo digestibilities of several diets fed to cattle and deer. However, our work and that of Cowan has largely been done on high quality rations, such as alfalfa hay or pellets.

There is little question that, given the choice, deer select diets considerably different than those of cattle or sheep. Cattle are graminivores or grazers, whereas deer tend to be either concentrate selectors (browsers) or intermediate (adaptable, opportunistic) feeders. The differences in feeding strategies have been caused by or are a result of differences between species in nutritional requirements, habitat, feed availabilities, and digestive system characteristics, such as rumen size, rate of passage, papilli development, and microbial populations. That these latter differences exist is well documented (Hofmann 1985); their cause is open to some conjecture. In addition, authors disagree as to which cervids fall into which categories. In general, it is agreed that roe deer and muntjacs are browsers, whereas red deer, wapiti, and caribou are intermediate feeders (Kay 1985; Hofmann 1985). Most other species tend to overlap these two groups, depending on who is reporting, and while no cervids are classed as grazers, some such sika deer and sambar deer, *C. unicolor* are selective grass eaters, and others such as red deer and even white-tails may eat grass during certain times of the year (Kay et al. 1980; Klein 1985; Meyer et al. 1984). Generally speaking, cervids digest high quality diets as well as domestic ruminants; they digest low quality diets less well (Kay et al. 1980).

One of the most impressive mechanisms deer have developed in order to cope with fluctuations in feed quality and quantity is their voluntary reduction in feed intake during the winter. That deer literally starve themselves during the winter has been well documented in red deer (Clutton-Brock et al. 1982), reindeer (McEwan 1968), mule deer (Julander et al. 1961) and white-tailed deer in both northern (Ullrey et al. 1970) and southern (Wheaton & Brown 1983a) climates. One would suppose this reduction in intake is in response to the lack of available forage. The fact is that it continues to occur in penned animals with adequate feed available year-round. Red deer in the wild commonly lose 14–17% of their body weight over winter, and white-tails in pens lose similar amounts (Chao & Brown 1984; Cowan, personal communicaiton). This decline in intake usually begins with the rut, but continues well past the breeding season. Ryg (1983b) suggested this hypophagia was regulated by a seasonally changing body weight "set point." Clutton-Brock and Albon (1985) claim that this decline largely occurs in male deer, and may be due to differences in feeding behavior. While there is a general consensus that the decline in feed intake and body weight is more pronounced in the male (McMillin et al. 1980), Wheaton & Brown (1983a) found no statistical differences in the intake decline per unit metabolic body weight over winter in yearling male and open female white-tails.

The decline in intake in the wild no doubt evolved as a survival tool. During the rut, male deer are simply too preoccupied with chasing females to eat. After the rut, food supplies are limited, and the pregnant females possibly have first choice at what is available. Even though testosterone levels are falling, the males may still need to defend territory and harems. Exhausted from the rut, the buck's lowered "set points" may allow them to conserve their energy for more important tasks

(Mrosovsky & Sherry 1980). As long as feed availability does not fall below minimums needed to maintain the rumen microflora, feed digestibility, at least of standard diets, will not fall (Hershberger & Cushwa 1984; Wheaton & Brown 1983a). This is probably due to changes in gut capacity during this period (Kay 1985).

While the seasonal cycle of feed intake and body weight may make ecological sense for deer in the wild, its existence in deer raised in pens over several generations, and even in southern climates with mild winters, is still a puzzle. We have noticed that it is the older and dominant bucks that seem to suffer the most. These deer do most of the breeding and consequently most of the fighting to defend their harems. Younger bucks can fight and then go rest and perhaps eat; dominant, older bucks must be constantly on the alert and may be challenged daily. At the end of the winter, they look much the worse for wear as compared to their younger counterparts. A common mistake of deer managers is their effort to supplementally feed their herds during the winter in an attempt to override this phenomenon. While supplemental feeding in the wild may be necessary in unusually severe winters where widespread starvation and decimation of the herd is a possibility, it is not recommended as a regular policy. The normal cycle of weight loss is not something to try to overcome. In fact, "the wildlife nutritionist need not consider catabolism and weight loss as undesirable, but rather as essential components of the life strategies of many wild animals" (Lemano 1977; and Sherry et al. 1980, quoted in Robbins 1983). With this in mind, the question becomes how much of a weight loss and how long of a starvation period is acceptable before it affects not only the survival of the deer, but the production of next year's antlers.

Effect of Feed Restriction on Antler Growth

One must again be cautious in interpreting information on restricted feed studies in deer. Restricted feeding studies are reported here, under a section on energy requirements, simply because they are the only energy studies which have been done in relation to antler growth. As mentioned previously, no studies have been done to specifically determine the energy requirements of antler growth. Such studies would require adult animals, with no requirements for growth, and rations which were adequate in all nutrients, but diluted with a low energy filler such as sawdust. The work which has been done, and is reported here, concerns the feeding of diets nutritionally adequate in all nutrients, but restricted as to the amount offered and the length of the restriction. Thus the cautious reader should realize that the deer restricted in such a fashion may be suffering from not only an energy deficiency, but also from a deficiency in protein, minerals, etc.

In the northeastern United States, severe winters and resulting deer die-offs have led to numerous programs of supplemental winter feeding by state and private agencies. The efficacy of such programs was studied in a number of experiments at Pennsylvania State University in the early 1970s and reported by Hershberger and Cushwa (1984). In this work, a number of white-tails were fasted for different periods of time, subjected to various levels of stress during the fasting, and then refed with either alfalfa hay or corn. The researchers found that the rumen fermentation patterns of the deer changed within three days of fasting; the deer could no longer digest cellulose, and would develop lactic acidosis if refed corn. Deer survived only 2–3 weeks when stress was severe, but 3–4 weeks when stress was minimal. The most important factor was the condition of the deer prior to the fasting. Deer with the least fat reserved did the worst. The authors concluded that white-tailed deer could survive a 20% weight loss with refeeding only if they were not stressed, and if they had not suffered irreversible damage to their digestive systems.

The earlier work at Penn State by French et al. (1955, 1956) also included restricted diets. Male fawns were weaned onto complete diets or restricted (one-half control) rations and fed for a year. At the end of the first year the four control deer averaged 70.3 kg, having gained 49 kg since weaning. The two deer on the restricted ration weighed 36.3 and 70.3 kg, respectively, suggesting feeder dominance by one of the bucks. The restricted deer seemed to have somewhat stunted antlers, and pelage changes were delayed about one month. In the second year of these studies, one deer of each group was switched to the control ration in April. At the end of the second year, the stunted buck did not catch up with his control counterparts. Although the sample size was distressingly small, the worst antlers were produced on the restricted diets.

Based on intake measurements, French and his coworkers concluded that 22.7–27.7 kg deer require 0.9 kg of air dry forage per day, or about 3,600 kcal of gross energy, 45.3 kg deer require 1.4–1.8 kg or 6,300 kcal, 68.2 kg deer require 2.3–2.7 kg, or 9,900 kcal, and 68.2–90.9 kg deer require 1.4–2.3 kg or 9,000 kcal. The latter figure is of course less than that required for growing yearlings. Magruder et al. (1957) continued this work confirming French's findings and concluded that 3.5-year-old deer weighing 68.2–90.9 kg require only 1.8 kg of air dry forage (3.6 kg of natural browse), which was about 0.5 kg less than that required by 2-year-olds and 1 kg less than yearlings. They concluded that due to the seasonal fluctuation in feed intake in male deer, the spring and summer period, from April to September was by far the most critical nutritional period for antler growth.

This hypothesis was to be tested by Long et al. (1959) in the next series of experiments at Penn State. In this work, 22 fawns were weaned onto a control ration. While one group served as controls, another group was restricted to one-half of the ration for 5 weeks in March and April, while another group was restricted for 10 weeks, March through May. At the end of the restriction, all the deer were allowed full feed. Summer pelage change was delayed in both of the restricted groups; velvet shedding was delayed about 1 day for each week of restriction. Both restricted groups increased their feed consumption above that of the controls once back on full feed. The 5-week restricted group ended the year weighing the same as the controls, while the 10-week restricted group was 4.5 kg lighter. There was no effect on the time of antler casting or on the length or number of points of the antlers. The group restricted for 10 weeks produced antlers somewhat lower in density than those in the other two groups.

This entire series of experiments was put into perspective by Cowan & Long (undated). They admitted that because the small number of animals only broad generalizations could be deduced from the Penn State work. They also reported further experiments which apparently were never published elsewhere. They expanded the previous work by lengthening the periods of feed restriction. Restriction began January 7 and continued until April or May 1. The deer were allowed to lose 30% of their pre-rut weight. Again antler growth was delayed until the deer were returned to the complete ration. Restriction had no effect on antler size, branching, or velvet shedding. Thus the critical period of nutrition for antler growth was narrowed to May–September.

Species differences often become apparent when comparing the effect of feed restriction on the timing of the antler cycle. White-tailed deer on restricted rations generally cast their antlers earlier than deer on complete rations (Cowan, personal communicaiton). Jacobsen and Griffin (1982) reported that spike white-tailed deer in Mississippi generally shed their velvet later and cast their antlers earlier than the larger antlered, and supposedly better fed, counterparts. Ozoga and Verme (1982) reported that a supplementally fed herd of white-tails in Michigan cast their antlers in March, two to three months later than deer in the wild. The opposite seems to happen in red deer. Watson (1971) reported that after either mild or severe winters, stags in Scotland which had not cast their antlers by April were in poorer condition than those which had cast. Data collected over several years showed 434 stags which had died of natural causes in April and 52 found dead in May all still carried hardened antlers. He suggested the use of the date of antler casting as a nutritional condition index.

Wika (1980) reported a series of observations on reindeer, a species in which both the males and females grow antlers. He found that deer foraging around a garbage dump had markedly larger antlers than deer in the surrounding area. He found that male calves and yearlings had greater antlers per unit body weight than did females. In an experiment, three males each were put on an ad lib ration, a restricted ration, and an intermediate ration (the rations were undefined). The resulting body weights and antler sizes corresponded to the rations, and only the ad lib group had fully cast their antlers by July. Wika cites other examples of underfed reindeer which kept their hardened antlers longer than normal.

With the financial impetus of the antler velvet marker, some of the most intensive antler studies have been done in New Zealand. Fennessy & Moore (1981) observed the casting dates and later velvet yield of red deer in New Zealand fed either alfalfa hay or hay plus a pelleted supplement. The supplemented deer cast their antlers earlier and produced more velvet than the deer fed only hay. In a series of experiments at the Invermay Research Centre in New Zealand, Fennessey and Suttie (1985) fed diets of different qualities and amounts to a variety of red deer of different ages. They found that pedicle development in the fawn was dependent on the onset of puberty, which depended on the body weight (41–56 kg), and which was irrespective of the age of the deer. That is, the weight of the deer on high or low planes of nutrition were nearly identical at the time of pedicle initiation, antler initiation and velvet shedding, even though these events occurred at different ages. The deer on the high plane naturally grew heavier and longer antlers. They felt that light cycles largely controlled the date of antler casting, but nutrition had a moderating influence, in that deer on low planes of nutrition held their antlers longer. In an experiment with ten 4-year-old deer, half were fed ad lib and half were restricted to 80% of ad lib for the first 65 days of antler growth. The difference in the weights of antlers produced by the two groups were not significant at this level of restriction.

Work in Scotland with red deer produced similar results. Suttie et al. (1984) fed stag calves ad lib or 70% ad lib November to May and found that in the ad lib group testes and pedicles developed two months ahead of the restricted group. These studies were continued until the deer were three years-of-age. During the first cycle, the restricted group developed pedicles at 31 weeks of age on an increasing daylength of 8.3 hours, while the ad lib group developed pedicles at only 19 weeks on a decreasing photoperiod of 10.9 hours. Antler initiation and velvet shedding were also earlier in the ad lib group, but both groups reached the same event at the same body weight. Between velvet shedding and antler casting the restricted group caught up to the ad lib group in body weight but still tended to cast their antlers later. After the first year, the cycles coincided with photoperiods rather than with nutritional regimes. There were little differences in antler sizes between the groups. It was suggested by Suttie and Kay (1982), and later by G. Bubenik (1982) that deer have a threshold weight that they must attain before pedicles can develop. Lincoln (1971a) reported that Scot's red deer may go two to three years before pedicles develop, and he suggested that Scottish hummels, deer which do not grow antlers, may be simply undernourished. All of these authors agree that undernourishment can lead to poor development of the testes, the periosteium or the central nervous system which can lead to either hypogonadism or hyposensitivity of the central nervous system to hormonal control. It may be that an adequate diet is needed to allow young deer to reach a critical body weight which will allow them to respond to the light cycles via a pineal-hypothalamus-pituitary-gonadal/adrenal axis.

Conclusions and Management Implications

The statement that "the head grows according to the pasture, good or otherwise" (Dryden 1908, in Fennessey & Suttie 1985) is true enough. The great works of Franz Vogt (1936, 1948, 1951), recently popularized and eulogized in North America by V. Geist (1986a) proved that beyond a doubt. Through extensive feeding trials with red deer and roe deer, Vogt showed that pregnant hinds and does, supplemented with high protein-high energy rations could, over several generations, produce clearly superior antlered bucks and stags, even if the original genetic stock was mediocre. One group of his deer which averaged 180 kg whole body weight with 6 kg antlers produced 300–350 kg deer with 11–14 kg antlers just three generations later. While this work shows the potential of superior nutrition on the antler development of deer, it leaves few questions answered for the game manager.

The questions the manager wants to answer depends on his goals. The government biologist monitoring deer on public land wants healthy deer in adequate numbers to match hunting, predatory and other losses without over- or underutilization of his forest or range resources. The private land owner hoping to gain income from the sale of hunting rights has a similar interest in protecting his land, but with the added economic impetus of management for trophy antler size and body weights. The deer farmer producing velvet antler and venison for sale also has the economic incentive, but with an entirely different end product in mind. While the simplicity of Vogt's approach may benefit all three managers, it may not be feasible for any of them. Supplementing deer on public or even private game lands is generally not recommended, as it domesticates the deer and detracts from the original purpose of wildlife management—conserving the habitat and the species with minimal interrruption of its natural state. Even for the farmed or commercially hunted deer, a similar feeding program may lead to improved antlers, but the gain in body or antler weight may not be worth the extra cost of the feed. In the final analysis, our knowledge of the nutrient requirements of antler growth are woefully inadequate in relation to our needs of such information to properly manage the cervidae family.

Only continued interest and endeavor in the physiology of this fascinating phenomenon will earn us the rights of stewardship we presently claim.

17
Social Status and Antler Development in Red Deer

Luděk Bartoš

Introduction

Close correlations between social dominance and levels of some hormones modulated mainly by agonistic behavior have been reported in mammals. The hormone changes which accompany agonistic interactions appear to be more dramatic and longer lasting than those associated, for example, with sexual interactions (Harding 1981). Clearly, dominant animals generally have lower pituitary/adrenocortical activities than submissive animals living with them. Thus, the dominant position usually based on aggressive behavior often tends to be related to elevated androgen level, while subordinate status seems to be associated with lower androgen secretion and increasing levels of glucocorticoids (Brain 1980; Leshner 1980). Increased chronic ACTH/glucocorticoid production in subordinate animals suppresses androgens (Brain 1980). In *Cervidae* the evidence for these relationships has been obtained in white-tailed deer (Bubenik, A. & Bubenik, G. 1976b; Forand et al. 1985), red deer (Short 1979), and reindeer (Stokkan et al. 1980).

In theory, the mentioned hormones may be involved in both antler cycle timing and antler growth. In antlerogenesis the main role seems to be played by androgens. Typically, the more masculine a mammal male appears, the higher testosterone concentration in his blood (Crenshaw 1983). Deer are not an exception. Since the beginning of antler growth, androgens are probably the leading hormones affecting this process (Bubenik, G. 1982). Crenshaw (1983) injected GnRH into immature male white-tailed deer and then determined testosterone release. He obtained high correlation coefficients between testosterone response ratio and various antler measurements. In another experiment, Brown et al. (1978) found positive correlations between increasing serum androgen concentrations and increasing antler mass in bucks. Contrary to physiological levels of androgens, glucocorticoids were found to suppress antler growth (Bubenik, A. et al. 1976).

The above brief review allows one to put forward the hypotheses that dominance in a male deer is related to: (a) the timing of his antler cycle, i.e., dates of antler casting and/or cleaning; and (b) the process of his antler growth.

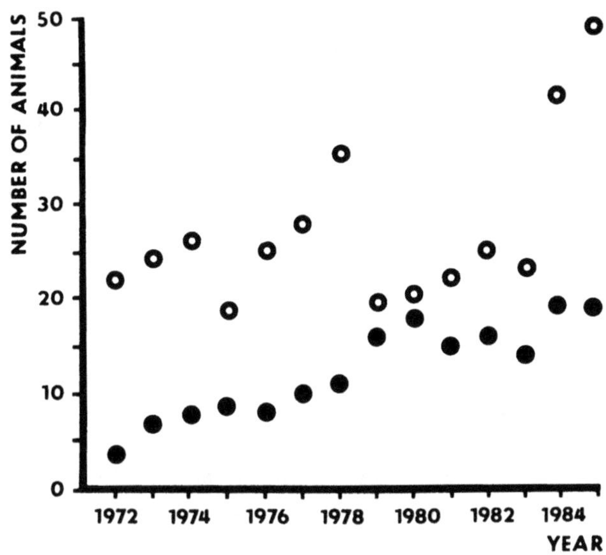

FIGURE 1. Number of deer in the observed population. *Open circles* = hinds. *Full circles* = stags.

To test these hypotheses, we studied an enclosed population of "white" red deer. The following text is based on published results of the study with some recent data added.

Red Deer Population Studied

The study was carried out in the Žehušice Game Reserve, Central Bohemia, Czechoslovakia, a fenced park of 2.42 square km, divided into two enclosures. In this report we used records from the population kept in the main enclosure (1.26 square km).

The subjects were male members of a herd of red deer, *Cervus elaphus*, containing many white and partially colored individuals (Bartoš 1980). No red deer in the reserve are culled, except for wild-colored male progeny. Fig. 1 shows the number of hinds and stags present in the main enclosure over the study period. From birth, all stags are identified individually from coat color and physical variation, so their ages are known exactly. Our analysis has involved all stags older than 2 years.

Observations were made between 1972 and 1985 (once every 4 weeks), except during the time of antler casting or cleaning, when they were made daily. The animals were fed almost every day during the whole year, and so could be inspected by an observer seated on a tractor during feeding at a distance of approximately

20 m without any apparent disturbance. The records of stag dominance hierarchy were made at the time of feeding. The observations of feeding deer lasted 10–70 min, until the animals left the feeding area. During observations, the food was always deposited in one place to induce competition among the stags. All animals encountered each other regularly, and if one animal moved away when approached by another, this was taken as an indication of subordinance. The outcome of such encounters was invariably clear. The rank order was based on the encounters of single stags with each animal of the bachelor group. The dates of antler casting and cleaning were recorded daily by the deerkeeper.

For the analysis, several relative values are used: *dominance index* (DI) (Bartoš & Hyánek 1982a)—calculated so that the position in a hierarchy (alpha=1, etc.) was divided by the number of males present. In those parts of the study, where it was necessary to study social position of stags in detail, DI is expressed in two forms (Bartoš & Perner 1985): *general dominance index* (GDI)—the position in the hierarchy of all the stags living within the same enclosure divided by their total number. When used for a single observation of social groups, a measure called *relative dominance index* (RDI) was used, i.e., DI within the group being just observed.

Once the rut was over, stags were separated temporarily from the main herd of hinds. The bachelor group tended to separate and/or disintegrate, mainly in summer, toward the period of antler cleaning. Although the mean age of the stags monitored was rather low, the alpha stags were not usually the oldest ones (Bartoš 1986a). There were very stable social relationships among the stags throughout the year in our study herd. Changes in the stag dominance hierarchy increased with the number of individuals present. The frequency of rank changes during the period with hard antlers was significantly higher than that during the velvet period (Bartoš 1986b). Between 1972 and 1983 stags developed a typical linear hierarchy. Triangular relationships occurred seldomly, being usually a temporal result of rare changes in the hierarchy during the period of antler casting. Only during and after the antler casting of 1984 the stability of the hierarchy decreased and the first permanent triangular relationships appeared. By the end of the year the hierarchy was stabilized again, this time with numerous nonlinear dominance relationships (Fig. 2).

Increasing the size of a social group affected animals at the two extremes of the hierarchy (the alphas and omegas). Increasing group size elicited an increase in agonistic activity in the former and suppressed it in the latter. The stags which occupied the middle range of the hierarchy showed a decrease in number of agonistic interactions when group size increased (Bartoš 1986a).

▶

FIGURE 2. Structure of dominance hierarchies in three different seasons related to calculated rank, date of antler casting, and ages of stags.

Hierarchy 1982	Rank x	Date y	Age z	Hierarchy 1983	Rank x	Date y	Age z	Hierarchy 1985	Rank x	Date y	Age z
Pišta ↓	1	3.3	7.5	Kramle ↓	1	14.3	5.5	Jubinal ↓	1	9.3	7.5
Bohumil ↓	2	4.3	9.5	Pišta ↓	2	10.3	8.5	Pekel ↓	2	10.3	7.5
Větrník ↓	3	18.3	4.5	Bohumil ↓	3	12.3	10.5	Pišta ↓	7	7.3	10.5
Kramle ↓	4	25.3	4.5	Špičkin ↓	4	22.3	6.5	Bohumil ↓	4	+	12.5
Špičkin ↓	5	22.3	5.5	Pekel ↓	5	20.3	5.5	Mozart ↓	3	17.3	6.5
Jubinal ↓	6	30.3	4.5	Jubinal ↓	6	20.3	5.5	Bedřich ↓	4	16.3	6.5
Pekel ↓	7	5.4	4.5	Košilka ↓	7	27.3	5.5	Jarmilin ↓	6	11.3	6.5
Košilka ↓	8	8.4	4.5	Ambrož ↓	8	27.3	4.5	Košilka ↓	13	10.3	7.5
Ambrož ↓	9	13.4	3.5	Bedřich ↓	9	8.4	4.5	Fahele ↓	8	1.4	4.5
Bedřich	10	15.4	3.5	Jarmilin ↓	10	9.4	4.5	Tolar ↓	12	13.4	3.5
Mozart =	11.5	20.4	3.5	Mozart ↓	11	11.4	4.5	Cyril ↓	11	10.4	4.5
Jarmilin	11.5	23.4	3.5	Fahele ↓	12	22.4	2.5	Jáchym ↓	10	11.4	4.5
				Cyril ↓	13	24.4	2.5	Kramle ↓	9	9.3	7.5
				Jáchym X	14	25.4	2.5	Markýz X	14.5	12.4	3.5
								Bělka	14.5	12.4	3.5

TABLE 1. Partial correlation coefficients between dominance and antler-testing dates, and between dominance index and antler-cleaning dates in individual seasons (with standardized age)

Season	Correlation between dominance index and	
	Antler-casting dates	Antler cleaning dates
1975	0.80	0.87
1976	0.91	0.33
1977	0.97	0.71
1978	0.95	0.36
1979	0.89	0.76
1980	0.67	0.87
1981	0.54	0.77
1982	0.95	0.39
1983	0.79	0.69
Total*	0.86	0.67

*According to Anděl (1978), the hypothesis was tested that the correlation coefficients are equal. The hypothesis could not be rejected ($P = 0.05$) for both antler casting and cleaning—hence, the estimation using z transformation for all seasons was made.

Social Status and Antler Cycle Timing

The first step of the investigation was to relate rank position to dates of antler casting and cleaning. The initial study (Bartoš 1980) indicated that the antler casting time of individual stags was dependent primarily on social status and that the influence of age was of secondary importance. The stags of higher rank also tended to shed velvet earlier. After a longer period of time, partial correlation coefficients were calculated for each season of the period when the stags established linear hierarchy in Žehušice (x = DI, y = date of antler casting/cleaning, z = age; Table 1). (Incidental asynchronous cast was calculated as a mean date of casting of the left and right antlers.)

Both relationships (between DI and casting/cleaning) when influence of age was eliminated, reached high statistically significant values. The first hypothesis advanced has been confirmed. The next step was to estimate social factors which could influence the relationships.

Social Structure and Antler Cycle Timing

The relationships between antler casting and social position under the situation of stabilized linear and nonlinear hierarchy were compared according to nonlinear social relationships occurring since the fall of 1984. For this purpose we used the data from 1982 (typical linear hierarchy), 1983 (linear hierarchy with an unusually young stag in the alpha position), and 1985 (nonlinear hierarchy; see Fig. 2). Data for spikers (deer with first antlers) were not included. For the pur-

TABLE 2. Comparison of correlations existing within different types of hierarchy

Correlation coefficients between:		1982	1983	1985
Rank and date of casting	(r_{xy})	0.98	0.97	0.68
Rank and age	(r_{xz})	−0.78	−0.82	−0.58
Date of antler casting and age	(r_{yz})	−0.87	−0.86	−0.92
Rank and date of casting with standardized age	$(r_{xy.z})$	0.98*	0.91*	0.45

*$P < 0.001$.

pose of this part of the study, the hierarchy of 1985 was estimated according to Clutton-Brock et al. (1982). The values were ranked. Order represented a rank position of a stag ("rank"). The bachelor groups of stags of 1982, 1983, and 1985, respectively, did not differ in mean (± SE), dates of antler casting (March 31 ± 5 days, March 31 ± 4 days, March 24 ± 4 days, ANOVA, $F(2,37) = 0.90$, NS), or in mean ages (4.92 ± 0.53, 5.21 ± 0.60, 6.00 ± 0.55 years, ANOVA, $F(2,37) = 0.98$, NS). The data for each season were calculated separately using partial correlations (x = DI, y = date of casting, z = age of the stag). The results are shown in Table 2.

In seasons with linear hierarchy (1982, 1983), correlations between rank and date of antler casting (with standardized age) reached high, significant values, while in the season with nonlinear hierarchy it did not show such a close relationship. Thus, the linearity of a hierarchy seems to be one of the important factors that allows a close relationship between social position of a stag and his antler-casting time.

Agonistic Activity and Antler Cycle Timing

Years lasting observations allowed more general analysis using data of cycle timing and social characteristics per unit calendar year. It was found that the strength of the relationship between stag rank and casting and cleaning order under the situation of stabilized hierarchy was significantly correlated with most of the indicators of general aggression (such as the number of killed stags, incidence of broken antlers, etc.; Bartoš 1986b). The higher the level of aggression within the herd of stags, the closer the relationship between rank and the timing of the antler cycle indicating that the process of antler cycle timing can be modified by an aggressive behavior of a stag related to his rank. To prove directly this suggestion, agonistic activity of selected stags was monitored in detail during competition at feeding before casting and before cleaning. Animals of the top, middle, and bottom rank were included. There were significant correlations between the casting time and the most frequent aggressive acts prior to casting. The more aggressive a stag, the earlier the date of casting. On the other hand, there were no statistically significant correlations between agonistic activities recorded during a velvet period and the date of antler cleaning (Bartoš 1985). This is consistent with the results presented in Table 1. It can be seen there that although both antler casting and cleaning correlated highly with DI, the correlation of the former was markedly higher than that of the latter. Is there any difference between the periods preceding the casting and cleaning?

Before antler casting (i.e., since the end of the rut), the stags of the studied herd lived usually in one large group, whereas after casting they tend to disperse into numerous unstable small groups not separated exclusively from a company of hinds. Therefore, the stags may be in a different social environment in the two periods. Presumably, in a situation of changing sizes and memberships of groups, animals of lower rank may increase their rank temporarily if they are in a group of the lowest-ranking individuals. On the other hand, dominant animals separated from the company of others need not have sufficient social stimulation which would influence their internal environment. If it is so, then sample monitoring of individual stags at that time could thus hardly record the same facets of their aggressive activity. To solve the problem we conducted detailed observations of the composition of individual social groups of stags throughout the velvet period (Bartoš & Perner 1985). That is, from the time just after antler casting of all stags to the time when the last stag is cleaned. The observations were made approximately every other day within the velvet period. As in other red deer populations (Appleby 1983; Bützler 1974; Clutton-Brock et al. 1982; Darling 1937), our stags tended to associate with animals of similar rank and age. RDI values were calculated for each stag for each observation and were compared with antler cleaning dates. The correlation coefficients increased toward the time of cleaning. For the last 2 weeks of the period the coefficients reached levels similar to those between stags' rank and antler casting. To complete the analysis, the association between stags was defined. The closest associates of each stag (determined after Appleby 1983) in the weeks preceding each individual's antler cleaning date were identified. The higher-ranked associates of the same age cleaned significantly earlier than the lower-ranked individual. It was concluded that both antler casting and cleaning are regulated by hormones modulated by agonistic behavior related to rank (Bartoš & Perner 1985).

Antler Casting in Different Cervid Species

Until now we have discussed the relationships between rank position and antler cycle timing in red deer. Forand et al. (1985), observing captive herds of white-tailed deer, found a highly significant inverse correlation between rank and order of casting antlers, indicating that dominant white-tailed deer bucks retained their antlers longer than subordinates. This result is in apparent contrast to what has been found in our red deer herd (Bartoš 1980, 1986b). The authors (Forand et al. 1985) presented a brief review giving literary evidence that in northern areas of the United States where antler casting is early and relatively short (from mid-December and to late January), older, larger, and presumably dominant bucks cast antlers earlier than their subordinates. For the Midwest, where antler casting extends from January to late March, white-tailed bucks with large antlers generally retain them longer than bucks with small antlers. The authors suggested that dominant white-tails in northern ranges may experience considerable stress and physical exhaustion due to a short but intensive rutting season. Then they related the suggested stress with increased levels of corticoids reducing levels of testosterone. They concluded that testosterone levels of subordinate animals may stay longer above the threshold for antler casting, resulting in longer retention of antlers.

Are the presented contradictory results based on a methodological mistake, or is there any species-specific differences? The latter seems to be more likely. It is our suggestion that the difference between red deer and white-tails in the relationship between rank position and the order in which individuals cast their antlers lies in the species-specific response of antler casting to seasonal pattern of testosterone. Brown et al. (1983a) have shown that the endocrine control of the antler cycles of two deer species may differ. It has been well documented that there are species-specific differences in seasonal patterns of casting and new antler growth. Deer species with seasonally determined antler cycles may be divided into two basic groups. *Group A*—those in which casting of old antlers is followed immediately by growing of a new antler, such as red deer, wapiti, sika, fallow, roe, and Pere David's deer; *Group B*—those in which an interval exists between antler casting and new antler growth such as white-tailed deer, mule-deer, moose, reindeer, and caribou (Bubenik, A. 1966; Goss 1983; Jaczewski 1981a; Sempéré & Boissin 1982). In red deer, antler casting and regrowth are closely interdependent. Our hypothesis about the positive correlation between antler casting and rank position of a male fits well (Bartoš 1980, 1986b). It is presumed that this might be the case for all species belonging to Group A. On the other hand, in species in Group B, antler casting and starting of new antler growth are well separated in time, so that they may have different relationships to rank position. While high rank position may associate with the delay of antler casting, as has been found by Forand et al. (1985), starting of new antler growth should be enhanced. In other words, the time between antler casting and new antler growth should be shortest in the highest-ranking animals of Group B. To support the above hypothesis, we can submit some empirical data. For species in Group A: in red deer (Bützler 1974; Lincoln 1972; Lydekker 1898; Nečas 1959), and wapiti (Bubenik A. 1982a), it is well established that the older and stronger males cast earlier. For species in Group B: Kozhukchov (1973) reported in farmed bull moose that young animals cast antlers 1–3 months earlier than old animals.

How can we explain the relationships between the behavior and antler casting and the difference between the two groups of deer species on a hormonal basis? While males of the species belonging to the Group A need some stimulation of a new antler growth, it is the decline of testosterone itself after the rutting season that seems to be responsible for the casting of the antlers which may occur in early winter in males of Group B (Brown et al. 1983a; Mirarchi et al. 1977b). White-tailed bucks treated with antiandrogen cyproterone acetate immediately after casting their old antlers did not renew growth of antlers (Bubenik, G. 1982). It can be predicted that the treatment of males in Group A with antiandrogen before casting might respond in retaining their antlers. Red deer stags that die in spring

FIGURE 3. An overaged stag (in the foreground) who failed to cast antlers in the spring is sparring with a mature stag carrying fully developed velvet antlers.

in Scotland are almost invariably still carrying their hard antlers (Mitchel in Lincoln 1971a; Lincoln & Bubenik, G. 1985). In 1985, the oldest stag of the Žehušice herd failed to cast his antlers (Fig. 3). He had retained them until next October, when unfortunately he was shot. It was found that he had fully regressed testes. The evident fall of testosterone levels was not sufficient to induce casting. On the other hand, van Ballenberghe (1982) reported several cases of bull moose which were handicapped by an injury and which cast earlier than others, suggesting that the fall of testosterone levels was potent enough to initiate casting. The presumed species-specific response to seasonal testosterone variation may be reflected also in the cast antlers. The longer dead antler is attached to the pedicle, the more it dies back (Lincoln 1984). The species which cast soon after the rut tend to have a convex casting surface at the base of the cast antlers compared with the concave casting surface in the species which cast later. This is also apparent when red deer are castrated soon after the rut (Lincoln 1984; Lincoln & Bubenik 1985).

It has been hypothesized that new antler growth may be initiated by a small reactivation of sexual function and hence testosterone pulse (Bartoš 1980; Bubenik, G. 1982; Goss 1983; Sempéré & Boissin 1982). This may correspond to the initiation of pedicle formation within a male's ontogeny. It has been shown that during puberty the deer testes must be activated for a short time to induce the growth of pedicles (Brown et al. 1983a, 1983b; Lincoln 1971; Sempéré & Boissin 1982; Suttie et al. 1984). The initial suggestion that new antler growth may be induced by the short-term pulse of testosterone was based more or less on indirect results (Bartoš 1980). However, now there are rather more direct indications available. Testosterone titers were determined to increase twice a year in red deer (Blaxter et al. 1974; Suttie et al. 1984), wapiti (Haigh et al. 1984), white-tailed deer (Bubenik, A. 1984; Bubenik, G. et al. 1982a; Brown et al. 1983b; Mirarchi et al. 1977b), and sika deer (Brown et al. 1983a). The effect of testosterone, modulated by behavior, on the initiation of antler regrowth could probably be determined by measuring the absolute hormone level. Low amounts of testosterone can stimulate bone growth, and higher levels may be inhibiting (Brown et al. 1978b). The more dominant males have earlier, higher, and more frequent testosterone pulses within the range of Brown et al.'s 'low amount'; a new antler bone growth may thus be initiated more vigorously and the antler casting in species in Group A may occur earlier (Bartoš 1980). In males in Group B, new antler growth of dominants may start earlier even though antler casting had occurred later.

In general, fighting stimulates the level of glucocorticoids (see Introduction). Increased corticoid levels inhibit antler growth (Bubenik, A. et al. 1976). Adrenal hypertrophy may occur in deer under conditions of stress (Bubenik, G. & Bubenik, A. 1965; Hughes and Mall 1958), a reaction which may delay antler casting in fallow and red deer (Fig. 4) (Bubenik, G. 1982; Topiński 1975), representatives of species Group A. Hypothetically, the species of Group B should be affected in just the opposite way by stress, enhancing antler casting. A role of some other hormones may be expected. Nevertheless, this possibility is not included in this simplified model.

If our earlier hypothesis is correct, how does it fit to antler casting caused by castration? From the point of view of the hypothesis, we should expect a sudden fall in testosterone levels after castration, followed by temporary restoration of androgen levels (of adrenal origin?) which afterward definitely declines. To our knowledge, the detailed pattern of testosterone decline after castration has not been investigated. However, there are some indirect data. After orchidectomy,

FIGURE 4. Asynchronously casting stags are on average older and higher ranking than those casting synchronously. Among the asynchronously casting stags, the higher a stag ranked, the shorter interval between the dates of casting of both his antlers (Bartoš and Perner 1987).

LH levels increased more than four times in red deer (Lincoln & Kay 1979) and more than four times in white-tailed bucks (Bubenik, G. et al. 1982a). In theory, testosterone secretion from the adrenal gland may be facilitated by this way to a short-term surge. Under such circumstances the difference between the two groups of Cervids should be diminished. This seems to be the case. As reported by Goss (1983), after castration, renewed antler growth occurs soon after the old antlers have been lost, even in those species in which there is normally a lag between these two events. However, the time of antler regrowth after castration is season-dependent in both groups of the deer species (Lincoln 1984; Bubenik, G., personal communication).

How would the above hypothesis explain that some solitary-living old individuals may cast their antlers rather early in the period of antler casting when compared to most of other conspecifics in Žehušice while performing minimal social interactions? In experiments with artificially altered daylight and its influence on the antler cycle, it was shown that older males can sometimes express endogenous annual antler growth, irrespective of artificial light conditions (Goss 1969a). This may be the case in solitary-living stags. On the other hand, the absence of social stimulation may cause the observed fact that these solitary-living stags always cast antlers later than the younger top dominant ones living in social groups (Bartoš 1980).

It must be emphasized, however, that general good nutritional status of a population seems to be an essential factor allowing an expression of the behavior in antlerogenesis. Suttie (1980a) found in red deer that the quality of nutrition influences the seasonal levels of testosterone and prolactin. Good nutrition caused even rutting of his stags twice a year, in spring and fall. In this respect, it is important to note that there is also sufficient evidence confirming a species-specific dependence of antler casting on nutrition. In red deer, Darling (1937) and A. Bubenik (1966) argued that inadequate nutrition after the rut significantly delays casting. Also, Lincoln (1971a) stated that red deer stags in poor condition cast their antlers later. Watson (1971) showed that antler casting among red deer was delayed by food restriction brought about by severe weather conditions, but supplemental feeding could reverse this trend. Similarly, in experiments with farmed deer, Suttie & Kay (1982) and Fennessy & Suttie (1985) found a trend for antlers of a nutritionally unrestricted group to be cast 1–2 weeks before the restricted group. In contrast, Long et al. (1959) showed that nutritional deprivation of white-tailed deer in the spring hastened antler casting. Also Lincoln & G. Bubenik (1985) stated that white-tailed deer cast their antlers earlier than normal in winter in response to poor feeding and loss of condition, while the regrowth of new antlers in such animals occurs later than normal in spring. Ozoga & Verme (1982) reported an influence of improved nutrition on antler casting in white-tailed deer. Supplementally fed bucks in the enclosure retained their antlers several months longer than did the bucks in that area. West & Nordan (1976a) found similar relations in mule-deer. Better fed captive bucks often cast antlers later than wild bucks.

Conversely some other data do not support the above hypothesis, either for Group A or Group B. Gibson & Guinness (1980) reported, for example, that red deer stags with the highest reproductive success (and hence of the highest rank) during the rut cast their antlers later than others. Presumably poor condition negatively affects the spring testosterone pulsation of these animals as in Suttie's (1980a) restricted group [see also the above-mentioned experience of Darling (1937) and A. Bubenik (1966)]. The same explanation may account for all the discrepancies in casting of white-tailed deer of various geographical origin cited earlier (Forand et al. 1985). In northern areas, one may expect worse nutrition of the deer than in southern areas. Dominant bucks are not able to maintain levels of testosterone above the threshold for antler casting because they are significantly more exhausted after rutting activity than their subordinates. Earlier casting in dominant bucks thus may occur. Moreover, seasonal fall in testosterone levels under conditions of naturally restricted nutrition cannot be stimulated substantially by dominant related behavior, which may also explain why the casting time in white-tailed deer is more synchronized and occurs earlier in the season in northern than in southern areas as reported by Forand et al. (1985). Under better general conditions, in the Midwest, such behavior may lead to maintenance of elevated testosterone levels of dominant bucks causing the delay of antler casting.

Antler Cleaning and Deer Species

Contrary to casting, antler cleaning seems to follow the same pattern in both groups of deer species in relation to social position. A tendency for a positive correlation between rank and order in which males clean antlers was found not only in our red deer (Table 1), but also in captive white-tailed deer (Forand et al. 1985). The stimulatory effects of social interactions among dominant males probably elevate levels of testosterone, while the interactions elevate glucocorticoids and depress testosterone levels in subordinates. As a result, antler cleaning may occur earlier in dominants and later in subordinates.

Many authors have suggested that antler cleaning dates are fully dependent on age, such as in red deer (Bützler 1974; Darling 1937; Nečas 1959) and in other cervids (e.g., Hirth 1977). Spike-antlered deer clean antlers later than fork-antlered males in red deer (Bubenik, A. 1966; Darling 1937; Jaczewski 1981a; Lincoln 1971b; Nečas 1959), white-tailed deer (Hirth 1977; Jacobson & Griffin 1982; Scanlon 1977), and moose (van Ballenberghe 1982). Nevertheless, there are also contradictory reports, mainly from captive populations. Both the earliest and the latest cleaning dates were observed among yearlings in fallow deer (Chapman & Chapman 1975; Štěrba & Klusák 1984), in white-tailed deer (Jacobson & Griffin 1982), and in our red deer herd. Exceptionally early cleaning by yearlings may be observed under natural conditions, too [e.g., in moose (van Ballenberghe 1982)]. Here again, different opportunities for social grouping and hence for differential social stimulation of the process, as well as differential opportunity to be stressed, may be involved.

Social Status and Antler Growth

The possibility of a relationship between social position and antler size in Cervids has been widely discussed.

Some authors have suggested that antlers advertise an individual's dominance status (Beninde 1937; Bubenik, A. 1968, 1982b; Geist 1966b; Henshaw 1969), but variable results have been obtained in field studies designed to assess this. Espmark (1964) in reindeer and Lincoln (1972) in red deer found that after the loss of antlers, either naturally or artificially, individuals became less effective in competition with other males, resulting in loss of social rank in the bachelor group. Bützler (1974) measured the lengths and weights of cast antlers from stags of known social position. There was a positive correlation between these measurements and social position but the author questioned whether the relationship was genuine. Clutton-Brock et al. (1979) observed more than 100 rutting fights between red deer on the isle of Rhum in Scotland and found a weak correlation between the number of points on the antlers and fighting ability. No relationship between antler length and fighting success was apparent. Appleby (1982) also observed red deer of the same population and found that the rank of mature stags in winter was correlated to antler length. Winter rank in mature stags was, however, correlated significantly with the weight of their antlers in one of two study years. Suttie (1980b) found in a group of farmed stags that antler weight but not antler length or number of points was positively correlated with social position. Miura (1984) reported for male sika deer that large antlers were related to dominance.

All the mentioned studies were based on observations at the time when the male deer had completed their antler growth. That is, the studies compared the relationship between males' fighting abilities and size of their grown antlers. The criticism of the suggested behavioral significance of grown antlers was made in red deer during the rutting season. The fighting ability of individual stags changes during the course of the rut as their body condition declines, and individuals vary in the timing of their declines. Consequently, a stag that assessed its opponents on criteria that did not vary with changes in body condition during the rut would take many incorrect decisions (Clutton-Brock et al. 1979, 1982). In our previous studies we have already suggested that social position and related agonistic activity of stags during the velvet period influence the antler weight, length and number of points, and therefore the size of grown antlers are a consequence of previous social position and not vice versa (Bartoš & Hyánek 1982a, 1982b). Also Wölfel (1983) claimed possible role of rank position of a male yearling red deer in antler development. The aim of the following study was to assess in detail how the social position of a growing red deer stag is related to various measures of antler development.

In red deer, the gain in antler growth or development correlates in the first 5–6 years almost linearly with body growth and weight. Afterwards, i.e., after the stags have reached their mature body size, there is always a substantial variation

TABLE 3. Relationships of general and relative dominance indices to antler weight and length, and number of antler points

	Antler weight	Antler length	No. of antler points
General dominance index (GDI)	−0.77**	−0.63*	−0.93***
Relative dominance index (RDI)	−0.87**	−0.78**	−0.99***

*$P < 0.05$.
**$P < 0.01$.
***$P < 0.001$.

(Bubenik, A. 1966, 1982; Huxley 1931). Hence we have simplified the antler growth into a model that states that antlers increase linearly during a stag's ontogeny up to 5 years of his age, while afterwards there is not regular increase. [The same pattern has been found also in a red deer stag's social position by Appleby (1980) and Bartoš & Hyánek (1982a).] According to the model proposed, a linear regression of growth of all the antler measurements and dominance indices (DI's) during the period of antler growth was estimated for each stag. Then values for two extreme ages of a stag's ontogeny were calculated (2 years—the beginning of the first regular branched antler growth, and 5 years—the end of body development). Such calculated values of the DI were correlated with those of antler measurements. The results of the analysis (Bartoš et al. 1988) showed that high-ranking stags of both extreme ages had heavier, longer and more branched antlers. The social position involved (DI) had always been estimated during the period of formation of the future antler. However, as mentioned earlier, in Žehušice the bachelor group tended to disintegrate during the velvet period. Thus DI does not reflect detailed changes in social environment. Hence we further presumed: If we compared GDI and RDI throughout the period of antler growth of a stag, then correlation between antler size characteristics and the indices should fit better to RDI than to former one. To solve the problem we used the data of 1983 when we followed the distribution of all stags in individual social groups. We used the records of the period between antler casting and cleaning for each stag. It represented 59.71 ± 2.70 different observations for a stag. Cast antlers of 11 individuals were collected during the following spring and measured. The values of GDI of the velvet period of these stags were equal to those of RDI in one case, while they were not equal in ten cases (Sign test, $P = 0.01$). All the values used in the analysis (GDI, RDI, antler weight and length, and number of points) were adjusted for age using an analysis of linear regression (Snedecor & Cochrane 1965). Correlation coefficients between the indices and selected antler size characteristics are shown in Table 3.

The correlation coefficients between antler characteristics and GDI were lower than those between antler characteristics and RDI in all three cases. So the presumption has been fully supported.

The results presented should be taken as applying to a model. The living conditions in Žehušice differ in several ways from those elsewhere. The situation in

FIGURE 5. A mature stag who cast one antler is "flailing" in an attempt to hold the rank over a subadult, antler-carrying stag.

this herd is difficult to compare with the free-living individuals. The deer could not leave the fenced area. So that the animals were unable to separate fully from others and interactions with conspecifics could not be avoided. From Suttie's (1985) observations of farmed red deer stags, it was apparent that although a stable hierarchy existed in bachelor herds, it did not prevent aggression. Suttie found that the level of aggression was much higher than in the wild. He concluded that members of a hierarchy may be stressed to an extent not seen in the wild. Hence physiological consequences of aggression related to rank position under conditions of restricted space may be expressed more markedly than those under natural conditions. On the other hand, it seems to be likely that a population such as ours may be a relevant model for free-living populations, since the consequences of high density in the pen which may influence antler structure (and/or antler cycle timing) are more evident under these defined living conditions. The behavioral significance of the antlers may be of secondary importance depending on social background of a studied herd and/or on previous social experience of a male deer. In Žehušice, the bachelor group of stags is constant throughout the year. The same animals encounter each other during the velvet period as during the rest of the year. This results in a very stable social hierarchy. Social status remains almost constant even after antler casting (Fig. 5) or antler breakage (Bartoš 1986b). It is no wonder that there is a close correlation between rank position and antler length, antler weight and the number of tines found also out of the velvet period (Bartoš & Hyánek 1982b). In free-living populations, reduction in gregariousness immediately after antler casting is known among red deer stags (Bützler 1974; Geist 1982; Lincoln et al. 1970). This may result in a less pronounced relationship between rank position and antler size. Just before the rut, bachelor groups completely disintegrate and stags widely disperse (Bützler 1974; Clutton-Brock et al. 1982; Darling 1937; Lincoln et al. 1970; Nečas 1959). During the rutting season, strange stags may be encountered. Sexual competition during that time brings a strong motivation for fighting. This may encourage stags to ignore experience gained in a bachelor group during the antler growth that taught them to avoid an interaction with larger antlered individuals. That is probably why Clutton-Brock et al. (1979) did not find any simple relationship between success in rutting fights and antler size. After the rut, stags usually return to their original ranges (Bützler 1974; Clutton-Brock et al. 1982; Darling 1937; Lincoln et al. 1970). Winter groups thus consist mainly of stags which had been present during antler development. Hence Appleby (1982) was able to detect at least partly some relationships between rank and antler characteristics in his winter observations. The behavioral meaning of fully grown antlers and the physiological consequence of the stag's behavior on antler growth seem to have quite a different basis. Behavioral meaning of the antler size of red deer stags probably depends on social background of the studied population and previous experience of an individual. Hence, there are a number of studies suggesting no consistent tendency for males to avoid fighting individuals with larger antlers, at least in red deer (Appleby 1982; Clutton-Brock et al. 1979; Krzywiński 1978; Lydekker 1898), while under some circumstances there is evidence of an advantage to bear large antlers (Bartoš & Hyánek 1982b; Bubenik, A. 1982b). The physiological consequence of the stag's behavior on antler growth may act since the beginning of the velvet period. The more dominant a stag, the higher the seasonally attained levels of androgens, the greater the enhancement of antler tissue formation. This suggestion has been supported by Shilang & Shanzi (1985) who found that a small amount of androgens to sika deer stag's food during the velvet period stimulated their antler growth. On the other hand, the lowest ranking stags may lack this androgen stimulation and the presumably elevated glucocorticoid levels may actually suppress antler growth (see Introduction). A fall in social hierarchy in aged stags may be an initiation of their antlers "going back." A. Bubenik (1982a) stated that in disorganized populations, a stag could be "over the hill" at 10–11 years, as opposed to 16–18 years in stags of well-organized herds. In young male deer, the size of antlers increases with increasing body size in succeeding seasons (Bubenik A. 1966; Goss 1983; Jaczewski 1981a). The seasonal peaks of testosterone also increase in parallel with antler and body growth, as was shown in red deer (Bubenik, A. 1984; Lincoln 1971a), roe deer (Sempéré & Lacroix 1982), and white-tailed deer (Bubenik & Schams 1986). Bubenik, G. et al. (unpublished, shown in Bubenik, A. 1982a) found in red deer that young animals had the highest levels of cortisol which then declined in prime stags (4–10 years) and then rose again in old ones (11+ years).

Cameron (1892, cited in Goss 1983) was probably the first who reported the primary importance of body weight in a combat of deer. He said that not males with heaviest antlers are favored in fights, but those with the heaviest body weight. Clutton-Brock (1982) has argued that antler size is related to individual differences in body size and weight. To support it, there is a large amount of evidence for a relationship between body size and weight in red deer and antler weight (Appleby 1982; Clutton-Brock et al. 1979, 1982; Huxley 1926, 1931; Hyvärinen et al. 1977). Initially, there was no body weight data available for our "white" stags. However, there were some indications that body weight did not have exclusive influence on the relationship between rank position and antler development (Bartoš et al. 1988). In the 1985 season, we succeeded in obtaining the data of the relative body weight of the stags under study. A close correlation between body weight of a stag and his rank position was found when the age was statistically standardized ($r = 0.79$, $P<0.01$). On the other hand, a nonsignificant ($P>0.05$) correlation was apparent between body weight of a stag and the size of his antlers (with eliminated influence of age) under the living conditions of the park, while rank did not correlate with several antler characteristics such as antler length, etc. When body weight was controlled by partial correlation, rank correlated with antler length, number of tines, number of points on the royal, bez tine, third point of the royal, and length of all the royal points. On the other hand, when rank was controlled by partial correlation, there was still no significant correlation between body weight and antler characteristics (Bartoš et al. 1988). This was contrary to results of the above-mentioned authors. Nevertheless, it has already been observed in Scotland that live weight need not correlate with antler size in red deer (Suttie 1980b).

It may be concluded that the relationship between rank position of a stag during the velvet period and intensity of his antler growth does exist regardless of his body weight. Those studies showing relationship between body weight and antler size of stags living in groups should be analyzed also from the point of view of the stag's rank position in the same time, since under certain conditions rank position may be even more influential on the antler growth than the body weight.

It can be concluded that the advanced hypothesis was confirmed. Dominance in male red deer was found to be related to timing of antler cycle and antler growth. The dominant individuals of socially stabilized group tend to cast and/or clean antlers earlier and produce larger antlers than subordinate ones.

Acknowledgments. I gratefully acknowledge the excellent field assistant I have had over many years from V. Perner, the former deerkeeper at Žehušice. I am indebted to A.B. Bubenik, G.A. Bubenik, R.N.B. Kay, and G.A. Lincoln for stimulative criticism of an earlier draft of the whole manuscript. Helpful comments of P.T. Brain, T.H. Clutton-Brock, and V. Geist on the part of this chapter originally prepared for other publication are also very much appreciated.

18
Genetic Variability and Antler Development

KIM T. SCRIBNER AND MICHAEL H. SMITH

Introduction

Body and antler size in *Cervidae* are important qualities contributing to dominance among males and to their value as trophies. Variation in these characteristics has occupied the interest of many people over the centuries and much speculation about the underlying causes of this variation has resulted. The basic importance of genetics to antler development is inferred primarily from the differences in antler expression among species but not from any specific information about genetic differences between cervids. The genetic basis of variation in antlers among individuals within a species is just beginning to be understood. Information on genetic variability has only recently been mentioned relative to antler development (Smith et al. 1983). The major difficulty in establishing an understanding of the genetic basis of antler development is the lack of quantitative data on the inheritance of antler characteristics. Supplying this information is a monumental task because of the need to maintain a large number of breeding animals in captivity.

Evolutionary Significance of Antler Development

Examination of the factors affecting variability in morphological characteristics, such as antlers, is important to understanding the adaptive nature of an individual's interactions with its environment. The major approach used in studying the genetic basis of morphological characters is to partition individual variability into several components including that of additive genetic variance (i.e., heritability; Falconer 1981). Many characters have been studied in the laboratory and their heritabilities estimated. The hypothesis that antler size must be heritable to some degree is suggested by studies of white-tailed deer, *Odocoileus virginianus*, under penned conditions (Harmel 1982; Templeton et al. 1982). Antler size also varies in response to environmental fluctuations (Severinghaus et al. 1951; Severinghaus & Moen 1983; Ullrey 1983). Morphological characters may be optimized by different genetic characteristics depending upon environmental conditions. Estimates of phenotypic variability and heritability vary for traits in populations experiencing spatial and temporal environmental variation (Falconer & Latyszewski 1952; Falconer 1960; MacKay 1981; Lasslo et al. 1985). The results of these studies and theoretical analyses of the underlying causes of character variability (Felsenstein 1976; Slatkin 1978; Via & Lande 1985) suggest important hypotheses that are testable in natural populations. The variability in antler characteristics among individuals can be related to spatial and temporal fluctuations in environmental quality and differences in the genetic constituency of individuals, especially their level of genetic variability.

A number of empirical studies have documented the relationship between expression of characters and heterozygosity. As a measure of the genetic variability of an individual, heterozygosity is calculated as the percentage of loci for which an individual received a different allele from each of its parents. Studies correlating growth rates (Singh & Zouros 1978; Pierce & Mitton 1982; Cothran et al. 1983; Chesser & Smith 1987) and overall body size (Boyer 1974; Garten 1976; Smith & Chesser 1981; Koehn & Gaffney 1984; Chesser & Smith 1987) with heterozygosity suggest the importance that different levels of genetic variability may have in varying morphological characters. Antler size shows strong positive relationships with body size within a given herd (McCullough 1982; Smith et al. 1982). A positive association between genetic and environmental variability exists across a variety of species (Nevo et al. 1984), but the relationship between these variables and the expression of traits in natural populations is not well substantiated. Analysis of the effects of heterozygosity or environmental variability on the expression of morphological traits is confounded by environmental-genetic interactions.

The evolutionary significance of the phenotypic expression of morphological characteristics is determined by their contribution to individual fitness. Heritabilities for characteristics closely associated with fitness generally will be low (Fisher 1930). Structures related to the ability of an animal to utilize available resources to maximize reproductive output will be selected for and preferentially expressed to a greater degree in subsequent generations. Male reproductive success in cervids is associated with access to females and directly related to an individual's dominance (DeVos et al. 1967). Social rank in a variety of species is correlated with age, body size, antler size (Severinghaus & Cheatum 1956; Hirth 1977; Clutton-Brock et al. 1979) and antagonistic behavior or avoidance thereof (Geist 1974a; Clutton-Brock et al. 1980). The relationship between heterozygosity and morphological characteristics suggests that heterozygosity may also affect dominance status. Mice with higher heterozygosities have greater dominance and more effectively compete for limited resources than those less heterozygous (Garten 1976). Antler size may be a reflection of an animal's heterozygosity as well as part of the strategy by which animals achieve dominance. Since antlers are regrown each year, they are ideal structures to test for the importance of heterozygosity in varying environments. However, the only cervid studied in this regard is the white-tailed deer, which will be emphasized in this presentation. Demographic, body condition and size, and genetic characteristics of white-tailed deer can be used to illustrate several important points concerning antler growth. Antler size and shape change are a function of age, but the problem is to understand how important environmental and genetic factors are in altering growth and how the pattern is changed at specific ages during the life of cervids. Determining the relative importance of these factors is a statistical problem analogous to that of determining heritability and involves partitioning variability of antler characteristics associated with such variables as age, habitat, year, body condition, and genetic characteristics. Resource availability as reflected in body condition (Johns et al. 1982, 1984) and heterozygosity (Cothran et al. 1983; Scribner et al. 1984; Chesser & Smith 1987) are correlated with life history characteristics of white-tailed deer. Heterozygosity is calculated from heritable variations in protein banding patterns as demonstrated by electrophoresis (Manlove et al. 1975).

Model of Antler Growth

In mammals, growth is normally a sigmoidal function of age (Zullinger et al. 1984). Antler growth differs in several important ways from this general pattern. In white-tailed deer it is generally initiated during the second half of the first year of life, and the antlers are shed and regrown each year. Since animals are normally collected after antlers have completed their yearly growth, the function describing changes in antlers with age calculated from these data is curvilinear but not sigmoidal (Fig. 1). A good mathematical function for describing changes in particular antler characteristics (Y) *vs* age (X) appears to be a quadratic equation: $Y = a + b_1 (X - c) + b_2 (X - c)^2$. The constant c is species specific and is the average time at which antlers are measured after completing their first year of growth. Environmental ($E_{1\text{-}N}$; Fig. 1) and genetic effects are expressed uniquely each year as well as incrementally by influencing body size and its allometric relationships with antler characteristics (Smith et al. 1982). The general quadratic model of antler development may be affected by environmental and genetic factors and their interactions by changes in the intercept or initial expression (a), and linear (b_1) and quadratic (b_2) slope parameters of the model. Variability in the antler relationships with age may be observed among populations and/or temporally within a population. This variation may be due to resource quantity and quality affecting body condition and also to levels of genetic variability.

The quadratic model seems to hold for body weight as well as antler characteristics when both types of variables are examined over the same range of ages (Fig. 2). Over this range, maximum variation in the antler size/body weight relationship occurs at the youngest and oldest ages (Fig. 2). The effects at younger ages are primarily reflected by variation in a or at the oldest ages in b_2. Maternal effects on antler development are not documented but would be expected to primarily influence the value of a. Maternal effects could be mediated through the mother's genetic characteristics or body condition. Fetal growth rates vary relative to the level of maternal genetic variability (Cothran et al. 1983). The

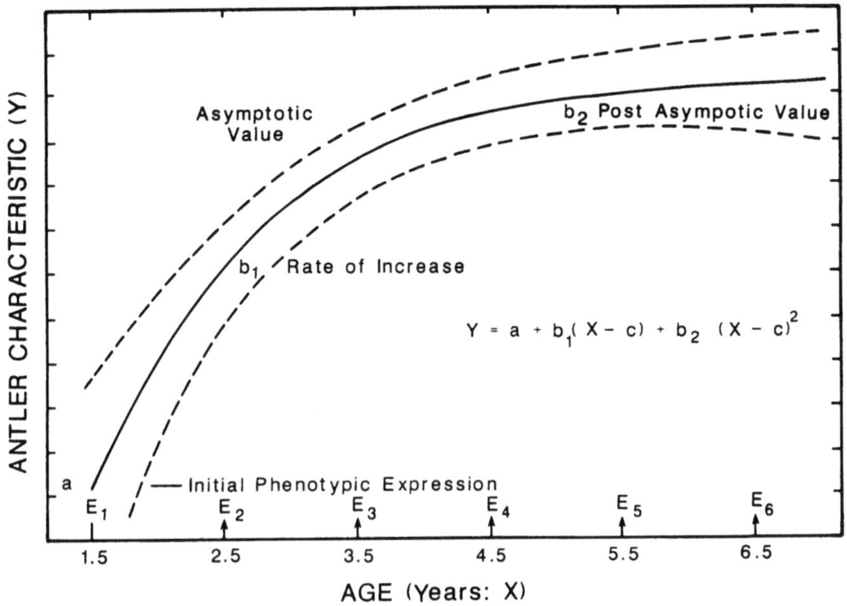

FIGURE 1. Curvilinear model describing the changes of a general antler characteristic (Y) as a function of age (X). Variation in the general quadratic model may be affected by environmental ($E_{1\text{-}N}$) and genetic factors each year. The constant c is species specific and is the average time at which antlers are measured after completion of the first full year of growth. Time to asymptote varies within and across species.

value of b_1 may be relatively invariant and more a reflection of species specific allometric growth.

The general form of the curvilinear function used for describing antler development is well known for various species, but it has seldom been specific or parameterized for individual populations or species. This function has been used infrequently for age-related changes in antler size/body weight relationships (Fig. 2; Gould 1974). We analyzed antler and age data for ten herds of white-tailed deer in the southeastern United States. The quadratic model accounts for 29–90% of the variation in main beam length, 27–88% for beam diameter, and 37–63% for number of points. The values of b_1 are always positive while those for b_2 are mostly negative. Differences among populations are observed for intercept and slope values for the number of points and beam diameter though the allometric relationships between antler characteristics have no relationship with latitude. That locality effects are important in altering the portion of the growth curves most affected by the quadratic term (X^2) supports the need for further analyses of potential environmental and genetic effects. The negative value commonly observed for b_2 probably indicates the difficulties older males have in maintaining body condition while trying to maximize the number of females

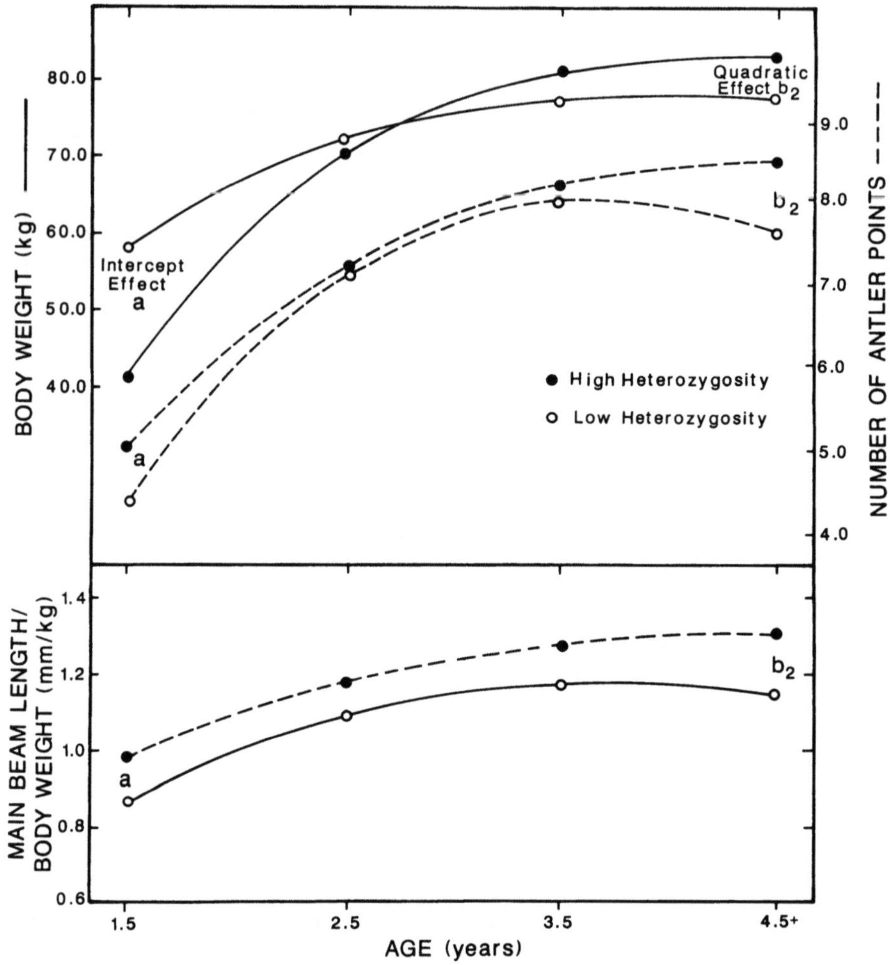

FIGURE 2. Differences in growth of number of antler points, body weight, and main beam length/unit body weight between white-tailed deer of high and low levels of heterozygosity. The greatest variation in antler size and body weight relationships occurs in the earliest (intercept effect: a) or oldest (quadratic effect: b_2) ages.

impregnated. Fat levels in white-tailed males during the rut are sometimes reduced to levels below those thought to be minimal for survival in mammals (Pond 1978; Johns et al. 1982). The value of b_2 should become more negative as resource availability or the genetic capability to acquire and process resources decline, or as the need for resources increases. High positive values might be expected only in the most ideal environments. The application of the quadratic model to the evaluation of the importance of genetics, local environmental conditions and their interaction to antler development has been accomplished for white-tailed deer on the Savannah River Plant (SRP) in South Carolina (Scribner et al. 1989).

TABLE 1. Analysis of variables associated with antler characteristics of white-tailed deer (N = 588) collected during 1978 and 1979 on the Savannah River Plant, South Carolina

Source of variation	Antler characteristics[a] No. of points	Beam diameter	Width of inside spread	Beam length	Incidence of spiking*
Body weight	0.170	0.191	0.180	0.182	
	<0.0001	<0.0001	<0.0001	<0.0001	<0.0001
Kidney fat	0.008	0.007	0.005	0.004	>0.05
index	0.061	0.004	0.023	0.036	
Age	0.057	0.045	0.048	0.051	
	<0.0001	<0.0001	<0.0001	<0.0001	<0.0001
Year	0.017	0.004	0.050	0.016	<0.01
	0.008	0.039	<0.0001	<0.0001	
Region	0.008	0.002	<0.001	0.001	
	0.059	0.111	0.681	0.229	<0.01
Heterozygosity	0.001	0.003	0.003	<0.001	
	0.491	0.052	0.079	0.859	<0.05

[a] Upper values represent the proportion of the overall model sums of squares attributed to each source of variation. Lower values designate level of significance.

*Significance levels of the relationships between incidence of spiking and body weight and kidney fat were tested using a one-way analysis of variance. Differences attributable to age, year, region (swamp *vs* upland), and heterozygosity (number of loci heterozygous/individual) were tested using Chi-square analysis.

White-Tailed Deer on the Savannah River Plant

Demographic, nutritional, genetic, body weight, and antler data have been collected during fall hunts for white-tailed deer on the SRP for over 20 years (Ramsey et al. 1979; Dapson et al. 1979; Johns et al. 1984). This site (approx. 800 km^2) has a variety of habitat types (Whipple et al. 1981) and has generally been divided into upland and swamp regions based on vegetation and stream drainage patterns (Urbston 1967). The extensive scope and diversity of the data facilitates analyses addressing potential physiological, environmental, and genetic factors affecting expression of antler traits in this herd. Our conclusions are presumably applicable to other cervids, especially for populations with diverse age structures that occur in heterogeneous environments.

Most of the variation in antler development is related to individual age and body weight (Table 1; Smith et al. 1983). Within an age class, antler expression is significantly related to body weight. Thus, age-specific body weight may be the most important indicator of an individual's condition. Some animals in each class show unbranched or spiked antlers and the incidence of this trait declines with increasing age and body weight (Table 1; Scribner et al. 1984). Antler development is also related to levels of body fat although the amount of variation in antler development accounted for is relatively low. While the data on fat levels are taken after the time of antler growth, they are an indicator of overall nutritional condition (Finger et al. 1981; Johns et al. 1984). The relationship between antler size and fat levels is probably stronger than indicated in Table 1 because of the difference in time between initiation of antler growth and the time at which fat measurements were taken. It is still apparent that as animals become larger and/or in better body condition their antlers increase in size and branching.

White-tailed deer on the SRP show spatial and annual differences in antler expression and growth. Antler expression differs among deer harvested in different years. Deer with low antler quality also have lower levels of body fat and body weight (Scribner et al. 1989). Deer from the swamp area of the SRP have a higher incidence of spiking than those inhabiting upland areas (Scribner et al. 1984). The relationships between antler characteristics measured at different ages also vary over years as does the detectability of significant antler-heterozygosity relationships.

The presence of detectable heterozygosity effects on antler development in some years and not others raises important questions about the nature of the genetic effects. The interaction of environmental and genetic factors might be responsible for the observed annual patterns, or some methodological problem might confound our ability to detect heterozygosity effects. For example, the genetically related differences in antler characteristics may be of a constant magnitude, but the total variability in these characteristics could fluctuate over years and modify our ability to detect statistically the genetic effects. Fortunately, the extensive historical record for the SRP herd facilitates testing of hypotheses concerning annual fluctuations in levels of variance in antler size. The variance and standard deviation of the number of antler points declines with an increase in their mean (Fig. 3). Thus, the level of variability in the number of antler points and of other correlated antler characteristics is likely to be different over years. Heterozygosity-related effects are apparent in all years of the antler study but significant for general antler characteristics during years in which the deer were in the best body condition (Scribner et al. 1989). Fat reserves of males vary among years on the SRP (Johns et al. 1982) and vary as a function of heterozygosity (Scribner et al. 1989). During years of optimal environmental conditions, antler development is maximized as illustrated by the number of points and other correlated antler characteristics (Smith et al. 1982). In terms of the overall quadratic model (Fig. 1), a becomes more positive and b_2 becomes less negative or may even assume positive values. High-quality environments may lead to decreased variance in a number of characters closely associated with essential functions and allow the detection of heterozygosity effects because of the decreased variance to be partitioned within the statistical model. Heterozygosity effects seem to be present in deer populations every year and are probably of relatively constant magnitude. The ability to statistically document these effects seems to vary as a function of the amount of environmental variation.

Variation across locations has occurred between swamp and upland portions of the SRP deer herd relative to demographic (Dapson et al. 1979), reproductive (Urbston 1967), and body condition characteristics (Johns et al. 1984) during the expansion of the herd in the 1960s and early 1970s. Differences in body size

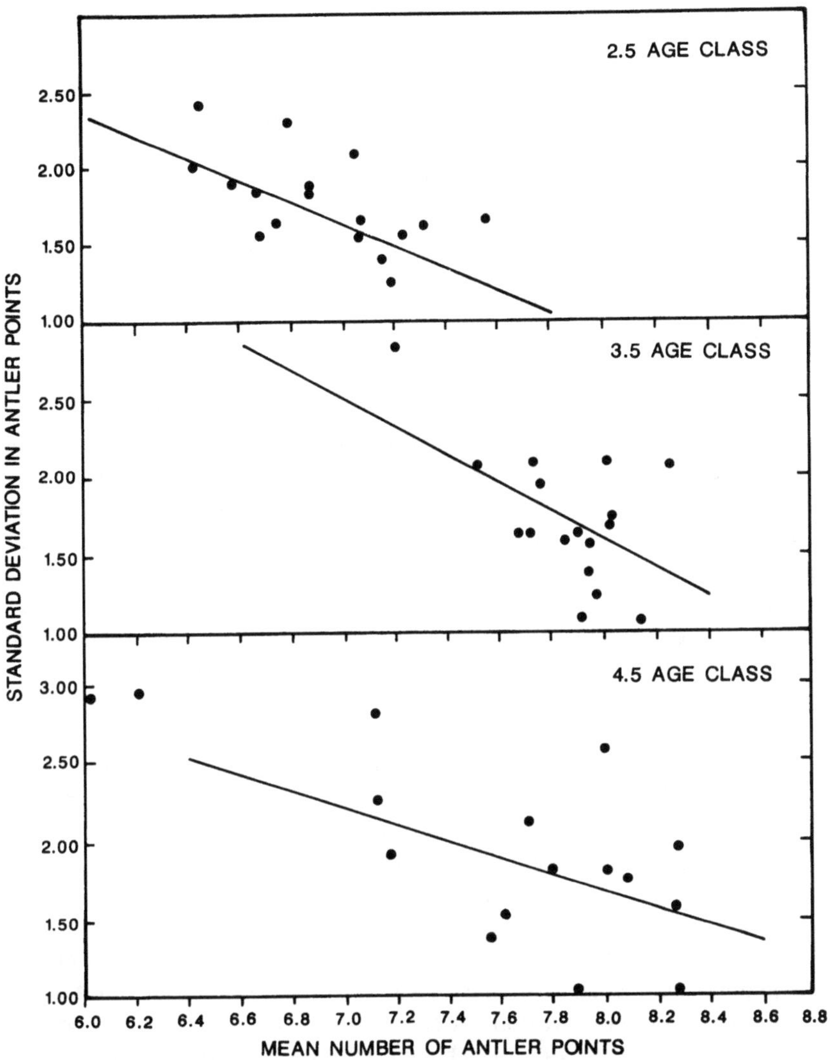

FIGURE 3. Relationship between the standard deviation and mean number of antler points derived from data collected in each of 17 years (1968–84) for white-tailed deer 2.5, 3.5, and 4.5 years of age.

(Brisbin et al. 1984), body condition (Johns et al. 1984), and reproductive (Urbston 1967) variables seem to be the result of historical differences in herd densities in these areas. Deer from these two areas also show variation in gene frequency, suggesting that spatial differentiation is related to breeding structure and historical separation of the subpopulations of the herd (Ramsey et al. 1979; Chesser et al. 1982). Incidence of spiking is different between deer from swamp and upland areas. However, other antler characteristics studied after the early period (which was characterized by demographic and nutritional variation between the swamp and upland subpopulations) does not support the importance of environmental differences associated with this dichotomy. Since the effects associated with the swamp-upland category have varied over years but have included at some time most of the characteristics measured for the deer, the explanation of the swamp and upland effects is confounded relative to the importance of environmental and genetic factors affecting antler characteristics.

Heterozygosity and Antler Development

Heterozygosity is correlated with body growth and development of antler characteristics. Deer 2.5 years of age or older with higher heterozygosities have a lower incidence of spiked antlers (Fig. 3; Scribner et al. 1984). Body growth (Chesser & Smith 1987), number of antler points (Fig. 2; Scribner et al. 1989), width of inside spread, beam diameter, and beam length (Scribner et al. 1989) vary among deer with different levels of heterozygosity. The strongest of these associations is between body weight and heterozygosity. Thus, antler size is most affected by body weight and both sets of variables are independently related to genetic variability, which may affect antler size most by its effect on body size.

Antler expression per unit body weight is curvilinear with increasing age (Fig. 2). The greatest incremental increase in antler size is observed between 1.5 and 2.5 years of age (Fig. 2). Antler growth continues through the 4.5-year age class (Fig. 2) and probably beyond. However, antler size/kg of body weight has its highest value at 3.5 years of age. The differential antler expression observed among deer with different heterozygosities is reflected further in antler size/kg. Deer with higher levels of genetic variability have greater antler size/kg than those with lower levels of genetic variability (Fig. 2).

A large number of studies document relationships between levels of heterozygosity and morphological or life history traits. Variations in growth rates (Singh & Zouros 1978; Pierce & Mitton 1982; Chesser & Smith 1987) and overall body size (Boyer 1974; Garten 1976; Koehn & Gaffney 1984) are related to heterozygosity, and individuals exhibit different adaptive strategies depending on their levels of genetic variability. Several hypotheses concerning the underlying properties related to heterozygosity that account for differences in individual performance are summarized by Mitton and Grant (1984) as follows: (a) enzymes are controlled by genes which occur in linkage groups on chromosomes, and these genes may singly or in various combinations influence growth; (b) enzymes represent loci which can be used as indicators of past population breeding structure with different levels of heterozygosity representing a continuum from highly inbred (low-heterozygosity) to randomly outbred (high-heterozygosity) individuals; (c) there are different kinetic and physiological properties of enzyme polymorphisms, and the superiority of heterozygosity is inferred as a result of the multiple molecular pathways animals can employ in processing energy. A partial test for the importance of these explanations has been proposed by Smouse (1986). His model tests for the cumulative effects of heterozygote superiority across loci. There are individual locus effects on antler characteristics (Smith et al. 1983), but differences in antler characteristics between genotypic categories are not prevalent, and only white-tailed deer that are heterozygous for the phosphoglucomutase-2 locus have larger antlers.

The lack of consistent deviations from expected genotypic proportions (Ramsey et al. 1979) and high levels of heterozygosity for deer on the SRP does not implicate inbreeding as a cause of individual variation in antler size. Levels of heterozygosity may be indicative of variability of loci involved in the regulation of calcium deposition and antler development. Levels of parathyroid hormone, calcitonin, and alkaline phosphatase are related to the regulation of serum Ca at various stages of the antler cycle as discussed elsewhere in this book. Variations in antler size among individuals of different heterozygosities could reflect shift in metabolic efficiency (Garton et al. 1984). These metabolic effects would be particularly important during the spring at which time resources are needed, both for body and antler growth and to replace body reserves lost during the rut of the previous autumn.

Levels of genetic variability are significantly related to antler growth (Fig. 2), mean differences in antler size within an age class (Fig. 4), mean differences in antler size, and overall size and symmetry as reflected in Boone and Crockett (1982) scores (Fig. 5). These effects are correlative in nature; however, the consistency and direction of these relationships suggest that heterozygosity is an important factor for antler development.

The effect of heterozygosity on antler growth and expression appears most notable in the first year of growth (1.5 age class) and in the post asymptotic ($\geq$3.5 age class) portion of the antler-age relationships. Deer with higher heterozygosities produce antlers of greater size in the first year of development than do deer of lower heterozygosity (Fig. 2; Smith et al. 1983). Antler size/kg of body weight is also greater in deer with higher heterozygosities (Fig. 2), though individuals with high heterozygosities exhibit lower body weight in the 1.5 age class. This apparent contradiction to the general trend can be explained because the males with higher heterozygosities have mothers of higher heterozygosities which have more than one offspring (Chesser & Smith 1987). Animals with higher heterozygosities may be physiologically better able to partition resources into secondary structures, such as antlers at a time of rapid body growth. Greater antler expression in the 1.5 age class could enhance reproductive success and individual fitness, especially in populations which have a low mean age structure as on the SRP (mean male age is 1.77 years; Scribner et al. 1984).

Deer with higher heterozygosities exhibit significantly greater asymptotic antler size (Fig. 2) and size per unit body weight (Fig. 2B) than individuals with lower heterozygosities. Greater antler size in older, more heterozygous males could also infer a competitive reproductive advantage during the rut when body

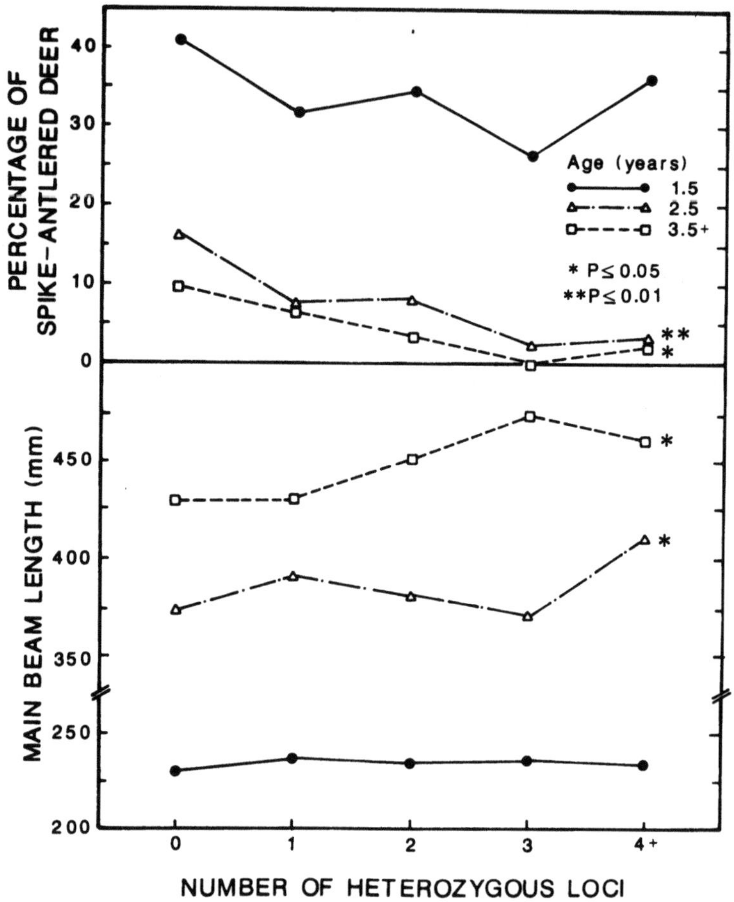

FIGURE 4. Age-specific expression of the percentage of spike-antlered deer and main beam length in white-tailed deer as a function of the number of loci heterozygous per individual.

fat is used extensively. Individuals with high levels of heterozygosity have higher levels of stored body fat than those of lower heterozygosity, and their pattern of fat utilization may also be different in males than it is in females (Cothran et al. 1987). Since males greatly reduce feeding during the rut and lower their fat reserves to starvation levels (Johns et al. 1984), the amount of stored fat and its rate of utilization should determine the length of time a male can effectively participate in the rut.

Age and body weight account for most of the measurable variation in antler expression on the SRP (Table 1; Smith et al. 1983). Temporal variation in antler size occurs, with the larger antlers being expressed at a time when body condition and size are the highest (Table 1; Scribner et al. 1989). The variance in number

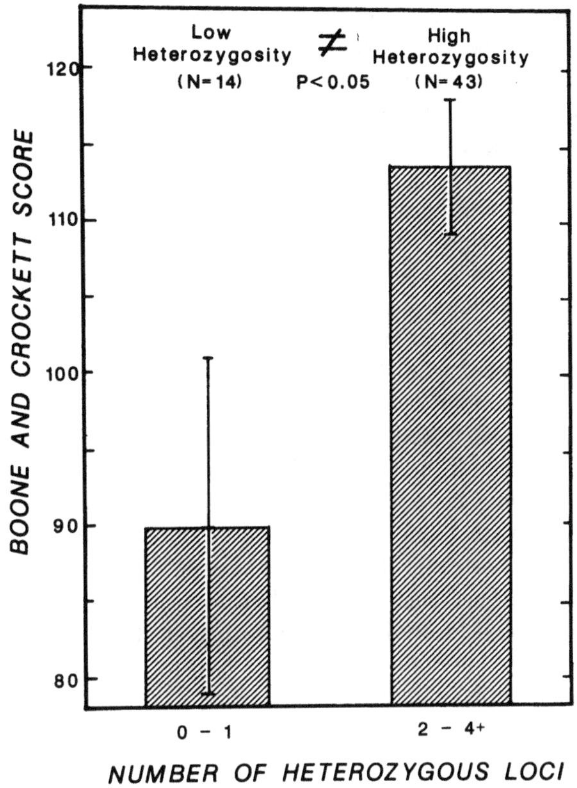

FIGURE 5. Differences in mean Boone and Crockett scores (±2 SE) of white-tailed deer relative to the number of loci heterozygous per individual.

of antler points measured annually each year for 20 years on the SRP is negatively related to the mean number of points (Fig. 3) and thus to resource quality. In good years antler size and number of points increase, and the uniformity of individual expression of these traits also increases. Yearly differences in the predictability of the relationships between antler expression and heterozygosity appear to be due to environmental influences. The ability to statistically detect differences in antler size relative to levels of heterozygosity is enhanced under more optimal environmental conditions which result in larger antlers and lower variability among individuals.

The variability in antler expression attributable to heterozygosity alone is low relative to that due to age and body condition effects (Table 1). However, within an age class, heterozygosity is responsible for 10–15% of the differences in main beam length, beam diameter, inside beam spread, number of antler points, and incidences of spiking (Fig. 4; Scribner et al. 1984; Scribner et al. 1989). A 20% difference in Boone and Crockett (1982) scores occurs among individuals with high and low levels of heterozygosity (Fig. 5). Boone and Crockett scores are a composite measure of phenotypic expression and thus reflect the cumulative effects of heterozygosity over numerous antler characteristics.

From an evolutionary perspective, all animals compete to maximize the number of their offspring in subsequent generations. Small male cervids in natural populations breed with very few females (Hirth 1977). Antler size seems to be a major factor contributing to the success of male mating behavior. As animals increase their body weight with age, their antlers increase in a correlated manner. Variation among individuals in the a and b_2 parameters of the antler growth curve suggest different individual strategies in optimizing fitness. The more genetically variable animals grow faster than those less genetically variable (Cothran et al. 1983) and reach higher adult body weights (Chesser & Smith 1987). In some species, the more homozygous males may achieve a rapid growth rate by having mothers who give birth to fewer offspring and invest more resources in each one than would be possible if the female raised a greater number. Males with higher genetic variability may be more aggressive (Garten 1976) and better able to control home ranges with higher-quality food resources than more homozygous males. Individuals with high levels of heterozygosity exhibit 10–20% greater antler size within an age class than do those with lower heterozygosities. This differential expression for individual characteristics (Fig. 2; Fig. 4) and overall expression as depicted in Boone and Crockett scores suggests that heterozygosity is an important genetic characteristic.

Conclusion

Antler size increases in a curvilinear manner with increasing age in white-tailed deer. Antler size and growth vary between individuals relative to age, levels of heterozygosity, body size, and body condition. Environmental variability, manifested in year-to-year variation in resource availability and differences in age-specific body size and condition, have a significant effect on antler size. Differences in habitat and heterozygosity between deer inhabiting swamp and upland regions of the SRP may be partially responsible for the spatial variation in antler size.

Body size, nutritional condition, and size and symmetry of many antler characteristics are postively correlated with heterozygosity. Variation in the antler-age and body weight-age relationship among individuals of high and low heterozygosities could imply different means of achieving fitness. Individuals with lower levels of heterozygosity mature at an earlier age and potentially minimize generation length by becoming large at an early age and reproducing as yearlings. Individuals with higher levels of heterozygosity potentially maximize reproductive output by maintaining high levels of postasymptotic body and antler size, dominance, and reproductive output. Older-age (≥ 3.5) males with higher levels of heterozygosity have higher body weight, fat levels, and antler size, suggesting greater utilization or efficiency of use of available resources. The knowledge about genetic effects and the interrelationships between various characteristics of cervids should play an important role in the formulation of effective management programs for antler quality.

19
The Antler as a Model in Biomedical Research

George A. Bubenik

Introduction

The growing antler is a unique phenomenon, as it is the only completely regenerating organ found in mammalian species (Goss 1984). The growing antler is an extension of the pedicle, the nonrenewable protuberance of the frontal bone. Similarly to the skull bones, growing antler is enclosed in hair-covered skin (velvet), contains subdermal layers of mostly collagenous tissues and has a bony core composed of compact and spongious components (Fig. 1). In the past several decades, deer antlers have become recognized as an excellent model for a number of biomedical studies, including tissue differentiation, skin and bone regeneration, collagen formation, bone growth and mineralization, and the trophic influences of the endocrine and the nervous systems (Goss 1961, 1983; Banks 1973, 1974; Banks & Newbrey 1982a,b; Banks et al. 1968a,b; Bubenik, A. & Pavlansky 1965; Frasier & Banks 1973; Frasier et al. 1975; Newbrey & Banks 1974, 1975, 1983; Newbrey et al. 1982; Bubenik, G. et al. 1982b; Speer 1982; Suttie et al. 1988; Suttie & Fennessy 1985). Detailed results of some of these studies are also available in Chapters 9, 11, and 12.

Why Antlers?

Despite some minor differences in the process of osteogenesis (Lojda 1956; Banks & Newbrey 1982b), growing antler tissue is very similar to a developing skeletal bone (Banks 1973, 1974). However, there are numerous advantages in using the antler as model: (1) Its external position allows simple observation and measurement as well as an easy access to blood vessels and nerves; (2) lack of muscles favors the bone biopsy; (3) an extremely fast growth rate (up to 2.0 cm/day) (Goss 1970) leads to complete development from preosseous tissues to bone in 120–150 days); (4) the massive mineralization process is concentrated in the last 2–3 weeks of antlerogenesis; (5) the deciduous nature of antler growth allows different studies to be performed on the same animal; (6) a rapid growth of bony tissue (antler) occurs in the mature animal (skeletal bones grow only in

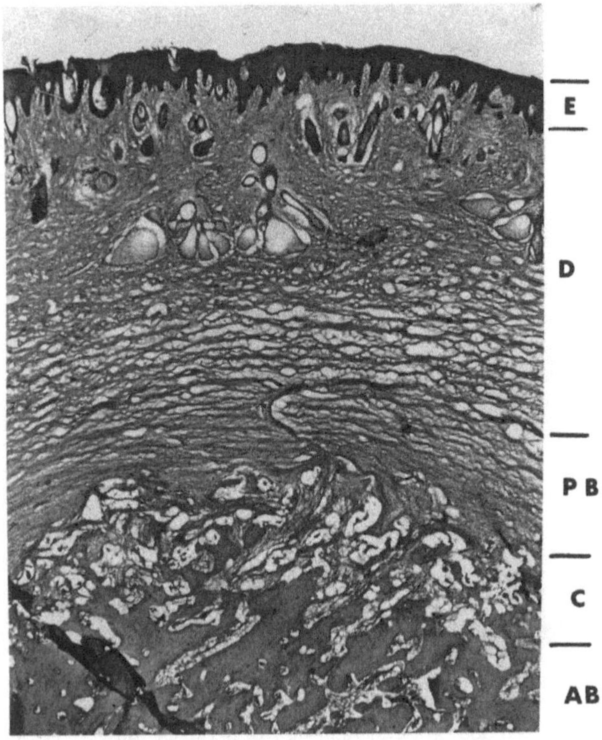

FIGURE 1. Histological section of the tip of a growing white-tailed deer antler. *Staining*: hematoxylin and eosin. *Magnification*: 20x E = epidermis, D = dermis, PB = prochondral blastema, C = cartilage, AB = antler bone.

juvenile mammals); (7) the two antlers develop in a mirror image but have partly independent regulatory mechanisms (one can serve as an experimental organ, the other as a control) (Bubenik, G. et al. 1982b); (8) antlers exhibit an extreme sensitivity to neuronal and endocrine manipulation; (9) except for gravitational forces, the antler is free of stresses that influence the development and turnover of skeletal bones (such as tensions caused by muscular contraction); and (10) the life span of larger cervids is 12 to 20 years; therefore, the influences of maturation and aging processes can be studied on the same animal for a number of years (Bubenik, G. & Schams 1986).

Endocrine Studies

Steroids and Reproduction

The close relationship of the antler cycle to the reproductive cycle of the male deer provides a unique indicator of blood androgen levels. Antler tissue is a target

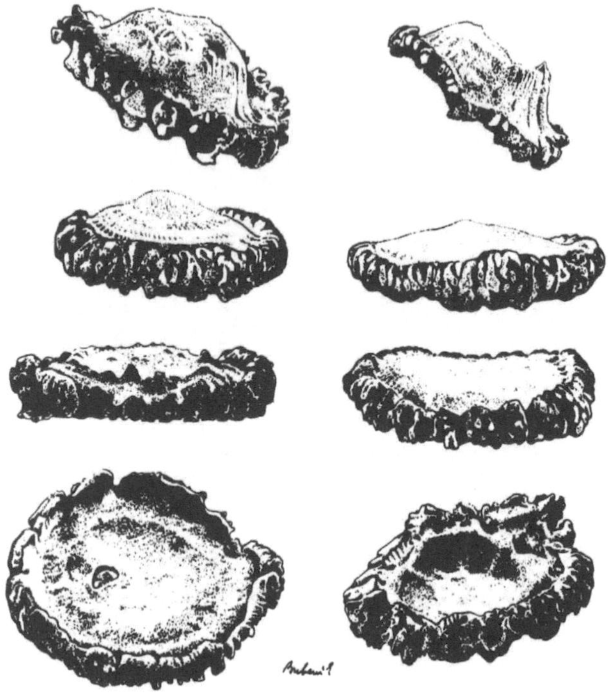

FIGURE 2. "Seals" of red deer antlers, ranging from convex shape of strong stags (*top*) to concave shape of weak or overaged ones (*bottom*). (From Bubenik, A. 1966.)

organ for steroidal hormones (Bubenik, G. et al. 1974; Morris & Bubenik, G. 1982). The antler growth period is associated with minimal seasonal levels of testosterone (T). The rapid mineralization phase (accompanied by the desiccation of the velvet) is associated with a slow rise of T levels. Rapid increases of androgens in blood (occurring several months before the rut) result in shedding of velvet and the death of antlers. Finally, a rapid reduction of T levels after the rut is manifested by the casting of antlers (Bubenik, G. 1982a).

These characteristics of individual phases of the antler cycle make it possible to investigate various treatments affecting the reproductive cycle, such as immunization (Lincoln et al. 1982), stimulation and blockade of receptors (Bubenik, G. et al. 1975b; Bubenik, G. & Bubenik, A. 1978a; Bubenik, G. et al. 1985a; Schams & Barth 1987), application of steroids (Lincoln et al. 1970; Bubenik, A. et al. 1976; Semperé & Boissin 1982; Morris & Bubenik, G. 1982), or modification of photoperiodicity (Jaczewski 1954; Goss 1969a,b, 1976; Bubenik, G. 1983; Bubenik, G. et al. 1987a). In the majority of these studies a qualified estimate of blood T levels, derived from visual observation of developmental stages of antlers, was very close to hormonal concentrations determined later in blood samples. For instance, premature mineralization of antlers in bromocriptine-treated bucks indicated an early rise in T concentration in blood (Bubenik, G.

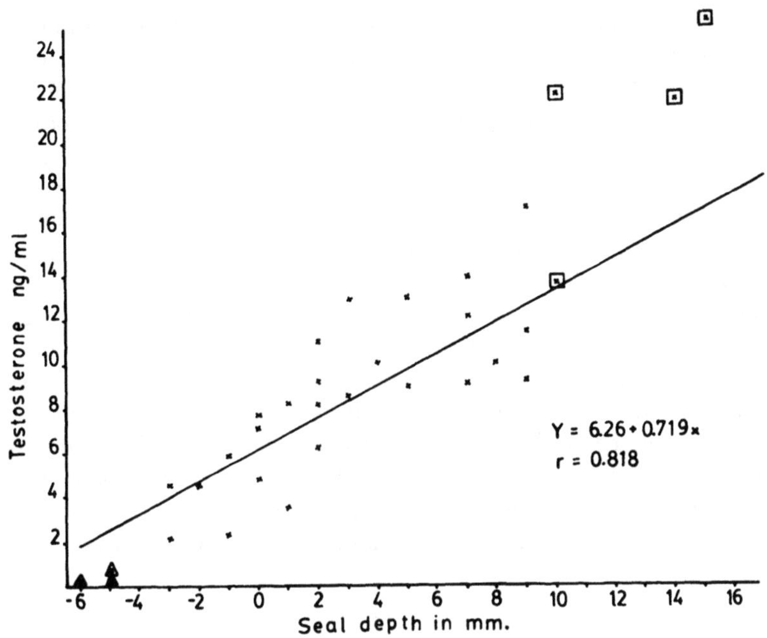

FIGURE 3. Correlation of plasma testosterone levels and the depth of antler seal in white-tailed deer. *Triangles* = castrates. *Squares* = very aggressive buck number 24.

et al. 1985a). Similar premature mineralization of antlers in melatonin-treated deer also indicated a shift in T secretion (Bubenik, G. 1983). Conversely, a premature casting of antlers in light-treated bucks indicated a sudden decrease in levels of T. These observational conclusions were later confirmed by blood analyses (Bubenik, G. et al. 1985a, 1987a).

The shape of the proximal surface of cast antlers the so-called "seal" (convex in an antlerogenetically promising deer and concave in an antlerogenetically poor one) (Fig. 2) correlates very well with maximal T levels in blood achieved in the male deer in the previous rut (Fig. 3) (Bubenik, G. & Schams, 1990). Therefore, evaluation of cast seals can reveal the endocrine status of deer without exposing them to the stress of anesthesia and blood sampling. As the stress is a well-established cause of male reproductive malfunctions (Ducharme et al. 1982), the noninvasive technique such as the seal evaluation has a major advantage in endocrine studies. This technique has been utilized for a long time by European hunters, who use seal shape as an indicator of future antlerogenetic potential.

Role of Hormones and Nerves in Bone Growth

The role of hormones in the normal and the pathological development of skeletal bones is only poorly understood. Studies performed on antler tissues could advance our knowledge in this important field. The antlers of castrated and intact deer have been used to investigate the effect of various steroidal hormones (e.g., androgens, estrogens, corticoids) on the development, ossification, and mineralization of bony tissues (Tachezy 1956; Bubenik, A. et al. 1976; Goss 1983; Morris & Bubenik, G. 1982; Schams et al. 1987). The antler is an ideal model, due to extremely fast antler response to exogenous steroids (transformation of preosseous tissues into fully developed osteons in 2–3 weeks), an easy biopsy technique, and the lack of complications from the biopsy (Morris & Bubenik, G. 1982; G. Schwartz, personal communication). Recently, the growing antler also has been used for detection and elucidation of function of various growth factors, such as the epidermal growth factor (EGF) (Ko et al. 1986) and the insulin-like growth factor (IGF–1) (Suttie et al. 1988).

The "trophic role" of nerves in the regulation of growth of bony tissues has been suggested for many years, but it is difficult to prove. Studies on antlers indicate that nerves might be involved in the regulation of antlers shape and the speed of antler growth (Bubenik, G. et al. 1982b; Suttie & Fennessy 1985). However, despite some progress in this field (for details see Chapters 9 and 10), more studies are necessary to clarify individual components of this suspected neural regulatory mechanism.

Tissue Metabolism

Because the antler is a terminal bony organ with an isolated blood and nerve supply, it can be used to study the changes of blood parameters (such as proteins, carbohydrates, minerals, hormones, and enzymes) occurring during the flow through antler tissues. Differences between inflow and outflow values will indicate which factors are produced in antler bone, which are utilized there, and which are not affected at all. Similar studies on skeletal bones are rather difficult, as a major surgical procedure is required to obtain blood from bone vessels. On the other hand, superficial antler blood vessels are relatively large, easily visible, and accessible (Fig. 4). Blood sampling from antlers is practically without risk, and the amount of blood taken is almost unlimited. Differences in arteriovenous concentration of calcium in reindeer antlers were reported by Hove and Steen (1978); they obtained arterial blood from the carotid artery and venous blood from the superficial antler vein. A technique for a repeated sampling from the carotid artery of a deer has been developed (Suttie et al. 1986); however, it requires extensive surgery and carries some risks. A safer procedure is blood sampling from small arteries supplying growing antlers. The arterial blood supply to growing antlers is provided mostly by the temporal artery (a branch of the carotid artery), which divides extensively in the pedicle (Fig. 4). The branches first run under the velvet and then penetrate the antler bone cortex (Suttie et al. 1985). As we were rather unsuccessful in obtaining larger amounts of blood from these arteries and repeated sampling caused damage which resulted in reduction of growth, we decided to circumvent this problem. Instead of measuring arterio-

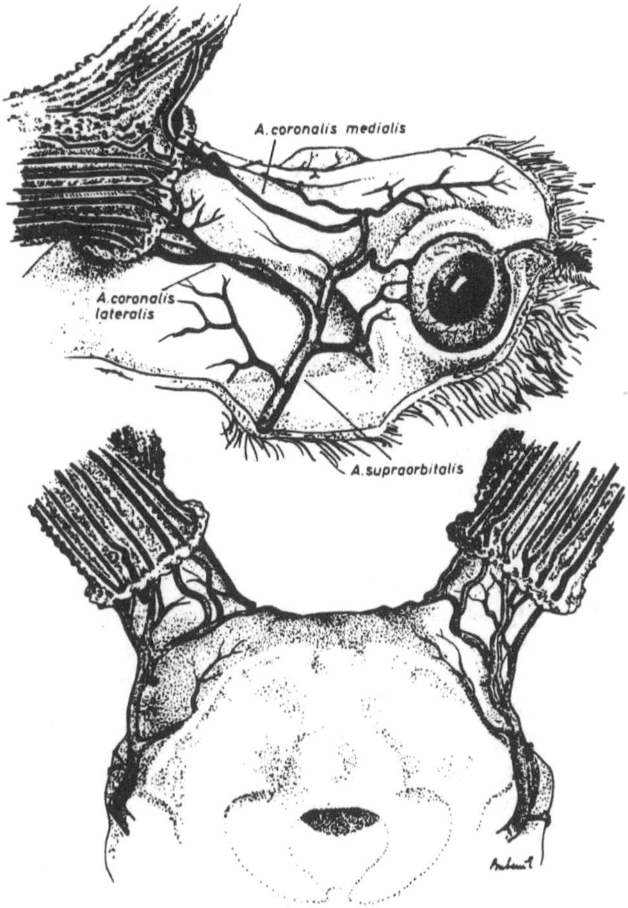

FIGURE 4. Diagram of blood supply to growing antlers. *Top*: lateral view. *Bottom*: posterior view. (From Bubenik, A. 1966.)

venous differences, we decided to compare concentrations of blood components in veins leaving antlers, with concentrations in veins located further away. We have chosen two easily accessible veins—the vena jugularis and the vena saphena. The external jugular vein in deer drains blood from the whole head area, except the internal organs (mostly brain tissues). Therefore, the jugular blood contains the venous blood from antlers. On the other hand, blood in the saphenous vein is drained from the lower part of the front leg (not directly connected to antler circulation) and as such should provide rather contrasting results.

Blood samples taken from the three previously described veins (velvet, jugular, and saphena) during the three important phases of the antler cycle (rapid growth,

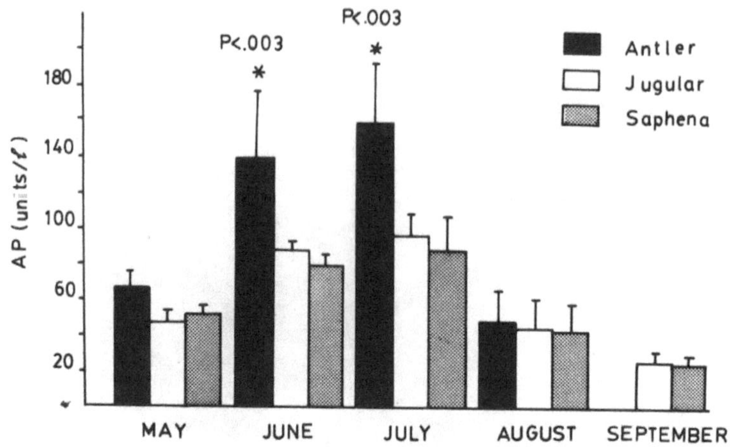

FIGURE 5. Levels of alkaline phosphatase in three white-tailed deer veins during the antler growth period. (From Bubenik, G. et al. 1987b.)

ossification, and mineralization) have thus provided clues of metabolic activity in the antler tissues during these periods. During the most intense antler growth period in July, the level of alkaline phosphatase (AP) in the antler vein was 63% higher than levels in the jugular vein and 83% higher than values of AP detected in the saphena (Fig. 5). On the other hand, there was no significant difference in AP levels from the three veins during the mineralization period in August (Bubenik, G. et al. 1987b). Surprisingly, values from the jugular vein were not different from concentrations found in the saphena. These findings seem to confirm the well-supported hypothesis that elevated levels of AP found in deer blood during the antler growth period is due to AP production in the antler tissue (Lojda 1956; Kuhlman et al. 1963) and that AP is involved mainly in antler growth (Semperé et al. 1986), but not much in antler mineralization as suggested by Eiben et al. (1984). On the other hand, 37% lower levels of triiodothyronine (T_3) detected in the antler vein during the May growth period indicates that T_3 is utilized in rapidly growing tissues of the developing antler (Bubenik, G. et al. (1987b) (Fig. 6).

Bone Growth–In Vitro Studies

Thin slices of growing antler tip biopsies were kept alive up to 48 hours in incubation medium. Two parameters indicating the activity of bone metabolism, calcium, and hydroxyproline were measured at various intervals (Ramirez & Brown 1988). This successful experiment indicates that the rapidly growing antler tip can serve as a model for detailed studies of bone metabolism during individual developmental stages.

The tissue of the developing antler has also been grown successfully as cell suspension in tissue cultures. Unlike the skeletal bones (which cease to divide after

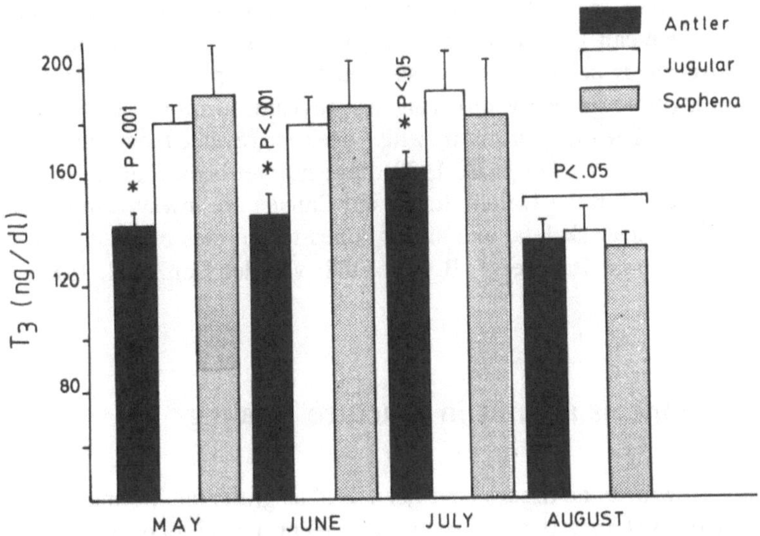

FIGURE 6. Levels of T_3 in three white-tailed deer veins during the antler growth period. (From Bubenik, G. et al. 1987b.)

only few generations) the preosseous tissues of the growing antler (such as the reserve mesenchyme, prechondroblasts, and chondroblasts) has been grown for several weeks without any sign of weakening mitotic activity (C. Grant, personal communication). In that respect, the growing antler bone exhibited the properties of embryonic or neoplastic tissues, which are used classically for tissue culture studies.

The process of endocrine regulation of bone development and mineralization at the tissue level is only poorly understood. It is suspected that this process requires a sequential action of many growth factors as well as several steroidal (androgenic, estrogenic, and corticoidal) and nonsteroidal (thyroidal and hypophyseal) hormones (for more details see Chapter 8). It is obvious, therefore, that to understand bone growth and development in vitro studies on cultured antler bone have to supplement our data obtained in vivo.

Studies of Osteoporosis

As mentioned earlier, the antler growth represents a very fast developmental process. It takes only a few weeks to transform an embryonic-like blastemic and cartilagenous tissue to fully developed Haversian lamellae. During that short period a huge amount of calcium and phosphate is transferred into the antler bone, a large proportion of which is taken from skeletal deposits in ribs, sternum, and vertebrae (Banks et al. 1968b; Brown et al. 1978). The demineralization of skeletal bones in deer is a process that resembles osteoporosis, a crippling disease occurring often in postmenopausal women (Davis 1987). However, unlike humans, deer can replenish the calcium deposits in the skeleton after antler mineralization is achieved. It is assumed that the process involves hormonal regulation (most probably involving parathyroid hormone (PTH), calcitonin, and steroids). The mechanisms by which deer can reverse the noticeable changes in bone density (Pritzker et al. 1988) have not yet been found. However, several laboratories in the United States and Canada are involved in research projects utilizing deer skeletal and antler bones for studies of osteoporosis (Baksi & Newbrey 1988; Ramirez & Brown 1988; van der Ems et al. 1988; Pritzker et al. 1988).

Antler Bone as a Graft in Fracture Healing

Rapidly growing bone tissues have been used as grafts facilitating the healing of bone fractures (Lance 1985). Two exceptional properties of developing antlers—the rapid growth and the undifferentiated blastemic character which resembles an embryonic or bone sarcoma tissue (Olt 1927a,b; Modell 1969)—would make growing antlers very useful as a bone graft. Fast-growing tissues contain a great amount of growth-supporting factors, such as the epidermal growth factor, fibroblast growth factor, nerve growth factors, and others (Centrella & Canalis, 1985), which are known to dramatically alter metabolic processes of mammalian cells. Tissues which have embryonic character have been used as successful grafts because of their very low antigenicity reducing the chance of rejection (Arendt et al. 1988). The antigenicity of growing antler tissue exhibiting blastemic character has not been tested to our knowledge, but the presence of one of the growing factors (epidermal growth factor) has been detected in the growing antler of the sika deer (Ko et al. 1986). In addition, the antler appears to be a target organ for IGF-1 (Suttie et al. 1988). Finally, preliminary tests in my laboratory indicate that extract from the growing antler tip facilitates healing of epidermal wound on the dorsum of albino rats.

Growing antler is composed of several distinct zones (Fig. 1) (Bubenik, G. et al. 1974). The "growing top" is made of rapidly growing epidermis supported by a thick layer of tissue of mesenchymal and blastemic character. The middle segment is formed by a hyaline cartilage which in the central regions is replaced by a bone through the process of modified endochondral ossification (Banks & Newbrey 1982a).

Each of the antler zones has special characteristics which could be utilized to support bone growth. Therefore, each of the four tissues—the prochondral blastema, cartilage, primary osteoid, and mineralized antler bone—should be tested separately. The tissues could be used in the natural stage, or they could be denatured by autoclaving, which is used in other homo- or xenographic implants (Lance 1985).

Death of the Velvet–a Model of Myocardial Infarct?

If it seems probable that death of the antler is caused by death of the velvet (Bubenik, G. 1982), not enough data exist on the cause of this event. The drying of velvet is a process that takes several days. The final restriction of blood flow is rather fast (Mollelo et al. 1963). Large deposits of lipid substances were observed in arteries of mineralizing antlers and antler velvet after treatment with a synthetic steroid, diethylstilbestrol (DES) (Bubenik, A., personal communication). Therefore, it might be speculated that in many aspects, the death of antlers can be compared to the process which leads to a myocardial infarct. Arteriosclerosis, the hardening of arteries by deposition of mineral salts into arterial plaques, restricts the flow of blood to the heart muscle. When only a small lumen is left open, any major stress (e.g., the reduction of oxygen supply) induces contraction of smooth muscle in the arterial wall. A complete occlusion of blood vessels, even for several minutes, results in stagnation thrombosis and in the death of tissues supplied by the affected arteries. A similar situation might develop in mineralization of antlers. When arteries in the velvet are practically closed by heavy deposits of minerals shortly before shedding (Wislocki & Singer 1946), blood flow is shut off by the sympathetic nerves found in the walls of the velvet blood vessels (Nodl, cited by Vacek 1955; Rayner & Ewen 1981). The stimulus for the vasoconstriction of smooth muscles in the arterial walls could originate in the local tissues. A novel, potent vasoconstrictor peptide secreted by endothelial cells was recently isolated and characterized (Yanagisawa et al. 1988). Blood vessels of the velvet should be investigated as a possible source of this peptide.

Another possibility is that the stimulus for vasoconstriction might have come from the hypothalamic sympathetic center which control peripheral blood flow. The hypothalamus might react to a certain threshold of androgen levels which are associated with the death and desiccation of the velvet. That threshold is usually a little below the levels needed for initiation of antler shedding. Various thresholds exist for individual target tissues. Variable thresholds of target organs for androgens in deer were postulated by Lincoln (1971a). The role of sexual hormones in development of arteriosclerosis is well established and the changes in velvet arteries caused by androgens or their metabolites might serve one day as a model for studies of heart attacks. The easy access to velvet blood vessels, the simplicity of biopsies, and a quick recovery of sampled tissues are some of the advantages provided by the developing antler.

Antler–a Model for Cancer Studies?

The rapidly proliferating tissues have a primitive morphological character resembling a bone cancer, osteosarcoma (Olt 1927a). However, despite the fast growth, the process is remarkably orderly and the bilateral shape and size of antlers are

FIGURE 7. Tumor-like peruke of the roe-deer antlers. (From Bubenik, A. 1966.)

maintained with great precision. As the process of bone development requires interactions of numerous growth factors (Centrella & Canalis 1985), the control of individual phases of growth and differentiation must be very effective. As the rapid proliferation of antler tissues is limited to the so-called "growing tops" (composed of tips of beams and points) which solely maintain the blastemic character, positional memory mechanisms must exist which limit the boundary of that fast proliferation. In that respect the prochondral blastemic layer of the growing antler top represents an organ which is analogous to the epiphysial plate of the long skeletal bone. Cartilagenous tissue very rarely acquires neoplastic character. Similarly, despite the tumor-like appearance of growing antler bone, no true antler tumors were ever recorded. It can be argued that there is not enough time for neoplasmic mutation in a short life span (3–4-months) it takes to grow antlers. However, no true tumors have ever been detected in the perennially growing antlers of castrated deer, which can persist and grow for many years (Goss 1983).

On the other hand, bizarrely shaped, sarcoma-like structures (Fig. 7) called peruke (wig) or antleromas (Goss 1983) develop regularly in castrated roe-deer or muntjac. However, the peruke never forms metastases (Goss 1983), despite the fact that micromorphology of this tissue closely resembles malignant tumour (Olt 1927a) and in roe-deer invariably causes death of the buck within 1 or 2 years from the time of castration.

In other deer species, castration or blockade of androgen receptors will cause impairment of antler growth and prevent petrification; however, the development of perukes is extremely rare (Bubenik, A. 1966; Goss 1983) (Fig. 8). As was

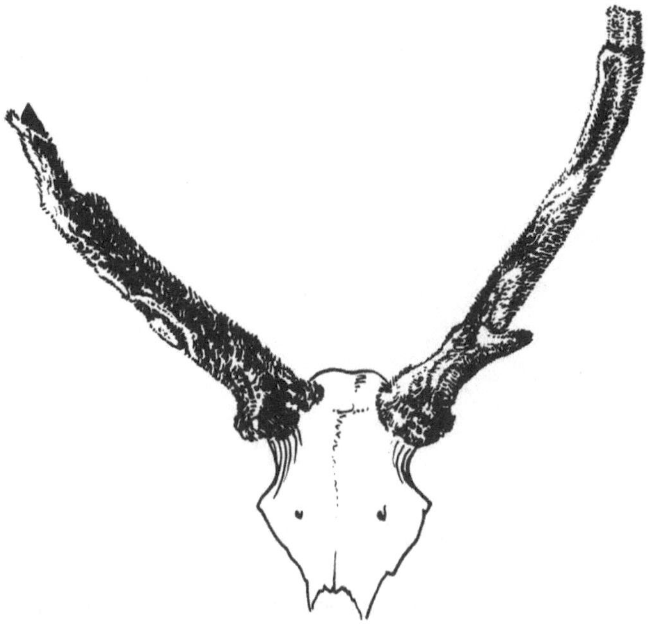

FIGURE 8. Well-controlled growth of the red deer peruke. (From Bubenik, A. 1966.)

mentioned in Chapter 8, antler tissue is extremely sensitive to effects of steroidal hormones, particularly testosterone and estradiol. Whereas in cases of cancerous growth in tissues of the reproductive system (such as the mammary gland, uterus, or prostate) the steroids promote tumor growth, in the cases of the deer peruke the steroids keep the proliferation under control. However, similar to target organs of reproductive tissues, growing antlers also contain high concentrations of receptors for androgens (Plotka et al. 1983) and estrogens (Barrell et al. 1987). It is suspected that in both cases steroids are acting on the same genetic material, the so-called "oncogenes." The name is a historical misnomer, since in a normal cell these genes are controlling the production of autocrine and paracrine factors needed for differentiation and growth of tissues. In cases of neoplastic mutations, the control of production of growth factors is impaired, which results in unlimited proliferation of tissue and the development of the cancer.

One of the tissue growth factors (epidermal growth factor) has already been detected in growing antler tissue (Ko et al. 1986), and the search for more of these factors is under way. These tissue growth factors are implicated in promotion of cancerous growth. Elucidation of mechanisms of control of their production is extremely important for our understanding of oncogenesis. Perhaps one day an understanding of mechanisms that prevent the development of deer perukes might help us prevent the development of some cancers in humans.

FIGURE 9. Antler development of a young white-tailed buck in which the right antler bud was treated with the all-trans-retinoic acid.

Morphogenesis and Positional Memory

The cellular mechanisms of morphogenesis and positional memory of developing organs is mostly unknown, but recent research data indicate that positional memory of the proximodistal axis of developing and regenerating vertebrate limbs can be manipulated by retinoic acid (RA) Tickle et al. 1982; Kim & Stocum 1986; Scadding & Maden 1986). Retinoic acid has been detected in developing chicken limb buds with a concentration gradient highest in the proximal end (Thaller & Eichele 1987).

So far, morphogenetic action of RA has not been observed in mammalian tissues. Because of the rapid proliferation of antler tissue and the blastemic character of its growing antler tip a deer antler can be utilized as a unique model for testing the morphogenic action of RA in nonfetal mammalian tissue.

The first results indicate that a single subperiosteal injection of all-trans-retinoic acid (10 mg in 0.5 ml of olive oil) deposited in the lateral region of antler prochondral blastema (Fig. 1) was followed by a positional displacement and acceleration of antler growth on the RA-treated side. The experimentally treated antler grew initially at the 90° angle as compared to the control side; in a later phase it developed a slight curvature upward (Fig. 9). The RA treatment was effective only in the deer treated in the early developmental stage (antler length 1.5 cm). In the later growing stage (antler length around 10 cm), the RA injection was without effect, presumably because RA was delivered into tissue which already lost the embryonic mesenchymal character.

Further studies of this phenomenon are in progress to confirm the reported findings and to elucidate the possible role of RA in the positional memory of growing antler. I hope that the data obtained in studies utilizing RA can also contribute to our understanding of the largely unexplained phenomenon of the trophic memory (Bubenik, A. & Pavlansky 1965), which is discussed in detail in Chapter 11.

Social Studies

Finally, deer antler is also an indicator of the social status of an individual male (Bubenik, G. 1982) (for more details, see Chapters 1 and 17). Therefore, factors affecting the strength of the animal (such as social and environmental stress, nutritional standard, or disease) which might result in the change of the social rank will be reflected in the change of the size and the branching pattern of antlers. In a well-balanced deer population, each individual male carries antlers of an optimal size. Therefore in many European countries, records of individual antler development obtained from cast antlers collected over the years became an important tool of modern wildlife management.

Summary

It is my conviction that the data presented here indicate the potential of growing antlers to serve as a unique model for biomedical studies investigating the roles of humoral and neuronal factors in regulation of tissue growth, differentiation, and metabolism. The major advantages of using developing antlers, as compared to presently used animal models, include: their phenomenal speed of growth; the fact that the differentiation, development, and mineralization phases are compressed into a 3–4-month period; and the exterior position of a complex organ that provides for easy access to the vascular system as well as to integumental, preosseous, and osseous tissues.

The progress in recognition of the antler as a uniquely suited model for biomedical studies is painfully slow. It is my hope that this chapter as well as several others in this book will facilitate this process and lead to a further utilization of antler tissues in experimental as well as in clinical research.

References

Acharjyo, L.N. 1982. Observations on aspects of antler casting in captive *Sambar* deer, pp. 23–28. *In*: R.D. Brown (ed.), Antler Development in *Cervidae*. Caesar Kleberg Wildl. Res. Inst., Kingsville, TX.

Acharjyo, L.N. & A.B. Bubenik. 1982. The structural peculiarities of antler bone in genera *Axis, Rusa*, and *Rucervus*, pp. 195–209. *In*: R.D. Brown (ed.), Antler Development in *Cervidae*. Caesar Kleberg Wildl. Res. Inst., Kingsville, TX.

Adams, D.B. 1979. Brain mechanisms for offense, defense and submission. Behav. Brain Sci. 2:201–241.

Adams, J.L. 1979. Innervation and blood supply of the antler peduncle in the red deer. N.Z. Vet. J. 27:200–201.

Aitchison, J. 1946. Hinged teeth in mammals: A study of the tusks of muntjacs (*Muntiacus*) and Chinese Water deer (*Hydropotes inermis*). Proc. Zool. Soc. London 116:329–338.

Aitken, R.J. 1981. Aspects of delayed implantation in the roe deer (*Capreolus capreolus*). J. Reprod. Fertil. (Suppl.) 29:83–95.

Alberch, P., S.J. Gould, G.F. Oster & D.B. Wake. 1979. Size and shape in ontogeny and phylogeny. Paleobiol. 5:296–317.

Aleyev, Yu.G. 1986. Ekomorphologiya. (Ecomorphology). Naukova Dymka, Kiev. (In Russian.)

Aleyev, Yu.G. & V.D. Bubrak. 1984. Ekologo-morphologicheskye konvergentsie i edinaya ekomorphologicheskaya sistema organizmov. Ekologiya Morya. 17:3–17. (In Russian.)

Allen, G.M. 1939. Mammals of China and Mongolia. American Museum of Natural History.

Alston, E.R. 1879. On female deer with antlers. Proc. Zool. Soc., London: 296–299.

Anděl, J. 1978. Matematická statistika. SNTL, Praha.

Anderson, A.E. & D.E. Medin. 1969. Antler morphology in a Colorado mule deer population. J. Wildl. Manage. 36:579–594.

Anderson, J.L. 1979. Reproductive seasonality of the Nyala *Tragelaphus angasi*; the interaction of light, vegetation phenology, feeding style and reproductive physiology. Mammal Rev. 9:33–46.

Andrews, P., J.M. Lord & E.M.N. Nesbit Evans. 1979. Patterns of ecological diversity in fossil and modern mammalian communities. Biol. J. Linn. Soc. 11:177–205.

Anonymous. 1983. Buck antelope has three horns. S.D. Cons. Digest 50:31.

Ansell, W.F.H. 1971. Artiodactyla, Part 15, pp. 1–84. *In*: J. Meester & H.W. Setzer (eds.), The Mammals of Africa: An Identification Manual. Smithsonian Institution Press, Washington, D.C.

Appleby, M.C. 1980. Social rank and food access in red deer stags. Behaviour 74:294–309.

Appleby, M.C. 1982. The consequences and causes of high social rank in red deer stags. Behaviour 80:259–273.

Appleby, M.C. 1983. Competition in a red deer stag social group-rank, age and relatedness of opponents. Anim. Behav. 31:913–918.

Arambourg, C. 1941. Antilopes nouvelles du Pleistocene ancien de l'Omo (Abyssinie). Bull. Mus. Natn. Hist. Nat. Paris 2 13:339–347.

Arambourg, C. 1947. Contribution à l'étude géologique et paléontologique du bassin du lac Rodlophe et de la basse vallée de l'Omo. Mission scient. Omo 1932–1933, Paris, I Geol. Anthrop. 3:232–562.

Arambourg, C. 1959. Vertébrés continentaux du Miocène supérieur de l'Afrique du Nord. Mem. Carte géol. Algerie (n.s. Paléont.) 4:1–159.

Arambourg, C. 1979. Vertébrés Villafranchiens d'Afrique du Nord. Singer-Polignac, Paris.

Arambourg, C., M. Boule, H. Vallois & R. Verneau. 1934. Les grottes paléolithiques des Beni Segoual (Algerie). Archs. Inst. Paléont. hum. Paris 13:1–242.

Arendt, T., Y. Allen, J. Sinden, M.M. Schugens, R.M. Marchbanks, P.L. Lantos & J. Gray. 1988. Cholinergic-rich brain transplants reversed alcohol-induced memory deficit. Nature 332:448–451.

Aristotle. 1910. "Historia Animalum." (Engl. transl. by D'A.W. Thompson), Oxford University Press, London.

Arman, P., R.N.B. Kay, E.D. Goodall & G.A.M. Sharman. 1974. The composition and yield of milk from captive red deer (*Cervus elaphus* L.). J. Reprod. Fertil. 37:67–84.

Asdell, S.A. 1944. The genetic sex of intersexual goats and a probable linkage with the gene for hornlessness. Science 99:124.

Asdell, S. 1964. Patterns of Mammalian Reproduction. 2nd ed. Cornell University Press, Comstock Publ. Associates, Ithaca, NY.

Asher, G.W., J.L. Adam, W. Otway, P. Bowmar, G. van Reenan, C.G. Mackintosh & P. Dratch. 1988. Hybridization of Père David's deer (*Elaphurus davidianus*) and red deer (*Cervus elaphus*) by artificial insemination. J. Zool. 215:197–203.

Atkeson, T.D., V.F. Nettles, R.L. Marchinton & M.V. Branan. 1988. Nasal glands in the *Cervidae*. J. Mammal. 69:153–156.

Atzkern, J. 1923. Zur Entwicklung des Os cornu der Cavicornier. Anat. Anz. 57:125–130.

Autenrieth, R.E. & E. Fichter. 1975. On the behavior and socialization of pronghorn fawns. Wildlife Monogr. 42:1–111.

Avise, J.C. 1986. Mitochondrian DNA and the evolutionary genetics of higher animals. Philos. Transact. R. Soc. London B 312:325–342.

Axelrod, D.I. 1939. A Miocene flora from the western border of the Mohave Desert. Carnegie Inst. Washington Publ. 516.

Axmacher, H. & R.R. Hofmann. 1988. Morphological characteristics of the masseter muscle of 22 ruminant species. J. Zool. 215:463–473.

Azzaroli, A. 1948. Revisioni della fauna dei terreni fluvio-lacustri del Valdarno Superiore – III. I cervi fossili della Toscana – con particolare riguardo alle specie villafranchiane. (Fossil deer of Toscana—with special respect to species of the Villafranchian). Paleontogr. Ital. 43:47–82. (In Italian.)

Azzaroli, A. 1979. Critical remarks on some giant deer (genus *Megaceros* Owen) from the Pleistocene of Europe. Palaeontogr. Ital. 71:5–16.

Baber, D.W. 1987. Gross antler anomaly in a California mule deer: the cactus buck. The Southwestern Naturalist. 33:404–406.

Bahnak, B.R., J.C. Holland, L.J. Verme & J.J. Ozoga. 1981. Seasonal and nutritional influences on growth hormone and thyroid activity in white-tailed deer. J. Wildl. Manage. 45:140–146.

Bailey, V. 1920. Old and new horns of the prong-horned antelope. J. Mammal. 1:128–130.

Bailey, V. 1932. The Oregon antelope. Proc. Biol. Soc. 45:45–46.

Baker, D.E., D.E. Johnson, L.H. Carpenter, O.C. Wallmo & R.B. Gill. 1979. Energy requirements of mule deer fawns in winter. J. Wildl. Manage. 43:162–169.

Baksi, S.N. & J.W. Newbrey. 1988. Plasma calcemic hormones in mature female reindeer. Gen. Comp. Endocrinol. 69:262–266.

Balon, E.K. 1985a. The theory of saltatory ontogeny and life history models revisited, pp. 13–30. *In*: E.K. Balon (ed.), Early Life Histories of Fishes: New Developmental, Ecological and Evolutionary Perspectives. Dr. W. Junk Publishers, Dordrecht.

Balon, E.K. 1985b. Reflections on epigenetic mechanisms: Hypotheses and case histories, pp. 239–270. *In*: E.K. Balon (ed.), Early Life Histories of Fishes: New Developmental, Ecological and Evolutionary Perspectives. Dr. W. Junk Publishers, Dordrecht.

Balon, E.K. 1986. Saltatory ontogeny and evolution. Rivista di Biologia—Biology Forum 79:151–190.

Balon, E.K. 1988. Tao of life: universality of dichotomy in biology. Biology Forum 81:185–230.

Bamberg, F.B. 1985. Untersuchungen von gefangenschaftsbedingten Verhaltensänderungen beim Damwild (*Cervus dama* Linné, 1958) Beitrage zur Wildbiologie, Heft 5. Verlag Gunter Hartmann, Kronshagen, West Germany.

Banfield, A.W.F. 1960. The use of caribou antler pedicles for age determination. J. Wildl. Manage. 24:99–102.

Banks, W.J. 1973. Histological and ultrastructural aspects of cervinae antler development. Anat. Rec. 175:487.

Banks, W.J. 1974. The ossification process of the developing antler in the white-tailed deer (*Odocoileus virginianus*). Calc. Tiss. Res. 14:257–274.

Banks, W.J. & R.W. Davis. 1966. Observations on the relationship of antlerogenesis to bone morphology and composition in the Rocky Mountain mule deer (*Odocoileus hemionus hemionus*). Anat. Rec. 154:312.

Banks, W.J. & J.W. Newbrey. 1982a. Light microscopic studies of the ossification process in developing antlers, pp. 231–260. *In*: R.D. Brown (ed.), Antler Development in *Cervidae*, Caesar Kleberg Wildl. Res. Inst., Kingsville, TX.

Banks, W.J. & J.W. Newbrey. 1982b. Antler development as a unique modification of mammalian endochondral ossification, pp. 279–306. *In*: R.D. Brown (ed.). Antler Development in Cervidae, Caesar Kleberg Wildl. Res. Inst., Kingsville, TX.

Banks, W.J., G.P. Epling, H.A. Kainer & R.W. Davis. 1968a. Antler growth and osteoporosis. I. Morphological and morphometric changes in the costal compacta during the antler growth cycle. Anat. Rec. 162:387–398.

Banks, W.J., G.P. Epling, H.A. Kainer & R.W. Davis. 1968b. Antler growth and osteopo-

rosis. II. Gravimetric and chemical changes in the costa compacta during the antler growth cycle. Anat. Rec. 162:399–405.
Barrell, G.K., P.D. Muir, & A.R. Sykes. 1985. Seasonal profiles of plasma testosterone, prolactin, and growth hormone in red deer stags, pp. 185–190. *In*: K.R. Drew & P.F. Fennessy (eds.), Biology of Deer Production. The Royal Soc. of New Zealand, Bull. 22.
Barrell, K.G., C.M. Lengoc & F.D. Muir. 1987. Receptors for estradiol in growing antlers of red deer stags, pp. 17–18. Abstr. of the 18. Congress of I.U.G.B., Krakow, Poland.
Barrette, C. 1977. Fighting behavior of muntjac and the evolution of antlers. Evolution 31:169–176.
Barry, J.C., E.H. Lindsay & L.L. Jacobs. 1982. A biostratigraphic zonation of the Middle and Upper Siwaliks of the Potwar plateau of northern Pakistan. Palaeogeogr. Palaeoclimatol. Palaeoecol., Amsterdam 37:95–130.
Barry, J.C., N.M. Johnson, S.M. Raza & L.L. Jacobs. 1985. Neogène mammalian faunal change in southern Asia. Geology, Boulder, CO. 13:637–640.
Barth, D., T. Gimenez, B. Hoffman & H. Karg. 1976. Testosteron konzentration im peripheren Blut beim Rehbock (*Capreolus capreolus*). Z. Jagdwiss. 22:134–148.
Bartke, A. 1980. Role of prolactin in reproduction in male mammals. Fed. Proc. 39:2577–2582.
Bartoš, L. 1980. The date of antler casting, age and social hierarchy relationships in the red deer stag. Behav. Process. 5:293–301.
Bartoš, L. 1985. Social activity and the antler cycle in red deer stags, pp. 269–272. *In*: K.A. Drew & P.F. Fennessy (eds.), Biology of Deer Production. The Royal Soc. of New Zealand. Bull. 22.
Bartoš, L. 1986a. Dominance and aggression in various sized groups of red deer stags. Aggress. Behav. 12:175–182.
Bartoš, L. 1986b. Relationships between behaviour and antler cycle timing in red deer. Ethology 71:305–314.
Bartoš, L. & J. Hyánek. 1982a. Social position in the red deer stag. I. The effect on developing antlers, pp. 451–461. *In*: R.D. Brown (ed.), Antler Development in Cervidae. Caesar Kleberg Wildl. Res. Inst., Kingsville, TX.
Bartoš, L. & J. Hyánek. 1982b. Social position in the red deer stag. II. The relationship with developed antlers, pp. 463–466. *In*: R.D. Brown (ed.), Antler Development in Cervidae. Caesar Kleberg Wildl. Res. Inst., Kingsville, TX.
Bartoš, L. & V. Perner. 1985. Integrity of a red deer stag social group during velvet period, association of individuals, and timing of antler cleaning. Behaviour 95:314–323.
Bartoš, L. & V. Perner. 1987. Asynchronous antler casting in red deer, pp. 18–19. *In*: XVIIIth Congress IUGB, Abstracts, Jagiellonian University, Kraków.
Bartoš, L., V. Perner & S. Losos. 1988a. Red deer stags rank position, body weight and antler growth. Acta Theriol. 33:209–217.
Bartoš, L., V. Perner & B. Procházka. 1988b. On the relationship between social rank during the velvet period and antler parameters in a growing red deer stag. Acta Theriol. 32:403–412.
Bateson, G. 1963. The role of somatic change in evolution. Evolution 17:529–539.
Bateson, G. 1979. Mind and Nature—A Necessary Unity. E.P. Dutton, New York, N.Y.
Bayern, A.v. & J.v. Bayern. 1975. Über Rehe in einem Steirischen Gebirgsrevier. Johannes Bauer, Hamburg.
Beach, F.A. 1975. Sexual attractivity, proceptivity and receptivity in female mammals. Horm. Behav. 7:105–138.

Becker, R.O. 1961. The bioelectric factors in amphibian-limb regeneration. J. Bone & Joint Surgery 43A:643–656.

Bejšovec, J. 1955. Špatný vyvoj parůžků u srnčí zvěře. (Poor antler development in roe deer.) Acta Soc. Zool. Bohemosloven. 19:119–137. (In Czech.)

Belyaev, D.K. 1979. Destabilizing selection as a factor in domestication. J. Heredity 70(5):301–308.

Beninde, J. 1937. Zur Naturgeschichte des Rothirsches. Monographien der Wildsäugetiere 4. Dr. P. Schöps Verlag, Leipzig.

Bentley, A. 1978. An introduction to the deer of Australia with special reference to Victoria, Melbourne. The Koetong Trust Service Fund, Forest Commission Victoria.

Berger, J. 1985. Interspecific interactions and dominance among wild Great Basin ungulates. J. Mammal. 66:571–573.

Bernhard, R. 1963. Specific gravity, ash, calcium and phosphorus content of antlers of Cervidae. Can. Field-Nat. 90:310–322.

Bernor, R.L. 1983. Geochronology and zoogeographic relationships of Miocene Hominoidae, pp. 21–64. *In*: R.L. Ciochon & R.S. Corruccini (eds.), New interpretations of ape and human ancestry. Plenum Press, New York, N.Y.

Berthold, A.A. 1831. Ueber das Wachstum, den Abfall und die Wiedererzeugung der Hirschgeweihe. Beitrag Anatomie, Zootomie u. Physiologie. Goettingen: 39–96.

Beyer, C.K., F. Larsson, J. de la Torre & G. Perez-Palacios. 1975. Synergistic actions of estrogens and androgens on the sexual behavior of castrated male rabbit. Horm. Behav. 6:301–306.

Bigalke, R.C. 1963. A note on reproduction in the Steenbok *Raphicerus campestris* Thunberg. Ann. Cape Provinc. Mus. 3:64–67.

Billingham, R.E., R. Mangold & W.K. Silvers. 1959. The neogenesis of skin in the antlers of deer. Ann. N.Y. Acad. Sci. 83:481–498.

Bird, R.D. 1933. A three-horned wapiti (*Cervus canadensis canadensis*). J. Mammal. 14:164–166.

Bischoff, T.L. 1854. Entwicklungsgeschichte des Rehes. J. Ricker's Buchhandlung, Giessen.

Blasius, W. 1894–95. Über den Schädel einer gehörnten Ricke. Jahresberichte d. Vereins f. Naturwiss. zu Braunschweig, IX, Sitzunsberichte: 11–13.

Blasius, W. 1903. Gehörnte Ricke. Jahresber. Verein f. Naturwiss. Braunschweig 9:11–13.

Blauel, G. 1935. Beobachtungen über die Entstehung der Perücke beim Rehbock. Endokrinologie 15:321–329.

Blauel, G. 1936. Beobachtungen über die Entstehung der Perücke beim Rehbock, 2. Mitt. Endokrinologie 17:369–372.

Blaxter, K.K. 1962. The energy metabolism of ruminants. Thomas, Springfield, IL.

Blaxter, K.L., R.N.B. Kay, G.A.M. Sharman, J.M.M. Cunningham & W.J. Hamilton. 1974. Farming the red deer. Department of Agriculture and Fisheries for Scotland, Edinburgh, Her Majesty's Stationary Office.

Boas, J.C.V. 1892. Beitrag zur Kenntnis des Hermaphroditismus beim Rehwild. Der Waidmann.

Boas, J.E.V. 1917. Das Gehörn von Antilocapra und sein Verhältnis zu den anderen Cavicornia und der Hirsche. Vidensk. Selsk. Biol. Med. 1:1–23.

Bohlin, B. 1935. *Tsaidamotherium hedini*, n.g. n.sp., ein Einhoerniger Ovibovine aus den tertiären Ablagerungen aus der Gegend der Tossun Nor, Tsaidam. Geogr. Annaler 17.

Festschr. d. Svenska Saellsk. f. Antropol u. Geogr. zu Sven Hedin's 70. Geburtstag: 66–74.

Bohlin, B. 1953. *Triceromeryx*: an American immigrant to Europe. Bull. Geol. Inst. Uppsala 35:1–6.

Bolk, L. 1926. Das Problem der Menschwerdung. G. Fischer, Jena.

Boone & Crockett. 1982. Official Measurer's Manual. Boone & Crockett Club, Alexandria, VA.

Borgens, R.B., J.W. Vanable, Jr. & L.F. Jaffe. 1977. Bioelectricity and regeneration. 1. Initiation of frog limb regeneration by minute currents. J. Exp. Zool. 200:403–416.

Bosnjak, C. 1898. Ein seltener Kümmerer. Wild und Hund 4:457–458.

Bouchud, J. 1966. Essai sur le renne et la climatologie du Paléolithique Moyen et Supérieur. C.N.R.S. Paris.

Bouvrain, G. 1982. Revision du genre *Prostrepsiceros* Major 1891. Palaont. Z. Berlin 56:113–124.

Bouvrain, G. & L. de Bonis. 1984. Le genre *Mesembriacerus* (Bovidae, Artiodactyla, Mammalia): un oviboviné primitif du Vallésien (Miocène supérieur) de Macédoine (Grèce). Palaeovertebrata 14:201–223.

Bower, R.T. 1983. Osteophagia and antler breakage among Roosevelt elk. Calif. Fish and Game 69:84–88.

Boyd, R.J. 1970. Elk of the White River Plateau, Colorado. Colorado Div. Game, Fish and Parks. Tech. Bull. No. 25.

Boyer, J.F. 1974. Clinal and size dependent variation at the Lap locus in *Mytilus edulis*. Biol. Bull. 147:535–549.

Brain, P.F. 1980. Adaptive aspects of hormonal correlates of attack and defence in laboratory mice—a study of ethobiology, pp. 391–413. *In*: P.S. McConnell, G.J. Boer, H.J. Romijn, N.E. Vandepoll & M.A. Corner (eds.), Adaptive Capabilities of the Nervous System. Elsevier North-Holland Biomedical Press, Amsterdam.

Bramley, P.S. 1970. Territoriality and reproductive behaviour of the deer. J. Reprod. Fertil. Suppl. 11:43–70.

Branan, W. & R.L. Marchinton. 1981. Deer populations and management in Suriname: A preliminary report. 4th. Ann. Meeting Southeast Deer Study Group. Panama City Beach, Florida.

Brandt, K. 1882. Die "dritte" Rose und Stange beim Rehgehörn und ihre Beziehungen zum normalen Rosenstock, Illustrierte Jagdzeitung, Organ für Jagd, Fischerei und Naturkunde, IX: 185–188, 197–200, 207–209.

Brandt, K. 1901. Das Gehörn und Entstehung monströser Formen. Paul Parey, Berlin.

Breedlove, S.M. 1985. Hormonal control of the anatomical specificity of motor neurons-to-muscle innervation in rats. Science 227:1357–1359.

Brisbin, I.L., Jr. & M.S. Lenarz. 1984. Morphological comparisons of insular and mainland populations of southeastern white-tailed deer. J. Mammal. 65:44–50.

Brockstedt-Rasmussen, H., P. Leth Sorensen, H. Ewald & F. Melson. 1987. The rhythmic relation between antler and bone porosity in Danish deer. Bone 8:19–22.

Brody, S. 1945. Bioenergetics and Growth. Hafner, New York, NY.

Brokx, P.A. 1972. Ovarian composition and aspects of the reproductive physiology of Venezuelan white-tailed deer (*Odocoileus virginianus Gymnotis*). J. Mammal. 53:760–773.

Bromley, P.T. 1967. Pregnancy, birth, behavioral development of the fawn, and territoriality in the pronghorn (*Antilocapra americana* Ord) on the National Bison Range, Moiese, Montana. M.S. Thesis, University of Montana, Missoula, MT.

Bromley, P.T. 1969. Territoriality in pronghorn bucks on the National Bison Range, Moiese, Montana. J. Mammal. 50:81–89.

Bromley, P.T. 1977. Aspects of the behavioral ecology and socio-biology of the pronghorn. Ph.D. Thesis, University of Calgary, Calgary, Alberta, Canada.

Bromley, P.T. & D.W. Kitchen. 1974. Courtship in the pronghorn (*Antilocapra americana*), pp. 356–364. *In*: V. Geist & F. Walther (eds.), The Behavior of Ungulates and its Relation to Management, Vol. I. I.U.C.N. Publ. 24, Morges, Switzerland.

Brooke, V. 1878. On the classification of the Cervidae with a synopsis of the extinct species. Proc. Zool. Soc. London: 883–928.

Brooks, R.V. 1975. Androgens. Clinics in Endocrinology and Metabolism 4:503–520.

Brown, K.I. & K.E. Nestor. 1973. Some physiological responses of turkeys selected for high and low adrenal response to cold stress. Poultry Sci. 52:1948–1954.

Brown, R.D. (ed.). 1982. Antler Development in *Cervidae*. Caesar Kleberg Wildl. Res. Inst., Kingsville, TX.

Brown, R.D. 1985. Observations on the genetics of antler growth. Caesar Kleberg Wildl. Res. Inst., Ann. Rpt., Kingsville, TX.

Brown, R.D. & K.P.H. Pritzker. 1985. Osteoporosis in white-tailed deer. Caesar Kleberg Wildl. Res. Inst., Ann. Rpt., Kingsville, TX.

Brown, R.D., R.L. Cowan & J.F. Kavanaugh. 1978a. Effect of pinealectomy on seasonal androgen titers, antler growth and feed intake in white-tailed deer. J. Anim. Sci. 47:435–440.

Brown, R.D., R.L. Cowan & L.C. Griel. 1978b. Correlation between antler and long bone relative bone mass and circulating androgens in white-tailed deer (*Odocoileus virginianus*). Am. J. Vet. Res. 39:1053–1056.

Brown, R.D., R.L. Cowan & J.F. Kavanaugh. 1981. Effect of parathyroidectomy on white-tailed deer. Texas J. Sci. 33:113–120.

Brown, R.D., C.C. Chao & L.W. Faulkner. 1983a. The endocrine control of the initiation and growth of antlers in white-tailed deer. Acta Endocrinol. 103:138–144.

Brown, R.D., C.C. Chao & L.W. Faulkner. 1983b. Hormone levels and antler development in white-tailed deer and sika fawns. Comp. Biochem. Physiol. 75A:385–390.

Brown, W.L. 1957. Centrifugal speciation. Quart. Rev. Biol. 32:247–277.

Brüggemann, J., A. Adam & H. Karg. 1965. ICSH-Bestimmungen in Hypophysen von Rehböcken (*Capreolus capreolus*) und Hirschen (*Cervus elaphus*) unter Berücksichtigung des Saisoneinflusses. Acta Endocrinol. 48:569–580.

Bruhin, H. 1953. Zur Biologie der Stirnaufsätze bei Huftieren. Physiol. Comp. et Oecol. B. III:63–127.

Bruns, E.H. 1969. A preliminary study of behavioral adaptations of wintering pronghorn antelopes. M.S. Thesis, University of Calgary, Calgary, Alberta, Canada.

Bryden, H.A. (ed.) 1899. The great and small game of Africa. Rowland Ward, London.

Bubenik, A.B. 1956. Eine seltsame Geweihentwicklung beim Ren (*Rangifer tarandus* L.). Z. Jagdwiss. 2:21–24.

Bubenik, A.B. 1959a. Der Feinbau der Geweihe von *Cervus* (*Dama*) Linné, 1758, und *mesopotamicus* Brooke, und ihre Entwicklungsstufe. Säugetierkdl. Mitt. 7:90–95.

Bubenik, A.B. 1959b. Grundlagen der Wildernährung. Deutscher Bauernverlag, Berlin.

Bubenik, A.B. 1959c. Ein weiterer Beitrag zu den Besonderheiten der Geweihtrophik beim Ren. Z. Jagdwiss. 5:51–55.

Bubenik, A.B. 1962. Geweihmorphogenese im Lichte der neurohumoralen Forschung. Symp. Theriolog. ČSAV, Brno 1960:59–66.

Bubenik, A.B. 1963. Rund um die "Perückengeweihe." St. Hubertus 6:81–85.

Bubenik, A.B. 1966. Das Geweih. Paul Parey Verlag, Hamburg.
Bubenik, A.B. 1968. The significance of antlers in the social life of the Cervidae. Deer 1:208–214.
Bubenik, A.B. 1971. Geweihe und ihre biologische Funktion. "n+m" (Naturwiss. u. Medizin) 8:33–51.
Bubenik, A.B. 1973. Hypothesis concerning the morphogenesis in moose antlers. Alces 9:195–231.
Bubenik, A.B. 1975. Taxonomic value of antlers in genus Rangifer H. Smith. 1st. Intern. Reindeer-Caribou Symposium College, Alaska: 9–11.
Bubenik, A.B. 1982a. Physiology of wapiti, pp. 125–179. *In*: J.W. Thomas & D.E. Toweill (eds.), Elk of North America, Ecology and Management. Wildl. Management Institute, Washington, D.C.
Bubenik, A.B. 1982b. The behavioral aspects of antlerogenesis, pp. 389–449. *In*: R.D. Brown (ed.), Antler Development in Cervidae, Caesar Kleberg Wildl. Res. Inst., Kingsville, TX.
Bubenik, A.B. 1982c. Taxonomy of the Pecora in relation to morphophysiology of their cranial appendages, pp. 163–185. *In*: R.D. Brown (ed.), Antler Development in Cervidae. Caesar Kleberg Wildl. Res. Inst., Kingsville, TX.
Bubenik, A.B. 1982d. Proposals for standardized nomenclature for bony appendages in Pecora, pp. 187–194. *In*: R.D. Brown (ed.), Antler Development in Cervidae. Caesar Kleberg Wildl. Res. Inst., Kingsville, TX.
Bubenik, A.B. 1984. Ernährung, Verhalten und Umwelt des Schalenwildes. BLV Verlagsgesellschaft, München, Wien, Zürich.
Bubenik, A.B. 1985. Reproductive strategies in cervids, pp. 367–374. *In*: P.F. Fennessy & K.R. Drew (ed.). Biology of Deer Production. Royal Soc. of New Zealand, Bull. 22.
Bubenik, A.B. 1986a. Grundlagen der soziobiologischen Hirschwildbewirtschaftung, pp. 97–159. *In*: H. Reuss (ed.), Das Rotwild – Cerf rouge – Red Deer. Proc. C.I.C. Sympos. in Graz, Austria, Juni 1986.
Bubenik, A.B. 1986b. Taxonomic position of *Alcinae* Jerdon 1874 and the history of the genus *Alces* Gray 1821. Alces 22:1–67.
Bubenik, A.B. 1987a. The behavior of the moose of North America. 2nd. Intern. Symp. on Moose 1984, Uppsala, Swedish Wildl. Res. Suppl. 1:333–365.
Bubenik, A.B. 1987b. Die Brunftstrategien und Brunft-Taktiken unserer Geweihträger. Der Anblick 10:382–389.
Bubenik, A.B. 1989. The sociobiological versus hunters viewpoints on antlers and horns, pp. 355–380. *In*: W. Trense (ed.), The Big Game of the World and Its Trophies. P. Parey, Hamburg-Berlin.
Bubenik, A.B. & R. Pavlansky. 1956. Von welchem Gewebe geht der eigentliche Reiz zur Geweihbildung aus? II. Mitteilung: Operative Eingriffe auf den Rosenstöcken der Rehböcke, *Capreolus capreolus* (Linné, 1758). Säugetierkundl. Mitt. 4:97–103.
Bubenik, A.B. & R. Pavlansky. 1959. Von welchem Gewebe geht der eigentliche Reiz zur Geweihbildung aus? III. Mitteilung: Operative Eingriffe am Bastgeweih. Säugetierkundl. Mitt. 7:157–163.
Bubenik, A.B. & R. Pavlansky. 1965. Trophic responses to trauma in growing antlers. J. Exp. Zool. 159:289–302.
Bubenik, A.B. & V. Munkačevič. 1967. Peculiarities of antler trophic in red deer (*Cervus* elaphus) in the region of Belje, Yugoslavia. 7th Congr. IUGB Beograd-Ljubljanna: 255–259.

Bubenik, A.B. & G.A. Bubenik. 1976a. New, non-traumatic, disposable automatic injection dart. C.A.L.A.S./A.C.T.A.L. Proc., pp. 48–53.

Bubenik, A.B. & G.A. Bubenik. 1976b. Pigmentation in scrotal and other hairs in deer as indication of sexual activity and social status. Ann. Meet. Anim. Behav. Soc. No. 241.

Bubenik, A.B. & R. König. 1985. Morphology of antlers of the genus *Capreolus* (Grey 1821), pp. 273–278. *In*: P.F. Fennessey & K.R. Drew (eds.), Biology of Deer Production. Royal Soc. New Zealand, Bull. 22.

Bubenik, A.B. & C. Weber-Schilling. 1986. Perücken der Geweihträger und das Phänomenon des Abwerfens oberhalb der Rose. Z. Jagdwiss. 32:158–171.

Bubenik, A.B., R. Pavlansky & J. Řeřabek. 1956. Untersuchungen des Mineralstoffwechsels bei Geweihträgern mittels radioaktiver Isotopen. Z. f. Jagdwiss. 2:119–123.

Bubenik, A.B., R. Tachezy & G.A. Bubenik. 1976. The role of the pituitary-adrenal axis in the regulation of antler growth processes. Säugetierkundl. Mitt. 24:1–5.

Bubenik, A.B., F.L. Raymond & P. Meile. 1977. Morphometry of the horns of chamois (*Rupicapra rupicapra* L.). Preliminary study. Transact. 13th. Cong. IUGB Atlanta, GA, USA, pp. 351–364.

Bubenik, G.A. 1982. Endocrine regulation of the antler cycle, pp. 73–107. *In*: R.D. Brown (ed.), Antler Development in Cervidae. Caesar Kleberg Wildl. Res. Inst., Kingsville, TX.

Bubenik, G.A. 1983. Shift of seasonal cycle in white-tailed deer by oral administration of melatonin. J. Exp. Zool. 225:155–156.

Bubenik, G.A. 1986. Regulation of seasonal endocrine rhythms in male boreal cervids, pp. 461–474. *In*: I. Assenmacher & J. Boisson (eds.), Endocrine Regulation as Adaptive Mechanisms to the Environment. CNRS-CEBAS, France.

Bubenik, G.A. & A.B. Bubenik. 1967. Adrenal glands in roe deer (*Capreolus capreolus* L.). 7th Congr. des Biol. du Gibier, Beograd, Ljubljana, pp. 93–97.

Bubenik, G.A. & A.B. Bubenik. 1978a. The role of sex hormones in the growth of antler bone tissue: Influence of an antiestrogen therapy. Säugetierkdl. Mitt. 26:284–291.

Bubenik, G.A. & A.B. Bubenik. 1978b. Thyroxine levels in male and female white-tailed deer (*Odocoileus virginianus*). Can. J. Physiol. Pharmacol. 56:945–949.

Bubenik, G.A. & J.F. Leatherland. 1984. Seasonal levels of cortisol and thyroid hormones in intact and castrated mature male white-tailed deer. Can. J. Zool. 62:783–787.

Bubenik, G.A. & A.B. Bubenik. 1985. Seasonal variations in hair pigmentation of white-tailed deer and their relationship to sexual activity and plasma testosterone. J. Exp. Zool. 235:387–395.

Bubenik, G.A. & A.B. Bubenik. 1986. Phylogenic and ontogenic development of antlers and neuroendocrine regulation of the antler cycle—a review. Säugetierkdl. Mitt. 33:97–123.

Bubenik, G.A. & D. Schams. 1986. Relationship of age to seasonal levels of LH, FSH, prolactin and testosterone in male, white-tailed deer. Comp. Biochem. Physiol. 83A:179–183.

Bubenik, G.A. & A.B. Bubenik. 1987. Recent advances in studies of antler development and neuroendocrine regulation of the antler cycle, pp. 99–109. *In*: Ch. Wemmer (ed.), Biology and Management of Cervidae, Smithsonian Inst. Press, Washington, DC.

Bubenik, G.A. & D. Schams. 1990. Mineralization of the antler pedicle correlates with plasma testosterone levels in deer. Säugetierkdl. Mitt. (In press.)

Bubenik, G.A. & J.S. Smith. 1986. The effect of thyroxine (T_4) administration on plasma levels of triiodothyronine (T_3) and T_4 in male white-tailed deer. Comp. Biochem. Physiol. 83A:185–187.

Bubenik, G.A. & P.S. Smith. 1987. Circadian and circannual rhythm of melatonin in adult male white-tailed deer; the effect of oral administration of melatonin. J. Exp. Zool. 241:81–89.
Bubenik, G.A., G.M. Brown, G.A. Bubenik & L.J. Grota. 1974. Immunohistochemical localization of testosterone in the growing antlers of white-tailed deer (*Odocoileus virginianus*). Calc. Tiss. Res. 14:121–130.
Bubenik, G.A., A.B. Bubenik, G.M. Brown, A. Trenkle & D.A. Wilson. 1975a. Growth hormone and cortisol levels in the annual cycle of white-tailed deer (*Odocoileus virginianus*). Can. J. Physiol. Pharmacol. 53:787–792.
Bubenik, G.A., G.M. Brown, A.B. Bubenik & D.A. Wilson. 1975b. The role of sex hormones in the growth of antler bone tissue. I. Endocrine and metabolic effects of antiandrogen therapy. J. Exp. Zool. 194:349–358.
Bubenik, G.A., A.B. Bubenik, G.N. Brown & D.A. Wilson. 1977a. Sexual stimulations and variations of plasma testosterone in normal, antiandrogen and antiestrogen treated white-tailed deer (*Odocoileus virginianus*) during the annual cycle, pp. 377–386. 13th Congress of Game Biologists, Atlanta, GA.
Bubenik, G.A., A.B. Bubenik, A. Trenkle, A. Sirek, D.A. Wilson & G.M. Brown. 1977b. Short-term changes in plasma concentration of cortisol, growth hormone and insulin during the annual cycle of a male white-tailed deer (*Odocoileus virginianus*). Comp. Biochem. Physiol. 58A:387–391.
Bubenik, G.A., A.B. Bubenik & J. Zamečnik. 1979. The development of circannual rhythm of estradiol in plasma of white-tailed deer (*Odocoileus virginianus*). Comp. Biochem. Physiol. 62A:869–872.
Bubenik, G.A., J.M. Morris, D. Schams & C. Klaus. 1982a. Photoperiodicity and circannual levels of LH, FSH, and testosterone in normal and castrated male, white-tailed deer. Can. J. Physiol. Pharmacol. 60:788–793.
Bubenik, G.A., A.B. Bubenik, E.D. Stevens & A.G. Binnington. 1982b. The effect of neurogenic stimulation on the development and growth of bony tissues. J. Exp. Zool. 219:205–216.
Bubenik, G.A., A.B. Bubenik, D. Schams & J.F. Leatherland. 1983. Circadian and circannual rhythms of LH, FSH, testosterone (T), prolactin, cortisol, T_3 and T_4 in plasma of mature, male, white-tailed deer. Comp. Biochem. Physiol. 76:37–45.
Bubenik, G.A., D. Schams & J.F. Leatherland. 1985a. Seasonal rhythms of prolactin and its role in the antler cycle of white-tailed deer, pp. 257–262. *In*: K. Drew & F. Fennessy (eds.), Biology of Deer Production. The Royal Soc. of New Zealand, Bull. 22.
Bubenik, G.A., A.B. Bubenik & A. Frank. 1985b. Aussersaisonale Brunft beim Edelhirsch (*Cervus elaphus maral*). Z. Jagdwiss. 21:129–133.
Bubenik, G.A., P.S. Smith & D. Schams. 1986. The effect of orally administered melatonin on the seasonality of deer pelage exchange, antler development, LH, FSH, prolactin, testosterone, T_3, T_4, cortisol and alkaline phosphatase. J. Pineal Res. 3:331–349.
Bubenik, G.A., D. Schams & G. Coenen. 1987a. The effect of artificial photoperiodicity and antiandrogen treatment on the antler growth and plasma levels of LH, FSH, testosterone, prolactin and alkaline phosphatase in the male white-tailed deer. Comp. Biochem. Physiol. 87A:551–559.
Bubenik, G.A., D. Schams & A. Semperé. 1987b. Assessment of reproductive and antler performance of male white-tailed deer by Gn-RH stimulation test. Comp. Biochem. Physiol. 86A:767–771.
Bubenik, G.A., D. Pomerantz & D. Schams. 1987c. The role of androstenedione and testosterone in the reproduction and antler growth of a male white-tailed deer. Acta. Endocrinol. 114:147–152.

Bubenik, G.A., A. Semperé & J. Hamr. 1987d. Developing antler, the model for endocrine regulation of bone growth. 1. Concentration gradient of T_3, T_4 and alkaline phosphatase in the antler-, jugular- and saphenous veins. Calcif. Tiss. Int. 41:38–43.
Bubenik, G.A., Smith, J.H. & Flynn, A. 1988. Plasma levels of β-endorphin in white-tailed deer: Seasonal variation and the effect of thyroxine, GnRH, Dexamethasone and ACTH administration. Comp. Biochem. Physiol. 90A:309–313.
Burchardt, B. 1978. Oxygen isotope palaeontemperatures from the Tertiary period in the North Sea area. Nature 275:121–123.
Buechner, H.K. 1950a. Life history, ecology, and range use of the pronghorn antelope in Trans-Pecos, Texas. Am. Midl. Nat. 43:257–354.
Buechner, H.K. 1950b. Range ecology of pronghorn on the Wichita Mountains Wildlife Refuge. Trans. North Am. Wildl. Conf. 15:627–644.
Buss, I.O. & J.D. Solf. 1959. Record of an antlered female elk. J. Mammal. 40:252.
Bützler, W. 1974. Kampf- und Paarungsverhalten, soziale Rangordnung und Aktivitätsperiodik beim Rothirsch. Beiheft Z. Tierpsychol. No. 16, Paul Parey Verlag, Hamburg, Berlin.
Byers, J.A. 1987. Why deer and the antelope play. Nat. History 15:54–60.
Byers, J.A. & D.W. Kitchen. 1988. Mating system shift in a pronghorn population. Behav. Ecol. Sociobiol. 22:355–360.
Cameron, A.G. 1892. The value of antlers in the classification of deer. Field 79:625.
Care, A.D., R. Ross, G.K. Barrell, P.D. Muir, J. Aaron & C. Oxby. 1985. Effects of long-term thyroparathyroidectomy on antler growth in red deer, pp. 251–254. *In*: D.R. Drew & P.F. Fennessy (eds.), The Biology of Deer Production, The Royal Soc. of New Zealand, Bull. 22.
Carr, D.E. 1972. The Forgotten Senses. Doubleday and Co., Inc., Garden City, NY.
Carr, S.M., S.W. Ballinger, J.N. Derr, L.H. Blankenship & J.W. Bickham. 1986. Mitochondrial DNA analysis of hybridization between sympatric white-tailed deer and mule deer in west Texas. Proc. Natl. Acad. Sci. USA 83:9576–9580.
Caton, J.D. 1874. On the structure and casting of the antlers of the deer. Am. Natur. 8:348–353.
Caton, J.D. 1877. The Antelope and Deer of America. Forest and Stream Publishing Co., New York, NY.
Centrella, M. & E. Canalis. 1985. Local regulators of skeletal growth: A perspective. Endocrine Rev. 6:544–551.
Chaddock, T.T. 1940. Chemical analysis of deer antlers. Wisc. Cons. Bull., Madison 5:42.
Chao, C.C. & R.D. Brown. 1984. Seasonal relationships of thyroid, sexual and adrenocortical hormones to nutritional parameters and climatic factors in white-tailed deer (*Odocoileus virginianus*) of south Texas. Comp. Biochem. Physiol. 77A:299–305.
Chao, C.C., R.D. Brown & L.J. Deftos. 1984a. Seasonal levels of serum parathyroid hormone, calcitonin and alkaline phosphatase in relation to antler cycle in white-tailed deer. Acta Endocrinol. 106:234–240.
Chao, C.C., R.D. Brown & L.J. Deftos. 1984b. Effects of xylazine immobilization on biochemical and endocrine values in white-tailed deer. J. Wildl. Dis. 20:328–332.
Chapman, D.I. 1981. Antler structure and function—a hypothesis. J. Biomech. 14:195–197.
Chapman, D.L. 1975. Antlers, bones of contention. Mammal Review 5:121–172.
Chapman, N. & D. Chapman. 1975. Fallow deer. The British Deer Society.
Chard, J.S.R. 1958. The significance of antlers. Bull. Mammal. Soc. Br. Isles 10:9–14.

Charlesworth, D., B. Charlesworth, J.J. Bull, A. Grafen & L. Cairns. 1988. Origin of mutants disputed. Nature 336:525–528.

Chen, G. 1988. Remarks on the *Oioceros* species (Bovidae, Artiodactyla, Mammalia) from the Neogene of China. Vertebr. Palasiat., Beijing 26:169–172.

Chen, G. & W. Wu. 1976. Miocene mammalian fossils of Jiulongkou, Ci Xian district, Hebei. Vertebr. Palasiat., Beijing 14, 1:6–15.

Chesser, R.K. & M.H. Smith. 1987. Relationship of genetic variation to growth and reproduction in the white-tailed deer, pp. 168–177. *In*: C.M. Wemmer (ed.), Biology and Management of the Cervidae. Smithsonian Institution Press, Washington, DC.

Chesser, R.K., M.H. Smith, P.E. Johns, M.N. Manlove, D.O. Straney & R. Baccus. 1982. Spatial, temporal and age-dependent heterozygosity of Beta-hemoglobin in white-tailed deer. J. Wildl. Manage. 46:983–990.

Christian, J.J. 1975. Hormonal control of population growth, pp. 205–274. *In*: B.E. Eleftheriou & R.L. Sprotts (eds.), Hormonal correlates of behavior, Vol. 1. Plenum Press, New York, NY.

Churcher, C.S. 1970. Two new Upper Miocene Giraffids from Fort Ternan, Kenya, East Africa: *Palaeotragus primaevus* n.sp. and *Samotherium africanum* n.sp., pp. 1–105. *In*: L.S.B. Leakey & R.J.G. Savage (eds.), Fossil Vertebrates of Africa, Vol. 2. Academic Press, London.

Churcher, C.S. 1978. Giraffidae, Chapter 25, pp. 509–535. *In*: V.J. Maglio & H.B.S. Cooke (eds.), Evolution of African Mammals. Harvard University Press, Cambridge, MA.

Churcher, C.S. & J.D. Pinsof. 1987. Variation in the antlers of North American *Cervalces* (Mammalia: Cervidae): Review of new and previous recorded specimens. J. Vertebr. Paleontol. 7:373–397.

Clark, C.M.H. & C.L. Batcheler. 1972. A high incidence of antler malformation among red deer in Cupola Basin, Nelson. New Zealand Forestry Service Reprint No. 615, ODC 149.6.

Clark, J., J.R. Beerbower & K. Kietzke. 1967. Oligocene sedimentation, stratigraphy, paleoecology and paleoclimatology in the Big Badlands of North Dakota. Fieldana (Geology) 5.

Clutton-Brock, T.H. 1982. The function of antlers. Behaviour 79:108–125.

Clutton-Brock, T.H. & S.D. Albon. 1979. The roaring or red deer and the evolution of honest advertisement. Behaviour 69:145–170.

Clutton-Brock, T.H. & S.D. Albon. 1985. Reproductive success in wild Red deer, pp. 205–212. *In*: P.F. Fennessy & K.R. Drew (eds.), Biology of Deer Production. Royal Soc. of New Zealand, Bull. No. 22.

Clutton-Brock, T.H., S.D. Albon, R.M. Gibson & F.E. Guinness. 1979. The logical stag: Adaptive aspects of fighting in red deer (*Cervus elaphus*). Anim. Behav. 27:211–225.

Clutton-Brock, T.H., S.D. Albon & P.H. Harvey. 1980. Antlers, body size and breeding group size in the Cervidae. Nature 285:565–567.

Clutton-Brock, T.H., F.E. Guinness & S.D. Albon. 1982. Red deer: behavior and ecology of two sexes. The University of Chicago Press, Chicago, IL.

Colbert, E.S. 1933. A skull and mandible of *Giraffokeryx punjabiensis* Pilgrim. Amer. Mus. Novitates No. 632.

Colbert, E.S. 1935. Siwalik Mammals in the American Museum of Natural History. Trans. Amer. Phil. Soc., Philadelphia 26:1–401.

Colbert, E.S. 1936. Tertiary deer discovered by American Museum Asiatic Expedition. Amer. Museum Novit. No. 854:1–21.

Colbert, E.S. 1938. The relationships of the Okapi. J. Mammal. 19:47–64.
Colbert, E.S. 1940. Some cervid teeth from the Tung Gur formation of Mongolia and additional notes on the genera *Stephanocemas* and *Lagomeryx*. Amer. Mus. Novitates, No. 1062:1–6.
Colbert, E.S. 1958. Morphology and behavior, pp. 27–48. *In*: A. Roe & G.G. Simpson (eds.), Behavior and Evolution. Yale University Press, New Haven, CT.
Cole, G.F. 1956. The pronghorn antelope—its range use and food habits in central Montana with special reference to alfalfa. Mont. Fish and Game Dep., Helena, and Mont. State Coll. Agric. Exp. Station, Bozeman. Tech. Bull. 516.
Cole, G.F. & B.T. Wilkins. 1958. The pronghorn antelope—its range use and food habits in central Montana with special reference to wheat. Mont. Fish and Game Dep., Helena. Tech. Bull. 2.
Colenbrander, R., S.J. Dieleman & G.J.G. Wensing. 1977. Changes in serum LH concentration during normal and abnormal sexual development in the pig. Biol. Reprod. 17:506–513.
Collias, N.E. 1956. The analysis of socialization in sheep and goats. Ecology 37:228–239.
Collinson, M.E. & J.J. Hooker. 1987. Vegetational and mammalian faunal changes in the Early Tertiary of southern England, pp. 259–304. *In*: E.N. Friis, W.G. Chaloner & P.R. Crane (eds.), The Origin of Angiosperms and their Biological Consequences. Cambridge University Press, Cambridge, MA.
Collinson, M.E., K. Fowler & M.C. Boulter. 1981. Floristic changes indicate a cooling climate in the Eocene of Southern England. Nature 291:312–317.
Cook, T.A. 1914. The Curves of Life. Constable and Co., London.
Cooke, H.B.S. & S.C. Coryndon. 1970. Pleistocene mammals from the Kaiso Formation and other related deposits in Uganda, pp. 107–224. *In*: L.S.B. Leakey & R.J.G. Savage (eds.), Fossil Vertebr. Afr., London 2.
Cope, E.D. 1874. Report on the stratigraphy and Pliocene vertebrate paleontology of Northern Colorado, pp. 9–28. *In*: F.V. Hayden (ed.), Bull. of the U.S. Geol. and Geograph. Survey of the Territories, Vol. I.
Cope, E.D. 1887. The Origin of the Fittest. Essays on Evolution. D. Appleton, New York. (Reprinted edition, 1974, Arno Press, New York, NY.)
Corner, R.G. 1977. A late Pleistocene-Holocene vertebrate fauna from Red Willow County, Nebraska. Nebraska Transact. Nebraska Acad. Sci. 4:77–93.
Coslovsky, R. & R.S. Yalow. 1974. Influence on the hormonal forms of ACTH on the pattern of corticosteroid secretion. Biochem. Physiol. Res. Commun. 60:1351–1356.
Cothran, E.G., R. Chesser, M.H. Smith & P.E. Johns. 1983. Influences of genetic variability and maternal factors on fetal growth in white-tailed deer. Evolution 37:282–291.
Cothran, E.G., R.K. Chesser, M.H. Smith & P.E. Johns. 1987. Fat levels in female white-tailed deer during the breeding season and pregnancy. J. Mammal. 68:111–118.
Cowan, I.McT. 1946. Antlered doe mule deer. Can. Field-Nat. 60:10–12.
Cowan, R.L. & A.C. Clark. 1981. Nutritional requirements, pp. 72–86. *In*: W.A. Davidson (ed.). Diseases and Parasites of White-tailed Deer. Southeastern Coop. Wildl. Disease Study, University of Georgia, Athens, GA.
Cowan, R.L. & T.A. Long. (undated). Studies on antler growth and nutrition of white-tailed deer. Pennsylvania Coop. Wildl. Res. Unit Paper No. 107:1–8.
Cowan, R.L., E.W. Hartsook & J.B. Whelan. 1968. Calcium-strontium metabolism in white-tailed deer as related to age and antler growth. Proc. Soc. Exp. Biol. Med. 129:733–737.
Cowan, R.L., E.W. Hartsook & J.B. Whelan. 1969. Deer antler growth—ideal test for study of bone metabolism. Science in Agr. 17:1.

Cowan, W.M. 1979. The development of the brain, pp. 56–69. *In*: The Brain. Scientific American, September 1979.

Crenshaw, D.B. 1983. Antler quality prediction by testosterone radioimmunoassay after GnRH challenge in immature white-tailed bucks. Caesar Kleberg Wildl. Res. Inst., Annual Report, 8-9.

Crispens, C.G. & J.K. Doutt. 1970. Studies of the sex chromatin in the white-tailed deer. J. Wildl. Manage. 34:642–644.

Crispens, C.G., Jr. & J.K. Doutt. 1973. Sex chromatin in antlered female deer. J. Wildl. Manage. 37:422–423.

Cronin, M.A., E.R. Wyse & D.G. Cameron. 1988. Genetic relationships between mule deer and white-tailed deer in Montana. J. Wildl. Manage. 52:320–328.

Crusafont, P.M. 1952. Los jirafídos fosiles de España Disp. Prov. Barcelona. Mem. y Com. Inst. Geol., C.S.I.C., Barcelona 8:1–239. (In Spanish.)

Cubitt, G. & G. Mountfort. 1985. Wild India. The wildlife and scenery of India and Nepal. Collins, London.

Cunningham, J.T. 1900. Sexual Dimorphism in the Animal Kingdom. Adam and Charles Black, London.

Curlewis, J.D., A.S. Loudon & A.P.M. Coleman. 1988. Oestrous cycles and the breeding season of the Pere David's deer hind (*Elaphurus davidianus*). J. Reprod. Fert. 82:119–126.

Dagg, A.I. & J.B. Foster. 1976. The Giraffe, its biology, behavior and ecology. Van Nostrand Reinhold, New York, NY.

Dansie, O. 1983. The Muntjac. British Deer Soc. Booklets, 2nd ed.

Dapson, R.W., P.R. Ramsey, M.H. Smith & D.F. Urbston. 1979. Demographic differences in contiguous populations of white-tailed deer. J. Wildl. Manage. 43:889–898.

Darling, F.F. 1937. A herd of red deer. Oxford University Press, Humphrey Milford, London.

Darwin, C. 1859. The origin of species by means of natural selection, or the preservation of favored races in the struggle for life. J. Murray, London.

Darwin, C. 1871. The Descent of Man, and Selection in Relation to Sex. 2 vols. J. Murray, London.

Darwin, E. 1794. Zoonomia or the Laws of Organic Life. J. Johnson, London.

Davies, A.M., C. Bandtlow, R. Heumann, B. Korsching, H. Rohrer & H. Thoenen. 1987. Timing and site of nerve growth factor synthesis in developing skin in relation to innervation and expression of the receptor. Nature 326:353–358.

Davis, M.R. 1987. Screening for post menopausal osteoporosis. Am. J. Obst. Gynecol. 156:1–5.

Davis, R.W. 1962. Studies on antler growth in mule deer (*Odocoileus hemionus hemionus Rafinesque*), pp. 61–64. *In*: Proceedings of the First National White-tailed Deer Disease Symposium, Athens, GA, University of Georgia, Center for Continuing Education.

Davis, S.L., D.L. Ohlson, M.S. Anfinson & J. Klindt. 1977. Episodic growth hormone secretory patterns in sheep: Relationship to gonadal steroid hormones. Am. J. Physiol. 6:E519–E523.

Davis, T.A. 1982. Antler assymetry caused by limb amputation and geophysical forces, pp. 223–230. *In*: R.D. Brown (ed.), Antler Development in *Cervidae*. Caesar Kleberg Wildl. Res. Inst., Kingsville, TX.

Dawkins, R. 1986. The Blind Watchmaker. Longman, London.

Deblinger, R.D. & J.E. Ellis. 1976. Aspects of intraspecific social variations in pronghorns. Proc. Pronghorn Antelope Workshop 7:26–48.

Dehm, R. 1944. Frühe Hirschgeweihe aus dem Miozan Süddeutschlands. Neues Jb. Miner. Monatshefte, Abt. B.: 81–98.

Demment, M. & P. Van Soest. 1985. A nutritional explanation for body-size patterns of ruminant and non-ruminant herbivores. Am. Nat. 125:641–672.

De Vos, A., P. Brokx & V. Geist. 1967. A review of social behavior of the North American Cervids during the reproductive period. Amer. Midl. Nat. 77:390–417.

De Vos, J. & M.D. Dermitzakis. 1986. Models of the development of Pleistocene deer on Crête (Greece). Modern Geology 10:243–248.

Dharamjaygarh, P.O. & Chur Chandra. 1950. A doe cheetal with horns. J. Bombay Nat. Hist. Soc. 49:547.

Dietrich, W.O. 1950. Fossile Antilopen und Rinder Äquatorialafrikas. Palaeontographica, Stuttgart 99A:1–62.

Dixon, J.S. 1927. Horned does. J. Mammal. 8:289–291.

Dixon, J.S. 1934. A study of the life history and food habits of mule deer in California: Part I—life history. Calif. Fish and Game. 20:181–282.

Dobroruka, L.J. 1960. Hornwachstum bei der Nilgau-Antilope, *Boselaphus tragocamelus* Pallus 1766. Zool. Anz. 165:145–146.

Dobroruka, L.J. 1966. Periodisches Auswechseln der Apicalhaut an Ossiconen bei der Giraffe. Zool. Anz. 179:230–232.

Dollman, G. & J.B. Burlace. 1935. Rowland Ward's records of big game. 10th ed. Rowland Ward, London.

Donaldson, J.C. & J.K. Doutt. 1965. Antlers in female white-tailed deer: A 4-year study. J. Wildl. Manage. 29:699–705.

Dorfman, R.A. & R.A. Shipley. 1956. Androgens. John Wiley & Sons, Ltd., New York, NY.

Dorst, J. & P. Dandelot. 1970. A Field Guide to the Larger Mammals of Africa. Collins, London.

Doutt, J.K. & J.C. Donaldson. 1959. An antlered doe with possible masculinizing tumor. J. Mammal. 40:230–236.

Dove, W.F. 1935. The physiology of horn growth: A study of the morphogenesis, the interaction of tissues, and the evolutionary processes of a Mendelian recessive character by means of transplantation of tissues. J. Exp. Zool. 69:347–405.

Dubois, E. 1891. Voorlopig Bericht omtrent het Onderzock naar de Plestocene en Tertiaire Vertebraten-Fauna van Sumatra en Java, gedurende het jaar 1890. Natuurk. Tijdschr. v. Nederl. Indie. 51:93. (In Dutch.)

Dubost, G. 1971. Observations éthologique sur le Muntjak (*Muntiacus muntjac* Zimmermann 1780, et *M. reevesi* Ogilby 1839) en captivité et semiliberté. Z. Tierpsychol. 27:387–427.

Dubost, G. 1975. Le comportement du Chevrotain africain *Hyemoschus aquaticus* Ogilby (Artiodactyla, Ruminantia). Z. Tierpsychol. 37:403–501.

Dubost, G. & R. Terrade. 1970. La transformation de la peau des Tragulidae en bouclier protecteur. Mammalia 34:505–513.

Ducharme, J.R., Y. Tache, G. Charpenet & R. Collu. 1982. Effects of stress on the hypothalamic-pituitary-testicular function in rats. pp. 305–318. *In*: E. Collu, R.A. Barbeau, J.R. Ducharme & G. Tolis (eds.), Brain Peptides and Hormones. Raven Press, New York, NY.

Duerst, L. 1902. Versuch einer Entwicklungsgeschichte der Hörner der Cavicornier nach Untersuchungen am Hausrinde. Festschr. f. KRAEMER, Forschung a. d. Gebiete d. Landwirtsch. Frauenfeld, pp. 1–47.

Duerst, J.U. 1926. Das Horn der Cavicornia. Denkschr. Schweiz. naturf. Ges. 63, 1:1–160.

Dulverton, L. 1970. Red hind with antler. Deer 2:578.

Duvernoy, G.L. 1851. Nôte sur une espèce de buffle fossile (*Bubalus (Arni) antiquus*), découverte en Algérie. C.r. hebd. Séanc. Acad. Sci. 33:595–597.

Eiben, B., S.T. Scharla, K. Fischer & H. Schmidt-Gayk. 1984. Seasonal variations of serum 1,25-dihydroxyvitamin D_3 and alkaline phosphatase in relation to the antler formation in the fallow deer (*Dama dama* L.). Acta Endocrinol. 107:141–144.

Einarsen, A.S. 1948. The pronghorn antelope and its management. Wildl. Manage. Inst., Washington, DC, Monumental Printing Co., Baltimore, MD.

Eisenberg, J.F. 1987. The evolutionary history of the *Cervidae* with special reference to the South American radiation, pp. 60–64. *In*: C.M. Wemmer (ed.), Biology and Management of the Cervidae. Smithsonian Institution Press, Washington, DC.

Eldredge, N. 1985. Time Frames. Simon and Schuster, New York, NY.

Engelmann, C. 1938. Über die Gross-Säuger Szetschwans, Sikongs und Osttibets. Z. Säugetierkunde 13:1–76.

Erbach-Erbach, F. 1986. Die Erbacher Hirschgalerie—eine Kollektion weltstärkster Trophäen, pp. 25–32. *In*: H. Reuss (ed.), Das Rotwild—Cerf rouge—Red Deer. C.I.C. Symposium Graz, Austria.

Espmark, Y. 1964. Studies in dominance-subordination relationship in a group of semi-domestic reindeer (*Rangifer tarandus* L.). Anim. Behav. 12:420–426.

Estes, R.D. 1974. Social organization of the African Bovidae, pp. 235–246. *In*: V. Geist & F. Walther (eds.), The Behaviour of Ungulates and its Relation to Management. I.U.C.N. Pub. No. 24, Morges, Switzerland.

Etches, R.J. 1976. A radioimmunoassay for corticosterone and its application to the measurement of stress in poultry. Steroids 28:763–773.

Evans, E.M.N., J.A.H. Van Couvering & P. Andrews. 1981. Palaeoecology of Miocene sites in western Kenya. J. Human Evol., London 10:99–116.

Ewer, R.F. 1958. Adaptive features in the skulls of African suidae. Proc. Zool. Soc. London 131:135–155.

Ewer, R.F. 1960. Natural selection and Neoteny. Acta Biotheoretica 13:161–184.

Falconer, D.S. 1960. Selection of mice for growth in high and low planes of nutrition. Genet. Res. 1:91–113.

Falconer, D.S. 1981. Introduction to Quantitative Genetics. Longman Inc., New York, NY.

Falconer, D.S. & M. Latsyzewski. 1952. The environment in relation to selection for size in mice. J. Genet. 51:67–80.

Fambach, R. 1909. Geweih und Gehörn. Z. f. Naturwiss. 81:225–264.

Felsenstein, J. 1976. The theoretical population genetics of variable selection and migration. Ann. Rev. Gen. 10:253–280.

Fennessy, P.F. 1982. Growth and nutrition, pp. 105–114. *In*: D. Yerex (ed.), The Farming of Deer. Agr. Promo. Assoc. Ltd., Wellington, New Zealand.

Fennessy, P.F. & H. Moore. 1981. Red deer velvet antler growth and harvesting. Aglink 1/5000/7/81:FFP 261. MAF, Wellington, New Zealand.

Fennessy, P.F. & J.M. Suttie. 1985. Antler growth: Nutritional and endocrine factors, pp. 239–250. *In*: P.F. Fennessy & K.R. Drew (eds.), Biology of Deer Production. Royal Soc. of New Zealand, Bull. 22.

Fennessy, P.F., G.H. Moore & I.D. Corson. 1981. Energy requirements of red deer. Proc. New Zealand Soc. Anim. Prod. 41:167–173.

Filhol, H. 1890. Etudes sur mammifères fossiles de Sansan. Bibl. de l'Ecole des Hautes Etudes, Sect. d. Sciences Natur. 37, G. Masson, Paris.

Finger, S.E., I.L. Brisbin, M.H. Smith & D.F. Urbston. 1981. Kidney fat as a predictor of body condition in white-tailed deer. J. Wildl. Manage. 45:964–968.

Fisher, R.A. 1930. The Genetic Theory of Natural Selection. Clarendon Press, Oxford.

Fisher, R.A. 1958. Genetical Theory of Natural Selection, 2nd rev. ed. (first ed. 1930). Dover, New York.

Flerov, K.K. 1952. Kabargi i Oleni. Fauna SSSR—Mlekopitajushchie. (Moschidae and Cervidae, Fauna USSR), Vol. 1: An SSSR Moskva. (In Russian.)

Fletcher, T.J. 1978. The induction of male sexual behavior in red deer (*Cervus elaphus*) by the administration of testosterone to hinds and estradiol-17β to stags. Horm. Behav. 11:74–88.

Fletcher, T.J. & R.V. Short. 1974. Restoration of libido in castrated red deer stag (*Cervus elaphus*) with estradiol-17β. Nature 248:616–617.

Folker, R.V. 1956. A preliminary study of an antelope herd in Owyhee County, Idaho. M.S. Thesis, University of Idaho, Moscow, ID.

Forand, K.J., R.L. Marchington & K.V. Miller. 1985. Influence of dominance rank on the antler cycle of white-tailed deer. J. Mammal. 66:58–62.

Foster, D.L., I.A. Nickelson, K.D. Ryan, G.A. Coon, R.A. Drondowski & J.A. Holt. 1978. Ontogeny of pulsatile LH and testosterone secretion in male lambs. Endocrinology 102:1137–1146.

Fowler, G.H. 1984. Notes on some specimens of antlers of the fallow deer, showing continuous variation, and the effects of total or partial castration. Proc. Zool. Soc. London 1984:485–494.

Frädrich, H. 1981. Ein schopftragender Muntjak (Muntiacus spec.). Bongo, Berlin 5:57–60.

Frankenberger, Z. 1951. Prvé začátky vývoje parohů u Cervidů (The first stages of the development of antlers in Cervidae). Biolog. Listy 1 (Suppl. 2):127–147. (In Czech.)

Frankenberger, Z. 1954a. Interstitialní buňky jelena (*Cervus elaphus* L.) (The interstitial cells of the red deer, *Cervus elaphus* L.). Československa Morfologie 2:36–41. (In Czech.)

Frankenberger, Z. 1954b. Pučnice a paroh. (The pedicle and the antler). Čs. Morfologie 2:89–95. (In Czech.)

Frankenberger, Z. 1961. K otázce mechanismu shozů parohů u jelenů. (Some remarks on the mechanism of the shedding of the antlers in the deer). Čsl. Morfologie 9:41–45. (In Czech.)

Franklin, W.L. 1974. The social behavior of the vicuña, pp. 477–497. *In*: V. Geist & F. Walther (eds.), The Behaviour of Ungulates and its Relation to Management. I.U.C.N. Pub. No. 24, Morges, Switzerland.

Franzmann, A.W., A. Flynn & P.D. Arneson. 1975. Serum corticoid levels relative to handling stress in Alaskan moose. Can. J. Zool. 53:1424–1426.

Fraser Stewart, J.W. 1981. A study of wild rusa deer (*Cervus rusa timorensis*) in the Tonda Wildlife Management area, Papua, New Guinea. F.A.O. PNG/78/040 Field Document No. 1.

Frasier, M.B. & W.J. Banks. 1973. Characterization of antler mucosubstances by selected histochemical techniques. Anat. Rec. 175:323.

Frasier, M.B., W.J. Banks & J.W. Newbrey. 1975. Characterization of developing antler cartilage matrix. I. Selected histochemical and enzymatic assessment. Calcif. Tiss. Res. 17:273–288.

French, C.E., L.C. McEwen, N.D. Magruder, R.H. Ingram & R.W. Swift. 1955. Nutritional requirements of white-tailed deer for growth and antler development. Pennsylvania Agr. Exp. Station, Bull. 600.

French, C.E., L.C. McEwen, N.D. Magruder, R.H. Ingram & R.H. Smith. 1956. Nutrient requirements for growth and antler development in the white-tailed deer. J. Wildl. Manage. 20:221–232.

Frenzel, B. 1967. Die Klimaschwankungen des Eiszeitalters. F. Vieweg und Sohn, Braunschweig.

Frick, C. 1937. Horned Ruminants of North America. Bull. Am. Mus. Natur. History, New York. Bull. 69.

Friend, T.H., C.E. Polan, F.C. Gwazdauskas & C.W. Heald. 1977. Adrenal glucocorticoid response to exogenous adrenocorticotropin mediated by density and social disruption in lactating cows. J. Dairy Sci. 60:1958–1963.

Fulkerson, W.J. & P.A. Jamieson. 1982. Pattern of cortisol release in sheep following administration of synthetic ACTH or imposition of various stressor agents. Aust. J. Biol. Sci. 35:215–222.

Fuschleberger, H. 1939. Das Gamsbuch, Naturgeschichte, Hege und Jagd des Gams und etwas von seiner Umwelt. F.C. Mayer, München.

Gadow, H. 1902. The evolution of horns and antlers. Proc. zool. Soc. Lond. I:206–222.

Gairdner, K.G. 1914. Note on two rare mammals, Berdmore's Rat (*Hapalomys longicaudatus*) and Fea's Muntjac (*Cervulus feae*). J. Siam. Soc., Nat. Hist. Suppl. 1:115–116.

Gairdner, K.G. 1915. Additions to the mammalian fauna of Ratburi. J. Nat. Hist. Soc. Siam. 1:252–255.

Garrod, A.H. 1877. Notes on the visceral anatomy and osteology of the Ruminants. Proc. Zool. Soc. London 1:2–18.

Garten, C.T., Jr. 1976. Relationships between aggressive behavior and genic heterozygosity in the oldfield mouse *Peromyscus polionotus*. Evolution 30:59–72.

Garton, D.W., R.K. Koehn & T.M. Scott. 1984. Multiple-locus heterozygosity and the physiological energetics of growth in the coot clan, *Mulinia lateralis*, from a natural population. Genetics 108:445–455.

Gates, C.C. & R.J. Hudson. 1978. Energy costs of locomotion in wapiti. Acta Theriol. 23:365–370.

Geist, V. 1966a. The evolution of horn-like organs. Behaviour 27:175–214.

Geist, V. 1966b. Horn-like structures as rank symbols, guards and weapons. Nature 220:813–814.

Geist, V. 1968. Horn-like structures as rank symbols, guards and weapons. Nature 220:813–814.

Geist, V. 1971. Mountain Sheep. A Study in Behavior and Evolution. The University of Chicago Press, Chicago-London.

Geist, V. 1974a. On fighting strategies in animal combat. Nature 250:354–356.

Geist, V. 1974b. The relation of social evolution and dispersal in ungulates, with emphasis on the Old World deer and the genus Bison. Quartern. Res. 1:285–315.

Geist, V. 1978a. On weapons, combat, and ecology, pp. 1–30. *In*: L. Kramer, P. Pliner & T. Alloway (eds.), Aggression, Dominance and Individual Spacing. Plenum Publ. Co., New York, NY.

Geist, V. 1978b. Life Strategies, Human Evolution, Environmental Design. Springer-Verlag, New York, NY.

Geist, V. 1982. Adaptive behavioral strategies, pp. 219–277. *In*: J.W. Thomas & D.E. Toweill (eds.), Elk of North America, Ecology and Management. Stackpole Books, Harrisburg.

Geist, V. 1983. On the evolution of Ice Age mammals and its significance to an understanding of speciation. ASB Bull. 30:109–133.

Geist, V. 1985. On Pleistocene bighorn sheep: some problems of adaptation, and relevance to today's American megafauna. Wild. Soc. Bull. 13:351–359.

Geist, V. 1986a. Super antlers and pre-World War II European research. Wildl. Soc. Bull. 14:81–84.

Geist, V. 1986b. On speciation in ice age mammals, with special reference to cervids and caprids. Can. J. Zool. 65:1067–1084.

Geist, V. 1986c. New evidence of high frequency of antler wounding in cervids. Can. J. Zool. 64:380–384.

Geist, V. 1987a. On the evolution of optical signals in deer. A preliminary analysis, pp. 235–255. *In*: C.M. Wemmer (ed.), Biology and Management of the Cervidae. Smithsonian Institution Press, Washington, DC.

Geist, V. 1987b. On the evolution and adaptation of *Alces*. Swed. Vildl. Res. Viltrevy, Suppl. 1, Part 1:11–24.

Geist, V. & P.T. Bromley. 1978. Why deer shed antlers. Z. Säugetierkd. 43:223–231.

Gentry, A.W. 1970. The bovidae of the Fort Ternan fossil fauna, pp. 243–323. *In*: L.S.B. Leakey & R.J.G. Savage (eds.), Fossil Vertebrates of Africa, Vol. 2. Academic Press, London.

Gentry, A.W. 1978a. The fossil Bovidae of the Baringo area, Kenya, pp. 293–308. *In*: W.W. Bishop (ed.) Geological background to fossil man. Geol. Soc. Spec. Publ. no. 6. Scottish Academic Press, Edinburgh.

Gentry, A.W. 1978b. Bovidae, pp. 540–572. *In*: V.J. Maglio & H.B.S. Cook (eds.), Evolution of African Mammals. Harvard University Press, Cambridge, MA.

Gentry, A.W. 1980. Fossil Bovidae from Langebaanweg, South Africa. Ann. S. Afr. Mus., Cape Town 79:213–337.

Gentry, A.W. 1981. Notes on Bovidae from the Hadar Formation, Ethiopia. Kirtlandia, Cleveland, OH 33:1–30.

Gentry, A.W. 1987. Pliocene Bovidae from Laetoli, pp. 378–408. *In*: M.D. Leakey & J.M. Harris (eds.), The Pliocene Site of Laetoli, Northern Tanzania. Clarendon Press, London.

Gentry, A.W. & A. Gentry. 1978. Fossil Bovidae of Olduval Gorge, Tanzania. Bull. Br. Mus. Nat. Hist. (Geol.), London 29:289–445; 30:1–83.

George, A.N. 1956. An important stage in the development of the os cornu (bone core) of the horns of the sheep. Am. J. Vet. Res. 582–587.

Geraads, D. 1981. Bovidae et Giraffidae du Pleistocène de Ternifine (Algérie). Bull. Mus. Natn. Hist. Nat. Paris 4, 3, C:47–86.

Geraads, D. 1986. Remarques sur la systématique et la phylogénie des Giraffidae (Artiodactyla, Mammalia). Geobios (Paléontologie, Stratigraphie, Paléoécologie), No. 19, Fasc. 4:465–477. Lyon.

Gibson, R.M. & F.E. Guinness. 1980. Differential reproduction among red deer (*Cervus elaphus*) stags on Rhum. J. Anim. Ecol. 49:199–208.

Gilbert, B.K. 1973. Scent marking and territoriality in pronghorn (*Antilocapra americana*) in Yellowstone National Park. Mammalia 37:25–33.

Gimenez, T., D. Barth, B. Hoffmann & H. Karg. 1975. Blook levels of testosterone in the roe deer (*Capreolus capreolus*) in relationship to the season. Acta Endocr. Suppl. 193:59.

Gingerich, P.D. 1977. Patterns of evolution in the mammalian fossil record, pp. 469–500. *In*: A. Hallam (ed.), Patterns of Evolution. Elsevier, Amsterdam.

Gingerich, P.D. 1981. Variation, sexual dimorphism and social structure in the early Eocene horse, *Hyracotherium* (Mammalia, Perissodactyla). Paleobiology 7:443–455.

Gingerich, P.D. 1984. Pleistocene extinctions in the context of orignation-extinction equilibria in Cenzoic mammals, pp. 211–222. *In*: P.S. Martin & R.G. Klein (eds.), Quaternary Extinctions. University of Arizona Press, Tucson, AR.

Ginsburg, L. 1963. Les mammifères fossiles recoltés à Sansan au cours du XIX siècle. Bull. Soc. Geol. France 5:3–15.

Ginsburg, L. 1968. L'évolution du climat au cours du Miocène en France. Bull. l'Assoc. d. Natural. Orléans et de la Loire Moyenne. Nouv. S. XLI–Geologie.

Ginsburg, L. & E. Heintz. 1966. Paléontologie–Sur les affinités du genre Palaeomeryx (Ruminant du Miocène Européen). C.R. Acad. Paris 262:979–982.

Ginsburg, L. & F. Crouzel. 1976. Contribution à la connaissance d'heteroprox larteti (Filhol) cervidé du Miocène européen. Bull. Mus. Nat. d'Hist. Natur. 58:345–357.

Goldschmidt, R. 1940. The Material Basis of Evolution. Yale University Press, New Haven, CO.

Gorbman, A., W.W. Dickhoff, S.R. Vigna & C.L. Ralph. 1983. Comparative Endocrinology, John Wiley and Sons, Toronto.

Goldman, E.A. 1945. A new pronghorn antelope from Sonora. Proc. Biol. 58:3–4.

Goss, R.J. 1961. Experimental investigations of morphogenesis in the growing antler. J. Embryol. Exp. Morph. 9:342–354.

Goss, R.J. 1964a. The role of skin in antler regeneration, pp. 194–207. *In*: W. Montagna & R.E. Billingham (eds.), Advances in Biology of Skin Wound Healing. Pergamon Press, New York, NY.

Goss, R.J. 1965a. Mammalian regeneration and its phylogenetic relationships, pp. 33–38. *In*: V. Kiortsis & H.A.L. Trampusch (eds.), Regeneration in Animals and Related Problems. North Holland Publ. Co., Amsterdam.

Goss, R.J. 1965b. The functional demand theory of growth regulation, pp. 444–451. *In*: V. Kiortsis & H.A.L. Trampusch (eds.), Regeneration in Animals and Related Problems. North Holland Publ. Co., Amsterdam.

Goss, R.J. 1968. Inhibition of growth and shedding of antlers by sex hormones. Nature 220:83–85.

Goss, R.J. 1969a. Photoperiodic control of antler cycles in deer. I: Phase shift and frequency changes. J. Exp. Zool. 170:311–324.

Goss, R.J. 1969b. Photoperiodic control of antler cycles in deer. II: Alterations in amplitude. J. Exp. Zool. 171:223–234.

Goss, R.J. 1969c. Principles of Regeneration. Academic Press, New York, NY.

Goss, R.J. 1970. Problems of antlerogenesis. Clin. Orthopaed. 69:227–238.

Goss, R.J. 1972. Wound healing and antler regeneration, pp. 219–228. *In*: H.I. Maibach & D.T. Rovee (eds.), Epidermal Wound Healing. Yearbook Medical Publishers, Chicago, IL.

Goss, R.J. 1976. Photoperiod control of antler cycles in deer. III: Decreasing versus increasing day-lengths. J. Exp. Zool. 197:307–320.

Goss, R.J. 1980a. Prospects for regeneration in man. Clin. Orthop. Relat. Res. 151:270–282.

Goss, R.J. 1980b. Is antler assymetry in reindeer and caribou genetically determined? pp. 364–372. *In*: E. Reimers, E. Garre & S. Skjenneberg (eds.), Proc. 2nd Int. Reindeer/Caribou Symp., Roros, Norway, 1979.

Goss, R.J. 1983. Deer Antlers. Regeneration, Function and Evolution. Academic Press, New York, NY.

Goss, R.J. 1984. Epimorphic regeneration in mammals, pp. 554–573. *In*: T.K. Hunt, R.B. Heppenstall, E. Pines & D. Rovee (eds.), Soft and Hard Tissue Repair. Surgical Sci. Series, Vol. 2. Praeger Scientific, New York, NY.

Goss, R.J. 1985. Tissue differentiation in regenerating antlers, pp. 229–238. *In*: P.F. Fennessy & K.R. Drew (eds.), Biology of Deer Production. Royal Soc. of New Zealand Bull. 22.

Goss, R.J. 1987. Induction of deer antlers by transplanted periosteum. II. Regional competence for velvet transformation in ectopic skin. J. Exp. Zool. 244:101–111.

Goss, R.J. & J.K. Rosen. 1973. The effect of latitude and photoperiod on the growth of antlers. J. Reprod. Fert. Suppl. 19:111–118.

Goss, R.J. & R.S. Powel. 1985. Induction of deer antlers by transplanted periosteum 1. Graft size and shape. J. Exp. Zool. 235:359–373.

Goss, R.J., C.W. Sveringhaus & S. Free. 1964. Tissue relationship in the development of pedicles and antlers in the Virginia deer. J. Mammal. 45:61–68.

Gould, S.J. 1974. The origin and function of "bizarre" structures: Antler size and skull size in the "Irish Elk", *Megaloceros giganteus*. Evolution 28:191–220.

Gould, S.J. 1977. Ontogeny and Phylogeny. Harvard University Press, Cambridge, MA.

Gould, S.J. & E.C. Levontin. 1979. The spandrels of San Macro and the Panglossian paradigm: A critique of the adaptationist programme. Proc. R. Soc. London Ser. B 205:581–598.

Gould, S.J. & E.S. Vrba. 1982. Exaptation—a missing term in the science of form. Paleobiology 8:4–15.

Gower, D.B. 1975. Regulation of steroidogenesis, pp. 127–147. *In*: H.L.J. Makin (ed.), Biochemistry of Steroid Hormones. Blackwell Scientific Publications, Osney Mead, Oxford.

Grafflin, A.L. 1943. Further observations upon the parathyroid gland in the Virginia deer specimens taken throughout the year. Endocrinology 30:571–580.

Gray, A.P. 1954. Mammalian hybrids. A check-list with bibliography. Technic. Comm. No. 10. Commonwealth Bureau of Animal Breeding and Genetics. Farnham Royal Bucks, England.

Gray, J.E. 1821. On the natural arrangement of vertebrose animals. London Med. Reposit., Vol. 15, Part 1:296–310.

Green, M.J.B. 1985. Aspects of the ecology of the Himalayan musk deer. Ph.D. Thesis, University of Cambridge, Cambridge, MA.

Greer, K.R. 1968. Special collections–Yellowstone elk study, 1967–68. Job Compl. Rep., Fed. Aid Proj. No. W-83-R-11. Montana Fish and Game Dept.

Gregg, H.A. 1955. Summer habits of Wyoming antelope. Ph.D. Thesis, Cornell University, Ithaca, NY.

Groves, C.P. 1974. A note on the systematic position of the muntjac (Artiodactyla, Cervidae). Z. Säugetierkunde 39:369–372.

Groves, C.P. 1989. A Theory of Human and Primate Evolution. Oxford University Press.

Groves, C.P. & P. Grubb. 1982. The species of muntjac (genus Muntiacus) in Borneo: unrecognised sympatry in tropical deer. Zool. Meded. Leiden 56:203–216.

Groves, C.P. & P. Grubb. 1987. Relationships of living Cervidae, pp. 21–59. *In*: C. Wemmer (ed.), Biology and Management of Cervidae. Smithsonian Institution Press, Washington, DC.

Grubb, P. 1977. Notes on a rare deer, *Muntiacus feae*. Ann. Mus. Civ. Stor. Nat. Genoa 81:202–207.

Grubb, P. 1978. Patterns of speciation in African mammals. Bull. Carnegie Mus. nat. Hist., Pittsburgh 6:152–167.

Grzimek, B. 1968. Grzimek's Tierleben, Bd. 1–13: Säugetiere 4. Kindler Verlag, München.

Hadjouis, D. 1985. Les bovidés du gisement Atérien des phacocheères (Alger.). Ph.D. Thesis, Univ. Pierre et Marie Curie, Paris.

Hafez, E.S.E. & M.R. Jainudeen. 1966. Intersexuality in farm mammals. Anim. Breed. Abstr. 34:1–15.

Haigh, J.C. 1982. Reproductive seasonality of male wapiti. M.S. Thesis, University of Saskatchewan, Saskatoon, Sask., Canada.

Haigh, J.C., W.F. Cates, G.J. Iover & N.C. Rawlings. 1984. Relationships between seasonal changes in serum testosterone concentrations, scrotal circumference and sperm morphology of male wapiti (*Cervus elaphus*). J. Reprod. Fert. 70:413–418.

Hall, B.K. 1978. C. Epigenetic control. I. Hormones–The growth of antlers, pp. 213–215. *In*: B. Hall (ed.), Developmental and Cellular Skeletal Biology, Academic Press, New York, NY.

Hall, T.C., W.F. Ganong, E.B. Taft & J.C. Aub. 1960. Endocrine control of deer antler growth. Acta Endocrinologica, Suppl. 51:525.

Hall, T.C., W.F. Ganong & E.B. Taft. 1966. Hypophysectomy in the Virginia deer; technique and physiologic consequences. Growth 30:383–392.

Hamilton, A.C. 1982. Environmental history of east Africa. Academic Press, London.

Hamilton, W.J., R.J. Harrison & B.A. Young. 1960. Aspects of placentation in certain Cervidae. J. Anat. 94:1–33.

Hamilton, W.R. 1973. On the lower Miocene ruminants of Gebel Zelten, Libya. Bull. Br. Mus. nat. Hist. (Geol.) 21:73–150.

Hamilton, W.R. 1978a. Fossil giraffes from the Miocene of Africa and a revision of the phylogeny of the Giraffoidea. Phil. Trans. Roy. Soc. London, Ser. B 283:165–229.

Hamilton, W.R. 1978b. *Cervidae* and *Palaeomerycidae*, Chapter 24, pp. 496–508. *In*: V.J. Maglio & H.B.S. Cooke (eds.), Evolution of African Mammals. Harvard University Press, Cambridge, MA.

Harding, C.F. 1981. Social modulation of circulating hormone levels in the male. Amer. Zool. 21:223–231.

Hardy, A.C. 1965. The Living Stream. Collins, London.

Harmel, D.E. 1977. Antler formation in white-tailed deer. Job Perf. Rept., Job No. 20, Fed. Aid Proj. No. W-76-R-20. Texas Parks and Wildl. Dept., Austin, TX.

Harmel, D.E. 1978. Antler formation in white-tailed deer. Job Perf. Rept., Fed. Aid Proj. No. W-109-R-1, Texas Parks and Wildl. Dept., Austin, TX.

Harmel, D.E. 1979. Antler formation in white-tailed deer. Job. Perf. Rept., Fed. Aid Proj. No. W-109-R-2, Texas Parks and Wildl. Dept., Austin, TX.

Harmel, D.E. 1982. Effects of genetics on antler quality and body size in white-tailed deer, pp. 339–348. *In*: R.D. Brown (ed.), Antler Development in Cervidae. Caesar Kleberg Wildl. Res. Inst., Kingsville, TX.

Harrington, R. 1985. Evolution and distribution of the Cervidae, pp. 3–12. *In*: P.F. Fennessy & K.R. Drew (eds.), Biology of Deer Production. Royal Soc. New Zealand, Bull. 22.

Harris, J.M. 1976. Pleistocene giraffidae (mammalia, artiodactyla) from East Rudolf, Kenya. *In*: R.J.G. Savage (ed.), Fossil Vertebrates of Africa, Vol. 4, pp. 283–332. Academic Press, London.

Harris, J.M. 1978. Palaeontology, pp. 32–63. *In*: M.G. Leakey & R.E.F. Leakey (eds.), Koobi Fora Research Project. Clarendon Press, London.

Hartwig, H. 1967. Experimentelle Untersuchungen zur Entwicklungsphysiologie der Stangenbildung beim Reh (*Capreolus C. Capreolus L.* 1758). Roux Archiv für Entwicklungsmechanik 158:358–384.

Hartwig, H. 1968. Durch Periostverlagerung experimentell erzeugte, heterotope Stirnzapfenbildung beim Reh. Z. Säugetierkd. 33:246–248.

Hartwig, H. 1969. Versuch zur Analyse der Entwicklunsbedingungen, die zu einer Doppelstangen-Bildung beim Reh führten. Z. Jagdwiss. 15:167–169.

Hartwig, H. 1972. "Fegeverhalten" bei einem gehörnlosen Rehbock. Z. Jagdwiss. 18: 166–168.

Hartwig, H. 1979. Komplizierte Gehörnrestitution nach Rosenstockfraktur. Z. Jagdwiss. 25:4–8.

Hartwig, H. 1985. Der einseitig gehörnte Braunschweiger Rickenschädel und das "Dornröschen"-Prinzip. Niedersächsischer Jäger 6:300–303.

Hartwig, H. & J. Schrudde. 1974. Experimentelle Untersuchungen zur Bildung der primären Stirnauswüchse beim Reh (*Capreolus capreolus L.*). Z. Jagdwiss. 20:1–13.

Hartwig, H., J. Schrudde, H. Pade & E. Ueckermann. 1968. Verhinderung der Rosenstock-und Stangenbildung beim Reh, *Capreolus capreolus*, durch Periostausschaltung. Zool. Garten 35:252–255.

Haugen, A.O. & E.W. Mustard, Jr. 1960. Velvet-antlered pregnant white-tailed doe. J. Mammal 41:521–523.

Hediger, H. 1976. Proper names in the animal kingdom. Experientia 32:1357–1364.

Heimer, W.E. 1987. Publication of Dall sheep findings and development of future research direction. Alaska Dept. Fish and Game, Div. Game. Project W-22-5. Mimeo, 33 pp.

Heintz, E. 1969. Le dimorphisme sexuel des appendices frontaux chez *Gazella deperdita* Gervais (Bovidae, Artiodactyla, Mammalia) et sa signification phylogénique. Mammalia 33:626–629.

Heintz, E. 1970. Les Cervidés Villafranchiens de France et d'Espagne. Mémoires Mus. Nat. d'Hist. Natur. Nouv. Serie C. Tom. 22, Vols. I and II.

Heintz, E. & M. Brunet. 1982. Rôle de la Thetys et de la chaîne alpine asiatique dans la distribution spatio-temporelle des Cervidés. (Role of the Thetys and of the Asiab Alpine Range in the space-time distribution of Cervids). C.R. Acad. Sci. Paris 294: 1095–1098.

Hendey, Q.B. 1968. The Milkbos site . . . in the south-western Cape Province. Ann. S. Afr. Mus., Cape Town 52:89–119.

Hendey, Q.B. 1978. Preliminary report on the Miocene vertebrates from Arrisdrift, South West Africa. Ann. S. Afr. Mus., Cape Town 76:1–41.

Henshaw, J. 1969. Antlers—the bones of contention. Nature 224:1036–1037.

Hepworth, W. & F. Blunt. 1966. *In*: Research findings on Wyoming antelope, pp. 24–29. Wyoming Wildlife Special Antelope Issue.

Hershberger, T.V. and C.T. Cushwa. 1984. The effects of fasting and refeeding white-tailed does. Bull. 846, The Penn State Univ. Ag. Extp. Stn., University Park, PA.

Hershkowitz, P. 1969. The evolution of mammals on Southern Continents. VI. Recent mammals of the Neotropical region: A zoogeographic and ecological review. Quart. Rev. Biol. 44:1-70.
Hershkowitz, P. 1982. Neotropical deer (*Cervidae*), Part I. Pudus, Genus *Pudu* Gray. Fieldiana Zool. New Ser. 11.
Herzog, A. 1986. Biochemisch-genetische Untersuchungen an Rotwild (*Cervus elaphus* L.), pp. 432-434. *In*: H. Reuss (ed.), Das Rotwild—Cerf rouge—Red Deer. CIC Symp. Graz.
Heude, P.-M. 1884. Catalogue des cerfs tachetés (sikas) du Musée de Zi-ka-wei, ou notes préparatoires à la monographie de ce group. Mémoires concernant l'Histoire Naturelle de l'Empire Chinois 1:1-12.
Heude, P.-M. 1894. Catalogue revisé des cerfs tachetés (sikas) de la Chine centrale. Mémoires concernant l'Histoire Naturelle de l'Empire Chinois. 2:146-163.
Heumann, R. 1987. Regulation of the synthesis of nerve growth factor. J. Exp. Biol. 132:133-150.
Hillis, D.M. 1987. Molecular versus morphological approaches to systematics. Ann. Rev. Ecol. Syst. 18:23-42.
Hillman, J.R., R.W. Davis & Y.Z. Abdelbaki. 1973. Cyclic bone remodeling in deer. Calc. Tiss. Res. 12:323-330.
Hilzheimer, M. 1905. Eine kleine Sendung chinesischer Säugetiere. Abh. Mus. Naturk. Magdeburg 1:165-184.
Hilzheimer, M. 1922. Ueber die Systematik einiger fossilen Cerviden. Centralbl. f. miner. Geol. u. Palaentol. 19227:712-749.
Hirth, D. 1977. Observations on loss of antler velvet in white-tailed deer. Southwestern Nat. 22:269-296.
Hirth, D.H. 1977. Social behavior of white-tailed deer in relation to habitat. Wildlife Monogr. No. 53:5-55.
Ho, M.W. 1988. How rational can rational morphology be? A post-Darwinian rational taxonomy based on a structuralism of process. Biology Forum 81(1):11-55.
Ho, M.W. & P.T. Saunders. 1980. Adaptation and natural selection, and mechanism and teleology. Sympos. The Dialectics of Biology and Society in the Production of Mind. Bresannone, Padua Univ. pp. 85-102.
Hoffman, R.A. & P.F. Robinson. 1966. Changes in some endocrine glands of white-tailed deer as affected by season, sex and age. J. Mammol. 47:266-280.
Hoffmann, H. 1956. Zur Verästelung der Edelhirschgeweihe. Deutsche Jägerztg. 20:418-419.
Hoffstetter, R. 1952. Les mammifères Pleistocènes de la République de l'Equateur. Mémoires Soc. Geol. de France 66:1-391.
Hofman, R. 1978. Die Verdauugsorgane des Rehes und ihre Anpassung an die besondere Ernährungsweise, pp. 101-112. *In*: Wildbiologische Information für den Jäger, Bd. I. Enke-Verlag, Stuttgart.
Hofmann, A. 1893. Die Fauna von Göriach. Abhandl. d.k.k. Geol. Reichsanstalt 15:63-72.
Hofmann, C. 1901. Zur Morphologie der Geweihe der rezenten Hirsche. P. Schettlers Erben, Coethen.
Hofmann, R.R. 1985. Digestive psychology of the deer—their morphophysiological specialization and adaptation, pp. 393-407. *In*: P.F. Fennessy & K.R. Drew (eds.), Biology of Deer Production. Royal Soc. of New Zealand Bull. 22.
Holding, R.E. 1896. Bemerkungen über Spaltung der rechten Geweihstange von *Cervus dama*. Proc. Zool. Soc. London. p. 855.

Holter, J.B., W.E. Urban, Jr. & H.H. Hayes. 1977. Nutrition of northern white-tailed deer throughout the year. J. Anim. Sci. 45:365–376.

Holter, J.B., W.E. Urban, Jr. & H.H. Hayes. 1979a. Predicting energy and nitrogen retention in young white-tailed deer. J. Wildl. Manage. 43:880–888.

Holter, J.B., H.H. Hayes & S.H. Smith. 1979b. Protein requirement of yearling white-tailed deer. J. Wildl. Manage. 43:872–879.

Hoogerwerf, A. 1970. Udjung Kulon. The land of the last Javan rhinoceros. E.J. Brill, Leiden.

Hoover, R.L., C.E. Till & S. Ogilvie. 1959. The antelope of Colorado. Colo. Game and Fish Dep., Denver, Tech. Bull. 4.

Hopkins, D.M. 1959. Some characters of the climate in forest and tundra regions in Alaska. Arctic 12:215–220.

Hopkins, D.M., J.V. Matthews, C.E. Schweger & S.B. Young. 1982. Paleoecology of Beringia. Academic Press, New York, NY.

Hove, K. & J.B. Steen. 1978. Blood flow, calcium deposition and heat loss in reindeer antlers. Acta Physiol. Scand. 104:122–128.

Hsu, T.C. & K. Benirschke. 1971. *Elaphurus davidianus*. An atlas of mammalian chromosomes. 5:2411.

Hudson, R.J. & R.G. White. 1985. Bioenergetics of wild herbivores. CRC Press, Boca Raton, FL.

Hudson, R.J., W.G. Watkins & R.W. Pauls. 1985. Seasonal bioenergetics of wapiti in western Canada, pp. 447–452. *In*: P.F. Fennessy & K.R. Drew (eds.), Biology of Deer Production. Royal Soc. of New Zealand.

Hughes, E. & R. Mall. 1958. Relation of the adrenal cortex to condition of deer. Calif. Fish Game 44:191–196.

Hutt, F.B. 1964. Animal Genetics, p. 546. Ronald Press, New York.

Hutton, D.A. 1972. Variations in skull and antlers of wapiti. M.Sc. Thesis, University of Calgary, Calgary, Alberta, Canada.

Huxley, J.S. 1926. The annual increment of the antlers of the red deer (*Cervus elaphus*). Proc. Zool. Soc. Lond. 67:1021–1036.

Huxley, J.S. 1931. The relative size of antlers in deer. Proc. Zool. Soc. Lond. 19:819–863.

Huxley, J.S. 1932. Problems of relative growth. Methuen, London.

Hyvarinen, H., R.N.B. Kay & W.J. Hamilton. 1977. Variation in weight, specific gravity and composition of antlers of red deer (*Cervus elaphus*). Br. J. Nutr. 38:301–311.

Ingold, D.A. 1969. Social organization and behavior in a Wyoming pronghorn population. Ph.D. Thesis, University of Wyoming, Laramie, WY.

Innis, A. 1958. The behaviour of the giraffe in the Eastern Transvaal. Proc. Zool. Soc. Lond. 131:245–278.

Jackson, J. 1985. Behavioural observations on the Argentinian pampas deer (*Ozotoceros bezoarticus* Cabrera, 1943). Z. Säugetierkd. 50:107–116.

Jackson, L. 1977. To see or not to see. The special world of equine vision. Equus 16: 24–31.

Jacobson, H.A. 1984. Investigation of phosphorus in the nutritional ecology of white-tailed deer. Prog. Rept., Fed. Aid in Wildl. Restoration Proj. W-48-31, Study XXIII.

Jacobson, H.A. & R.N. Griffin. 1982. Antler cycles of white-tailed deer in Mississippi, pp. 15–22. *In*: R.D. Brown (ed.), Antler Development in Cervidae. Caesar Kleberg Wildl. Res. Inst. Kingsville, TX.

Jaczewski, Z. 1954a. The effect of altered daylight on the growth of antlers in the deer (*Cervus elaphus* L.). Folia Biologica II:133–143.

Jaczewski, Z. 1954b. Regeneracja rogow u losia (*Alces alces L.*). Kosmos. 4:460.

Jaczewski, Z. 1955. Regeneration of antlers in red deer (*Cervus elaphus L.*). Bull. Acad. Polon. Sci., Cl. II, 3:273–278.

Jaczewski, Z. 1956a. Free transplantation of antler in red deer (*Cervus elaphus L.*). Bull. Acad. Polon. Sci., Cl. II, 4:107–110.

Jaczewski, Z. 1956b. Further observations on transplantation of antler in red deer (*Cervus elaphus L.*). Bull. Acad. Polon. Sci., Cl. II, 4:289–291.

Jaczewski, Z. 1958. Free transplantation and regeneration of antlers in fallow deer (*Cervus dama L.*). Bull. Acad. Polon. Sci., Cl. IV, 6:179–182.

Jaczewski, Z. 1961. Observations on the regeneration and transplantation of antlers in deer, Cervidae. Fol. Biol. 9:47–99.

Jaczewski, Z. 1967. Regeneration and transplantation of antlers in deer, *Cervidae*. Z. Säugetierkd. 32:215–233.

Jaczewski, Z. 1976. The induction of antler growth in female red deer. Bull. de l'Académie Polonaise des Sciences. 21:61–65.

Jaczewski, Z. 1977. The artificial induction of antler cycles in female red deer. Deer. 4:83–86.

Jaczewski, Z. 1981a. Poroze jeleniowatych (Deer Antlers). Panstwowe Wydawnictwo Rolnicże i Lesne, Warsaw. (In Polish.)

Jaczewski, Z. 1981b. Further observations on the induction of antler growth in red deer females. Folia Biol. 29:131–140.

Jaczewski, Z. 1982. The artificial induction of antler growth in deer, pp. 143–162. *In*: R.D. Brown (ed.), Antler Development in Cervidae. Caesar Kleberg Wildl. Res. Inst., Kingsville, TX.

Jaczewski, Z. 1985. Hormonal regulation of antler casting in red deer. Fortschritte der Zoologie 30:167–171.

Jaczewski, Z. & B. Galka. 1967. Effect of administration of testosterone propionicum on the antler cycle in red deer. Finnish Game Res. 30:303–308.

Jaczewski, Z. & B. Galka. 1970. Effect of human chorionic gonadotrophin on the antler cycle in red deer, pp. 217–218. Transactions of the 9th International Congress of Game Biologists, Moscow.

Jaczewski, Z. & T. Jasiorowski. 1974. Observations on the electro-ejaculation in red deer. Acta Theriol. 19:143–151.

Jaczewski, Z. & K. Krzywinska. 1974. The induction of antler growth in a red deer male castrated before puberty by traumatization of the pedicle. Bull. Acad. Polon. Sci., Cl. V, 32:67–72.

Jaczewski, Z. & K. Krzywinski. 1975. The effect of testosterone on the behavior of castrated females of red deer (*Cervus elaphus L.*). Pr. Mater. Zool. 8:37–45.

Jaczewski, Z., L. Zaniewski & W. Zurowski. 1965. Observations on the circulation in the pedicle arteries of red deer (*Cervus elaphus L.*). Trans. VIth. Congr. Intern. Union Game Biol. Bournemouth 1963, London. pp. 145–155.

Jaczewski, Z., T. Doboszynska & A. Krzywinski. 1976. The induction of antler growth by amputation of the pedicle in red deer (*Cervus elaphus L.*) males castrated before puberty. Fol. Biol. 24:299–307.

Janis, C.M. 1976. The evolutionary strategy of the Equidae, and the origins of rumen and cecal fermentation. Evolution 30:757–774.

Janis, C.M. 1979. Aspects of the evolution of herbivory in ungulate mammals. Unpublished Ph.D. Thesis, Harvard University, Cambridge, MA.

Janis, C.M. 1982. Evolution of horns in ungulates: ecology and paleoecology. Biol. Rev. 57:261–318.

Janis, C.M. 1984. The use of fossil ungulate communities as indicators of climate and environment, pp. 85–104. *In*: P. Brenchley (ed.), Fossils and Climate. John Wiley and Son, London.

Janis, C.M. 1987. Grades and clades in hornless ruminant evolution: the reality of the Gelocidae and the systematic position of *Lophiomeryx* and *Bachitherium*. J. Vertebr. Paleo. 7:200–216.

Janis, C.M. 1988. An estimation of tooth volume and hypsodonty indices in ungulate mammals, and the correlation of these factors with dietary preferences, pp. 367–387. *In*: D.E. Russell, J.-P. Santoro & D. Sigogneau-Russell (eds.), Teeth Revised: Proceedings of the VIIth. International Symposium on Dental Morphology, Paris 1986. Mem. Mus. Natl. Hist. Paris (C) 53.

Janis, C.M. (in press a). Correlation of cranial and dental variables with dietary preferences: a comparison of macropodoid and ungulate mammals. *In*: S. Turner, R.A. Thulborn & R. Molnar (eds.), Problems in Vertebrate Biology and Phylogeny, and Australian Perspective. Mem. Qld. Mus. Special Pub.

Janis, C.M. (in press b). Correlation of cranial and dental variables with body size in ungulates and macropodoids. *In*: J. Damuth & B.J. MacFadden (eds.), Body Size Estimation in Mammalian Paleobiology. Cambridge University Press, Cambridge, MA.

Janis, C.M. & D. Ehrhardt. 1988. Correlation of relative muzzle width and relative incisor width with dietary preferences in ungulates. Zool. J. Linn. Soc. 92:267–284.

Janis, C.M. & A.M. Lister. 1985. The morphology of the lower fourth premolar as a taxonomic character in the Ruminantia (Mammalia, Artiodactyla) and the systematic position of *riceromeryx*. J. Paleontol. 59:405–410.

Janis, C.M. & K.M. Scott. 1987. The interrelationships of higher ruminant families, with special emphasis on the members of the Cervoidea. Am. Mus. Novit. 2893:1–85.

Jarman, P.J. 1974. The social organization of antelope in relation to their ecology. Behaviour 48:213–267.

Jarman, P.J. 1983. Mating systems and sexual dimorphism in large terrestrial mammalian herbivores. Biol. Rev. 58:485–520.

Jehenne, Y. 1986. Les ruminants primitifs du Paléogène et du Néogène inférieur de l'Ancien Monde: Systématique, phylogénie, biostratigraphie. Ph.D. Thesis, University of Poitiers.

Johns, P.E., M.H. Smith, E.G. Cothran & R.K. Chesser. 1982. Fat levels in male white-tailed deer during the breeding season, pp. 454–462. *In*: Proc. Ann. Conf. Southeast. Assoc. Fish and Wildl. Agencies, No. 36.

Johns, P.E., M.H. Smith & R.K. Chesser. 1984. Annual cycles of the kidney fet index in a southeastern white-tailed deer herd. J. Wildl. Manage. 48:969–973.

Johnson, E.M. & H.K. Yip. 1985. Central nervous system and peripheral nerve growth factor provide trophic support critical to mature sensory neural survival. Nature 314:751–752.

Johnson, G.L. 1901. Contributions to the comparative anatomy of the mammalian eye, chiefly based on ophtalmoscopic examination. Philosophic. Transact. Royale Soc. London, Series B 194:1–82.

Johnson, J.D. & R.B. Buckland. 1976. Response of male holstein calves from seven sires to four management stresses as measured by plasma corticoid levels. Can. J. Anim. Sci. 56:727–732.

Johnston, R.E. 1983. Chemical signals and reproductive behavior, pp. 3–37. *In*: J.G. Vandenbergh (ed.), Pheromones and Reproduction in Mammals. Academic Press, New York, NY.

Julander, O., L.W. Robinette & D.A. Jones. 1961. Relation of summer range condition to mule deer herd productivity. J. Wildl. Manage. 25:54–60.

Kahlke, H.D. 1951. Der altpleistozäne Verticornis-Kreis und die Frage der Entstehung der Riesenhirsche (*Megaceros*). Hallesches Jhrb. Mitteldeutsche Erdgeschichte 1: 174–179.

Kalb, J.E. et al. 1982. Vertebrate faunas from the Awash Group, Afar, Ethiopia. J. Vertebr. Paleont., Normal, Oklahoma 2:237–258.

Kapherr, G. 1924. Das Hirschgeweih. J. Neumann, Neudam.

Karg, H., D. Schams & V. Reinhardt. 1973. Prolactin secretion in animals, pp. 414–434. *In*: P. Franchimont (ed.), Some Aspects of Hypothalamic Regulation of Endocrine Function. F.K. Schattauer Verlag, Stuttgart.

Kass, E.H., O. Hechter, I.A. Macchi & T.W. Mou. 1954. Changes in patterns of secretion of corticosteroids in rabbits after prolonged treatment with ACTH. Proc. Soc. Exp. Biol. Med. 85:583–587.

Kay, R.F. & H.H. Covert. 1983. Anatomy and behavior of extinct primates, pp. 467–508. *In*: D.J. Chivers, B.A. Wood & A. Bilsborough (eds.), Food Acquisition and Processing in Primates. Plenum Press, New York, NY.

Kay, R.N.B. 1985. Body size, pattern of growth, and efficiency of production in Red deer, pp. 411–421. *In*: P.F. Fennessy & K.R. Drew (eds.), Biology of Deer Production. Royal Soc. of New Zealand Bull. 22.

Kay, R.N.B., W.V. Englehardt & R.G. White. 1980. The digestive physiology of wild ruminants, pp. 743–761. *In*: Y. Rucklebusch & P. Thivend (eds.), Digestive Physiology and Metabolism of Wild Ruminants. AVI Pub. Co., Inc., Westport, CT.

Kay, R.N.B., M. Phillippe, J.M. Suttie & G. Wenham. 1981. The growth and mineralization of antlers. J. Physiol. 332, Abstract 4 B.

Kelch, R.P., M.R. Jenner, R. Weinstein, S. Kaplan & M.M. Grumbach. 1972. Estradiol and testosterone secretion by human, simian, and canine testes, in male with hypogonadism and in male pseudohermaphrodites with the feminizing testes syndrome. J. Clin. Inv. 51, 824–830.

Kelly, R.W., K.P. McNatty, G.M. Moors, D. Ross & M. Gibb. 1982. Plasma concentrations of LH, prolactin, oestradiol and progesterone in female red deer (*Cervus elaphus*) during pregnancy. J. Reprod. Fert. 64:475–483.

Kendrick, K.M. & B.A. Baldwin. 1987. Cells in temporal cortex of conscious sheep can respond preferentially to the sight of faces. Science 236:448–450.

Keqing, C. 1987. Gradual increase of Mi-lu deer *Elaphurus davidianus* in China. Investigatio et Studium Naturae: 1–5.

Khan, I.F., C.H. Biddulph & E.A. D'Abreu. 1936. Horn growth as observed in black buck and nilgai. Bombay Natural Hist. Soc. J. 39:171–173.

Khomenko, J. 1913. La faune méotique du village Taraklia du district de Bendery. 1. Les ancêtres des Cervinae contemporains et fossiles. 2. Giraffinae et Cavicornia. Ann. Geol. et Miner. de la Russie 15:107–143.

Kierdorf, U. 1985. Gehörnte Ricken. WJSC Blatter 61:1–12.

Kiernik, M.E. 1913/1914. Ueber ein Dicrocerus-Geweih aus Polen. Bull. Intern. de l'Acad. Sci. de Cracovie: 449–464.

Kiltie, R.A. 1984. Seasonality, gestation time, and large mammals extinctions, pp. 299–414. *In*: P.S. Martin & R.G. Klein (eds.), Quaternary Extinctions. University of Arizona Press.

Kiltie, R.A. 1985. Evolution and function of horns and horn-like organs in female ungulates. Biol. J. Linn. Soc. 24:299–320.

Kim, W.-S. & D.L. Stocum. 1986. Retinoic acid modifies positional memory in the anteriposterior axis of regenerating axolotl limbs. Dev. Biol. 195:170–179.

Kingdon, J. 1979, 1982. East African Mammals, Vol. 3, B, C, D. Academic Press, New York, NY.

Kirkpatrick, M. 1987. Sexual selection by female choice in polygynous animals. Ann. Rev. Ecol. Syst. 18:43–70.

Kitchen, D.W. 1974. Social behavior and ecology of the pronghorn. Wildl. Monogr. 38:1–96.

Kitchen, D.W. & P.T. Bromley. 1974. Agonistic behavior of territorial pronghorn bucks, pp. 365–381. *In*: V. Geist & D. Walthers (eds.), The Behavior of Ungulates and its Relation to Management, Vol. I. I.U.C.N. Publ. 24, Morges, Switzerland.

Kitchen, D.W. & F. Griep. 1978. Variation in breeding behavior of pronghorn bucks. Ann. Meet. Anim. Behav. Soc., Seattle, WA.

Kitchen, D.W. & B.W. O'Gara. 1982. Pronghorn (*Antilocapra americana*), pp. 960–972. *In*: J.A. Chapman & G.A. Fieldhammer (eds.), Wild Mammals of North America–Biology, Management and Economics. The John Hopkins University Press, Baltimore, MD.

Kitchener, A. 1986. Why do deer lose their antlers. Deer 6:403–406.

Kitchener, A. 1988. An analysis of the forces of fighting of the blackbuck (*Antilope cervicapra*) and the bighorn sheep (*Ovis canadensis*) and the mechanical design of the horns of bovids. J. Zool. Lond. 214:1–20.

Kleiber, M. 1961. The Fire of Life. John Wiley and Sons, New York, NY.

Klein, D.R. 1985. Population ecology: The interaction between deer and their food supply, pp. 13–22. *In*: P.F. Fennessy & K.R. Drew (eds.), Biology of Deer Production. Royal Soc. of New Zealand, Bull. 22.

Klein, R.G. 1980. Environmental and ecological implications of large mammals from Upper Pleistocene and Holocene sites in southern Africa. Ann. S. Afr. Mus., Cape Town 81:223–283.

Klein, R.G. 1984. Mammalian extinctions and Stone Age people in Africa, pp. 553–573. *In*: P.S. Martin & R.G. Klein (eds.), Quaternary Extinctions. University of Arizona Press, Tucson, AZ.

Klingel, H. 1974. Observations on social organization and behavior of African and Asiatic wild asses (*Equus africanas* and *E. hemionus*). Z. Tierpsychol. 44:323–331.

Klingel, H. 1977. Observations on social organization and behavior of African and Asiatic wild asses (*Equus africanus* and *E. hemionus*). Z. Tierpsychol. 44:323–331.

Ko, K.M., T.T. Yip, S.W. Tsao, S.W., K.C. Kong, P. Fennessy, M.C. Belew, & J. Porath. 1986. Epidermal growth factor from deer (*Cervus elaphus*) submaxillary gland and velvet antler. Gen. Comp. Endocrinol. 63:45–51.

Kobrin-Morizi, S. 1984. The co-evolution of cranial morphology and fighting behavior in deer, the *Cervidae*. Ph.D. Thesis, University of Pennsylvania, Philadelphia, PA.

Kocan, A.A., B.L. Glenn, T.R. Thedford, R. Doyle, K. Waldrup, G. Kubat & M.G. Shaw.

1981. Effects of chemical immobilization on hematologic and serum chemical values in captive white-tailed deer. J. Amer. Vet. Med. Assoc. 179:1153–1156.

Koehn, R.K. & P.M. Gaffney. 1984. Genetic heterozygosity and growth rate in *Mytilus edulis*. Mar. Biol. 5:1–9.

Koenigswald, G.H.R. 1933. Beitrag zur Kenntnis der fossilen Wirbeltiere Javas, I. Teil. Wetenschapplijke Mededeelingen 23:57–86.

Koford, C.B. 1957. The vicuna and the puna. Ecol. Monogr. 27:153–219.

Kohler, M. 1987. Boviden des türkischen Miozens (Kanozoikum und Braunkohlen der Türkei. 28). Paleont. Evoluc. Sabadell 21:133–246.

Kolda, J. 1951. Osteologicky atlas. (Atlas of Osteology). Zdravotnicke nakladatelstvi Prague. (In Czech.)

Komura, T. 1926. Transplantation der Hoerner bei Cavicornien, und Enthornung nach einfachster Methode. J. Jap. Soc. Vet. Sci. 5:69–85.

Kong, Y.C. & P.P. But. 1985. Deer–The ultimate medicinal animal (Antler and deer parts in medicine), pp. 311–324. *In*: P.F. Fennessy & K.R. Drew (eds.), Biology of Deer Production. Royal Soc. New Zealand, Bull. 22.

Koulischer, L., J. Tyskens & J. Mortelmans. 1972. Mammalian cytogenetics. VII. The chromosomes of *Cervus canadanesis, Elaphurus davidianus, Cervus nippon* (Temminck) and *Pudu pudu*. Acta Zool. et Pathol. Antverpiensia 56:25–30.

Kowalski, K. 1971. The biostratigraphy and paleoecology of Late Cenozoic mammals of Europe and Asia, pp. 465–476. *In*: K.K. Turekian (ed.), The Late Cenozoic Glacial Ages. Yale University Press, New Haven, CO.

Kozhukchov, M.B. 1973. Itogi 20-letnyey eksperimentalnoy raboty po odomashnivaniyu losya v Petchoroilytcheskom zapovednike, pp. 17–27. *In*: T.B. Sablina (ed.), Odomoshnivanie losya. Nauka, Moskva. (In Russian.)

Kraglievich, L. 1932. Neuvos apuntes para la geologia y paleontologa Uruguays. Anal. Mus. Hist. Natur. de Montevideo II:257–438.

Kretzoi, M. Wichtigere Streufunde in der Wirbeltiersammlung der ungarischen geologischen Anstalt. Magyar allami Foldtani intezet evi jelentese: 426–429.

Krieg, H. 1936a. Luxusbildungen bei den Tieren. Zool. Jahrb. 69:303–318.

Krieg, H. 1936b. Das Reh in biologischer Sicht-Gestalt, jahreszeitlichem Rythmus und Variation. Neumann, Neudam.

Krieg, H. 1937. Zur Frage der Degeneration und Kummermodifikation beim Rehwild. Biolog. Zentraebl. 57:225–228.

Krieg, H. 1944. Der Schädel einer Giraffe. Naturwissenschaften 32:148–156.

Krieg, H. 1948. Zwischen Anden und Atlantik. C. Hanser Verlag, München.

Krimbas, C. 1984. On adaptation, neo-Darwinian tautology, and population fitness, pp. 2–57. *In*: M.K. Hecht, B. Wallace & G.T. Prance (eds.), Evolutionary Biology, Vol. 17. Plenum Press, New York, NY.

Krumbiegel, I. 1965. Gabelungsspuren an Giraffenhörnern. Säugetierkundl. Mitt. 14:107–108.

Krzywiński, A. 1974. Wstepne obserwacje nad traumatyzacja poroza jelenia szlachetnego podczas wzrostu. (Observation of antler traumatization in deer during the growth period.) Biul. V. Zjazdu PTNW, AR-T Olsztyn 2:516. (In Polish.)

Krzywiński, A. 1978. Obserwacje nad sztucznym rozrodem jelenia szlachetnego (*Cervus* elaphus L.). (Observation on artificial breeding of the red deer.) Ph.D. Thesis, Institute of Genetics and Animal Production, Pol. Acad. Sci., Popielno. (In Polish.)

Kuhle, M. 1986. Die Vergletscherung Tibets und die Entstehung von Eiszeiten, pp. 42–54.

In: Spektrum der Wissenschaft, September.

Kuhlman, R.B., R. Rainey & R. O'Neill. 1963. Biochemical investigation of deer antler growth. J. Bone & Joint Surgery 45:345–350.

Kumar, R. 1984. Metabolism of 1,25-Dihydroxyvitamin D_3. Physiol. Rev. 64:478–504.

Kurnosov, K.M. 1962. Interfetal placental connections of the elk in embryonic parabiosis. Dokl. Akad. Nauk SSSR 142:253–256.

Kurt, F. 1968. Das Sozialverhalten des Rehes (*Capreolus capreolus*), pp. 1–102. Verlag P. Parey, Hamburg and Berlin.

Kurt, F. 1978. Socio-ecological organization and aspects of management in South Asian deer, pp. 219–239. *In*: Threatened Deer. Morges, I.U.C.N.

Kurtén, B. 1968. Pleistocene Mammals of Europe. Weidenfeld and Nicholson, London.

Kurtén, B. & E. Anderson. 1980. Pleistocene Mammals of North America. Columbia University Press, New York, NY.

Lacroix, A. & J. Pelletier. 1979. Short-term variations in plasma LH and testosterone in bull calves from birth to one year of age. J. Reprod. Fertil. 55:101–106.

Lake, F.T., R.W. Davis & G.C. Solomon. 1978. The effects of continuous direct current on the growth of the antler. Am. J. Anat. 153:625–630.

Lake, F.T., G.C. Solomon, R.W. Davis, N. Pace & J.B. Morgan. 1979. Bioelectric potentials associated with the growing deer antler. Clin. Orthopaed. and Related Res. 142:237–243.

Lake, F.T., R.W. Davis & G.C. Solomon. 1982. Bioelectric phenomena associated with the developing deer antler, pp. 317–328. *In*: R.D. Brown (ed.), Antler development in Cervidae, Caesar Kleberg Wildl. Res. Inst., Kingsville, TX.

Lamprecht. 1892. Abnormitäten von Hirschgeweihen. Der Weidmann, Blätter für Jäger und Jagdfreunde XXIII 1:3–4.

Lance, E.M. 1985. Some observations on bone graft technology. Clin. Orthopaed. Rel. Res. 200:115–124.

Landauer, W. 1925. Ergebnisse in der Erbanalyse der Behörnung von Rind, Schaf und Ziege. Z. f. inductive Abstammungs-u. Vererbungslehre 39:294–332.

Landois, H. 1904. Eine dritte Edelhirsch-Geweihstange über den mit der Hinterhauptschuppe verwachsenen Zwischenscheitelbeinen. Roux Arch. Entwickl. Mech. d. Organismen 18:290–295.

Lankester, E.R. 1902. On *Okapia*, a new genus of Giraffidae from Central Africa. Trans. Zool. Soc. London, 1903 16:279–307.

Lankester, E.R. 1905. Extinct Animals. Archibald Constable, London.

Lankester, E.R. 1907a. The origin of the lateral hornes of the giraffe in foetal life on the area of the parietal bones. Proc. Zool. Soc. London 1907:100–115.

Lankester, E.R. 1907b. On the existence of rudimentary antlers in the Okapi. Proc. Zool. Soc. London 1907:126–135.

Lankester, E.R. 1910. Monograph of the okapi. Atlas Br. Mus. Trustees.

Lankester, E.R. & W.G. Ridewood. 1910. Monograph of the Okapi. Atlas of 48 Plates. Brit. Museum (Nat. Hist.), London.

Lartet, E. 1839. Notice présentant quelques aperçus géologiques dans le département du Gers. Annuiaire du Département du Gers pour 1839. (Reimprinted 1951).

Lasslo, L.L., G.E. Bradford, D.T. Torell & B.W. Kennedy. 1985. Selection for weaning weight in Targhee sheep in two environments. I. Direct response. J. Anim. Sci. 61:376–386.

Laurie, A. 1982. Behavioural ecology of the greater one-horned rhinoceros (*Rhinoceros unicornis* Linn.). J. Zool. Lond. 196:307–341.

Lavasseur, M.C. 1976. Thoughts on puberty. Initiation of gonadotropic function. Ann. Biol. Anim. Biochem. Biophys. 17:345–361.

Lavocat, R. 1961. In: Choubert, G. & A. Faure-Muret (eds.), Le gisement de vertébrés Miocènes de Beni Mellal. Notes Mém. Serv. Mines Carte géol. Maroc 155:1–122.

Leader-Williams, N. 1978. Age-related changes in the testicular and the antler cycles of reindeer (*Rangifer*). J. Reprod. Fertil. 57:117–126.

Leakey, L.S.B. 1965. Olduvai Gorge 1951–61, Vol. I. Cambridge University Press, Cambridge, MA.

Lebedinsky, N.G. 1939. Beschleunigung der Geweihmorphogenese beim Reh (*Capreolus capreolus* L.) durch das Schilddrüsenhormon. Acta Biol. Latvica. 9:125–134.

Lehman, V.W. & J.B. Davis. 1942. Experimental wildlife management in the south Texas chaparral. P-R Quart. Rep. 1-R C1, Texas Game, Fish and Oyster Comm.

Lehmann, E. 1960. Das Problem der Grössenabnahme (Deminutionstendenz) beim Reh. Z. Jagdwiss. 6:41–51.

Lehmann, U. & H. Thomas. 1987. Fossil Bovidae from the Mio-Pliocene of Sahabi (Libya), pp. 323–335. *In*: N.T. Boaz, A. El-Arnauti, A.W. Gaziry, J. de Heinzelin & D.D. Boaz (eds.), Neogene Paleontology and Geology of Sahabi. Alan R. Liss, New York, NY.

Leinders, J.J.M. 1979. On the osteology and function of the digits of some ruminants and their bearing on taxonomy. Z. Säugetierkd. 44:305–319.

Leinders, J. 1984. Hoplitomerycidae fam. nov. (Ruminantia, Mammalia) from Neogene fissure fillings in Gargano (Italy). Part 1: The cranial osteology of *Hoplitomeryz* gen. nov. and a discussion on the classification of pecoran families. Scripta Geol. 70:1–51.

Leinders, J. & E. Heintz. 1980. The configuration of the lacrimal orifices in Pecorans and Tragulids (Artiodactyla, Mammalia) and its significance for the distinction between *Bovidae* and *Cervidae*. Beaufortia 30:155–162.

Lekagul, B. & J.A. McNeely. 1977. Mammals of Thailand. Association for the Conservation of Wildlife, Bangkok.

Lent, P.C. 1974. Mother-infant relationships in ungulates, pp. 14–15. *In*: V. Geist & F. Walther (eds.), The Behaviour of Ungulates and its Relation to Management. IUCN Publ. No. 24, IUCN Morges.

Leshner, A.I. 1980. The interaction of experience and neuroendocrine factors in determining behavioral adaption to aggression. pp. 427–438. *In*: P.S. McConnell, G.J. Boer, H.J. Romijn, N.E. Vandepoll & M.A. Corner (eds.), Adaptive Capabilities of the Nervous System. Elsevier North-Holland Biomedical Press, Amsterdam.

Leuthold, W. 1977. African Ungulates. A Comparative Review of Their Ethology and Behavioral Ecology. Springer-Verlag, Berlin.

Levine, S. 1985. A definition of stress?, pp. 51–69. *In*: G.P. Moberg (ed.), Animal Stress. Am. Physiol. Soc., Bethesda, MA.

Lewontin, R.C. 1978. Adaptation. Scient. Am. 239:212–230.

Lincoln, G.A. 1971a. The seasonal reproductive changes in the red deer stag (*Cervus elaphus*). J. Zool. (London) 163:105–123.

Lincoln, G.A. 1971b. Puberty in a seasonally breeding male, the red deer stag (*Cervus elaphus*). J. Reprod. Fert. 25:41–54.

Lincoln, G.A. 1972. The role of antlers in the behaviour of red deer. J. Exp. Zool. 182:233–250.

Lincoln, G.A. 1973. Appearance of antler pedicles in early foetal life in red deer. J. Embryol. Exp. Morph. 29:431–437.

Lincoln, G.A. 1975. An effect of the epididymis on the growth of antlers of castrated red deer. J. Reprod. Fertil. 42:159–162.

Lincoln, G.A. 1984. Antlers and their regeneration—a study using hummels, hinds and haviers. Proc. Roy. Soc. Edinburgh, 82B:243–259.

Lincoln, G.A. 1985. Seasonal breeding in deer, pp. 165–179. *In*: K. Drew & P. Fennessy (eds.), Biology of Deer Production. Royal Soc. of New Zealand, Bull. 22.

Lincoln, G.A. & G.A. Bubenik. 1985. Antler physiology, pp. 474–475. *In*: K.R. Drew & P.F. Fennessy (eds.), Biology of Deer Production. The Royal Soc. of New Zealand, Bull. 22.

Lincoln, G.A. & T.J. Fletcher. 1976. Induction of antler growth in a congenitally polled Scottish red deer stag. J. Exp. Zool. 195:247–262.

Lincoln, G.A. & T.J. Fletcher. 1984. History of a hummel. Part 7: Nature versus nurture. Deer 6:127–131.

Lincoln, G.A. & R.N.B. Kay. 1979. Effect of season on the secretion of LH and testosterone in intact and castrated red deer stags (*Cervus elaphus*). J. Reprod. Fertil. 55:75–80.

Lincoln, G.A., R.W. Youngson & R.V. Short. 1970. The social and sexual behaviour of the red deer stag. J. Reprod. Fertil. (Suppl.) 11:71–103.

Lincoln, G.A., H.M. Fraser & T.J. Fletcher. 1982. Antler growth in male deer (*Cervus elaphus*) after active immunization against LH-RH. J. Reprod. Fertil. 66:703–708.

Linnaeus, C. 1758. System naturae per regna naturae, secundum classes, ordines, genera, specium cum characteribus, differentiis, synonymis. Editio decima, reformata. Stockholm, Laurentii Salvii. (In Latin.)

Lister, A.M. 1984. Evolutionary and ecological origins of British deer. Proc. Roy. Soc. Edinburgh 82B:205–229.

Lister, A.M. 1987. Diversity and evolution of antler form in Quaternary deer, pp. 81–98. *In*: C.M. Wemmer (ed.), Biology and Management of the Cervidae. Smithsonian Institution Press, Washington, DC.

Lockerbie, R.O. 1987. The neuronal growth cone: a review of its locomotory, navigational and target recognition capabilities, and polled condition in cattle. J. Hered. 69:395–400.

Lojda, Z. 1956. Histogenesis of the antlers of our cervidae and its histochemical picture. Československa Morfologie, 4:43–65. (In Czech.)

Long, C.R. & K.E. Gregory. 1978. Inheritance of the horned, scurred, and polled condition in cattle. J. Hered. 69(6):395–400.

Long, T.A., R.L. Cowan, C.W. Wolfe, T. Radar & R.W. Swift. 1959. Effect of seasonal feed restriction on antler development of white-tailed deer. Penn. Agr. Exp. Sta. Progress Report 209.

Lorenz, K. 1941. Vegleichende Bewegungsstudien bei Anatiden. J. Ornith. 89:194–294.

Lorenz, K. 1964. Das sogenannte Böse. Dr. G. Bortha-Schoeler Verlag, Wien.

Lostroh, A.J. & C.H. Li. 1958. Effect of GH and thyroxine on body weight of hypophysectomized C3H mice. Endocrinology 62:484–492.

Loudon, A.S.I. & J.D. Curlewis. 1988. Cycles of antler and testicular growth in an aseasonal tropical deer (*Axis axis*). J. Reprod. Fertil. 83:729–738.

Lovari, S. 1984/1985. Behavioural repertoire of the Abruzzo chamois, *Rupicapra pyrenaica ornata* Neumann, 1899 (Artiodactyla: Bovidae). Säugetierkundl. Mitt. 32:113–136.

Løvtrup, S. 1988. Epigenetics, pp. 189–227. *In*: C.J. Humpries (ed.), Ontogeny and Systematics. Columbia University Press, New York.

Lowenstein, J.M. 1986. Molecular phylogenetics. Ann. Rev. Earth planet. Sci., Palo Alto, California 14:71–83.

Lu, Ho-gee & He-lin Sheng. 1984. Status of the Black muntjac, *Muntiacus crinifrons*, in eastern China. Mammal Rev. 14:29–36.

Ludwig, W. 1970. Das Rechts-Links-Problem im Tierreich und beim Menschen. Springer-Verlag, Berlin.

Lydekker, R. 1898. The Deer of All Lands. A History of the Family Cervidae Living and Extinct. Rowland Ward Ltd., London.

Lydekker, R. 1904. On the subspecies of *Giraffa camelopardalis*. Proc. Zool. Soc. London 1904:202–227.

Lydekker, R. 1915. Catalogue of ungulate mammals in the British Museum, Vol. 4. British Museum Trustees, London.

Lyon, M.W., Jr. 1908. Remarks on the horns and on the systematic position of the American antelope. Proc. U.S. Nat. Mus. 34:393–401.

Ma, S., Y. Wang & L. Xu. 1986. Taxonomic and phylogenetic studies on the genus *Muntiacus*. Acta Theriol. Sin. 6:191–209.

Ma, S., Y. Wang & C.P. Groves. 1988. Taxonomic notes on the subspecies of the Indian Muntjac (*Muntiacus muntak*) in Yunnan, China. Acta Theriol. Sin. 8:95–104.

MacInnes, D. 1936. A new genus of fossil deer from the Miocene of Africa. Linnean Soc. J. Zool. 34:521–530.

MacKay, T.F.C. 1981. Genetic variation in varying environments. Genet. Res. 37:79–93.

MacNamara, M. & W.D. Eldridge. 1987. Behavior and reproduction in captive pudu (*Pudu pudu*) and red brocket (*Mazama americana*), a descriptive and comparative analysis, pp. 371–387. *In*: C.M. Wemmer (ed.), Biology and Management of the Cervidae. Smithsonian Institution Press, Washington, DC.

Magruder, N.D., C.E. French, L.C. McEwen & R.W. Swift. 1957. Nutritional requirements of white-tailed deer for growth and antler development. II: Experimental results of the third year. Pennsylvania Agr. Expt. Sta. Bull. 628.

Manlove, M.N., J.C. Avise, H.O. Hillestad, P.N. Ramsey, M.H. Smith & D.O. Straney. 1975. Starch gel electrophoresis for the study of population genetics in white-tailed deer, pp. 392–401. *In*: Proc. Ann. Conf. Southeast. Wildl. and Fish Comm., No. 29.

Marburger, R.G., R.M. Robinson, J.W. Thomas, M.J. Andregg & K.A. Clark. 1972. Antler malformation produced by leg injury in white-tailed deer. J. Wildl. Dis. 8:311–314.

Marchinton, R.L. & D.A. Dobie. 1987. Genetic stock and environment as factors in production of record class antlers, p. 114. *In*: Abstracts of the XVIIIth Congress IUGB, Krakow.

Markowski, J. 1987. Epigenetic asymmetry as a biological indicator of population differentiation in roe deer, p. 115. *In*: Abstracts of the XVIIIth Congress IUGB, Krakow.

Martin, C.R. 1977. Status and ecology of the barasingha (*Cervus duvauceli branderi*) in Kanha National Park (India). J. Bombay Nat. Hist. Soc. 84:60–132.

Martin, C.R. 1978. The male reproductive system, pp. 241–264. *In*: Textbook of Endocrine Physiology. Oxford Univ. Press, New York, NY.

Martin, J.T. & T.B. van Wimersma-Greidanus. 1979. Imprinting behavior: influence of vasopressin and ACTH analogues. Psychoneuroendocrinology 3:261–269.

Martinka, C. 1966. The international antelope herd. Mont. Wildl. 28–30.

Matthew, W.D. 1904. A complete skeleton of Merycodus. Bull. Am. Mus. Nat. Hist. 20:101–129.

Matthew, W.D. 1908. Osteology of *Blastomeryx* and phylogeny of the American Cervidae. Bull. Am. Mus. Nat. Hist. 24:535–562.

Mautz, W.W. & J. Fair. 1980. Energy expenditure and heart rate for activities of white-tailed deer. J. Wildl. Manage. 44:333–342.

Mautz, W.W., P.J. Pekins & J.A. Warren. 1985. Cold temperature effects on metabolic rate of white-tailed, mule, and black-tailed deer in winter coat, pp. 453–457. *In*: P.F. Fennessy & K.R. Drew (eds.), Biology of Deer Production. Royal Soc. of New Zealand, Bull. 22.

Maynard-Smith, J., R. Burian, S. Kauffman, P. Alberch, J. Campbell, B. Goodwin, R. Lande, D. Raup & L. Wolpert. 1985. Developmental constraints and evolution. Quart. Rev. Biol. 60:265–287.

Mayr, B., J. Krutzler, H. Auer, M. Kalat & W. Schleger. 1987. NORs, Heterochromatin, and R-bands in three species of Cervidae. J. Heredit. 78:108–110.

Mazur, P.E. 1969. Location, mapping, and surgical removal of the parathyroid gland in the white-tailed deer (*Odocoileus virginianus*). M.Sc. Thesis, The Pennsylvania State University.

Mazur, P.E. 1973. Seasonal plasma androgen levels and its relation to antler growth and seasonal feed consumption in male white-tailed deer (*Odocoileus virginianus borealis*). Ph.D. Thesis, The Pennsylvania State University.

McCullough, D.R. 1969. The Tule elk: its history, ecology, and behavior. Univ. Calif. Publ. Zool. 88.

McCullough, D.R. 1971. The Tule Elk: Its History, Behavior and History, 2nd ed. University of California Press, Berkeley, CA.

McCullough, D.R. 1982. Antler characteristics of George Reserve deer. J. Wildl. Manage. 46:821–826.

McCullough, Y.B. 1980. Niche separation of seven North American ungulates on the National Bison Range, Montana. Ph.D. Thesis, University of Michigan, Ann Arbor, MI.

McEwan, E.H. 1968. Growth and development of barren ground caribou. II. Postnatal growth rates. Can. J. Zool. 46:1023–1029.

McEwan, E.H. 1970. Energy metabolism of barren ground caribou (*Rangifer tarandus*). Can. J. Zool. 48:391–392.

McEwen, B.S. 1980. The brain as a target organ of endocrine hormones, pp. 32–42. *In*: D.T. Krieger & J.C. Hughes (eds.), Neuroendocrinology. Sinauer Associates, Sunderland, MA.

McEwen, L.C., C.E. French, N.D. Magruder, R.W. Swift & R.H. Ingram. 1957. Nutrient requirements of the white-tailed deer. Trans. North Am. Wildl. Conf. 22:119–132.

McEwen, B.S., P.G. Davis, B. Parson & D.W. Pfaff. 1979. The brain as a target organ for steroid hormone action. Ann. Rev. Neurosci. 2:65–112.

McEwen, B.S., E.R. De Kloet & W. Rostene. 1986. Adrenal steroid receptors and actions in the nervous system. Physiol. Rev. 66:1121.

McFarland, W.N., F.H. Pough, T.J. Cade & J.B. Heiser. 1985. Vertebrate Life, 2nd ed. MacMillan Publishers, New York, NY.

McLean, D.D. 1944. The pronghorned antelope in California. Calif. Fish and Game 30:221–241.

McMillin, J.M., U.S. Seal, K.D. Keenlyne, A.W. Erickson & J.E. Jones. 1974. Annual testosterone rhythm in the adult white-tailed deer (*Odocoileus virginianus borealis*). Endocrinology 94:1034–1040.

McMillin, J.M., U.S. Seal & P.D. Karns. 1980. Hormonal correlates of hypophagia in white-tailed deer. Fed. Proc. 39:2964–2968.

McNay, M.E. 1980. Causes of low pronghorn fawn:doe ratios on the Sheldon National Wildlife Refuge, Nevada. M.S. Thesis, University of Montana, Missoula, MT.

Meile, P. & A.B. Bubenik. 1979. Zur Bedeutung sozialer Auslöser für das Sozialverhalten der Gemse, *Rupicapra rupicapra* (Linné, 1758). Säugetierkdl. Mitt. 27, Sonderheft.

Merkt, J.R. 1987. Reproductive seasonality and grouping pattern of the North Andean deer or Taruca (*Hippocamelus antisiensis*) in Southern Peru, pp. 399-401. *In*: C.M. Wemmer (ed.), Biology and Management of Cervidae. Smithsonian Institution Press, Washington, DC.

Meschaks, P. & M. Nordkvist. 1962. On the sexual cycle in the reindeer male. Acta Vet. Scand. 3:151–162.

Meunier, K. 1981. Die Geweihform der Hirscharten–ein Gestaltproblem, pp. 133–144. *In*: R.R. Hofmann (ed.), Wildbiologische Informationen für den Jäger. F. Enke Verlag, Stuttgart.

Meyer, M.W., R.D. Brown & M.W. Graham. 1984. Protein and energy content of white-tailed deer diets in the Texas Coastal Bend. J. Wildl. Manage. 48:527-534.

Meyer, P. 1979. Tumor am Kopf eines Rehes (*Capreolus capreolus L.*). Z. Jagdwiss. 25:239-241.

Mierau, G.W. 1972. Studies on the biology of an antlered female mule deer. J. Mammal. 53:403-404.

Miller, K.V., R.L. Marchinton, J.R. Beckwith & P.B. Busy. 1985. Variations in density and chemical composition of white-tailed deer antlers. J. Mammal. 66:693–701.

Mirarchi, R.E., P.F. Scanlon & R.L. Kirkpatrick. 1977a. Annual changes in spermatozoan production and associated organs of white-tailed deer. J. Wildl. Manage. 41: 92–99.

Mirarchi, R.E., P.F. Scanlon, R.L. Kirkpatrick & C.B. Schreck. 1977b. Androgen levels and antler development in captive and wild white-tailed deer. J. Wildl. Manage. 41:178–183.

Mirarchi, B.E., B.E. Howland, R.E. Scanlon, R.L. Kirkpatrick & L.M. Sanford. 1978. Seasonal variation in plasma LH, FSH, prolactin, and testosterone concentrations in adult male white-tailed deer. Can. J. Zool. 56:121–172.

Mitchell, G.J. 1965. Natality, mortality, and related phenomena in two populations of pronghorn antelope in Alberta, Canada. Ph.D. Thesis, Washington State University, Pullman, WA.

Mitchell, G.J. 1967. Minimum breeding age of female pronghorn antelope. J. Mammal. 48:489–490.

Mitchell, G.J. 1971. Measurements, weights, and carcass yields of pronghorns in Alberta. J. Wildl. Manage. 35:76–85.

Mitchell, G.J. 1980. The pronghorn antelope in Alberta. Alberta Dep. Lands and Forests, Fish and Wildl. Div., Edmonton, Alberta, Canada.

Mitton, J.B. & M.C. Grant. 1984. Associations among protein heterozygosity, growth rate and developmental homeostasis. Ann. Rev. Ecol. Syst. 15:479–499.

Miura, S. 1984. Social behavior and territoriality in male sika deer (*Cervus elaphus* Temminck 1883) during the rut. Z. Tierpsychol. 64:33–73.

Moberg, G.P. 1985. Influence of stress on reproduction: measure of well-being, pp. 245–267. *In*: G.P. Moberg (ed.), Animal Stress. Am. Physiol. Soc., Bethesda, MD.

Modell, W. 1969. Horns and antlers. Scientific Am. 220:114–122.

Moen, A.N. 1973. Wildlife Ecology. W.H. Freeman and Co., San Francisco, CA.

Moen, A.N. 1978. Seasonal changes in heart rates, activity, metabolism and forage intake in white-tailed deer. J. Wildl. Manage. 42:715–738.

Moen, A.N. 1985. Energy metabolism of deer in relation to environmental variables, pp. 439–445. *In*: P.F. Fennessy & K.R. Drew (eds.), Biology of Deer Production. Royal Soc. of New Zealand, Bull. 22.

Mohr, E. 1918. Biologie und Systematik der Sechsender-Hirsche Gattung Rusa. Arch. F. Naturgesch. 84:105–143.

Mohr, E. 1932. Materialien über die Hirschzuchten des ehemaligen Hamburger Zoo. Zool. Garten, NF V:3–14.

Mohr, E. 1962. Über Geweih und Geweihwechsel beim Milu, *Elaphurus davidianus*. M.-Edw. Wissenschaftl. u. Kulturelle Mitt. Tierpark Berlin 1:156–163.

Mohr, E. 1965. Besonderheiten an Cavicornier-Hörnern. Milu, Berlin 2:21–47.

Mollelo, J.A., C.P. Epling & R.W. Davis. 1963. Histochemistry of the deer antler. Am. J. Vet. Res. 24:573–579.

Montulet, J.-P. 1984. Les Cervidés du Monde Entier. Lechevalier, Paris.

Moritz, C., T.E. Dowling & W.M. Brown. 1987. Evolution of animal mitochondrial DNA: relevance for population biology and systematics. Ann. Rev. Ecol. Syst. 18:262–269.

Morris, J.M. & G.A. Bubenik. 1982. The effects of androgens on the development of antler bone, pp. 123–141. *In*: R.D. Brown (ed.), Antler Development in *Cervidae*. Caesar Kleberg Wildl. Res. Inst., Kingsville, TX.

Morrison, B.G. 1961. Some aspects of the histology and growth of the horns of *Antilocapra americana*. M.S. Thesis, University of Wyoming, Laramie, WY.

Morrison-Scott, T.C.S. 1960. Antler anomalies. J. Mammal. 41:412.

Moy, R.F. 1970. Histology of the subauricular and rump glands of the pronghorn (*Antilocapra americana* Ord). Am. J. Anat. 129:65–87.

Moy, R.F. 1971. Histology of the forefoot and hindfoot interdigital and median glands of the pronghorn. J. Mammal. 52:441–446.

Moya Sola, S. 1983. Los Boselaphini del Neogeno de la Peninsula Iberica. Univ. Barc. Publs. Geol., Barcelona 18:1–236.

Mrosovsky, N. & D.F. Sherry. 1980. Animal anorexias. Science 207:837–842.

Muir, P.D., G.K. Barrell & A.R. Sykes. 1982. Modification of antler growth in red deer stag by the use of synthetic progestagen. Proc. N.Z. Anim. Prod. 42:145–147.

Muir, P.D., A.R. Sykes & G.K. Barrell. 1985. Mineralization during antler growth in red deer, pp. 251–254. *In*: P.F. Fennessy & K.R. Drew (eds.), Biology of Deer Production. Royal Soc. of New Zealand, Bull. 22.

Muir, P.D., A.R. Sykes & G.K. Barrell. 1987. Growth and mineralisation of antlers in red deer (*Cervus elaphus*). N.Z. Agric. Res. 30:305–315.

Mulachius, G.A. 1843. Democritos Abderita Operum Fragmenta. Berlin. (In Greek and German.)

Muller, S. & H. Schlegel. 1839–1844. The deer of Indian Archipelago, pp. 209–237. *In*: Verhandelingen over Natuurlijke Geschiedenis der Nederlandsche Overzeesche Bezittingen door leden der Natuurkundige commissie in Indie en andere schrijvers, Leiden. (In Dutch.)

Murie, J. 1870. Notes on the anatomy of the prongbuck, *Antilocapra americana*. Proc. Zool. Soc. London: 334–368.

Murie, O.H. 1951. The Elk of North America. Stackpole Publ. Co., Harrisburg, Pennsylvania.

Murie, O.J. 1928. Abnormal growth of moose antlers. J. Mammal. 9:65.

Murphy, D.A. & J.A. Coates. 1966. Effects of dietary protein on deer. Trans. N. Am. Wildl. Nat. Resour. Conf. 31:129–138.

Naaktgeboren, C. 1969. Geburtskundliche Bemerkungen über die Hörner der neugeborenen Giraffen. Z. Säugetierkd. 34:375–379.

Nečas, J. 1959. Jelení zveř. (Red deer.) SZN, Praha. (In Czech.)

Neitzel, H. 1982. Karyotypenevolution und deren Bedeutung für den Speciationsprozess der Cerviden (Cervidae; Artiodactyla; Mammalia). Ph.D. Thesis, Freie Universität, Berlin.

Nellis, C.H. 1965. Antler from right zygomatic arch of white-tailed deer. J. Mammal. 46:108–109.

Nevo, E., A. Beiles & R. Ben-Shlomo. 1984. The evolutionary significance of genetic diversity: Ecological, demographic and life history correlations, pp. 13–212. *In*: G.S. Mani (ed.), Evolutionary Dynamics of Genetic Diversity, Springer-Verlag, Berlin.

Newbrey, J.W. & W.J. Banks. 1974. Ultrastructural characterization of mineralized cartilage by selected enzymatic digestions. Anat. Rec. 87:426.

Newbrey, J.W. & W.J. Banks. 1975. Characterization of developing antler cartilage matrix. II. An electron microscopic evaluation. Calcif. Tiss. Res. 17:289–302.

Newbrey, J.W. & W.J. Banks. 1983. Ultrastructural changes associated with the mineraliztion of antler cartilage. Am. J. Anat. 166:1–17.

Newbrey, J.W., D.F. Counts, W.J. Foreyet & W.W. Laegreid. 1982. Isolation of collagen by guanidine extraction and pepsin digestion from the growing deer antler. pp. 307–316. *In*: R.D. Brown (ed.), Antler Development in Cervidae. Caesar Kleberg Wildl. Res. Inst., Kingsville, TX.

Nitsche, H. 1898. Studien über Hirsche (Gattung *Cervus* im weitesten Sinne). Untersuchungen über mehrstangige Geweihe und die Morphologie der Huftier-Hörner im Allgemeinen, Bd. 72, Heft 1. W. Engelmann, Leipzig.

Notz, F.W. 1967. Geweihbildung bei weiblichem Wilde. Wild u. Hund 70:177–181.

Obergfell, F.A. 1957. Vergleichende Untersuchungen an den Dentitionen und Dentale altburdigaler Cerviden von Wintershof-West in Bayern und rezenter Cerviden. (Eine phylogenetische Studie.) Paleontographica (Stuttgart) 106 Abt. A 3/6:71–166.

O'Gara, B.W. 1968. A study on the reproductive cycle of the female pronghorn (*Antilocapra americana* Ord). Ph.D. Thesis, University of Montana, Missoula, MT.

O'Gara, B.W. 1969a. Unique aspects of reproduction in female pronghorn (*Antilocapra americana* Ord). Am. J. Anat. 125:217–231.

O'Gara, B.W. 1969b. Horn casting by female pronghorns. J. Mammal. 50:373–375.

O'Gara, B.W. & G. Matson. 1975. Growth and casting of horns by pronghorns and exfoliation of horns by bovids. J. Mammal. 56:829–846.

O'Gara, B.W. & R.F. Moy. 1972. Histology and morphology of scent glands and their possible roles in pronghorn behavior. Antelope States Workshop, Billings, MT 5: 192–208.

O'Gara, B.W., R.F. Moy & G.D. Bear. 1971. The annual testicular cycle and horn casting by the pronghorn (*Antilocapra americana*). J. Mammal. 52:537–544.

Ohtaishi, N. & K. Too. 1974. The possible thermo-regulatory function and its character of the velvety antlers in the Japanese deer *Cervus nippon*. J. Mamm. Soc. Japan 6:1–11.

Olt, A. 1927a. Die Perücke des Cervidengeweihes und ihre Bedeutung für die Krebsforschung. Brcht. Oberhess. Gscht. f. Natur. and Heilkunde zu Giessen 2:3–7.

Olt, A. 1927b. Die Perücke der Cerviden und das Karzinom. Deutsche Tierärztl. Wochenschr. 35:131–133.

Ortavant, R., J. Pelletier, J.P. Ravault & J. Thimonier. 1978. Annual cyclic variation in prolactin in sheep, pp. 75–78. *In*: I. Assenmacher & D.S. Farner (eds.), Environmental Endocrinology. Springer-Verlag, Berlin.

Osgood, W.H. 1932. Mammals of the Kelley-Roosevelts and Delacour Asiatic expeditions. Field Mus. Nat. Hist., Zool. Ser. 18:193–339.

Otsuka, H. 1972. *Elaphurus shikamai* Otsuka (Pleistocene cervid) from the Akarbi Formation of the Osaka Group, Japan, with special reference to the genus *Elaphurus*. Bull. Nat. Sci. Tokyo 15:197–210.

Otsuka, H. & Y. Hasegawa. 1976. On a new species of *Elaphurus* (Cervid, Mammal) from Akishima City, Tokyo. Bull. Nat. Sci. Mus., Ser. C (Geol.) 2:139–143.

Owen, R. 1840. Notes on the anatomy of the Nubian giraffe. Trans. Zool. Soc. London 2:217–248.

Owen, R. 1849. Notes on the birth of the giraffe at the Zoological Society gardens. Trans. Zool. Soc. London 3:21–28.

Owen, R. 1868. On the Anatomy of Vertebrates, Vol. 3. Mammalia. Longmans, Green and Co., London.

Owen-Smith, R.N. 1974. The social system of the white rhinoceros, pp. 341–351. *In*: V. Geist & F. Walther (eds.), The Behaviour of Ungulates and its Relation to Management. I.U.C.N. Pub. No. 24, Morges, Switzerland.

Owen-Smith, R.N. 1977. On territoriality in ungulates and an evolutionary model. Quart. Rev. Biol. 52:1–38.

Ozoga, J.J. & L.J. Verme. 1982. Physical and reproductive characteristics of a supplementally-fed white-tailed deer herd. J. Wildl. Manage. 46:281–301.

Packer, C. 1983. Sexual dimorphism: the horns of African antelopes. Science 221:1191–1193.

Pantič, V. & N. Stošič. 1966. Investigation of the thyroid of deer and roe-bucks. Acta Anatomica. 63:580–590.

Parker, K.C., C.T. Robbins & T.A. Hanley. 1984. Energy expenditure of locomotion for mule deer and elk. J. Wildl. Manage. 48:474–488.

Patterson, C. 1981. Methods of paleobiogeography, pp. 446–500. *In*: G. Nelson & D.E. Rosen (eds.), Vicariance Biogeography. A Critique. Columbia University Press, New York, NY.

Patton, T.H. & B.E. Taylor. 1971. The Synthetoceratinae (Mammalia, Tylopoda, Protoceratidae). Bull. Am. Mus. Natur. Hist. 145:119–218.

Patton, T.H. & B.E. Taylor. 1973. The Protoceratinae (Mammalia, Tylopoda, Protoceratidae) and the systematics of the Protoceratidae. Bull. Am. Mus. Natur. Hist. 150: 348–412.

Pavlanský, R. & A. Bubenik. 1955. Von welchem Gewebe geht der eigentliche Reiz zur Geweihentwicklung aus? I. Mitteilung: Versuch der Transplantation eines Geweihzapfens bei einem Damspieser, Dama dama dama (Linné 1758). Säugetierkdl. Mitt. 3:49–53.

Pavlanský, R. & A. Bubenik. 1960. Von welchem Gewebe geht der eigentliche Reiz zur Geweihentwicklung aus? IV. Mitteilung: Versuche mit Auto- und Homotransplantation des Geweihzapfens. Säugetierkdl. Mitt. 8:32–37.

Payne, J., C.M. Francis & K. Phillips. 1985. A field guide to the mammals of Borneo. Sabah Society and World Wildlife Fund, Malaysia. Kuala Lumpur.

Peacock, E.H. 1933. A game book for Burma and adjoining territories. H.F. & G. Witherby, London.

Peck, A.L. 1965. Aristotelis Historia Animalium. Libr. II. Cap. i. IX. W. Heinemann, London. (In Latin and English.)

Peterson, R.L. 1955. North American Moose. University of Toronto Press, Toronto.

Petrocchi, C. 1956. *Leptobos* di Sahabi. Boll. Soc. geol. Ital., Rome 75:206–238.

Phelps, J.S. 1981. Biological observations on the Sonoran pronghorn, pp. 28–33. *In*: The Sonoran Pronghorn. Spec. Rep. 10. Ariz. Fish and Game Dep., Phoenix, AZ.

Phillippo, M., G.A. Lincoln & C.B. Lawrence. 1972. The relationship between thyroidal calcitonin and seasonal reproductive change in the stag (*Cervus elaphus* L.). J. Endocrinol. 53:47–49.

Pickford, M. & H. Thomas. 1984. An aberrant new bovid in subrecent deposits from Rusinga Island, Kenya. Proc. K. Ned. Akad. Wet. B., Amsterdam 87:441–452.

Pierce, B.A. & J.B. Mitton. 1982. Allozyme heterozygosity and growth in the tiger salamander *Ambystoma tigrinum*. J. Hered. 33:250–253.

Pilgrim, G.E. 1937. Siwalik antelopes and oxen in the American Museum of Natural History. Bull. Am. Mus. Nat. Hist. New York 72:729–874.

Pilgrim, G.E. 1941a. The dispersal of the Artiodactyla. Biol. Rev. Cambridge 16:134–163.

Pilgrim, G.E. 1941b. The relationship of certain variant fossil types of 'horns' to those of the living Pecora. Ann. Mag. Nat. Hist. 7:182–184.

Plotka, E.D., U.S. Seal, M.A. Letellier, L.J. Verme & J.J. Ozoga. 1981. The effect of pinealectomy on seasonal phenotypic changes in white-tailed deer (*Odocoileus virginianus borealis*), pp. 45–46. *In*: C.D. Matthews & R.F. Seamark (eds.), Pineal Function. Elsevier/North Holland Biomedical Press, New York, NY.

Plotka, E.D., D.E. Koller, M.A. Letellier, L.J. Verme & J.J. Ozoga. 1983. Androgen receptors in prostate, neck and antler pedicle skin of fawns and adult intact and castrated white-tailed deer. Society for Study for Reproduction, Mtg. Cleveland, Ohio, 6–7 October.

Pocock, R.L. 1905. The effects of castration on the horns of a pronghorn (*Antilocapra americana*). Proc. Zool. Soc. London 1:191–197.

Pocock, R.I. 1923. On the external characters of Elaphurus, Hydropotes, Pudu and other Cervidae. Proc. Zool. Soc. Lond. 181–210.

Pocock, R.I. 1933. The homologies between the branches of the antlers of the Cervidae based on the theory of dichotomous growth. Proc. Zool. Soc. London 1933:377–406.

Pocock, R.I. 1943a. The larger deer of British India. Part III. The sambar (*Rusa*). J. Bombay Nat. Hist. Soc. 44:27–36.

Pocock, R.I. 1943b. The larger deer of British India. Part IV. The chital (*Axis*) and the hog-deer (*Hyelaphus*). J. Bombay Nat. Hist. Soc. 44:169–178.

Pollard, J.W. 1987. The moveable genome—Weismann's Doctrine and new models for speciation. Rivista di Biologia—Biology Forum 80:11–54.

Pomel, A. 1894. Les boeufs taureaux. Paleont. Monogr. Carte geol. Alger. 3:1–108.

Pond, C.M. 1978. Morphological aspects and the ecological and mechanical consequences of fat deposition in wild vertebrates. Ann. Rev. Ecol. Syst. 9:519–570.

Popp, J.W. 1985. Horn size and body size among antelope. Säugetierkdl. Mitt. 32:245–248.

Portmann, J. 1970. Überlegungen zur Entstehung von Korkenziehergeweihen und sogenannten Frostgehörnen. Z. Jagdwiss. 16:176–178.

Povilitis, A.J. 1983. Social organization and mating strategy of the huemul (*Hippocamelus bisulcus*). J. Mammal. 64:156–158.

Prenzlow, E.J., D.L. Gilbert & F.A. Glover. 1968. Some behavior patterns of the pronghorn. Colo. Dep. Game, Fish, and Parks. Spec. Rep. 17.

Presidente, P.J.A., J.H. Lumsden, K.R. Presnell, W.A. Rapley & B.M. McCraw. 1973. Combination of etorphine and xylazine in captive white-tailed deer. II. Effects on hematologic, serum biochemical, and blood gas values. J. Wildl. Dis. 9:342–348.

Prevost, M.F. 1869. De l'existance de cornes rudimentaires sur la tête des femelles de cerfs. Nouv. Arch. d. Mus. d'Hist. Nat. Paris V:271–275.

Prince, J.H. et al. 1960. Anatomy and histology of the eye and orbit in domestic animals. C.C. Thomas Publ., Springfield, IL.

Pritzker, K.P.H., M.D. Grynpas & R.D. Brown. 1988. Bone quality and bone kinetics in cyclic osteoporosis. 3rd Int. Conf. on the chemistry and biology of mineralized tissues (Abstr.). Chatham, MA, Oct. 16–21.

Pyrah, D. 1970. Antelope herd range in central Montana. Proc. Antelope States Workshop. 4:16–20.

Pyrah, D.B. 1987. American pronghorn antelope in the Yellow Water Triangle, Montana. Bull. Mont. Dep. Fish, Wildl. and Parks and U.S.D.I. Bur. Land Manage.

Qui, Z., D. Yan, H. Jia & B. Sun. 1985. Preliminary observations on the newly found skeleton of a *Palaeomeryx* from Shanwag, Shandong. Vert. Palasiat. 23:173–195. (In Chinese.)

Quinteros, I.R., A.O. Muller, W.J. Miller & J.R. Bischoff. 1971. Fenotipos de transferrinas en el venado argentino (*Ozotoceros bezoarticus* celer). (Phenotype of the transferrine of *Ozotoceros bezoarticus*). Analecta Vet. 33:107–114. (In Portuguese.)

Racey, P.A. & C. Skinner. 1979. Endocrine aspects in sexual mimicry in spotted hyena *Crocuta crocuta*. J. Zool. London 187:315–326.

Raesfeld, F. 1977. Das Rehwild, 8th ed. P. Parey, Hamburg und Berlin.

Raesfeld, F. & F. Vorreyer. 1978. Das Rotwild, 8th ed. P. Parey, Hamburg und Berlin.

Ralls, K. 1976. Mammals in which females are larger than males. Quart. Rev. Biol. 51:245–276.

Ralls, K., C. Barasch & K. Minkowski. 1975. Behaviour of captive mouse deer, *Tragulus napu*. Z. Tierpsychol. 37:356–378.

Ramirez, V. & R.D. Brown. 1988. A technique for *in vitro* incubation of deer antler tissue. Comp. Biochem. Physiol. 89A:279–282.

Ramsey, P.R., J.C. Avise, M.H. Smith & D.F. Urbston. 1979. Biochemical variation and genetic heterozygosity in South Carolina deer populations. J. Wildl. Manage. 43:136–142.

Rau, A. 1931. Das Rehgehörn. Neuman, Neudamm.

Ravault, J.F., B. Barenton, M. Blanc, A. Daveau, A., D.H. Garnier, R. Ortavant, J. Pelletier, M.M. de Reviers & M. Terqui. 1982. Influence of 2 BR-alpha-ergocryptine (CB 154) on the secretion of prolactin, LH, FSH and testosterone and on testicular growth in rams subjected to different photoperiods. Reprod. Nutr. Develop. 22:989–998.

Rawlings, N.C., H.D. Hafs & L.V. Swanson. 1972. Testicular and blood plasma androgens in Hollstein bull from birth through puberty. J. Anim. Sci. 34:435–440.

Rayner, V. & S.W.B. Ewen. 1981. Do the blood vessels of the antler velvet of the red deer have an adrenergic innervation? Quart. Rev. Exp. Physiol. 66:81–90.

Reiter, J. 1980. The pineal and its hormones in the control of reproduction in mammals. Endocrine Rev. 1:109–131.

Řeřábek, J. & Bubenik, A. 1963. The metabolism of phosphorus and iodine in deer. Transaction series AEC, 5631, U.S. Atomic Energy Commission: 51.

Reyel, L.A. 1963. The occurrence of certain anomalies in Michigan white-tailed deer. J. Mammal. 44:79–98.

Rhumbler, L. 1911. Über die Abhängigkeit des Geweihwachstums der Hirsche, speziell des Edelhirsches vom Verlauf der Blutgefässe im Kolbengewebe. Z. Forst- und Jagdwesen. 3:295–314.

Rhumbler, L. 1913. 1. Hat das Geweih des Damhirsches (*Dama dama* L.) eine morphologische Drehung erfahren? Zool. Anzeiger 42:577–586.

Rhumbler, L. 1932. Die Verschiedenheiten in der Stirnwaffenentwicklung bei Wiederkäuern und ihre Gründe, I. Teil. Jenaische Z. Naturwiss. 60:310–325.

Ringberg, T. 1979. The Spitzberger reindeer – a winter dormant ungulate? Acta Physiol. Scand. 105:268–273.

Ringberg, T., E. Jacobsen, M. Ryg & J. Krog. 1978. Seasonal changes in levels of growth hormone, somatomedin and thyroxine in free ranging semi-domesticated Norwegian reindeer (*Rangifer tarandus tarandus* L.). Comp. Biochem. Physiol. 60A: 123–126.

Robbins, C.T. 1983. Wildlife Feeding and Nutrition. Academic Press, New York, NY.

Robbins, C.T. & L.M. Koger. 1981. Prevention and stimulation of antler growth by injections of calcium chloride. J. Wildl. Manage. 45:733–737.

Robbins, C.T., Y. Cohen & B.B. Davitt. 1979. Energy expenditure by elk calves. J. Wildl. Manage. 43:445–453.

Robinette, W.L. & J.S. Gashwiler. 1955. Antlerless mule deer bucks. J. Mammal. 36: 202–205.

Robinette, W.L. & D.A. Jones. 1959. Antler anomalies of mule deer. J. Mammal. 40: 96–108.

Robinson, R.N., J.W. Thomas & R.G. Marburger. 1965. The reproductive cycle of male white-tailed deer in central Texas. J. Wildl. Manage. 29:53–59.

Rolf, H.J. & K. Fischer. 1987. Annual periodicity of blood testosterone and 5-alpha DHT levels in adult male fallow deer (*Dama dama* L.). Acta Endocrinol. 114: Suppl. 283, Abstr. 205.

Romer, A.S. 1966. Vertebrate Paleontology, 3rd ed. University of Chicago Press, Chicago, IL.

Roosevelt, T., T.S. Van Dyke, D.G. Elliot & A.J. Stone. 1902. The pronghorn antelope, pp. 46–130. *In*: The Deer Family. The Macmillan Company, Norwood, MA.

Rörig, A. 1899. Welche Beziehungen bestehen zwischen den Reproduktionsorganen der Cerviden und der Geweihbildung derselben? Roux Arch. Entw.-Mech. Org. 8:382–447.

Rörig, A. 1900. Ueber Geweihentwicklung und Geweihbildung. I. Abschnitt: Die phylogenetischen Gesetze der Geweihentwicklung. Roux Arch. Entw.-Mech. Org. 10:525–617.

Rörig, A. 1901. Ueber Geweihentwicklung und Geweihbildung. IV. Abschnitt: Abnorme Geweihbildungen und ihre Ursachen. Roux Arch. Entw.-Mech. Org. 11:66–148.

Rörig, A. 1907. Gestaltende Korrelationen zwischen abnormer Körperkonstitution der Cerviden und Geweihbilung derselben. Roux Arch. Entw.-Mech. Org. 23:1–150.

Rose, K.D. 1985. Comparative osteology of North American dichobunid artiodactyls. J. Paleontol. 59:1203–1226.

Roth, V.L. 1984. On homology. Biol. J. Lin. Soc. 22:13–29.

Roux, J. & K. Stott. 1948. Antler-bearing by a female Sika Deer. J. Mammal. 29:71.

Rush, W.M. 1932. North Yellowstone elk study. Montana Fish and Game Comm., Helena, MT.

Ruch, W.M. 1944. American pronghorn. Nat. Hist. 53:321-323.

Rutberg, A.T. 1987. Adaptive hypotheses of birth synchrony in ruminants: an interspecific test. Am. Nat. 130:692-710.

Ryel, L.A. 1963. The occurrence of certain anomalies in Michigan white-tailed deer. J. Mammal. 44:79-98.

Ryg, M. 1983a. Advances in the physiological studies in the reindeer/caribou in 1978-82. Acta Zool. Fenn. 175:77-80.

Ryg, M. 1983b. Relationships between hormone induced and compensatory weight changes in reindeer (*Rangifer tarandus tarandus*). Comp. Biochem. Physiol. 74A:33-35.

Ryg, M. 1984. Seasonal changes in the relationship between food intake and serum triiodothyronine in reindeer. Comp. Biochem. Physiol. 78A:427-429.

Ryg, M. & R. Langvatn. 1982. Seasonal changes in weight gain, growth hormone, and thyroid hormones in male red deer (*Cervus elaphus antlanticus*). Can. J. Zool. 60: 2577-2587.

Ryg, M. & E. Jacobsen. 1982. Seasonal changes in growth rate, feed intake, growth hormone, and thyroid hormones in young male reindeer (*Rangifer tarandus tarandus*). Can. J. Zool. 60:15-23.

Sallač, W. 1912. Die Kronenhirsche und die Mendel'schen Gesetze. II. Vereinsschrift f. Forst-, Jagd- u. Naturkunde, Prag: 73-112.

Sälzle, K. & H. Schedelmann. 1977. Deutsches Jagdmuseum München. Katalog. Stiftung Deutsches Jagdmuseum Munchen.

Sandiford, G. 1829. Over de vorming en ontwikkeling der Hornes von zogende Dieren in het algemeenen, en van die der Herten beesten in het bijzonder. (About the occurrence and development of horns in domestic animals). Nieuwe Verhandel. d. 1.Kl. v.h. Koninkl. Nederland. Inst. Van Wettenschappen II:67-106. (In Dutch.)

Sanford, L.M., B.E. Howland & W.M. Palmer. 1984. Seasonal changes in the endocrine responsiveness of the pituitary and testes of male sheep in relation to their patterns of gonadotropic hormone and testosterone secretion. Can. J. Physiol. Pharm. 62:827-833.

SAS Institute. 1985. SAS User's Guide: Statistics. Version 5. SAS Circle, Cary, N.C.

Savage, D.E. & D.E. Russell. 1983. Mammalian Paleofaunas of the World. Addison-Wesley Publishing Co., Reading, MA.

Scadding, S.R. & M. Maden. 1986. The effects of local application of retinoic acid on limbs and regeneration in tadpoles of *Xenopus laevis*. J. Embryol. Exp. Morph. 91: 55-63.

Scanlon, P.F. 1977. The antler cycle in white-tailed deer: a review of recent developments. Trans. N.E. Deer Study Group, 64-67.

Schaeffer, B. 1947. Notes on the origin and function of the artiodactyl tarsus. Am. Mus. Novit. 1356:1-24.

Schaffer, J. 1940. Die Hautdrüsenorgane der Säugetiere. Urban und Schwarzenberg, Berlin.

Schaffer, W.M. & C.A. Reed. 1972. The co-evolution of social behavior and cranial morphology in sheep and goats (Bovidae, Caprini). Fieldiana Zool. 61(1):1-63.

Schaller, G.B. 1967. The deer and the tiger. A study of wildlife in India. University of Chicago Press, Chicago, IL.

Schaller, G.B. 1977. Mountain Monarchs. University of Chicago Press, Chicago, IL.

Schams, D. & D. Barth. 1982. Annual profiles of reproductive hormones in peripheral plasma of the male roe deer (*Capreolus capreolus*). J. Reprod. Fertil. 66:463–468.

Schams, D. & D. Barth. 1987. Zur endokrinen Regulation des Geweihwachstums beim Rehbock (*Capreolus capreolus*). Forschung, Lehre, Praxis. Wissenschaftlicher Jahresbericht, TU München, Weihenstephan, FRG, pp. 21–22.

Schams, D., D. Barth & H. Karg. 1980. LH, FSH and progesterone concentrations in peripheral plasma of the female roe deer (*Capreolus capreolus*) during rutting season. J. Reprod. Fert. 60:109–114.

Schams, D., D. Barth & H. Karg. 1987. Is antler growth in adult roe deer (*Capreolus capreolus*) influenced by treatment with estradiol-17-β, an antiandrogen, or prolactin inhibitor, pp. 475–482. *In*: I. Assenmacher & J. Boissin (eds.), Endocrine Regulations as Adaptive Mechanisms to the Environment, CNRS-CEBAS, France.

Schams, D., S. Gombes, E. Schallenberger, V. Reinhardt & R. Claus. 1978. Relationships between short-term variations of LH, FSH, prolactin and testosterone in peripheral plasma of prepubertal bulls. J. Reprod. Fertil. 54:145–148.

Schlosser, M. 1924. Tertiary vertebrates from Mongolia. Pal. Sinica 1(1).

Schnare, H. & K. Fischer. 1987. Secondary sex characteristics and connected physiological values in male fallow deer (*Dama dama*) and their relationship to changes of the annual photoperiod: Doubling the frequency. J. Exp. Zool. 244:463–471.

Schneider, K. 1930. Über die Pupillengestalt bei einigen lebenden Säugetieren. Neue psycholog. Studien—Psychologische Optik 6:320–356.

Schmidt, F. 1985. Grundlagen der kybernetischen Evolution. Goecke & Evers, Krefeld.

Schönberg, F. 1928. Ein grosses Hauthorn bei einem Rinde. Z. Säugetierkd. 3:20.

Schonewald-Cox, C.M., J.W. Bayles & J. Schonewald. 1985. Cranial morphometry of Pacific Coast elk (*Cervus elaphus*). J. Mammal. 66:63–74.

Schuhmacher, S. 1939. Jagd und Biologie—Ein Grundriss der Wildkunde. Springer-Verlag, Berlin.

Schulte, B.A., J.A. Parsons, U.S. Seal, E.D. Plotka, L.J. Verme & J.J. Ozoga. 1980a. Heterologous radioimmunoassay for deer prolactin. Gen. Comp. Endocrin. 40:39–68.

Schulte, B.A., U.S. Seal, E.D. Plotka, L.J. Verme, J.J. Ozoga & J.A. Parsons. 1980b. Seasonal changes in prolactin and growth hormone cells in the hypophyses of white-tailed deer (*Odocoileus virginianus borealis*) studied by light microscopic immunocytochemistry and radioimmunoassay. Am. J. Anat. 159:369–377.

Schulte, B.A., U.S. Seal, E.D. Plotka, M.A. Letellier, L.J. Verme, J.J. Ozoga & J.A. Parsons. 1981a. The effect of pinealectomy on seasonal changes in prolactin secretion in the white-tailed deer (*Odocoileus virginianus borealis*). Endocrinology 108:173–179.

Schulte, B.A., U.S. Seal, E.D. Plotka, L.J. Verme, J.J. Ozoga & J.A. Parsons. 1981b. Characterization of seasonal changes in prolactin and growth hormone in the hypophyses of white-tailed deer (*Odocoileus virginianus borealis*) by ultrastructural and immunocytochemical techniques. Am. J. Nat. 160:277–284.

Scott, K.M. 1979. Adaptation and allometry in bovid postcranial proportions. Unpublished Ph.D. Thesis, Yale University, New Haven, CT.

Scott, K.M. 1983. Prediction of the body weight of fossil Artiodactyla. Zool. J. Linn. Soc. 77:199–215.

Scott, K.M. 1985. Allometric trends and locomotor adaptations in the Bovidae. Bull. Am. Mus. Nat. Hist. 179:197–288.

Scott, K.M. 1987. Allometry and habitat related adaptations in the postcranial skeleton of Cervidae, pp. 65–80. *In*: C. Wemmer (ed.), The Biology and Management of the Cervidae. Smithsonian Institution Publications, Washington, DC.

Scott, K.M. (in press). Skeletal dimensions of ungulates as predictors of body weight. *In*: J. Damuth & B.J. MacFadden (eds.), Body Size Estimation in Mammalian Paleobiology, Cambridge University Press, Cambridge, MA.

Scott, K.M. & C.M. Janis. 1987. Phylogenetic relationships of the Cervidae, and the case for a Superfamily "cervoidea," pp. 3–20. *In*: C.M. Wemmer (ed.). Biology and Management of the *Cervidae*. Research Symposia of the National Zool. Park, Smithsonian Institution, Smithsonian Institution Press, Washington, D.C.

Scott, W.B. 1885. *Cervalces americanus*, a fossil moose, or elk from Quaternary of New Jersey. Acad. Nat. Sci. of Philadelphia, Proceedings 37:174–202.

Scott, W.B. 1937. A History of Land Mammals in the Western Hemisphere. Hafner, New York, NY.

Scribner, K.T., M.H. Smith & P.E. Johns. 1984. Age, condition, and genetic effects on incidence of spiked bucks, pp. 23–32. *In*: Proc. Ann. Conf. Southeast. Fish and Wildl. Agencies, No. 38.

Scribner, K.T., M.H. Smith & P.E. Johns. 1989. Environmental and genetic components of temporal variability in white-tailed deer antler growth. J. Mammal.

Seal, U.S. & A.W. Erickson. 1969. Hematology, blood chemistry and protein polymorphisms in the white-tailed deer (*Odocoileus virginianus*). Comp. Biochem. Physiol. 30:695–713.

Seal, U.S., L.J. Verme, J.J. Ozogo & E.D. Plotka. 1983. Metabolic and endocrine responses of white-tailed deer to increasing population density. J. Wildl. Manage. 47:451–462.

Sebeok, T.A. 1985. On the phylogenesis of communication. Language and Speech, RSBI 5:361–367.

Sempéré, A. 1978. The annual cycle of plasma testosterone and territorial behavior in the roe deer. *In*: J. Assenmacher & D. Farner (eds.) Environmental Endocrinology. Springer-Verlag, Berlin.

Sempéré, A. 1982. Fonction de reproduction et caractères sexuels secondaires chez le Chevreuil (*Capreolus capreolus* L.): variations saisonnières et incidences sur l'utilisation du budget-temps-espace. Ph.D. Thesis, University of Tours.

Sempéré, A.J. & J. Boissin. 1982. Neuroendocrine and endocrine control of the antler cycle in roe deer, pp. 109–122. *In*: R.D. Brown (ed.), Antler Development in *Cervidae*. Caeser Kleberg Wildl. Res. Inst., Kingsville, TX.

Sempéré, A.J. & A. Lacroix. 1982. Temporal and seasonal relationships between LH, testosterone and antlers in fawn and adult male roe deer (*Capreolus capreolus*): a longitudinal study from birth to four years of age. Acta Endocrinol. 99:295–301.

Sempéré, A.J., J. Boissin, B. Dutourne, A. Lacroix, M.R. Blanc & J.P. Ravault. 1983. Variations de la concentration plasmatique en prolactine, LH, FSH et de l'activité testiculaire au cours de la première année de vie chez le Chevreuil (*Capreolus capreolus* L.). Gen. Comp. Endocrinol. 52:247–254.

Sempéré, A.J., G.A. Bubenik & J.H. Smith. 1986. Antler cycle and thermolabile and thermostable alkaline phosphatase in white-tailed deer; circannual and circadian rhythms and variation after thyroxine, dexamethasone and ACTH administration. Acta Endocrinol. 111:133–139.

Sempéré, A.J., R. Grimberg, C. Silve, C. Tau & M. Garabedian. 1989. Evidence for extrarenal production of 1,25-dihydroxyvitamin during physiological bone growth: In vivo and in vitro production by deer antler cells. Endocrinology 125:2312–2319.

Seton, E.T. 1927. Lives of Game Animals, Vol. III: Hooved Animals. Doubleday, Doran and Company, Inc., Garden City, NY.

Severinghaus, C.W. & E.L. Cheatum. 1956. Life and times of the white-tailed deer, pp. 57–186. *In*: W.P. Taylor (ed.), The Deer of North America. The Stackpole Co., Harrisburg, PA.

Severinghaus, C.W. & A.N. Moen. 1983. Prediction of weight and reproduction rates of a white-tailed deer population from records of antler beam diameter among yearling males. N.Y. Fish and Game J. 30:30–38.

Severinghaus, C.W., H.F. Maguire, R.A. Cookingham & J.E. Tanck. 1951. Variations by age class in the antler beam diameters of white-tailed deer related to range conditions, pp. 551–570. *In*: Trans. North Am. Wildl. Nat. Res. Conf., 15.

Sewertsow, S.A. 1951. Problemy ekologiyi zhivotnyh—Neopublikovanyie raboty. (Problems of animal ecology—unpublished papers.) Akademie SSSR, Moskva. (In Russian.)

Sheng, H. & H. Lu. 1980. Current studies on the rare Chinese Black Muntjac. J. Nat. Hist. 14:803–807.

Sheng, H. & H. Lu. 1981. Reproduction of the Chinese Black Muntjac. Acta Theriol. Sin. 1:14–18.

Sher, A.V. 1986. Istoriya evolyutciya losei. (History of evolution of alcins), pp. 6–35. *In*: V.E. Sokolov (ed.), Biologiya i ispolzovaniye losya. (Biology and Management of Moose). Nauka, Moscow. (In Russian.)

Shi, L. 1981. Unique cytogenetic characteristics of a Tufted Deer, *Elaphodus cephalophus*. Mamm. Chromosomes Newsletter, Dec.: 47–51.

Shi, L. 1983. Sex-linked chromosome polymorphism in Black Muntjac, *Muntiacus crinifrons*, p. 271. *In*: M.S. Swaminathan (ed.), Abstracts of the XV. Int. Congress of Genetics, New Delhi.

Shi, L., Y. Ye & X. Duan. 1980. Comparative cytogenetic studies on the Red Muntjac, Chinese Muntjac and their F1 hybrids. Cytogenet. Cell Genet. 26:22–27.

Shikama, T. 1937. Fossil Cervifauna of Syatin near Tainan, S.W. Taiwan. Rep. Tohoku Imp. Univ. Sendai. Ser. 2. XIX.

Shilang, Z. & W. Shanzi. 1985. Studies of velvet antler production of sika deer. pp. 154. *In*: K.R. Drew & P.F. Fennessy (eds.). Biology of Deer Production. The Royal Soc. New Zealand Bull. 22.

Short, R.V. 1967. Comparative endocrinology of early gestation, pp. 224–267. *In*: R.M. Wynn (ed.). Fetal Homeostasis. N.Y. Academy of Sciences, New York, NY.

Short, R.V. 1979. Sexual behavior in red deer, pp. 365–372. *In*: H.W. Hawk (ed.), Animal Reproduction. Beltsville Symposia, Vol. 3. Allanheld, Osman & Co., Montclair.

Short, R.V. & T. Mann. 1966a. Delayed implantation in the roe deer *Capreolus capreolus*. Symp. Zool. Soc. London 156:173–194.

Short, R.V. & T. Mann. 1966b. The sexual cycle of a seasonally breeding mammal, the roebuck (*Capreolus capreolus*). J. Reprod. Fertil. 12:337–351.

Silver, H., N.H. Colovos & H.H. Hayes. 1959. Basal metabolism of white-tailed deer—a pilot study. J. Wildl. Manage. 23:434–438.

Simpson, A.M., J.M. Suttie & R.N.B. Kay. 1984. The influence of artificial photoperiod on the growth, appetite and reproductive status of male red deer and sheep. Anim. Reprod. Sci. 6:291–299.

Simpson, G.G. 1945. The principles of classification and a classification of mammals. Bull. Am. Mus. Nat. Hist. 85:1–350.

Sinclair, A.R.E. 1979. The eruption of ruminants, pp. 82–103. *In*: A.R.E. Sinclair & M. Norton-Griffiths (eds.), Serengeti: Dynamics and Ecosystem. Chicago University Press, Chicago, IL.

Singer, M. 1964. The trophic quality of the neuron. Some theoretical considerations. Progr. Brain Res. 13:228–232.

Singer, R. & E. Boné. 1960. Modern giraffes and the fossil giraffes of Africa. Ann. S. Afr. Mus. 45:375–548.

Singh, S.M. & E. Zouros. 1978. Genetic variation associated with growth rate in the American oyster (*Rassostrea virginica*). Evolution 32:342–353.

Skinner, J.D. 1971. The effect of season on spermatogenesis in some ungulates. J. Reprod. Fert. (Suppl.) 13:29–37.

Skinner, J.D. & A.S. van Jaarsveld. 1987. Adaptive significance of restricted breeding in Southern African ruminants. S. Afr. J. Sci. 83:657–663.

Skinner, M.P. 1922. The pronghorn. J. Mammal. 3:82–105.

Slabý, O. 1962. The morphogenesis of the autopodium of our cervidae. Československ. Morfologie 10(1):94–106.

Slatkin, M. 1978. Spatial patterns in the distributions of polygenic characters. J. Theor. Biol. 70:213–228.

Slavkin, H.C. 1988. Genetic and epigenetic challenges in tooth development. J. Craniofacial Genetics and Developmental Biology 8:195–198.

Smelik, P.G. 1981. A concept of pituitary-adrenal regulation. Front. Horm. Res. 8:1–11.

Smith, J.H. & G.A. Bubenik. 1990. Plasma concentration of a glucocorticoids in white-tailed deer; The effect of acute ACTH and dexamethasene administration. Can. J. Zool. (in press).

Smith, M.C.T. 1974. Biology and management of the wapiti (*Cervus elaphus Nelsoni*) of Fiordland, New Zealand. New Zealand Deerstalkers Assn., Inc., Wellington.

Smith, M.H. & R.K. Chesser. 1981. Rationale for conserving genetic variation of fish gene pools. Ecol. Bull. (Stockholm) 34:13–20.

Smith, M.H., R.K. Chesser, E.G. Cothran & P.E. Johns. 1983. Genetic variability and antler growth in a natural population of white-tailed deer, pp. 365–387. *In*: R.D. Brown (ed.), Antler Development in Cervidae. Caesar Kleberg Wildl. Res. Inst., Kingsville, TX.

Smith, M.H., K.T. Scribner & P.E. Johns. 1987. Genetics and antler development, pp. 187. *In*: Abstracts of the XVIIIth Congr. IUGB, Krakow.

Smouse, P.E. 1986. The fitness consequences of multiple-locus heterozygosity under the multiplicative or dominance and inbreeding depression models. Evolution 40:946–958.

Snedecor, G.W. & W.G. Cochran. 1965. Statistical Methods Applied to Experiments in Agriculture and Biology, 5th ed. The Iowa State University Press, Ames, IA.

Snyder, D.L., R.L. Cowan, D.R. Hagen & B.D. Schanbacher. 1983. Effect of pinealectomy on seasonal changes in antler growth and concentrations of testosterone and prolactin in white-tailed deer. Biol. Reprod. 29:63–71.

Sohr, H.U. 1975. Vererbung beim Rotwild und extensive Tierhaltung unter besonderer Berücksichtigung land-, forst- und devisen-wirtschaftlicher Aspekte im Rahmen neuzeitlicher Umweltpolitik. University Valdivis, Chile.

Sokolov, I.I. 1957. On the Artiodactyl-fauna in the southern part of Yunnan Province (China). Zool. Zh. 36:1750–1760. (In Russian.)

Sokolov, V.E. & V.C. Gromov. 1985. Studies in hybridization of European and Asiatic roe deer. (Opyty po gibridizatsii evropeiskikh i aziatskikh kosul). Dokl. Acad. Nauk SSSR 285:1022–1024. (In Russian.)

Solounais, N. 1981. The Turolian fauna from the island of Samos, Greece. Contr. Vertebr. Evol. 6:1–232.

Soma, H., H. Kada, K. Mtayoshi, Y. Suzuki, C. Meckvichal, A. Mahannop & B. Vatanaromya. 1983. The chromosomes of *Muntiacus feae*. Cytogenet. Cell Genet. 35:156–158.

Sondaar, P.Y. 1977. Insularity and its effect on mammal evolution, pp. 671–707. *In*: M.K. Hecht, P.C. Goody & B.M. Hecht (eds.). Major Patterns in Vertebrate Evolution. Plenum Press, New York, NY.

Soyka, F. & J.D. Crawford. 1965. Antagonism by cortisone of the linear growth induced in hypopituitary patients and hypophysectomized rats by human growth hormone. J. Clin. Endocr. 25:469–475.

Speer, D.P. 1982. The collagenous architecture of antler velvet. pp. 273–278. *In*: R.D. Brown (ed.), Antler Development in Cervidae. Caesar Kleberg Wildl. Res. Inst., Kingsville, TX.

Spillmann, F. 1931. Die Säugetiere Ecuadors im Wandel der Zeit. (I. Teil). Quito–Ecuador, Verlag der Universidad Central.

Spinage, C.A. 1968a. The book of the giraffe. Collins, London.

Spinage, C.A. 1968b. Horns and other bony structures of the skull of the giraffe, and their functional significance. E. Afr. Wildl. J. 6:53–61.

Spinage, C.A. 1973. The role of photoperiodism in the seasonal breeding of tropical African ungulates. Mammal Review 3:71–84.

Spotorno, A.E. & R. Fernandez-Donoso. 1975. The chromosomes of the Chilean dwarf-deer 'pudu' (*Pudu pudu*–Molina). Mammalian Chromosomes Newsletter 16:17.

Spotorno, A.E., N. Brum & M. Di Tomaso. 1988. Comparative cytogenetics of South American deer. Fieldiana Zool. 39:473–483.

Stanley, S.M. 1979. Macroevolution: Pattern and Process. W.H. Freeman and Co., San Francisco, CA.

Stehlin, H.G. 1927. Bemerkungen über die Hirsche von Steinheim am Aalbuch. Ecl. Geol. Helv. Basel 21:245–256.

Stehlin, H.G. 1937. Bemerkungen über die miocänen Hirschgenera *Stephanocemas* und *Lagomeryx*. Verh. naturforsch. Ges. Basel 48:193–214.

Stehlin, H.G. 1939. *Dicroceros elegans* und sein Geweihwechsel. Ecl. Geol. Helvet. 32:162–169.

Stephenson, D.C. & R.D. Brown. 1984. Calcium kinetics in male white-tailed deer. J. Nutr. 114:1014–1024.

Štěrba, O. & K. Klusák. 1984. Reproductive biology of fallow deer, *Dama dama*. Acta Sci. Nat. Brno 18:1–46.

Stewart, J.W.F. 1985. Deer and development in South-west Papua, New Guinea, pp. 381–385. *In*: P.F. Fennessy & K.R. Drew (eds.). Biology of Deer Production. Royal Soc. New Zealand, Bull. 22.

Stieve, J. 1950a. Ist die Nebenbrunst der Tiere mit verlängerter Tragzeit eine Scheinbrunst? pp. 970–979. *In*: W. Herre (ed.), Neue Ergebnisse und Probleme der Zoologie. Festschrift f. B. Klatt. Suppl. Zool. Anz. 145.

Stieve, H. 1950b. Anatomische-biologische Untersuchungen über die Fortpflanzungstätigkeit des europäischen Rehes (*Capreolus capreolus* L.). Z. Mikrosk. Anat. Forsch. 55:425–530.

Stokkan, K.A., K. Hove & W.F. Carr. 1980. Plasma concentrations of testosterone and luteinizing hormone in rutting reindeer bulls (*Rangifer tarandus*). Can. J. Zool. 58: 2081–2083.

Stonehouse, B. 1968. Thermoregulatory function of growing antlers. Nature 218: 870–872.

Stringham, S.F. & A.B. Bubenik. 1974. Physical condition and the survival of chamois (*Rupicapra rupicapra* L.) as a function of maturity, sex class ratios in the population: implications for ungulate management, pp. 123–160. *In*: W. Schröder (ed.), Tagungsbericht: 1. Internal. Gamswild-Treffen. Inst. f. Wildforsch. und Jagdkunde, Oberammergau, FRG.

Stromer, E. 1907. Fossile Wirbeltier-Reste aus dem Uadi Faregh und Uadi Natrun in Ägypten. Abh. Senckenb. naturforsch. Ges., Frankfurt a.M. 19:99–132.

Stuart, A.J. 1982. Pleistocene vertebrates in the British Isles. Longman, London.

Stüve, M. 1985. Aspects of structure and reproduction of white-tailed deer populations, *Odocoileus virginianus* in Venezuela and Virginia. Säugetierkdl. Mitt. 32:137–141.

Sudre, J. 1984. *Cryptomeryx* Schlosser 1886, tragulidae de l'Oligocène d'Europe et l'origine des ruminants. Palaeovertebrata, Montpellier 14:1–31.

Suttie, J.M. 1980a. Influence of nutrition on growth and sexual maturation of captive red deer stags, pp. 341–349. *In*: E. Reimers, E. Gaave & S. Skjenneberg (eds.), Proc. 2nd Int. Reindeer/Caribou Symp., Roros, Norway.

Suttie, J.M. 1980b. The effect of antler removal on dominance and fighting behaviour in farmed red deer stags. J. Zool. 190:217–224.

Suttie, J.M. 1985. Social dominance in farmed red deer stags. Appl. Anim. Behav. Sci. 14:191–199.

Suttie, J. & P. Fennessy. 1985. Regrowth of amputated velvet antlers with and without innervation. J. Exp. Zool. 234:359–366.

Suttie, J.M. & P.F. Fennessy. 1987. Growth promoting hormones and antler development, pp. 194–195. *In*: Abstracts of the 18th Congress, IUGB–Krakow.

Suttie, J.M. & R.N.B. Kay. 1982. The influence of nutrition and photoperiod on the growth of antlers of young red deer, pp. 61–71. *In*: R.D. Brown (ed.). Antler Development in *Cervidae*. Caesar Kleberg Wildl. Res. Inst. Kingsville, TX.

Suttie, J.M. & R.N.B. Kay. 1985. Influence of plane of winter nutrition on plasma concentrations of prolactin and testosterone and their association with voluntary food intake in red deer stags (*Cervus elaphus*). Anim. Reprod. Sci. 8:247–258.

Suttie, J.M. & A.M. Simpson. 1985. Photoperiodic control of appetite, growth, antlers and endocrine status of red deer, pp. 429–432. *In*: K.R. Drew & P.F. Fennessy (eds.), Biology of Deer Production. Royal Soc. of New Zealand, Bull. 22.

Suttie, J.M., E.D. Goodall, K. Pennie & R.N.B. Kay. 1983. Winter food restriction and summer compensation in red deer stags (*Cervus elaphus*). Br. J. Nutrit. 50:737–747.

Suttie, J.M., G.A. Lincoln & R.N.B. Kay. 1984. Endocrine control of antler growth in red deer stags. J. Reprod. Fertil. 71:7–15.

Suttie, J.M., P.D. Gluckman, J.H. Butler, P.F. Fennessy, I.D. Corson & F.J. Laas. 1985a. Insulin-like growth factor 1 (IGF-1) antler stimulating hormone? Endocrinology 116:846–848.

Suttie, J.M., P.F. Fennessy, C.G. Mackintosh, I.D. Corson, R. Christie & S.W. Heap. 1985b. Sequential cranial angiography of young red deer stags, pp. 263–268. *In*: P.F. Fennessy & K.R. Drew (eds.), Biology of Deer Production. Royal Soc. of New Zealand, Bull. 22.

Suttie, J.M., I.D. Corson, M. Gray, P.D. Gluckman, H.J. Elgar & K.R. Lapwood. 1985c. Liveweight gain and insulin-like growth factor 1 (IGF1) levels in testosterone treated stags. Proc. Aust. Soc. Endocr. 28 (Suppl. 2):35.

Suttie, J.M., C.G. MacKintosh, I.D. Corson, P.F. Fennessy, R. Christie. 1986. Carotid artery exteriorisation in the red deer. New Zealand Vet. J. 34:214–216.

Suttie, J.M., P.F. Fennessy, P.D. Gluckman & I.D. Corson. 1988. Evidence for a true endocrine function for IGF-1 in anterless stag. Endocrinology 122:3005–3007.

Suttie, J.M., P.F. Fennessy, I.D. Corson, F.J. Laas, J.H. Crosbie, J.J. Butler & P.D. Gluckman. 1989. Pulsatile growth hormone, insulin-like growth factors and antler development in red deer (*Cervus elaphus*) stags. J. Endocr. 121:351–360.

Taber, R.D. & R.F. Dasman. 1958. The black-tailed deer of the chaparral: its life history and management in the North Coast range of California. California Dept. Fish and Game Bull. 8.

Tachezy, R. 1956. Über den Einfluss der Sexualhormone auf das Geweihwachstum der Cerviden. Säugetierkdl. Mitt. 4:103–112.

Tandler, J. 1910. Ueber den Einfluss der Geschlechtsdruesen auf die Geweihbildung bei Rentieren. Anz. d. Kaiserl. Akad. d. Wissenschaften 47:252–257.

Tandler, J. & S. Grosz. 1913. Die biologischen Grundlagen der sekundären Geschlechtscharaktere. Springer-Verlag, Berlin.

Taylor, C.R. 1966. The vascularity and possible thermoregulatory function of the horns in goats. Physiol. Zool. 39:127–139.

Tegner, H. 1961. Horn growth in infant roe deer. Proc. Zool. Soc. (Lond.) 137:635–637.

Teilhard de Chardin, P. 1939. The Miocene cervids from Shantung. Bull. Geol. Soc. China 19:269–278.

Teilhard de Chardin, P. & J. Piveteau. 1930. Les mammifères fossiles de Nihowan (Chine). Ann. Paleont. 19:3–134.

Teilhard de Chardin, P. & M. Trassaert. 1937. The Pliocene Camelidae, Giraffidae and Cervidae of South Eastern Shansi. Palaeont. Sinica (c) N.S. 1:1–56.

Templeton, J.W., R.M. Sharp, J. Williams, D. Davis, D. Harmel, B. Armstong & S. Wardroup. 1982. Single dominant major genes effect on the expression of antler point number in the white-tailed deer. p. 469. *In*: R.D. Brown (ed.), Antler Development in *Cervidae*. Caesar Kleberg Wildl. Res. Inst., Kingsville, TX.

Thaler, L. 1973. Nanisme et gigantisme insulaires. Recherche 4:741–750.

Thaller, C. & G. Eichele. 1987. Identification and spatial distribution of retinoids in the developing chick limb bud. Nature 327:625–628.

Thenius, E. 1948. Zur Kenntnis der fossilen Hirsche des Wiener Beckens, unter besonderer Berücksichtigung ihrer stratigraphischen Bedeutung. Ann. Naturhist. Mus. Wien 56:262–308.

Thenius, E. 1950. Die tertiären Lagomeryciden und Cerviden der Steiermark. Springer-Verlag, Wien.

Thenius, E. 1952. Die Boviden des steirischen Tertiärs. Sitzungsber. Ost. Akad. Wiss., Vienne Abt. I. 161:409–439.

Thenius, E. 1956. Zur Entwicklung des Knochenzapfens von *Protragocerus* Deperet aus dem Miozän. Geologie, Berlin 5:308–318.

Thenius, E. 1969. Stammesgeschichte der Säugetiere (einschliesslich der Hominiden). *In*: Handbuch der Zoologie, Bd. 8, Lief. 47–48. W. De Gruyter, Berlin.

Thenius, E. 1979. Niederösterreich zur Braunkohlenzeit. Landschaft, Klima, Tier- und Pflanzenwelt Niederösterreichs zur Tertiärzeit. Wiss. Schriften-R. No. 43/44:1–64. St. Polten.

Thenius, E. 1980. Grundzüge der Faunen- und Verbreitungsgeschichte der Säugetiere, 2nd ed. Gustav Fischer Verlag, Stuttgart.

Thenius, E. 1983. Niederösterreich im Wandel der Zeiten. Die Entwicklung der voreiszeitlichen Tier- und Pflanzenwelt von Niederösterreich, 3rd ed. Amt d. Niederösterr. Landesregiergung, Wien.

Thenius, E. & H. Hofer. 1960. Stammesgeschichte der Säugetiere. Springer-Verlag, Wien.

Thing, H., C.R. Olesen & P. Aastrup. 1986. Antler possession by west Greenland female caribou in relation to population characters. Rangifer, Special Issue No. 1:297–304.

Thomas, H. 1977. Un nouveau bovidé du Nagri, plateau du Potwar, Pakistan. Bull. Soc. geol. Fr., Paris 7, 19:375–383.

Thomas, H. 1979. Les bovidés miocènes des rifts est-africains: implications paléobiogéographiques. Bull. Soc. gèol. Fr., Paris 7, 21:295–299.

Thomas, H. 1980. Les bovidés du miocène supérieur des couches de Mpesida et de la formation de Lukeino (district de Baringo, Kenya), pp. 82–91. *In*: R.E.F. Leakey & B.A. Ogot (eds.), Proc. 8th. Panafr. Congr. Prehistory, Nairobi, 1977.

Thomas, H. 1981. Les bovidés miocènes de la formation de Ngorora du Bassin de Baringo (Kenya). Proc. K. Ned. Wet. B., Amsterdam 84:335–409.

Thomas, H. 1983. Les bovidés du miocène moyen de la formation Hofuf (Province du Hasa, Arabie Saoudite). Palaeovertebrata, Montpellier 13:157–206.

Thomas, H. 1984a. Les giraffoidea et les bovidae miocènes de la formation Nyakach (Rift Nyanza, Kenya). Palaeontographica, Stuttgart 183A:64–89.

Thomas, H. 1984b. Les origines africaines des bovidae miocènes des lignites de Grosseto (Toscane, Italie). Bull. Mus. Natn. Hist. nat., Paris 4, 6, C:81–101.

Thomas, H. 1984c. Les bovidae du miocène du sous-continent Indien, de la péninsule Arabique et de l'Afrique. Palaeogéogr. Palaeoclimat. Palaeoécol., Amsterdam 45:251–299.

Thomas, H. 1984d. Les bovides ante-hipparions des Siwaliks inferieurs (plateau du Potwar, Pakistan). Mem. Soc. geol. Fr., Paris NS 145:1–68.

Thomas, H. 1985. Les bovidae (Artiodactyla, Mammalia) du miocène du sous-continent Indien, de la péninsule Arabique et de l'Afrique: biostratigraphie, biogéographie et écologie. Palaeogéog. Palaeoclimatol. Palaeoécol. 45:251–299.

Thomas, H., R. Bernor & J.J. Jaeger. 1982. Origine du peuplement mammalien en Afrique du Nord durant le miocène terminal. Geobios, Lyon 15:283–297.

Thomas, H., S. Sen, M. Khan, B. Battail & G. Ligabue. 1982. The lower Miocene fauna of Al-Sarrar (Eastern province, Saudi Arabia). Atlal 5:109–136.

Thomas, J.W., R.M. Robinson & R.G. Marburger. 1970. Studies in hypogonadism in white-tailed deer. Park and Wildlife Dept. Tech. Ser. 5:1–50.

Thomas, O. & G. Doria. 1889. Diagnosi di una nova specie del genera Cervulus reccolta da L. Fea nel Tenasserim. Ann. Mus. Ci. Stor. Nat. Genova 10:913–944.

Thomas, P. 1884. Recherches stratigraphiques et paléontologiques sur quelques formations d'eau douce de l'Algérie. Mem. Soc. geol. Fr., Paris 3, 3, 2:1–51.

Thompson, C.B., J.B. Holter, H.H. Hayes, H. Silverr & W.E. Urban, Jr. 1973. Nutrition of white-tailed deer. I. Energy requirements of fawns. J. Wildl. Manage. 37:301–311.

Thompson, R.G., Rodriquez, A. Kowarski, C.J. Migeon & R.M. Blizzard. 1972. Integrated concentrations of growth hormone correlated with plasma testosterone and bone age in preadolescent and adolescent males. J. Clin. Endocrinol. Metab. 35:335–337.

Thompson d'Arcy, W. 1940. On Growth and Form. Cambridge University Press, Cambridge, MA.

Thun, R., E. Eggenberger, K. Zerobin, T. Lüscher & W. Vetter. 1981. Twenty-four-hour secretory pattern of cortisol in the bull: Evidence of episodic secretion and circadian rhythm. Endocrinology 109:2208–2212.

Thurley, D.C. & K.P. McNutty. 1973. Factors affecting peripheral cortisol levels in unrestricted ewes. Acta Endocrinol. 74:331–337.

Tickle, C., B. Alberts, L. Wolpert & J. Lee 1982. Local application of retinoic acid to the limb bond mimics action of the polarizing region. Nature 296:564–566.

Tilak, R. 1978. A case of a third antler in barking deer (*Cervidae; Mammalia*). Deer. 4:265.

Tinbergen, N. 1951. The Study of Instinct. Oxford University Press, Oxford.

Todd, N.B. 1975. Chromosomal mechanisms in the evolution of Artiodactyls. Paleobiology 1:175–188.

Topiński, P. 1975. Abnormal antler cycles in deer as a result of stress inducing factors. Acta Theriol. 20:267–279.

Topiński, P. 1978. Abnormal antler cycles in deer as a result of stress inducing factors. Acta Theriol. 21:267–279.

Trense, W., A.J.H. Boislambert & G.K. Whitehead. 1981. Die Jagdtrophäen der Welt. P. Parey, Hamburg und Berlin.

Trevino Fernandez, J.C. 1978. Number and distribution of pronghorn antelope in Chihuahua, Mexico. M.S. Thesis, New Mexico State University, Las Cruces, NM.

Truswell, E.M. & W.K. Harris. 1982. The Cainozoic paleobotanical record in arid Australia: fossil evidence for the origins of an arid-adapted fauna, pp. 57–76. *In*: W.R. Barker & P.J.M. Greenslade (eds.), Evolution of the Flora and Fauna of Arid Australia. Peacock Publications, Frewville, South Australia.

Turner, C.D. & J.T. Bagnara. 1976. General Endocrinology, W.B. Saunders, Co., Philadelphia, PA.

Ullrey, D.E. 1982. Nutrition and antler development in white-tailed deer, pp. 37–48. *In*: R.D. Brown (ed.), Antler Development in Cervidae. Caesar Kleberg Wildl. Res. Inst., Kingsville, TX.

Ullrey, D.E., W.G. Youatt, H.E. Johnson, L.D. Fay, B.L. Schoepke & W.T. Magee. 1970. Digestible and metabolizable energy requirements for winter maintenance of Michigan white-tailed does. J. Wildl. Manage. 34:863–867.

Ullrey, D.E., W.G. Youatt, H.E. Johnson, L.D. Fay, B.L. Schoepke, W.T. Magee, and K.K. Keahey. 1973. Calcium requirements of weaned white-tailed deer fawns. J. Wildl. Manage. 37:187–194.

Ullrey, D.E., W.G. Youatt, H.E. Johnson, A.B. Cowan, L.D. Fay, R.L. Covert, W.T. Magee & K.K. Keahey. 1975. Phosphorus requirements of weaned white-tailed deer fawns. J. Wildl. Manage. 39:590–595.

Umesono, K., V. Giguere, C.K. Glass, M.G. Rosenfeld & R.M. Evans. 1988. Retinoic acid and thyroid hormone induce gene expression through a common responsive element. Nature 336:262–265.

Urbain, A., J. Nouvel & P. Bullier. 1944. Neoformations cutannes et osseuses de la tete chez les Giraffes. Bull. Mus. Hist. Nat. Paris 16:91–95.

Urbston, D.F. 1967. Herd dynamics of a pioneer-like deer population, pp. 42–50. *In*: Proc. Ann. Conf. Southeast. Assoc. Fish Comm., 21.

Vacek, M.Z. 1955. Innervace lyči rostoucich parohů u *Cervidů*. (The innervation of the velvet of growing antlers of the *Cervidae*.) Čskl. Morfol. 3:249–264. (In Czech.)

Van Ballenberghe, V. 1982. Growth and development of moose antlers in Alaska. pp. 37–48. *In*: R.D. Brown (ed.), Antler Development in Cervidae. Caesar Kleberg Wildl. Res. Inst. Kingsville, TX.

Van Bemmel, A.C.V. 1949. Revision of the rusine deer in the Indo-Australian Archipelago. Treubia 20:191–262.

Van Bemmel, A.C. 1951. Some additions to a revision of the Rusine deer in the Indo-Australian Archipelago. Treubia 21:105–110.

Van Bemmel, A.C.V. 1952. Contributions to the knowledge of Muntiacus and Arctogalidia in the Indo-Australian archipelago. Beaufortia, 16:1–50.

Van Bemmel, A.C.V. 1973. The concept of superspecies applied to Eurasiatic Cervidae. Z. Säugetierkd. 38:295–302.

Van der Eems K.L., R.D. Brown & C.M. Gundberg. 1988. Circulating levels of 1,25 dihydroxyvitamin D, alkaline phosphatase, hydroxyprolin and osteocalcin associated with antler growth in white-tailed deer. Acta Endocrinol. 118:407–414.

Van Jaarsveld, A.S. & J.D. Skinner. 1987. Spotted hyaena monomorphism: an adaptive 'phallusy'? S. Africa J. Sci. 83:612–615.

Van Mourik, S. 1984. The use of Rompun for immobilizing rusa deer (*Cervus rusa timorensis*). Vet. Med. Rev. 2:163–167.

Van Mourik, S. 1985. The expression and relevance of dominance in formed rusa deer (*Cervus rusa timorensis*). Appl. Anim. Beh. Sci. 14:275–287.

Van Mourik, S. 1986. Maternal behaviour and reproductive performance of farmed rusa deer. (*Cervus rusa timorensis*). Appl. Anim. Beh. Sci. 15:147–159.

Van Mourik, S. & T. Stelmasiak. 1984a. The effect of immobilizing drugs on adrenal responsiveness to ACTH in rusa deer. Comp. Biochem. Physiol. 78C:467–471.

Van Mourik, S. & T. Stelmasiak. 1984b. Adrenal response to ACTH stimulation in rusa deer (*Cervus rusa timorensis*). Comp. Biochem. Physiol. 79A:581–584.

Van Mourik, S. & T. Stelmasiak. 1985. Seasonal variation in plasma prolactin concentrations in adult male rusa deer (*Cervus rusa timorensis*). Comp. Biochem. Physiol. 82A:323–328.

Van Mourik, S. & T. Stelmasiak. 1986. Rusa Deer—from past to present. Zool. Anz. 5/6: 309–320.

Van Mourik, S., T. Stelmasiak & K.H. Outch. 1985. Changes in plasma levels of cortisol and corticosterone after acute ACTH stimulation in rusa deer (*Cervus rusa timorensis*). Comp. Biochem. Physiol. 81A:545–550.

Van Mourik, S., T. Stelmasiak & K.H. Outch. 1986. Seasonal variation in plasma testosterone, luteinizing hormone concentrations and LH-RH responsiveness in mature male rusa deer (*Cervus rusa timorensis*). Comp. Biochem. Physiol. 83A:347–351.

Van Valkenburgh, B. 1985. Locomotor diversity within past and present guilds of large predatory mammals. Paleobiology 11:406–428.

Van Valkenburgh, B. 1988. Trophic diversity in past and present guilds of large predatory mammals. Paleobiology 14:155–173.

Vandal, D., C. Barrette & H. Jolicoeur. 1986. An ectopic antler in a male Woodland caribou (*Rangifer tarandus caribou*) in Quebec. Z. Säugetierkd. 51:52–54.

Vanoli, T. 1967. Beobachtungen an Pudus, Mazama Pudu (MOlina 1782). Säugetierkdl. Mitt. 15:155–163.

Vaughan, T.A. 1978. Mammalogy, 2nd ed. Saunders Publ. Co., Philadelphia, PA.

Vaurie, C. 1972. Tibet and its birds. Witherby, London.

Venkataseshu, G.K. & V.L. Estergreen, Jr. 1970. Cortisol and corticosterone in bovine plasma and the effect of adreno-corticotropin. J. Dairy Sci. 53:480–483.

Verme, L.J. & D.E. Ullrey. 1972. Feeding and nutrition of deer, pp. 275–291. *In*: D.C. Church (ed.). Digestive Physiology and Nutrition of Ruminants, Vol. 3: Practical nutrition. O & B Books, Corvallis, OR.

Vethamany-Globus, S., M. Globus & G. Milton. 1984. β-endorphins (β-EP) in Amphibians: High β-EP levels during regenerating stage of anuran life cycle and immunocytochemical localization of β-EP in regeneration blastemata. Exp. Zool. 232:259–267.

Via, S. & R. Lande. 1985. Genotype-environmental interaction and the evolution of phenotypic plasticity. Evolution 39:505–522.

Vodička, F. & J. Augusta. 1942. Karpatský jelen a jeho rod. (Carpatian red deer and its ancestors). V. Pavlik, Praha. (In Czech.)

Vogt, F. 1936. Neue Wege der Hege. Verlag J. Neumann-Neudamm, Vienna, Austria.

Vogt, F. 1947. Das Rotwild. Österr. Jagd- und Fischereiverlag, Vienna, Austria.

Vogt, F. & F. Schmidt. 1951. Das Rehwild. Österr. Jagd- und Fischereiverlag, Vienna, Austria.

Von Uexküll, T. 1982. Meaning and science in Jakob von Uexkull's Concept of Biology. Semiotica 42:1–24.

Voorhies, M.R. 1969. Taphonomy and population dynamics of an Early Pliocene vertebrate fauna, Knox County, Nebraska, pp. 36–42. *In*: Contribution to Geology, Special Paper No. 1. University of Wyoming, Laramie, WY.

Vrba, E.S. 1973. Two species of *Antidorcas* Sundevall at Swartkrans. Ann. Transv. Mus., Pretoria 28:287–352.

Vrba, E.S. 1975. Some evidence of chronology and palaeoecology of Sterkfontein, Swartkrans and Kromdraai from the fossil *Bovidae*. Nature 254:301–304.

Vrba, E.S. 1976. The fossil Bovidae of Sterkfontein, Swartkrans and Kromdraal. Transv. Mus. Mem., Pretoria 21:1–166.

Vrba, E.S. 1977. New species of *Parmularius* Hopwood and *Damaliscus* Sclater & Thomas from Makapansgat. Palaeont. afr., Johannesburg 20:137–151.

Vrba, E.S. 1979. Phylogenetic analysis and classification of fossil and recent Alcelaphini. Biol. J. Linn. Soc., London 11:207–228.

Vrba, E.S. 1980. Evolution, species and fossils: how does life evolve. S. Afr. J. Sci., Johannesburg 76:61–84.

Vrba, E.S. 1984. Evolutionary pattern and process in the sister-groups Alcelaphini-Aepycerotini, pp. 62–79. *In*: N. Eldredge & S.M. Stanley (eds.), Living Fossils. Springer Verlag, New York, NY.

Vrba, E.S. 1985. African Bovidae: Evolutionary events since the Miocene. S. Afr. J. Sci. 81:263–265.

Vrba, E.S. 1987. New species and new genus of Hippotragini (Bovidae) from Makapansgat Limeworks. Paleont. Afr. Johannesburg 26:47–58.

Waddington, C.H. 1957. The Strategy of the Genes. C. Allen and Unwin Ltd., London.

Waddington, C.H. 1961. Genetic assimilation. Adv. Genet. 10:257–293.

Waldo, C.M. & G.B. Wislocki. 1951. Observation on the shedding of the antlers of Virginia deer (*Odocoileus virginianus borealis*). Am. J. Anat. 88:351–396.

Waldo, C.M., G.B. Wislocki & D.W. Fawcett. 1949. Observations on the blood supply of growing antlers. Am. J. Anat. 84:27–61.

Wallace, B. 1982. Phenotypic variation with respect to fitness: the basis for rank-order selection. Biolog. J. Linnean Soc. 17:269–274.

Wallace, B. 1984. Adaptation, Neo-Darwinian Tautology, and population fitness: a reply, pp. 59–71. *In*: M.K. Hecht, B. Wallace & G.T. Prance (eds.), Evolutionary Biology. Plenum Press, New York, NY.

Walther, F. 1966. Mit Horn und Huf. P. Parey Verlag, Hamburg.

Walther, F.R. 1984. Communication and Expression in Hoofed Mammals. Indiana University Press, Bloomington, IN.

Watson, A. 1971. Climate and the antler shedding and performance of red deer in North-East Scotland. J. Appl. Ecol. 8:53–67.

Webb, J.W. & D.W. Nellis. 1981. Reproductive cycle of white-tailed deer of St. Croix, Virgin Islands. J. Wildl. Manage. 45:253–258.

Webb, S.D. 1973. Pliocene pronghorns of Florida. J. Mammal. 54:203–221.

Webb, S.D. 1977. A history of savanna vertebrates in the New World. Part I, North America. Ann. Rev. Ecol. Syst. 8:355–380.

Webb, S.D. 1978. A history of savanna vertebrates in the New World. Part II, South America and the great interchange. Ann. Rev. Ecol. Syst. 9:393–426.

Webb, S.D. 1981. *Kyptoceras amatorus*, new genus and species from the Pliocene of Florida, the last Protoceratid Artiodactyl. J. Vertebr. Paleontol. 1:357–365.

Webb, S.D. 1983a. A new species of *Pediomeryx* from the Late Miocene of Florida, and its relationship within the subfamily of Cranio-ceratinae (Ruminantia, Dromomerycidae). J. Mammal. 64:261–276.

Webb, S.D. 1983b. The rise and fall of the Late Miocene ungulate fauna in North America, pp. 267–306. *In*: M.H. Nitecki (ed.), Coevolution. University of Chicago Press, Chicago, IL.

Webb, S.D. 1984. Ten million years of mammalian extinctions in North America, pp. 189–210. *In*: P.S. Martin & R.G. Klein (eds.), Quaternay Extinctions: A Prehistoric Revolution. University of Arizona Press, AZ.

Webb, S.D. & B.E. Taylor. 1980. The phylogeny of hornless ruminants and a description of the cranium of *Archaeomeryx*. Bull. Am. Mus. Nat. Hist. 167:117–158.

Weber, M. 1904. Die Säugetiere. Einführung in die Anatomie und Systematik der recenten und fossilen Mammalia. G. Fischer, Jena.

Weiner, J. 1977. Energy metabolism of the roe deer. Acta Theriol. 22:3–24.

Weiner, J. 1985. Correlations and regression or on the mischievousness of calculators. Wied. Ecol. 31:67–84.

Weiss, R. 1989. A genetic gender gap. Science News 15 (May 20):312–315.

Weisburd, S. 1987. Tuning in to songbirds and their songs. Science News 131:182–183.

Welch, B.L. 1966. Projected assessment of new physiological indicators of population conditions in deer, pp. 157–160. *In*: Proc. 19th Ann. Conf. Southeastern Assoc. Game and Fish Comm., Oklahoma City, OK.

Wells, L.H. & H.B.S. Cooke. 1956. Fossil Bovidae from the Limeworks Quarry, Makapansgat, Potgiertersrus. Palaeont. Afr., Johannesburg 4:1–55.

Wemmer, C.M., L.R. Collins, B.B. Beck & B. Rettberg. 1983. The ethogram, pp. 91–125. *In*: B.B. Beck & C.M. Wemmer (eds.), The Biology and Management of an Extinct Species, Pere David's Deer. Noyes Publications, Park Ridge, NJ.

Wesson, J.A., P.F. Scanlon, R.L. Kirkpatrick, H.S. Mosby & R.W. Butcher. 1979. Influence of chemical immobilization and physical restraint on steroid hormone levels in blood of white-tailed deer. Can. J. Zool. 57:768–776.

West, N.O. & H.C. Nordan. 1976a. Hormonal regulation of reproduction and the antler cycle in the male Columbian black-tailed deer (*Odocoileus hemionus colombianus*). Part I: Seasonal changes in the histology of the reproduction organs, serum testosterone, sperm production and the antler cycle. Can. J. Zool. 54:1617–1636.

West, N.O. & H.C. Nordan. 1976b. Hormonal regulation of reproduction and the antler cycle in the male Columbian black-tailed deer (*Odocoileus hemionus colombianus*). Part II: The effects of methallibure and hormone treatment. Can. J. Zool. 54:1637–1656.

Wetzel, R.S. 1968. The seasonal utilization of calcium and phosphorus by white-tailed deer. M.S. Thesis, Pennsylvania State University, University Park, PA.

Wheaton, C. & R.D. Brown. 1983a. Feed intake and digestive efficiency of South Texas white-tailed deer. J. Wildl. Manage. 47:442–450.

Wheaton, C. & R.D. Brown. 1983b. Comparative digestive efficiency of white-tailed and sika deer. Tx. J. Sci. 1:89–92.

Whipple, S.A., L.H. Wellman & B.J. Good. 1981. A classification of hardwood and swamp forests of the Savannah River Plant, South Carolina. SRO-NERP-6 Savannah River National Environ. Res. Park, Aiken, SC.
White, J.A. & B.L. Keller. 1984. Evolutionary stability and ecological relationships of morphology in North American lagomorphs. Spec. Pub. Carn. Mus. Nat. Hist. 9:58–66.
White, M.J.D. 1978. Modes of Speciation. W.H. Freeman, San Francisco, CA.
White, R.G. & M.K. Yousef. 1977. Energy expenditure in reindeer walking in roads and on tundra. Can. J. Zool. 56:215–223.
Whitehead, G.K. 1970. Female deer with antlers. Deer. 2:638.
Whitehead, G.K. 1972. Deer of the world. Constable & Co., London.
Whitehead, P.E. & E.H. McEwan. 1973. Seasonal variation in the plasma testosterone concentration of reindeer and caribou. Can. J. Zool. 51:651–658.
Whitworth, T. 1958. Miocene ruminants of East Africa. Fossil Mamm. Afr., London 15:1–50.
Whybrow, P.J., M.E. Collinson, R. Daams, A.W. Gentry & B.A. McClure. 1982. Geology, fauna and flora from the early miocene of eastern Saudi Arabia. Tertiary Res., Leiden 4:105–120.
Wicke, I. 1989. Caterpillar disguise. You are what you eat. Science News 135:170.
Wika, M. 1978. *In vitro* studies of antler blood vessels. Acta Physiol. Scand. 70A:102–103.
Wika, M. 1980. On growth of reindeer antlers, pp. 416–421. *In*: E. Reimers, E. Garre & S. Skjenneberg (eds.), Proc. 2nd Int. Reindeer/Caribou Symp., Roros, Norway, 1979.
Wika, M. 1982. Foetal stages of antler development. Acta Zoologica (Stockh.) 63:187–189.
Wika, M. & J. Krog. 1980. Antler "disposable vascular bed", pp. 422–424. *In*: E. Reimers, E. Gaare & S. Skjenneberg (eds.), Proc. 2nd. Int. Reindeer/Caribou Symp., Roros, Norway, 1979.
Wika, M., J. Krog, P. Fjelheim, A. Blix & U. Rasmusen. 1975. Heat loss from growing antlers of reindeer (*Rangifer tarandus* L.) during heat and cold stress. Norw. J. Zool. 23:93–95.
Winans, W. 1913. Hirschzucht und Veredelung des Rotwildes. Drei- und vierfache Kreuzungen. P. Parey, Berlin.
Wislocki, G.B. 1943. Studies on growth of deer antlers, pp. 631–653. *In*: Essays in Biology in Honor of Herbert M. Evans. University of California Press, CA.
Wislocki, G.B. 1952. A possible antler rudiment on the nasal bones of a white-tailed deer (*Odocoileus virginianus borealis*). J. Mammal. 33:73–76.
Wislocki, G.B. 1954. Antlers in female deer, with a report of three cases in the genus Odocoileus. J. Mammal. 35:487–495.
Wislocki, G.B. 1956. Further notes on antlers in female deer of the genus Odocoileus. J. Mammal. 37:231–235.
Wislocki, G.B. & M. Singer. 1946. The occurrence and the function of nerves in the growing antlers of deer. J. Comp. Neurol. 85:1–19.
Wislocki, G.B., J.C. Aub & Ch.M. Waldo. 1947. The effect of gonadectomy and the administration of testosterone propionate on the growth of antlers in male and female deer. Endocrinology 40:202–224.
Wolfe, J.A. 1978. A paleobotanical interpretation of tertiary climates in the northern hemisphere. Am. Sci. 66:694–703.
Wolfe, J.A. 1985. Distribution of major vegetational types during the Tertiary, pp. 357–

375. *In*: The Carbon Cycle and Atmosphere Carbon Dioxide: National Variations Archean to Present. Geological Monographs 82 (American Geological Union).

Wölfel, H. 1983. Zur Jugendentwicklung, Mutter-Kind-Bindung und Feindvermeidung beim Rothirsch (*Cervus elaphus*). Z. Jagdwiss. 29:143–162.

Wright, P.L. & S.A. Dow, Jr. 1962. Minimum breeding age in pronghorned antelope. J. Wildl. Manage. 26:100–101.

Wurster, D.M. & N.B. Aitkin. 1972. Muntjac chromosomes: a new karyotype for Muntiacus muntjac. Experientia 28:972–973.

Wurster, D.M. & K. Benirschke. 1970. Indian Muntjac, Muntiacus muntjak: A deer with a low diploid chromosome number. Science 168:1364–1368.

Wurster, K., W. Hofman & W. Dondorf. 1983. Gehörnte Ricke–Symptom einer Tumorerkrankung. Z. Jadgwiss. 29:74–81.

Wurster-Hill, D.M. & B. Seidel. 1985. The G-banded chromosomes of Roosevelt's muntjac, *Muntiacus rooseveltorum*. Cytogenet. Cell Genet. 39:75–76.

Wurster-Hill, D.M., K. Benirschke & D.I. Chapman. 1983. Abnormalities of the X chromosome in mammals. Cytogenetics of the Mammalian X Chromosome. B:283–302.

Yanagisawa M., H. Kurihara, S. Kimura, Y. Tomobe, M. Kobayashi, Y. Mitsui, Y. Yazaki, K. Goto & T. Masaki. 1988. A novel potent vasoconstrictor peptide produced by vascular endothelial cells. Nature 32:411–415.

Yerex, D. 1979. Deer farming in New Zealand, pp. 73–76. D.F. Jones Ltd., Wellington.

Yoakum, J. 1972. Antelope-vegetative relationships. Antelope States Workshop, Billings, MT 5:171–177.

Young, E.A. & H. Akil. 1985. Corticotropin-releasing factor stimulation of adrenocorticotropin and β-endorphin release: Effects of acute and chronic stress. Endocrinology 117:23–30.

Yousef, M.K., R.D. Cameron & J.R. Luick. 1971. Seasonal changes in hydrocortisone secretion rate of reindeer, *Rangifer tarandus*. Comp. Biochem. Physiol. 40A:495–501.

Zhang, C., H. Sheng & H. Lu. 1984. On the Fea's muntjac from Xizang (Tibet), China. Acta Theriol. Sinica 4:88 and 106.

Zhanxiang, O., Y. Defa, J. Hang & A. Bo. 1985. Preliminary observations on the newly found skeletons of *Palaeomeryx* from Shanwang, Shandong. Vertebrata Palasiatica 23:173–195.

Zittel, K.A. 1925. Textbook of Palaeontology 3, Mammalia. Translated by A.S. Woodward. Macmillan, London.

Zullinger, E.M., R.E. Ricklefs, K.H. Redford & G.M. Mace. 1984. Fitting sigmoidal equations to mammalian growth curves. J. Mammal. 607–636.

Index of Extant and Extinct Species of Pecora

Addra gazelle (*Gazella dama*), 200
Aepycerotinae, 93, 207, 214, 217, 227
African reedbuck (*Redunca redunca*), 31, 169
Alcelaphinae, 93, 196, 204, 214–217, 221–223, 227
Alcinae, 107
Algamoceros (*Blastoceros*) *blicki*, 70, 72, 73
Antidorcas, 207, 219
Antifer, 96
Antilocapridae, 181, 196, 202, 218, 219, 225
Antilocapridae, *-nae*, 4, 5, 46, 52, 56, 58, 84, 91, 115, 129, 132
Antilopinae, 31, 33, 36–38, 195, 196, 204, 217, 218, 221–224
Argali (*Ovis ammon*), 33, 35
Artiodactyla, 3, 18, 181

Barasingha (*Rucervus duvauceli*), 70, 72, 73, 173
Barking deer, *see* Muntjac
Beatragus hunteri, 215
Benicerus, 205
Bighorn sheep (*Ovis canadensis*), 32, 35
Bison (*Bison bison*), 34, 35
Black-tailed deer (*Odocoileus hemionus columbianus*), 39, 100, 433
Black wildebeest (*Connochaetes gnou*), 23, 32, 199, 214, 216
Blastomerycidae, 4, 94
Blesbok (*Damaliscus albifrons*), 93
Blue wildebeest (*Connochoetes taurinus*), 199
Bontebok (*Damaliscus pyragus*), 93
Bovoids (*Bovoidae*), 28, 92, 96, 100, 101
Boselaphini, 195, 196, 202, 204–206, 210, 223, 225
Bovidae, 115, 180, 182, 222
Bovini, 195, 196, 204, 208, 221
Brabovis, 213
Bramatherium perimense, 191
Brontotheridae, 114
Bubalus, 220
Buffalo (*Syncerus caffer*), 197
Bushbuck (*Tragelaphus scriptus*), 197

Camelidae, 16, 114, 116, 124, 129, 130, 132
Candiacervus, 10
Cantumeryx sirtensis, 194
Capreolidae, 62, 97, 108, 109
Capridae, *-nae*, 15, 30, 180, 195, 206
Caprini, 195, 196, 204, 206, 208, 209, 219, 221, 222, 224, 225
Caprotragoides, 201, 205, 208, 224
Cephalophini, 196, 204, 211, 223, 225, 227
Cervalces scotti, 72, 73, 98, 99
Cervidae, 115, 135, 136, 138, 165, 180, 182, 191, 193
Cervinae, 39, 42, 88, 89
Cervoidae, 4, 13, 16, 26, 32, 37, 38, 43, 46, 52, 58, 81, 92, 100, 101, 102, 112

Chamois (*Rupicapra rupicapra*), 6, 29, 36
Chinese water deer (*Hydropotes inernis*), 136, 137
Chital (*Axis axis*), 46, 51, 78, 82, 91, 95, 169, 177, 371, 374
Climacoceridae, 8, 28, 43, 101, 192, 194
Common duiker (*Sylvicapra grimmia*), 197
Cranioceratidae, 7, 27, 28, 52
Cretan deer (*Candiacervus*), 10

Dall sheep (*Ovis dalli*), 33-35
Damalacra neanica, 214, 217
Damaliscus, 215, 216, 217
Damalops, 217
Dibatag (*Ammodorcas clarkei*), 200
Dicrocerotidae, 7, 38, 43, 44, 48, 53, 59, 61, 101, 125
Domestic sheep (*Ovis domestica*), 31, 32
Dremotherium, 201
Dromomerycidae, 26–28, 52, 115, 121, 125

Elaphin, *see Cervus*
Elaphuroidae, 95
Elk *see* Wapiti
Eotragus, 30, 201, 202, 205, 207, 208, 227
Equidae, 114, 116, 132, 225
Eucervidae, 62, 109–112
Eucladoceros, 98, 99
Eumeryx, 17
Euproxinae, 38, 43, 62, 94, 101

Fallow deer (*Damma damma*), 24, 38, 39, 72, 73, 74, 77, 82, 271, 276, 280, 302, 313, 321, 375, 382, 402, 409, 413, 449, 454

Gazella sp., 23, 30, 218, 219, 224, 227
Gelocidae, 196
Gemsbok (*Oryx gazella*), 23
Gerenuk (*Litocranius walleri*), 200
Giant duiker (*Cephalophus sylvicultor*), 34, 35
Giant fallow deer (*Megaceros solihacus*), 98, 99, 100
Giant Irish elk (*Megaceros* or *Megaloceros giganteus*), 72, 73, 98, 99
Giraffa sp., 5, 1, 19, 21, 22, 28, 29, 52, 180–182, 185, 187, 188, 189, 192, 195
Giraffoidae, 8, 13, 19, 22, 26–28, 32, 52, 101
Giraffokeryx, 28, 101, 191, 192
Goat (*Capra*), 31, 32, 89
Grant's gazelle (*Gazella granti*), 200
Greater kudu (*Tragelaphus strepticeros*), 31, 32, 197

Hartebeest (*Alcelaphus buselaphus*), 199
Helicoportax, 202
Hesperoceridas, 220
Heteroprox, 43, 47, 59, 61
Hidasperium, 191
Hipparion, 226
Hippotragini, 196, 204, 213, 214, 221, 222, 223, 227
Hog deer (*Hyelaphus porcinus*), 169, 170, 176
Homoidorcas, 202, 207, 218, 227
Homoioceros, 33
Hoplitomericidae, 11, 26, 28, 115
Huemul (*Hippocamelus bisulecus*), 72, 73, 95, 96
Hydropotinae, 4, 16, *see also* Chinese water deer
Hypocamelidae, 96

Ibex (*Capra ibex*), 32, 35
Illingoceros, 59, 84
Impala (*Aepiceros melampus*), 169

Kaama (*Alcelaphus buselaphus caama*), 34, 35
Kamchatka sheep (*Ovis nivicola*), 35
Kipsigicerus, 202

Klipspringer (*Oreotragus oreotragus*), 200
Kobus sp., 34, 35, 198, 213
Kubanotragus, 205, 208, 224

Lagomerycidae -inae, 27, 43, 48, 50, 58, 101, 192
Leptobos, 211
Lesser kudu (*Tragelaphus imbergis*), 197
Litocranium, 224, 225

Madoqua, 218
Maral (*Cervus elaphus maral*), 40
Marremia, 218
Marsh deer (*Dorcelaphus dichotomus*), 70, 72, 73, 91, 95
Mazama deer (*Mazama* sp.), 72, 78, 91, 94, 95, 96
Megaloceros (*Dolichodorycceros*) *susenbornensis*, 72, 73
Megalotragus, 216, 217, 220
Megalovis, 219
Menelikia, 223
Merycodontidae, 5, 7, 38, 42, 45, 47, 52–58, 60, 114, 120, 121
Mesembriportax acrae, 205
Miotragocerus, 202, 205
Moose (*Alces alces*), 40, 46, 56, 72, 73, 74, 77, 83, 98, 339, 445, 450
Mouflon (*Ovis musimon*), 23, 29, 32
Mule deer (*Odocoileus hemionus hemionus*), 64, 70, 72, 73, 88, 94, 100, 289, 364, 433, 436, 449
Muntjac, 17, 49, 59, 62, 65, 66, 78, 82, 94, 103, 128, 143–168, 365, 366, 434, 436
Muntjacidae, 112
Muntiacus atherodes, 49, 64, 138, 142, 143, 146, 147, 165
  crinifrons, 137–140, 143, 146, 152, 165
  feai, 135, 138, 139, 143, 146, 149–151
  lachrimans, 146, 148, 149
  muntiac, 138–141, 143, 145–147, 150, 154–165
  plaiharicus, 146, 147
  rafinesque, 146
  reevesi, 63, 138–142, 146, 147
  rooseveltorum, 137–140, 146, 149, 150
  sinensis, 148
Musk deer (*Moschus moschiferus*), 4, 16, 137
Muskox (*Ovibos moschatus*), 30, 31, 32, 34, 35
Mustelidae, 92

Nile lechwe (*Kobus megaceros*), 34, 35, 198
Neotragini, 196, 218
Nuomidocapra, 220
Nyanzameryx, 207

Odocoileidae, 39, 62, 106–108, 110, 135, 136
Oioceros, 30, 220
Okapi (*Okapia*), 5, 6, 28, 29, 92, 180, 182, 184, 185
Oreonagar, 215
Orthogonoceras, 42
Oryx (*Oryx gazella*), 23, 32, 33, 198, 225
Osbornoceros, 84
Ourebia, *218, 224*
Ovibovini, 196, 219
Ovis catclawensis, 35
Ozotoceras, 91

Pachyportax, 223
Pachytragus solignaci, 205–207, 223
Palaeomeryx, 125, 194, 201
Paleomerycidae, 4, 11, 13, 15, 26–30, 32, 52, 101, 125, 194, 201
Paleotraginae, 28, 191, 192
Pampas deer (*Ozotoceras bezoarticus*), 72, 73, 95, 96
Parantidorcas, 211, 220
Paratoceras wardi, 6
Parmularius, 215–217
Pelorovis, 210, 220
Pere David's deer (*Elaphurus davidianus*), 70, 71–74, 97, 171, 175, 449
Perrisodactyla, 131

Plioceros, 57
Pliocervinae, 62, 94, 102
Plisoaddax, 220
Praedamalis deturis, 213
Proamphibos, 210, 223
Procapreolidae, 88, 97, 112
Procervulidae, 26, 27, 40, 52, 94, 101, 201
Procopus, 138
Prolibytherium magnieri, 191, 192
Protoceratidae, 129, 130, 114, 115
Pronghorn (*Antilocapra americana*), 5, 16, 28, 30, 38, 39, 43, 53, 57, 59, 78, 83–85, 90, 92, 97, 119, 120, 231–264
Prostrepticeros, 208, 222, 224
Protoceratidae, 16
Protoryx carolinae, 205, 207
Protragoceros, 202, 208
Prox, 62
Pseudotragus, 30
Pudu (*Pudu pudu*), 72, 73, 83, 94

Rangiferinae, 107–108
Raphicerus, 218, 219, 224
Red deer (*Cervus elaphus hippelaphus*), 14, 39, 43, 50, 64, 73, 75, 77, 79, 80, 81, 83, 90, 173, 177, 272, 274, 280, 291, 296, 313, 320, 367, 368, 371, 403, 424, 428, 429, 433, 434, 436, 439, 440, 442, 443, 448, 453, 454, 455, 458, 476, 484
Red-flanked duiker (*Cephalophus rufiatus*), 197
Reduncini, 196, 204, 211, 215, 222, 224, 227
Reedbuck (*Redunca redunca*), 31, 169
Reindeer or caribou (*Rangifer tarandus*), 14, 16, 21, 23, 24, 39, 49, 62, 64, 77, 79, 83, 87, 89, 119, 248, 276, 288, 364, 366, 374, 392, 397, 402, 403, 406, 415, 433, 436, 439, 442, 449, 451, 455, 461, 478
Roe deer (*Capreolus capreolus*), 10, 13, 14, 24, 64, 72, 73, 76, 77, 78, 81, 89, 90, 96, 97, 272, 275, 279, 280, 288, 289, 291, 292, 371, 375, 377, 380, 382, 383, 396–415, 433, 436, 440, 458, 484
Rhinocerotidae, 114
Rhynotragus, 216, 220
Roan antelope (*Hippotragus equinus*), 23, 30
Royal antelope (Neotragus pygmaeus), 200
Rucervus, 91, 95
Rupicaprini, 196
Rusa (Javan) (*Cervus rusa timorensis*), 71–73, 100, 169, 172, 173, 176, 178, 316, 372, 374, 416, 417, 424, 425
Rusinogoryx primaevus, 193, 220

Sable antelope (*Hippotragus nigger*), 33, 34, 35
Sambar (*Cervus rusa unicolor*), 71, 72, 73, 75, 78, 82, 91, 335, 365, 436
Samoceros, 223
Samotherium boissieri, 190
Schomburg deer (*Rucervus schomburgki*), 172, 177
Selenoportax, 202, 223
Serow (*Capricornis sumatranis*), 93
Shou (*Cervus elaphus aphinis*), 170, 174
Sika deer (*Cervus nippon*), 23, 72, 73, 371, 380, 370, 406, 435, 436, 451, 482
Silvicapra, 211, 224
Simatherium, 210
Sinclairomeryx, 7
Sitanonga (*Tragelaphus spekei*), 34, 35
Sivatherinae, 101, 190
Sivoreas, 205
Soemmerring's gazelle (*Gazella soemmerringi*), 200
Sphenophalos, 56
Spiroceros, 220, 224
Spotted hyena (*Crocuta crocuta*), 13, 39
Stag-beetle (*Lucanidae*), 76
Stephanoceras, 125
Stockoceros, 59, 60
Subantilocapra, 56
Suidae, 13, 114
Suni (*Neotragus moschatus*), 220
Syndioceras, 194
Synthetoceras, 7, 11, 194

Taruca (*Hippocalemus antisiensis*), 95, 96
Taurotragus, 209
Tetracerus quadricornis, 21, 28, 192, 224
Thaleroceros, 212
Thamin (*Rucervus eldi*), 72, 73, 173
Tragelaphini, 196, 197, 204, 209, 221, 223, 224, 227
Tragoportax, 205
Tragulidae, 16, 115, 124, 125
Triceromeryx, 26, 52
Tsademotherium, 220
Tufted deer (*Elaphodus cephalophus*), 59, 138–140, 143–145, 165
Tylopoda, 7, 13, 17, 32, 38, 115, 124, 129, 130

Uintantheridae, 114
Urmiatherium, 220
Ursidae, 92

Vishnutherium, 191

Walangania, 201
Wapiti (*Cervus elaphus canadensis*), 72, 73, 89, 98, 343, 349, 371, 377, 392, 433, 436, 442, 449, 451
Waterbuck (*Kobus* sp.), 23, 31, 198
Wellsiana, 213
White-lipped deer (*Przewalskium albirostris*), 174
White-tailed deer (*Odocoileus virginianus*), 245, 277–280, 286, 288, 290, 291, 293, 296, 300, 352, 353, 367, 368, 371, 374, 377, 393, 396, 397, 402–404, 413, 415, 427, 429, 433, 435, 436, 442, 448, 449, 451, 453, 454, 458, 460, 462, 464, 471, 475, 477, 480, 481, 486

Zarafa zelteni, 191, 192, 194
Zebra fish (*Taeniopygia guttata*), 78

Index of General Terms

Adrenal, 275, 279, 286, 287, 290, 413, 440, 442, 451, 453
  androgen, 16, 19, 20, 39, 62, 66, 79, 90, 102, 107, 109, 287, 288, 291, 413, 449, 451
  testosterone, 288
  tumor, 39, 373
Adrenaline, 19, 355
Adrenocorticotropic hormone (ACTH), 286, 354, 442
Africa, 36, 91, 93, 129, 180, 190, 195, 201, 208, 210, 213, 218, 220, 223
Age, 31, 41, 51, 65, 66, 79, 86, 90, 92, 96, 103, 105, 109, 112, 135, 356, 409, 447, 454, 462
Aggression, *see* social aggression
Alkaline phosphatase (AP), 266, 294, 409–411, 415, 469, 480
Allele, 11–13, 16–18, 32, 38, 75, 90
  dormant (silent), 12, 13, 18, 38
  reactivation, 38
America
  North, 5, 8, 39, 52, 59, 100, 130, 195, 225
  South, 94, 96, 98, 100
Androgen, 6, 18, 19, 39, 62, 66, 79, 90, 102, 107, 109, 275–280, 288, 296, 347, 402, 407, 408, 410, 475, 481, 483
  receptor, 46, 278–280, 348
    blockade, 279, 285, 475, 484
  sensitivity, 18
Androgenization, 39
Androstanediol, 280
Androstenedione (A_4), 289, 290, 295
Androsterone, 281
Angiogenic factors, 336
Ankylosis, *see* Fusion
Antler
  angiography, 331, 332
    activation, 62
    anlagen, 4, 11, 12, 15, 17–21, 28, 30, 31, 38, 39, 62, 65, 70, 86, 101, 112
    dormant, 28
    female, 16, 28, 49
  anomaly, 12, 78
  artery, 20, 331, 332, 334–338, 478, 480, 481
  asymmetry, 78, 94, 102, 365
  beam, 8, 9, 26, 41, 68, 70, 78, 84, 106, 108, 109
  as biomedical model, 426
  biopsy, 474, 483
  /body ratio, 98
  branching pattern, *see* Branching
  cactus, 79, 81
  calcium, 79, 80, 427, 428, 429
  cartilage, 330, 475
  casting, 8, 9, 31, 43, 51, 59, 60, 65, 66, 78, 81, 83, 90, 97, 104, 105, 109, 139, 276, 284, 313, 314, 321, 331, 361, 364, 368, 345, 346, 348, 349, 351, 355, 369, 396, 397, 406, 409, 411, 413, 440, 442, 444, 446–449, 451
    desynchronization, 65, 66, 451
  cervicorn, 26, 39, 52

Antler (*cont.*)
characteristics, 465–468
composition, 18, 426, 427
construction, *see* Branching
convergence, 94
core, 46, 94, 103, 107, 474, 483
coronet, 8, 28, 43, 50, 53, 54, 59, 60, 61, 62, 64, 66, 70, 78, 81, 102–104, 108, 109
cortex, 5, 8, 43, 46, 53, 59, 65, 98, 102
cross-section structure, 70, 74, 75
cycle, 65, 67, 92–97, 104, 107, 109
asynchronous, 93, 94
biannual, 97
denervation, 278, 361, 362, 364
dry matter, 427
dummy, *see* Head-pole
ectopic, 13, 14, 28, 38, 307
electrical field (potential), 20, 53, 76, 364, 365
evolution, 460, *see* also Evolution
female, 277, 278, 348, 349, 371–373, 391
"frostbite", 43, 89
graft, 380
growth, 20, 32, 46, 66, 90, 266, 269, 273, 279, 280, 287–289, 291–293, 295, 299, 313, 315, 319, 320, 331, 333, 337, 339, 345, 359, 364, 365, 370, 379, 392, 394, 395, 397, 403, 406, 407, 409, 411, 413, 415, 426, 428–432, 434, 437, 439, 442, 451, 455, 456, 458, 462, 466, 469, 472, 474, 478, 480, 481
castrate, *see* Peruke or Velericorn
antler, 378
center (AGC), 19, 78, 90, 347, 348, 355–357, 368
desynchronization, 54, 93, 94, 348–350, 371, 381, 395
experimental, 380, 339
rate, 65, 98
timing, *see* Cycle
handicap organ, *see* Organ
Haversian system, 46
heterospecific, 342
histogenesis, 298, 300
histology, 429
hydroxyproline, 480
hypertrophy, 341, 342, 389
innervation, 301, 359, 369, *see also* Antler nerves
length, 143, 316, 319, 320, 398, 400, 401, 406–409, 412, 414, 464, 470, *see also* measurements
-less
male and female reindeer, 16, 39
male, 16, *see also* Hummel
locus, 11
macrostructure, 43, 44, 59, 60, 70, 75
malformation, 348, 353
measurements, 33
microstructure, 44, 59, 60
mineralization, 135, 142, 266, 276, 280, 284, 285, 286, 289, 292, 293, 295, 348, 349, 406, 408, 409, 411, 412, 415, 427, 429, 474, 477, 480, 483, *see also* petrification *or* ossification
minerals (Ca, P, Na, Mg, K, Fe, Mn, Zn, Ba, Sr, Al), 42, 427
deficiency, 78, 428
requirement, 428, 429
morphogenesis, 486
nerve, 38, 337, 339, 345, 348, 349, 369, 474
activity, recording, 351
facial, 359, 361
parasympathetic, 19
stimulation, 348, 350, 351, 352
trigeminal, 20, 359–361
nondeciduous, *see* Pseudoantler
ossification, 299, 301, 408, 482, *see also* mineralization *or* petrification
centrifugal, 43, 46, 49, 102
centripetal, 43, 46, 62, 78, 102
osteoma, 79, 84
osteon, *see* Bone
palm, palmicorn, 8, 74
pattern, 341, 345, 348, *see also* construction *or* species-specific
peruke, 79, 80, 288, 348, 378, 484, 485
petrification, 65, 278, 280, 286, 288, 484, *see also* mineralization
phylogeny, *see* Evolution
point, 8
number, 102

polishing, 280, 314, 361, 365, 368, 369, 444, 446, 448, *see also* velvet shedding
prong, *see* Sprout
protein, 430
  requirement, 430, 432
quality, 90, 292, 472
radiology, -graphy, 41, 48, 49, 59, 61, 64, 331, 430
regeneration, 307, 313, 315, 319, 320, 338, 361, 364, 365, 369, 371, 374, 378
regress, 86
royal, 8
rudiment, 14
sarcoma-like, 78
scent, 25
shield, 86
sparring tool, 66
seal, 9, 51, 55, 67, 476, 477
shaft, 8, 61
size, 66, 88, 318, 357, 360, 367–370, 455, 458, 459, 460, 463, 465, 469–472, *see* measurements
soft, *see* Panty *or* Velvet antler
species-specific, 319, 336, 346, 347, 352, 360, 362, 367, 369, 463
supernumerary, 350, 355, 356
surface, 26, 38, 39, 43, 51, 66
  enlargement, 27, 40, 79, 80, 85
terminology, 5–9
territory, 399, 303
tine, 9, 39, 40, 41, 42
  regeneration, 40
  regression, 42
transplantation, 346, 371
trauma, 344, 348, 373, 375, 383, 387, 394, 395, *see also* injury, wound *or* Velvet-antler trauma
trophy, 460
  scoring, (Boone & Crocket), 471, 472
tumour, 484, 485
vascularization, 19, 41, 46, 51, 55, 65, 98, 227, 478, 479
velvet, 39, 84
velveted, 43, 46, 48, 83
weapon, *see* Weaponry 9
weight, 98, 315, 320, 363, 366, 428–430, 432, 455, 457
Antlerogenesis, 20, 26, 46, 94, 97, 100, 298, 302, 303, 413, 442, 443, 477
Antleroma, 318, 319
Appendage
  cranial, 4, 6, 10, 15, 20, 22, 25, 26, 37, 39, 43, 81, 101, 102, 114–116, 119, 121, 128, 129, 131, 180, 181
    evolution, 123
  anlage, 15, 21, 28
  apophyseal, 3, 4, 9, 10, 12, 19, 20, 25, 27, 28, 38, 59, 123
  combined, 101
  epiphyseal, *see* Ossicone
  head adornment, 9
  luxurious (product), 9
  occipital, 32
  perennial, 5, 27, 43, 51
  cervicorn, 53, 59
  terminology, 3–8
  velericorn, 27, 43, 79, 81
Architecture, 13, 20, 26, 41, 75, 76, 85, 86, 90, 93
Artiodactyla
  evolution, 132
  locomotion, 124
  teeth, 124
Asia, 52, 190, 195
Assessment, *see* Discrimination
Atavism, 12, 30
Atherosclerosis, 482
Australia, 100

Bauplan, 41
Behavior
  aggressive, 82
  breeding, 23, 37, 69, 91, 95, 266, 268, 280
  greeting, 21
  significance, *see* Social
  strategy, 22, 94, 95, 97
  territorial
  traits, 100
Behavioral impact, 20, 90
Body, 8, 9, 20, 22, 25, 26, 28, 36, 66, 75, 78–80, 83, 86, 91, 95, 116, 117
  anal pole, 68, *see also* Rump patch
  axis, 26

Body (*cont.*)
  diminution, 4, 10
  growth, 469
  protein, 430
    requirement, 430, 433, 435, 437
  shoulder height, 36
  size, 9, 20, 116, 461, 467
  weight, 119, 430–432, 434, 435, 438–441
    loss, 436, 439
    /antler length ratio, 66, 69
    /gestation length ratio, 91
    /horn size ratio, 36
Bone, 115, 180, 185
  calcium, 78, 481
  compact, 5
  core
  graft, 482
  growth, 474, 477, 480
    apophyseal, *see* Appendage
    epiphyseal, *see* Appendage
    in vitro, 480, 481
  hydroxyapatite, 78
  Looser's zone, 43, 89
  matrix, 41, 46, 65
  metabolism, 479, 480
  occipital, *see* Skull
  osteon, 59, 102
  phosphorus, 78
Bottleneck effect, 13
Bovid, 4, 5, 10, 11, 13, 15, 17, 19, 20, 21, 22, 25, 30, 31, 36, 58, 86, 89, 195–227
  boreal, 31, 32, 36
  classification, 196, 204
  dispersal, 195
  evolution, 195
  teeth, 125, 192, 195, 201, 205, 206, 211, 214
  tropical, 10, 31
  hybrid, *see* Hybrid
Brain center, 78
Branching, 13, 20, 25, 39, 41, 43, 54, 55, 66, 68, 70, 73, 74, 75, 84, 86, 89, 98, 101, 102, 106, 108
  asymmetrical, 55
  atypic, 13, 38, 90
  heterospecific, 38
  pattern, 13, 20, 41, 66, 68, 101
    species-specific, 78
Breeder
  aseasonal, 94, 95, 97
  continuous, 93
  long-day, 93, 94
  seasonal, 94
  short-day, 91–93, 97, 98
Breeding
  period, 17, 91–94, 96–98, 101
  structure, 267, 268
Bromocriptine, 60, 268–270, 292, 404, 405, 475

Calcitonin, 292, 293, 295, 296
Calcium, *see* Antler Ca *or* Diet Ca
Cancer, 483, 484, 485
Carotid prosthesis, 331
Castrate(ion), 39, 46, 56, 78, 81, 82, 279, 280, 284, 287, 290, 278, 451
  psychological, 58
Cattle, 31, 32
Cavicornia, 4, 28, 30
Central Nervous System (CNS), 64, 339, 347, 348, 440
Chemoattractants, 18, 19
Chemoreception, 22
Chemotaxis, 19
Chewing, 17
Chromosomes, 11, 39, 90, 143, 144, 147, 150, 152, 153, 165, 468
Cladogram, 101, 102, 103
Classification, 4, 10, 37, 92, 100, 101, 103–106
Climate, 33, 88, 90, 91, 92, 95, 97, 123
  changes, 10, 91, 93, 122, 128, 131
  seasonal, 91
Computerized Axial Tomography (CAT), 321–330
Contest, *see* Fight
Co-optation, 21, 81
Cornuogenesis, 3, 20, 26, 29, 31, 93
Corticoids, 80, 286, 287, 295, 296, 478, 481
Cranial
  angiography, 330, 331, 334
  appendage, *see* Appendage
  ornaments, 180
Cranium, *see* Skull
Cryptorchid, *see* Testes

Cue, *see* Releaser
Cyproterone acetate, 46

Deer, 39, 46, 51, 55, 62, 65, 66, 86, 94
  equatorial, 65
  tropical, 56, 66, *see also* Index of species
Demographic, *see* Population
Dentition, 17, 26
Dermal condensation, 182
Dermis, *see* Skin
Desert, 124
Development, *see also* Appendage evolution or Antler or Horn evolution
Diapause (delayed implantation), 97, 392
Diet
  calcium, 292, 293, 427
  protein, 430, 431
Digestion, *see* Pecora
Digestive system, 114, 117, 427, 436
Dihydrotestosterone (DHT), 280, 284–286, 295
Dimorphism, 15, 31
Discrimination, 22, 25, 37, 86
Display, 21, 26, 87
Diversification, 93, 100, 122
DNA, 11–13, 42, 90, 100
  mtDNA, 100, 102
Dwarfism (insular), 10, 18

Ecology, 116, 119
Ecomorph, 9, 93
Electrical fields, 20
Electrophoresis, 462
Endorphin, 278, 354
Energetics, 93
Environmental factors, 10, 465
Enzyme, 18, 39, 90, 96, 468
Eocene, 180
Epidermal Growth Factor (EGF), 20, 478, 482, 485
Epidermis, *see* Skin
Epigenesis, 10, 11, 12, 13, 15, 20, 41, 42, 75, 92, 98
Epimorphic regeneration, 298
Epiphyseal disc, 187
Estrogen (E_2), 102, 275, 277, 378, 413, 478, 481, 485, 419
  blockade, 284, 287, 393
Estrus, 35, 91, 95, 97
  primer, 84, 85
Eubovids, 28, 30
Eupecora, 1, 4, 18, 19, 20, 21, 22, 24, 25, 28, 92, 100, 105
Eurasia, 10, 26, 38, 53, 59, 98, 129, 131, 180, 195, 201, 220, 223
Europe, 30, 95, 96, 218
Evolution, 3, 9, 1–13, 15–18, 20, 21, 24–26, 30, 31, 33, 38, 39, 42, 43, 68–76, 86, 92, 93, 96, 100–102, 112
  artiodactyla, 124
  bovidae, 114, 116, 121, 122
  theory, 3, 9, 12, 15
Exaptation, 10, 11, 12, 17, 18, 21, 22, 28
Exostosis, 182
  azyguous, 11, 28, 29
  cranial, *see* Appendage
  supraorbital, 29
Eye, 24

Fasting metabolic rate (FMR), 432, 434
Fat, 470
  kidney, 465, 466
    reserves, 466
Female's choice, 22, 24, 86, 88, 90
Fetalization, 20
Fetus/fetal
  development, 19, 39
  growth rate, 465
Fight (contest), 12, 16, 17, 21, 33, 55, 59, 82, 83, 86, 182, 188, 190–193
  females, 190
Fitness, 85, 88, 90, 92, 466
Flexibility, 92, 101
Follicle Stimulating Hormone (FSH), 267, 272, 273, 404, 405, 420
Follower, 21
Food (forage)
  fibrous, 9, 173
  intake, 78, 266, 430
    decline, 436, 437, 439
  resources, 16, 36, 132
  texture, 102
Freezing, *see* Antler frostbite

Frontoparietal sinus, 187, 188, 190, 193
Front-pole, *see* Head-pole
Fusion, 30, 31, 53, 92, 93

Game theories, 86
Gene, 18, 32, 41, 88, 90, 467
Genetic
  characteristics, 461, 462, 464
  heterozygosity, 461, 462, 464–466, 468, 472
  variability, 462, 468, 472
Genetics, 18, 31, 32, 75, 89, 96
Genocopy, 9, 13, 39, 41, 90
Genome, 11, 13, 15, 20, 41, 90, 92
Geophysical forces, 366, 367
Gestation, *see* Pregnancy
Giraffids dentition, 194, 195
Gland
  apocrine, 22
  preorbital, 138
  sebaceous, 22, 43
Glucocorticoids, 286, 287, 296, 413, 442, 451, 454, 458, *see also* Corticoids
Gonadotropin, 264, 266, *see also* LH and FSH
  Releasing Hormone (GnRH), 90, 273, 276, 295, 296, 397, 419–421, 442
Graft, *see* Periosteum or Antler
Growth
  asynchronous, 54
  synchronous, 32
Growth Hormone (GH), 270, 272, 292, 296, 302, 313, 321

Habitat, 9, 13, 20, 24, 25, 76, 95, 98, 132, 208, 472
  visibility, 76
Hair, 22, 32
Head-pole, 22, 24–26, 33, 66, 68, 86
  artificial, (dummy), 85, 86
Hermaphrodite, 32, 39, 89, 372
Hide, *see* Skin
Hider, 21
Holocene, 221
Holometacarpal, 62, 102
Homeostasis, 13
Hopeful monster, 98
Horn, 4, 20, 22, 25, 30, 32, 59, 88, 90
  anlagen, 30
  anomaly, 32
  asymmetry, 32
  azygous, 182
  core, 180, 181, 195, *see also* Os cornu
  cup, *see* sheath
  disk, 28
  display, 192
  ectopic, 28
  exfoliation, 58
  evolution, 3, 9, 13
  female, 15, 35, 36, 120, 170, 208
  growth, 31, 92, 93, 301
  length, *see* measurements
  -less female, 36
  measurements, 33, 36
  morphology, 26, 116
  necrosis, 189
  ontogeny, 196
  pairs, 21
  rejection, 29
  scent, 36, *see also* Olfaction
  shape, shaping, 32, 36, 76
  sheath, 4, 5, 32
  /shoulder height ratio, 37
  sinus, 193
  skin, *see* Scur
  ontogeny, 30, 86
  species-specific, 25
  spiral, 30, 32, 33
  stratum corneum, 32
  surface, 26, 33
  sparring tool, *see* weaponry
  thermoregulation, *see* Organ
  vascularization, 189
  velericorn, 28, 56, 81, 82
  visual releaser, 76, 77
  weaponry, 9, 26
Hummel, 64
Hybrid, -ization, 32, 89, 100
Hypermorph, 10, 93
Hypothalamus, 18, 483
Hypothesis, *see* Theory

Incisor, *see* Tooth
Indicator, *see* Social

Information, tactile, 19
Inheritance, 31, 32, 90
Injury, *see* Trauma
Innervation, *see* Antler or Velvet
Insulin-like Growth Factor (IGF), 272, 295, 296, 320, 321, 478, 482, *see also* Somatomedi
Intersex, *see* Hermaphrodite

Karyotype, 18, 94, 102, 139, *see also* Chromosome
Keratin, 28, 29, 30–33, 57, 115, 189
Keratovelvet, 57, 58
Knobler, *see* Hummel

Lactation, 392, 429, 433
Language, 13
Latitude, 123
Learning song, 78
Leg amputation, 365
Leydig cells, 268, 273
Life expectancy, 88
Lineage, 10, 13
Localization, 25
Luteinizing hormone (LH), 268, 269, 272–276, 295, 320, 396–398, 402–405, 414, 418–421

Malnutrition, 30, 70, 428, 435, *see also* Food intake
Management, 86, 441, 473, 487
Maturation, 81
Maxilla, 144
Melatonin, 266, 267, 269, 295
Midbrain, 20
Migration, 93, 97
Mimicry, 13, 18, 28, 36, 66, 96
Mineralocorticoids, 242
Mio-Bovid, 92, 94
Mio-Cervid, 94, 102
Mio-Giraffoid, 92
Miocene, 10, 30, 38, 39, 46, 51, 52, 59, 62, 70, 89, 91, 92, 97, 128, 129, 131, 190, 195, 200–202, 205, 207, 224–227
Mismanagement, *see* Management
Molars, *see* Tooth
Monogamy, 132
Monomorphism, 15, 21
Monoscopy, 24, 25
Mutation, 10, 12, 18, 32, 90, 92
Myocardial infarct, 483

Neck, 16, 22, 26
Neogene, 129, 130, 132
Neogeny, 20, *see also* Fetalization
Neonate, 15, 21
Neoteny, 10, 36, 93
Nerve, *see* Antler nerve *or* Pedicle nerve
Nerve Growth Factor (NGF), 19, 347
Nervous system, 339
Neuron, 11, 12, 13, 17–19, 20
Neurotransmitter, 19, 20
Niche, 12
Non-aptation, 11
Noradrenaline, 19
Nutrition, 9, 292, 426–441, 454

Odor, 22, 37, *see also* Olfactory *or* Scent
  dissipator, *see* Scent
Offensive defense, 12
Olfaction, 36, 83, 84
Oligocene, 17, 30, 38, 129, 225
Ontogeny, 15, 20, 21, 33, 36, 135, 192
Opportunism, 33
Optical accent, 22, 24, 33, 36
Organ
  development, 17
  handicap, 86, 88
  intimidating, 85
  thermoregulatory, 86, 88, 89
Os cornu, 5, 13, 30, 31, 32, 185
  activation, 31
  anlage, 28, 31
  boss, 5, 6, 30, 31
  function, 32
  growth, 31
  lamellae, 33
  polled, 31, 32, 89
  remodeling, 93
  surface, *see* Ossicone measurement
Ossicone, 5, 21, 26, 30, 37, 39, 93, 101, 180–190, 192, 193

Ossicone (*cont.*)
cavicorn, 5, 14, 31
circumferance, *see* measurements
female, 186, 187
horned, 12
genetics, 31
growth, 185
length, *see* measurements
maturation, 190
measurements, 33
ontology, 192
species-specific, 70, 74, 75
surface, 33
symmetry, 32
velericorn, 14
transplantation, 31
Ossification, *see also* Antler ossification
subperiosteal, 60
Osteochondrosis, 50, 51
Osteopathy, 89
Osteoporosis, 481, 482
Ovary, 29

Paleoecology, 130, 225
Panty, 65, 66, 79, 81, 83, 88, 89, 98
Parathyroid hormone, 292, 293, 295, 296, 469, 482
Paravelvet, 49, 52
Parturition period, 91, 93
Pecora, 3, 4, 115, 116, 121, 123, 125, 128
digestion, 225
Pedicle, 5, 7, 8, 19, 31, 32, 37, 46, 78, 101, 139, 276–279, 286, 295, 296, 298, 474
amputation, 375, 378, 380, 392
anlage, 15, 16, 30, 39
activation, 39
artery, 331–338, 478, 479
bone, 9, 300, 367, 380
graft, 300, 375, 382
cervicorn, *see* Antler cervicorn
core, 322
cortex, 53
demineralization, 66, 322
denervation, 347
development, 12, 21, 347, 355
embryology, 296
excission, 303, 375, 376
fracture, 341, 342, 377
growth, 20, 278, 355, 399
injury, *see* trauma
hair, 43
location, 59, 62, 63, 94
mineralization, 279
monomorphic, 39, 59
nerve, 337, 339, 340, 369, 370
periosteum (inductive), 38, 64, 301, 302, 311, 312, 347, 360
graft, 301–308, 380, 382
inverted, 304, 305
heterotopic, 309, 310
minced, 303, 304
orientation, 303
shape, 302
size, 302
injury, 350, 354, 374
regeneration, 39, 335, 319–321, 337
skin, 9, 11, 43, 299, 367, 380
graft, 299, 300
keratinization, 21
removal, 355
restructuring, 21
transplantation, 299
trauma, 21
splitting, 71
supernumerary, 355
transplantation, 19, 31, 346, *see also* periosteum graft
trauma, 313, 337, 367, 369, 370, 348, 349, 350, 354, 378
wound, 380
Pedomorph, 10, 93
Peduncle, 5, 7, 37, 38, 58
cavicorn, (horn protected), 5, 7, 16 , 56
pronghorned, *see* Pronghorn
skin, 43
split, 56
velericorn, 16, 52
Pelage exchange, 266, 414, 438
Penis, 83
Periosteum, *see* Pedicle *or* Velvet
Peruke, 79, 80, *see also* Antler peruke
proliferating, 79, 80
Phenotype, 9, 10
Pheromone, 66, 81, 83, 84, 85, 90

Phosphorus, 41
Photoperiodicity, 90, 91–93, 95, 266, 268, 278, 396, 397, 404, 413–415, 417, 418, 432
Phylogeny, 36, 41, *see also* Evolution
Pineal gland, 265, 266, 364, 440
Pituitary, 267, 379, 396, 397, 402, 413, 440, 442, 481
Pleisto-bovids, 93, 94
Pleistocene, 31, 51, 92, 208, 221, 227
  glacial, 91, 96
Pleisto-cervids, 97, 98, 100
Plesiometacarpal, 16, 62, 72, 94, 96, 134, 169
Plio-bovids, 93, 94
Pliocene, 10, 30, 38, 51, 70, 96, 97
  Villafranchian, 91, 94, 97
Plio-cervid, 94, 95, 96, 402
Plio-cervoid, 97
Pliomorphism, 11, 21
Polling, *see* Os cornu
Polygamy, 132
Polygyny, 88
Polymorphism, 26
Population, 12
  social organization, 12, 86, 88, 93
Positional memory, 486
Preantler, *see* Protoantler
Pregnancy, 91, 93, 94, 392
Premolar, *see* Tooth
Processus
  conuiformis, *see* Peduncle
  cornu cervi, *see* Pedicle
Prolactin (PRL), 276–269, 404, 405, 417–419, 453
Prong, *see* Sprout
Pronghorn (antelope)
  embryology, 97
  migration, 232, 233
  nutrients, 248, 264
  parturition, 234
  pelage, 232
  population, 231, 233
    hierarchy, 233, 252
    structure, 231, 232
  pregnancy, 233, 234
  protuberance, *see* Cranial apophysis
    pseudohermaphrodite, 241
    subspecies, 231, 236
  sex-hormone, 56
  taxonomy, 326
  territory, 238–252, 253–260
  testes, 243
  vocalization, 253, 264
Pronghorn (as appendage), 4, 38, 57, 58, 84, 97, 119, 120, 231–264
  androgen, 249, 264
  casting
    female, 249
    male, 236, 244–246, 264
  castrate, 248, 250
  characteristics, 58, 231
  core, 248
  cycle, 57, 92, 97
  embryology, 233
  growth
    female, 241, 242, 249
    male, 235, 236, 244, 246–248
    seasonal, 243
  hair, 235
  regeneration, 58
  sheath, 43, 57, 58
    hair, 58
  shedding, 236
  supernumerary, 239, 240
Propagule, 4, 11, 15, 52
Protoantler, 5, 38, 39, 42, 43, 50, 51, 78, 81
  asynchronous cycle, 60
  monomorphic, 59
Protobovid, 28, 30
Protocervidae, 5, 26, 43, 46, 52, 53, 58, 59, 66, 91, 93, 94
Protohorn, 30
Proto-ossicone, 5, 15, 28, 92
Protopedicle, 5, 50
Protovelvet, 49, 50
Protuberance, 4, 5
Pseudoantler, 5, 38, 42, 43, 51, 54, 78, 81
Pseudocervidae, 43, 46, 53, 66, 91
Pseudocoronet, 8, 50, 53, 54
Pseudohermaphrodite, 372, 373
Pseudopedicle, 5, 51, 52, 58
Pseudotine, *see* Sprout,
Pseudovelvet, 49, 52, 56, 57
  desiccation, 56
  infarction, 56

Pseudovelvet (*cont.*)
mummification, 56
regeneration, 56
Puberty, 84, 91, 98, 276, 397, 398

Radiation, 122
Radioisotope, 41
Rainfall, 123
Ramification, *see* Branching
Rank, *see* Social rank
Releaser, 25, 76, 85
Reproduction, 115, 117, 269, 273, 396, 414, 415, 461, 467, 475
Retinoic acid, (RA), 18, 19, 486
Ritualization, 20, 33, 36, 66, 76, 77, 85, 86
Rumen, 135
Rump patch, 22, 24
Rusa deer
evolution, 316
heredity, 424
nutrition, 424
reproduction, 396, 414

Savanna, 123, 124, 129
Scent, 58, 66, 81, 83, 84, *see also* Gland, Odor *or* Pheromone
Scur, 28, 29, 30, 31, 32, 58
Sebum, 22
Selection, 22, 88
Sensory receptor, 19
Sequestration, 7, 28, 43, 48, 50, 54, 55
Serotonin, 19
Sex-size relationships, 139
Sex(ual)
antagonism, 16
character, 13
dimorphism, 9, 15, 28, 31, 33, 114, 116, 117, 130–132, 190, 191
monomorphism, 15, 16, 21
selection, 11, 22, 36, 88
Shoulder height, *see* Body
Signal, *see* Releaser
Sinecorn(ua), 4, 16
Skeleton, 26, 32, 94, 100
Skin (hide), 88, 115, 145, 300, 301, *see also* Pedicle skin *or* Peduncle dermal ring, *see* Pseudocoronet
epidermis, 31, 46
hypodermis, 46
keratinized, 58
transformation, 46
Skull, 11, 12, 26, 29, 31, 38, 58, 81, 135, 137, 145, 148, 152, 183, 185, 190, 207
-angle, 37, 81
Social
agonistic activity, 447, 461
aggression, 401, 447, 472
avoidance, 461
behavior, gregarious, 95
dominance, 85, 442, 460
hierarchy, 22, 24, 444, 447, 457
index, 444, 446, 456
intimate zone, 35
inferiority, 85
rank, 85, 90, 401, 447, 454, 456, 457–459, 461
indicator, 36, 58, 81
significance, 33, 56, 58, 81
structure, *see* Population
subordinance, *see* inferiority
superiority, *see* dominance
well-being (welfare), 90
Socioecology, 129–131, 182
Somatomedins, 268
Space, time independent, 102
Sparring, 21, 25
strategy, 21, *see also* Fight
tool, 21
Speciation, 94
Spermatogenesis, 65, 92, 97, 272
Sprout, -ing, 8, 9, 39, 40, 41, 43, 60, 66, 101
Spur, 5, 31, 52
Stereoscopy, 24
Stilbestrol, 378, 392, 483
Stress, Stressor, 11, 12, 13, 38, 70, 89 295, 348, 354, 355, 395, 438, 449, 451, 477, 487
Superfetation, 97
Superior Cervical Ganglion (SCG), 364
Synapsis, 18

Tail, 22, 68
Target organ, 18, 20, 339, 347, 483
Taxon, 92, 94, 103, 134, 138, 139
Taxonomy, *see* Classification
Telemetacarpal, 16, 62, 72, 94, 95, 96, 100, 134
Temperature
  ambient, 13, 123, 187
  annual, 123
  seasonal, 123
Territory defense, 16
Testes, 39, 49, 50, 78, 66, 78, 81, 83, 98, 273, 276, 290, 396, 397
  fawn, 399
  size, 398
  volume, 398–400, 402
  weight, 402
Testosterone (T), 16, 19, 46, 51, 58, 60, 62, 65, 66, 82, 92, 98, 266–269, 273–280, 286, 288, 292, 295, 296, 299, 313, 315, 319–321, 337, 355, 378, 379, 391, 392, 394, 396, 397, 399, 401–403, 405, 409–414, 418–421, 427, 436, 442, 449, 475, 485
  adrenal, *see* Adrenal
  fetal, 39
  implant, 314
  pubertal, 31
  receptor, 357
    blockade or silenced, 46, 79
  testicular, 16, 60
Theory (Hypothesis), 33, 42, 43, 68, 78, *see also* Evolution
Thermoregulation, 16, 32, 33
Thyroidal hormones, 270, 291, 292, 296, 481
Thyroxine (T_4), 291, 292, 295, 296, 409, 410
Timer, *see* Zeitgeber
Timing, *see* Antler or Horn
Tine, *see* Antler tine
Tissue defense, 11
Tooth, 16, 101, *see also* Dentition
  canines upper, *see* tusk
  molar, 12, 17
  morphology, 10, 52, 102
  premolar, 52
  replacement, 135
  tusk, 12, 16, 17, 59, 82, 114
    -antler relationship, 17
    evolution, 17
    female, 17
    regression, 15, 17
Torso, 21, 22, 25, 36
Trauma, 12, 32, 37, 39, 78, 88
Triiodothyronine (T_3), 291, 292, 295, 296, 409, 410, 478
Trophic
  response, 346
  memory, 78, 354, 486
Tusk, *see* Tooth

Umwelt, 13
Urine jettisoning, 83

Velericorn, 5
Velvet, 43, 102, 278, 299
  arteries, 483
  death, 46, 65, 66
  desiccation, 52
  shedding, 51, 89, 94, 278, 280, 284, 288, 299, 302, 336, 351, 378, 391, 392, 394, 429, 430, 439, 440, 446
  survival, 49
Velvet-antler, 20, 321, 334, 415, 446, 475, 483
  age, 314, 319, 339, 420, 422, 423
  growth, 295, 422, 413
  infarction, 65
  innervation, 19, 43, 9, 345
  length, 319
  mummification, 46
  periosteum, 46
  regeneration, 46
  removal, 313–315, 337, 361, 362, 368, 369
  serial amputation, 315, 317, 319
  trauma, 318
  -wart, 79, 83
  weight, 314, 319, 422, 423
Villafranchian, *see* Pliocene
Visual
  assessment, *see* discrimination

Visual (*cont.*)
  discrimination, 25, 33
  perception, 22, 24
Vitamin D_3, 293–295, 409, 411, 415, 429

Weapon(ry), 17, 36, 55, 58, 76, 82, 86
Wind velocity, 76
Wound, *see* Trauma

Zeitgeber, 90, 91
Zone of Polarizing Activity (ZPA), 19

MIX
Papier aus verantwortungsvollen Quellen
Paper from responsible sources
FSC® C105338

If you have any concerns about our products,
you can contact us on
ProductSafety@springernature.com

In case Publisher is established outside the EU,
the EU authorized representative is:
Springer Nature Customer Service Center GmbH
Europaplatz 3, 69115 Heidelberg, Germany

Printed by Libri Plureos GmbH
in Hamburg, Germany